Cell Therapy for Brain Injury

David C. Hess
Editor

Cell Therapy for Brain Injury

Springer

Editor
Dr. David C. Hess
Department of Neurology
Georgia Regents University
Augusta, GA, USA

ISBN 978-3-319-15062-8 ISBN 978-3-319-15063-5 (eBook)
DOI 10.1007/978-3-319-15063-5

Library of Congress Control Number: 2015937202

Springer Cham Heidelberg New York Dordrecht London

Printed on acid-free paper

Springer is part of Springer Science+Business Media (www.springer.com)

Contents

About the Editor

Dr. David C. Hess MD is currently chairman and professor, Presidential Distinguished Chair, Department of Neurology, with a joint appointment at the Institute of Molecular Medicine and Genetics and the College of Graduate Studies at Georgia Regents University. He completed his MD at University of Maryland Medical College of Georgia in Baltimore and an internship and residency in internal medicine at Allegheny General Hospital in Pittsburgh, before completing an additional residency in neurology at Georgia Regent's University in Augusta. Dr. Hess has authored or co-authored more than 150 papers. This book reflects the intersection between Dr. Hess' clinical interest in stroke and his research interest in stem cells.

About the Editor

Contributors

Bhagelu R. Achyut Tumor Angiogenesis Laboratory, Department of Biochemistry and Molecular Biology, Cancer Center, Georgia Regents University, Augusta, GA, USA

Ali S. Arbab Tumor Angiogenesis Laboratory, Department of Biochemistry and Molecular Biology, Cancer Center, Georgia Regents University, Augusta, GA, USA

Ricardo L. Azevedo-Pereira Department of Neurosurgery and Stanford Stroke Center, Stanford University School of Medicine, Stanford, CA, USA

Oh Young Bang Department of Neurology, Samsung Medical Center, Sungkyunkwan University, Seoul, South Korea

Martin M. Bednar Pfizer Neuroscience Research Unit, Cambridge, MA, USA

Pallab Bhattacharya Department of Neurology, Leonard M. Miller School of Medicine, Miami, FL, USA

Tonya M. Bliss Department of Neurosurgery and Stanford Stroke Center, Stanford University School of Medicine, Stanford, CA, USA

S. Thomas Carmichael Department of Neurology, David Geffen School of Medicine, University of California, Los Angeles, CA, USA

James Carroll Department of Neurology, Georgia Regent's University, Augusta, GA, USA

Charles S. Cox, Jr. Department of Pediatric Surgery, University of Texas Health Sciences Center at Houston, Houston, TX, USA

Steven C. Cramer Department of Neurology, Anatomy and Neurobiology, and Physical Medicine and Rehabilitation, Institute for Clinical and Translational Science, Sue and Bill Gross Stem Cell Research Center, UC Irvine Medical Center, University of California, Orange, CA, USA

Nasir Fakhri Department of Neurology, Medical College of Georgia, Georgia Regent's University, Augusta, Georgia

David C. Hess Department of Neurology, Medical College of Georgia, Georgia Regent's University, Augusta, GA, USA

Md Nasrul Hoda Department of Neurology, Medical College of Georgia, Georgia Regent's University, Augusta, GA, USA

College of Allied Health Sciences, Georgia Regent's University, Augusta, GA, USA

Vikram Jadhav Neurology, Neurological Institute, University Hospitals Case Medical Center, Cleveland, OH, USA

Michael V. Johnston Departments of Neurology and Pediatrics, Kennedy Krieger Institute and Johns Hopkins University, School of Medicine, Baltimore, MD, USA

Brian H. Johnstone NeuroFx, Inc., Fishers, IN, USA

Yukiko Kasahara Department of Regenerative Medicine Research, Institute of Biomedical Research and Innovation, Hyogo, Japan

Vivian W. Lau Regenerative Bioscience Center, University of Georgia, Athens, GA, USA

Department of Small Animal Medicine and Surgery, University of Georgia, Athens, GA, USA

George Paul Liao Houston, TX, USA

Nathan C. Manley Department of Neurosurgery and Stanford Stroke Center, Stanford University School of Medicine, Stanford, CA, USA

Asterias Biotherapeutics, Inc., Menlo Park, CA, USA

Keith L. March Indiana Center for Vascular Biology and Medicine, Indianapolis, IN, USA

Vascular and Cardiac Center for Adult Stem Cell Therapy, Indiana University, Indianapolis, IN, USA

Tomohiro Matsuyama Institute for Advanced Medical Sciences, Hyogo College of Medicine, Hyogo, Japan

Robert W. Mays Athersys, Inc., Cleveland, Ohio, U.S.

Pouria Moshayedi Department of Neurology, David Geffen School of Medicine, University of California, Los Angeles, CA, USA

Keith W. Muir Institute of Neuroscience and Psychology, University of Glasgow, Southern General Hospital, Glasgow, Scotland, UK

Simon R. Platt Regenerative Bioscience Center, University of Georgia, Athens, GA, USA

Department of Small Animal Medicine and Surgery, University of Georgia, Athens, GA, USA

John D. Sinden ReNeuron Ltd, Surrey Research Park, Guildford, Surrey, UK

Gary K. Steinberg Department of Neurosurgery and Stanford Stroke Center, Stanford University School of Medicine, Stanford, CA, USA

Steven L. Stice Regenerative Bioscience Center, University of Georgia, Athens, GA, USA

Department of Animal and Dairy Science, University of Georgia, College of Veterinary Medicine, Athens, GA, USA

Akihiko Taguchi Department of Regenerative Medicine Research, Institute of Biomedical Research and Innovation, Hyogo, Japan

Masahiro Tsuji Department of Regenerative Medicine and Tissue Engineering, National Cerebral and Cardiovascular Center, Suita, Osaka, Japan

Franklin D. West Regenerative Bioscience Center, University of Georgia, Athens, GA, USA

Department of Animal and Dairy Science, University of Georgia, College of Veterinary Medicine, Athens, GA, USA

Dileep R. Yavagal Neurology and Neurosurgery, University of Miami and Jackson Memorial Hospitals, Miami, FL, USA

Chapter 1
Introduction to Cell Therapy in Brain Injury

David C. Hess

The failure of the promising neuroprotectant, NXY-059, in the SAINT II acute stroke clinical trial led to a sense of therapeutic nihilism in big pharma about the prospects of "neuroprotection" in acute stroke [1–4]. While important lessons about the quality of preclinical data have been learned from the development of NXY-059, the string of failures of neuroprotectants in acute stroke has continued with albumin, granulocyte colony stimulating factor (G-CSF), and magnesium [5, 6]. The recent failures of progesterone in TBI, despite strong preclinical data, have fueled the concern whether neuroprotection is a viable option in acute brain injury. The failures are not limited to acute stroke and brain injury; there have been similar failures of chronic neuroprotection in Parkinson's disease, amyotrophic lateral sclerosis, and Alzheimer's disease [7–9]. It is premature to conclude that "acute neuroprotecton" in stroke is dead as promising treatments are still under development; however, it is fair to say that big pharma have folded their tents and moved on.

However, there remains hope that "reparative" or "restorative" strategies for stroke and acute brain injury might be a more promising and ultimately successful space for therapeutic development. The seminal finding in 1998 that neurogenesis occurs in the dentate gyrus of the hippocampus in elderly subjects suggested that the brain possesses more plasticity and potential for recovery than previously thought and led to a revision of the dogma that neurons only died and never were born in the adult brain [10]. Moreover, the time window for therapeutic intervention in restorative approaches is longer allowing more patients to be better characterized and treated without the pressure of the "clock." There is also a large unmet clinical need with many disabled stroke patients with few options beyond stroke rehabilitation.

While the term "stem cell therapy" is often used, the terms "cell therapy" or "cell-based therapeutics" are more accurate as, in many cases, these cells may not be "true stem cells" but rather progenitor cells or other cells that do not fit the strict criteria for "stem cells." One interesting cell therapy-related approach is to use the conditioned media (CM) of cells rather than the cells themselves. An approach us-

D. C. Hess (✉)
Department of Neurology, Georgia Regent's University, Augusta, Georgia, USA
e-mail: dhess@gru.edu

D. C. Hess (ed.), *Cell Therapy for Brain Injury,* DOI 10.1007/978-3-319-15063-5_1

ing the CM of adipose stem cells is presented by Brian Johnstone and Keith March in Chap. 14.

Cell therapy for acute brain injury represents a promising therapeutic approach, a "third wave" of therapeutics. The "small molecule" or "chemical" approach has failed in acute stroke and brain injury and has not produced one Food and Drug Administration (FDA)-approved agent in acute stroke or stroke recovery (see Fig. 1.1). The "biologics" approach, so successful in cancer and in autoimmune diseases, has led to only one FDA-approved treatment, tissue plasminogen activator (tPA). It remains to be seen whether cell therapy will lead to FDA-approved treatments for stroke. Presently, most of the "players" in the cell therapy field are small biotechnology companies or groups of academic investigators. Big pharma remains on the sideline awaiting evidence of a signal of "activity" from early-phase clinical trials. The viewpoint of the pharmaceutical industry on cell therapy is captured by Martin Bednar in Chap. 2.

While there remains "hope" that cell therapies will be effective in stroke and brain injury there is an also an excess of "hype." The term "stem cells" often incites unrealistic expectations among our patients, their caregivers and families, and the press. Direct-to-consumer (DTC) marketing in this digital age with social media permits easier exploitation of patients and families [11]. Patient desperation and hyping of stem cells has led to "stem cell tourism" where patients and families

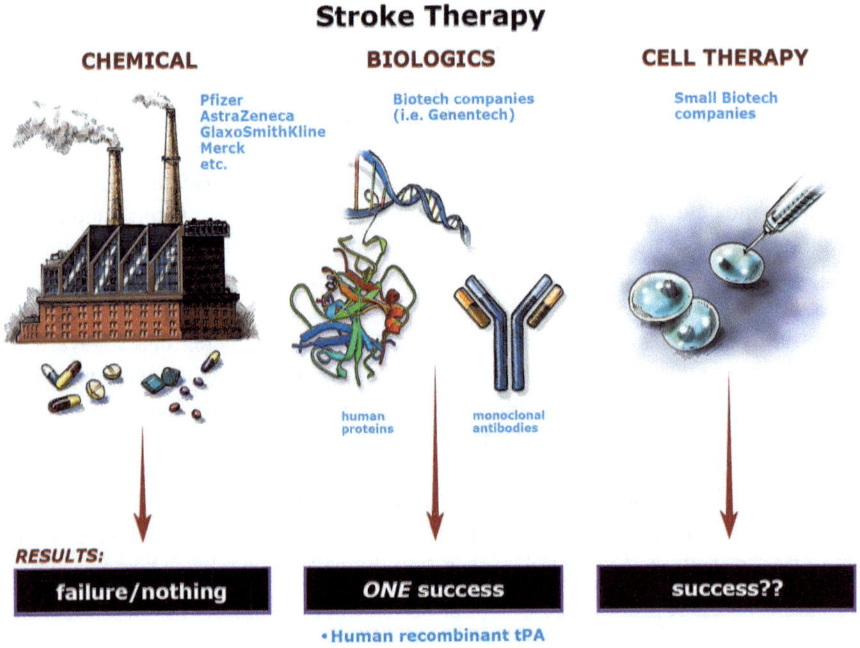

Fig. 1.1 Depiction of the three therapeutic approaches to acute brain injury. (Note that devices also have potential from clot-retrieval devices to devices that stimulate the sphenopalatine ganglia to blood pressure cuffs that deliver remote limb ischemic conditioning)

often travel overseas for expensive and unproven treatments with cost estimates of US$ 47,000 per treatment. In the USA, the FDA regulates cell therapies if they are "nonhomologous" and the cells are more than "minimally manipulated." "Homologous" function means that the stem cell has the same function in the donor and recipient. Hence, bone marrow cells for bone marrow transplant in a recipient with bone marrow failure are not regulated; however if bone marrow cells are used to repair the brain, then they are "nonhomologous" and under the purview of the FDA.

From a societal standpoint, only the oversight of regulatory agencies such as the FDA with the support of our professional societies will protect our patients and their families from modern "snake oil salesmen" and unscrupulous operators. Simultaneously, we need to uphold the highest standards for preclinical work and clinical trials. Only rigorous preclinical testing and randomized, blinded clinical trials will be the antidote to the hype of stem cell therapies and stem cell tourism. Due to the many failed clinical trials in stroke and other neurological diseases, there are now calls for more rigorous methodology and transparency in reporting of results of preclinical testing [12]. This includes randomization, concealment of allocation, blinding, sample size estimation, and reporting of negative studies. Organizations of clinicians and representatives from academia, industry, the National Institutes of Health (NIH), and the FDA such as The Stroke Academic Industry RoundTable (STAIR) and Stem Cells as an Emerging Paradigm for Stroke (STEPS) have published criteria to follow to help ensure that the best therapies are brought forward into clinical trial and that the clinical trials are conducted with the optimal design [13–17].

Types of Cells and Timing

With reparative and restorative cell therapies, a variety of types of cells, routes of administration, and "timing" of administration are proposed. The optimal timing and optimal routes of administration are not precisely known but likely will be related to the cell type used. The "golden time" of cell-based therapeutics is likely in the first week or weeks after injury when the brain is actively remodeling (Fig. 1.2). Intravascular (intravenous (IV) and intra-arterial (IA)) administration routes will likely be used in this time period. This early time point presents some challenges. There may not allow sufficient time to isolate and expand certain types of autologous stem cells and allow them to be transplanted in a short time window. In addition, the stroke patient may still be unstable and may be at risk for cerebral edema and brain herniation, complicating a clinical trial. While we lack extensive preclinical data on the optimal timing of cell therapy, one preclinical study suggests that IV mesenchymal stem cell (MSC) therapy may still have efficacy out to 1 month after stroke in rodents but there is a paucity of data at these later time points [18].

There are a wide variety of bone-marrow-derived stem cells (Fig. 1.3). To date, the most extensive preclinical support and clinical trial experience are with bone marrow mononuclear cells, MSC, and MultiStem, a proprietary plastic adherent

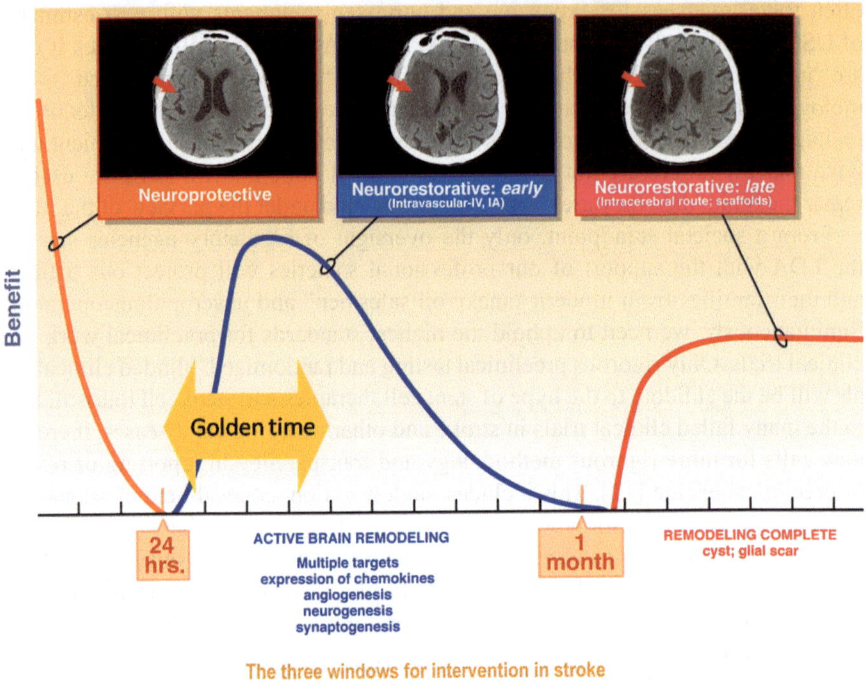

Fig. 1.2 The three main time windows for intervention: The acute neuroprotective window that closes rapidly in the first few hours, the "golden period" for reparative therapies where the brain is remodeling and there are immune system targets such as the spleen, and a later "chronic" period where the approaches will require intracerebral transplantation. From [15]

cell. Autologous bone marrow mononuclear cells can be rapidly isolated and do not require expansion and can be delivered back into the patient within 72 h and clinical trials to date in acute stroke demonstrate safety [19]. This approach is adopted by Charles Cox and colleagues (Chap. 15) in their use of intravenous autologous bone marrow mononuclear cells in pediatric patients with moderate to serve traumatic brain injury. Taguchi and colleagues transplant autologous bone marrow mononuclear cells at a later "subacute" period of 7–10 days after stroke (see Chap. 4). They isolate bone marrow mononuclear cells from stroke patients between 7 and 10 days from stroke and reinfuse these autologous cells the same day. In a later time window of 1–3 months after stroke, Oh Bang (see Chap. 3) has completed a clinical trial, the STARTING trial, of intravenously transplanted autologous MSC, where autologous MSC were expanded *ex vivo* in bovine sera. The treatment was safe with hints of activity and subjects were followed up for 5 years. Oh Bang now reports the ongoing STARTING 2 trial where the MSC are expanded in the sera of autologous stroke patients to "condition" them prior to transplantation.

An alternative approach is to use an "off-the-shelf" allogeneic cell. This cell type requires neither human leukocyte antigen (HLA) matching nor any isolation from the stroke patient. These cells from a healthy donor are stored and "ready to go" al-

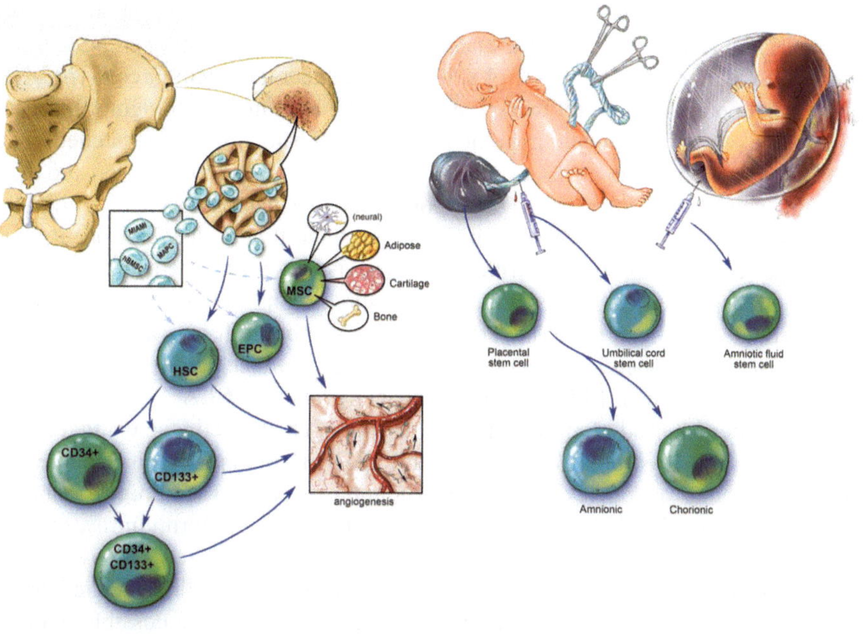

Fig. 1.3 Slide depicting the range of bone-marrow-derived cells and those derived from umbilical cord blood, placenta, and amniotic fluid

lowing early administration in stroke patients. Athersys, Inc. has launched an early-phase clinical trial of MultiStem, an allogeneic "off-the-shelf" plastic adherent cell distinct from the MSC. This multicenter, multinational randomized double blind, placebo-controlled, double blind trial of 136 patients incorporated a dose escalation phase and will finish enrollment in late 2014/early 2015 [20]. This trial utilizes a cell dose of 1.2 billion cells per patient, a higher dose than other intravascular trials. This trial targets a "homogeneous" group of moderately severe stroke patients with cortical involvement and a baseline NIHSS from 8 to 20 in the 24–48 h period after stroke [20]. This early time window targets the end of the neuroprotective window and the early portion of the recovery window. The development of MultiStem is reviewed in Chap. 5 by Robert Mays.

With the increasing use and availability of IA interventional therapy with clot-retrieval devices and the "positive" results of the MR CLEAN clinical trial from the Netherlands, there will be more opportunity to deliver IA cell therapy, an approach used in many interventional cardiology trials. Yavagal and colleagues have pioneered using IA delivery of MSC with testing in a canine model and discuss the IA approach in Chap. 6

Preclinical studies suggest that intracerebral delivery of neural stem cells (NSC) and iPS cell-derived neural progenitor may also be more effective in the first week after stroke than at later time points. There remain logistical challenges for these approaches in this early time period, and the approach will not be as scalable as intravenous delivery approaches to community hospitals. Steinberg's group at Stanford (Chap. 7) reviews the use of NSC and their own road to developing an NSC-based intracerebral therapy in human stroke. iPS-derived neural progenitor cells (iPS-NP) have great potential as a cell therapy and could be used in an autologous approach or more likely an allogeneic approach utilizing biobanks of HLA-matched cells. More work still needs to be done on the optimal timing and administration and there remains concern over the tumorigenic potential (see Chaps. 9 and 10). Unlike the intravascular approaches where significant cell engraftment does not occur, with these intracerebral approaches with NSC, there is cell engraftment with evidence that the transplanted cells integrate into brain circuitry as well as evidence that the cells provide beneficial paracrine effects on the host brain. It may also be possible to enhance the therapeutic effect of transplanted cells by using biodegradable matrices, discussed in Chap. 13 by Tom Carmichael's group at UCLA.

Later, after a period of months, the expression of chemotactic factors that attract stem cells such as stromal derived factor 1 (SDF-1) diminishes on endothelium and perivascular astrocytes reducing the homing of cells to the area of damage. Moreover, gliosis is often established and a cyst has started to form. Intravascularly delivered cells are less likely to "home" to tissue. Moreover, the "window" to target the spleen and to modulate the systemic immune system is likely closed. Since "homing" to the brain and the immune targets may no longer be possible, intracerebral, stereotactic transplantation is the best approach. The phase I and II trials of teratocarcinoma-derived NT 2 cells sponsored by Layton Biosciences beginning in the 1990s are reviewed by Nathan C. Manley from the Steinberg group in Chap. 7. They also briefly review the early-phase clinical trial of San Bio's human MSC line SB263. The development of the CTX human NSC line culminating in the PISCES clinical trial sponsored by Regeneron is reviewed in Chap. 8 by Keith Muir and John Sinden. One of the advantages of the later time points is that the stroke patient's clinical course has plateaued and the patient has a stable baseline allowing the effects of the intervention to be better measured and defined. In addition, patients have exhausted all other efforts at rehabilitation and have few options creating a large unmet clinical need and a population willing to travel and be enrolled in clinical trials.

Potency of Cells

A large number of cells die after transplantation and migrate poorly. Increasing the "potency" of stem cells, increasing their survival, homing, migration, engraftment potential, and capacity to secrete paracrine factors might be expected to improve patient outcome. Preconditioning stem cells with hypoxia and other factors such as SDF-1 and drugs such as minocycline have been shown to improve engraftment and outcome in some animal models and is discussed in Chap. 11. It is likely that some form of preconditioning of cells will become common in all cell therapy trials.

MRI Tracking of Cells

After transplantation, it is possible to track the in vivo migration, homing, and engraftment of cells after with MRI. This has been accomplished successfully in preclinical animal models. The use of MRI to track transplanted cells and the issues of cell labeling and safety are covered by Bhagelu R Achyut, and Ali S. Arbab, in Chap. 12

Neonatal Stroke and HI

Both neonatal stroke and hypoxic ischemic encephalopathy (perinatal asphyxia) present opportunities for cell therapy. The young brain has greater plasticity and potential for repair. Parents of children with cerebral palsy are desperate for new treatments and many are aware of "stem cell therapy." Cerebral palsy is a heterogeneous group of disorders and James Carroll addresses the particular challenges of use of cell therapy in "cerebral palsy" in Chap. 16. Neonatal stroke is specifically addressed by Masahiro Tsuji in Chap. 17.

Summary

With the recent interest in concussion and chronic traumatic encephalopathy in both athletes and soldiers with head injuries, there is intense interest in a wide range of therapies including cell therapies. While there are little data on cell therapy in these settings of mild traumatic brain injury (TBI), there are more data on patients with moderate to severe TBI as outlined by Charles Cox in Chap. 15. The secondary phase of TBI with its inflammatory cascade is a potential target of cell therapy. This chapter reviews the pathophysiology of TBI, past and current treatment approaches, and the development of a therapeutic approach with autologous bone marrow mononuclear cells that looks promising in early-phase clinical trials.

In the next few years, we should have safety, "activity," and some efficacy data from some of the ongoing and future planned trials. Clinical trials of iPS-based therapy in stroke will take longer to launch, as there are still issues about safety and logistics. It is critical to design cell therapy trials with rigor and with the appropriate outcome measures. This issue is addressed by Steven Cramer in Chap. 18. In the next decade, the mist of "stem cell therapy" should clear and we will begin to get answers whether cell therapy is a viable option for our patients and their families.

Acknowledgments The author wishes to acknowledge Michael Jensen, Department of Medical Illustration, Georgia Regent's University, for the medical illustrations of the three figures.

References

1 Diener HC, Lees KR, Lyden P, Grotta J, Davalos A, Davis SM, et al. NXY-059 for the treatment of acute stroke: pooled analysis of the SAINT I and II trials. Stroke. 2008;39(6):1751–8. doi:10.1161/STROKEAHA.107.503334. PubMed PMID: 18369171.
2 Dirnagl U, Macleod MR. Stroke research at a road block: the streets from adversity should be paved with meta-analysis and good laboratory practice. Br J Pharmacol. 2009;157(7):1154–6. doi:10.1111/j.1476-5381.2009.00211.x. PubMed PMID: 19664136; PubMed Central PMCID: PMC2743833.
3 Minnerup J, Sutherland BA, Buchan AM, Kleinschnitz C. Neuroprotection for stroke: current status and future perspectives. Int J Mol Sci. 2012;13(9):11753–72. doi:10.3390/ijms130911753. PubMed PMID: 23109881; PubMed Central PMCID: PMC3472773.
4 Savitz SI. Cosmic implications of NXY-059. Stroke. 2009;40(3 Suppl):S115–8. doi:10.1161/STROKEAHA.108.535112. PubMed PMID: 19064771.
5 Ringelstein EB, Thijs V, Norrving B, Chamorro A, Aichner F, Grond M, et al. Granulocyte colony-stimulating factor in patients with acute ischemic stroke: results of the AX200 for Ischemic Stroke trial. Stroke. 2013;44(10):2681–7. doi:10.1161/STROKEAHA.113.001531. PubMed PMID: 23963331.
6 Ginsberg MD, Palesch YY, Hill MD, Martin RH, Moy CS, Barsan WG, et al. High-dose albumin treatment for acute ischaemic stroke (ALIAS) Part 2: a randomised, double-blind, phase 3, placebo-controlled trial. Lancet Neurol. 2013;12(11):1049–58. doi:10.1016/S1474-4422(13)70223-0. PubMed PMID: 24076337; PubMed Central PMCID: PMC3929943.
7 Ahlskog JE. I can't get no satisfaction: still no neuroprotection for Parkinson disease. Neurology. 2007;69(15):1476–7. doi:10.1212/01.wnl.0000277645.60799.0e. PubMed PMID: 17923609.
8 Brew BJ. Lost in translation: again, another failed neuroprotection trial. Neurology. 2007;69(13):1308–9. doi:10.1212/01.wnl.0000277530.05450.ff. PubMed PMID: 17893290.
9 Gordon PH, Meininger V. How can we improve clinical trials in amyotrophic lateral sclerosis? Nature Rev Neurol. 2011;7(11):650–4. doi:10.1038/nrneurol.2011.147. PubMed PMID: 21947135.
10 Eriksson PS, Perfilieva E, Bjork-Eriksson T, Alborn AM, Nordborg C, Peterson DA, et al. Neurogenesis in the adult human hippocampus. Nature Med. 1998;4(11):1313–7. doi:10.1038/3305. PubMed PMID: 9809557.
11 Liang BA, Mackey TK. Stem cells, dot-com. Sci Transl Med. 2012;4(151):151cm9. doi:10.1126/scitranslmed.3004030. PubMed PMID: 22972840.
12 Landis SC, Amara SG, Asadullah K, Austin CP, Blumenstein R, Bradley EW, et al. A call for transparent reporting to optimize the predictive value of preclinical research. Nature. 2012;490(7419):187–91. doi:10.1038/nature11556. PubMed PMID: 23060188; PubMed Central PMCID: PMC3511845.
13 Savitz SI, Chopp M, Deans R, Carmichael T, Phinney D, Wechsler L, et al. Stem Cell Therapy as an Emerging Paradigm for Stroke (STEPS) II. Stroke. 2011;42(3):825–9. doi:10.1161/STROKEAHA.110.601914. PubMed PMID: 21273569.
14 Savitz SI, Cramer SC, Wechsler L, Consortium S. Stem cells as an emerging paradigm in stroke 3: enhancing the development of clinical trials. Stroke. 2014;45(2):634–9. doi:10.1161/STROKEAHA.113.003379. PubMed PMID: 24368562.
15 Albers GW, Goldstein LB, Hess DC, Wechsler LR, Furie KL, Gorelick PB, et al. Stroke Treatment Academic Industry Roundtable (STAIR) recommendations for maximizing the use of intravenous thrombolytics and expanding treatment options with intra-arterial and neuroprotective therapies. Stroke. 2011;42(9):2645–50. doi:10.1161/STROKEAHA.111.618850. PubMed PMID: 21852620.
16 Fisher M, Feuerstein G, Howells DW, Hurn PD, Kent TA, Savitz SI, et al. Update of the stroke therapy academic industry roundtable preclinical recommendations. Stroke.

2009;40(6):2244–50. doi:10.1161/STROKEAHA.108.541128. PubMed PMID: 19246690; PubMed Central PMCID: PMC2888275.
17 Saver JL, Albers GW, Dunn B, Johnston KC, Fisher M, Consortium SV. Stroke Therapy Academic Industry Roundtable (STAIR) recommendations for extended window acute stroke therapy trials. Stroke. 2009;40(7):2594–600. doi:10.1161/STROKEAHA.109.552554. PubMed PMID: 19478212; PubMed Central PMCID: PMC2761073.
18 Shen LH, Li Y, Chen J, Zacharek A, Gao Q, Kapke A, et al. Therapeutic benefit of bone marrow stromal cells administered 1 month after stroke. J Cereb Blood Flow Metab: Off J Int Soc Cereb Blood Flow Metab. 2007;27(1):6–13. doi:10.1038/sj.jcbfm.9600311. PubMed PMID: 16596121.
19 Savitz SI, Misra V, Kasam M, Juneja H, Cox CS, Jr., Alderman S, et al. Intravenous autologous bone marrow mononuclear cells for ischemic stroke. Ann Neurol. 2011;70(1):59–69. doi:10.1002/ana.22458. PubMed PMID: 21786299.
20 Hess DC, Sila CA, Furlan AJ, Wechsler LR, Switzer JA, Mays RW. A double-blind placebo-controlled clinical evaluation of MultiStem for the treatment of ischemic stroke. Int J Stroke: Off J Int Stroke Soc. 2014;9(3):381–6. doi:10.1111/ijs.12065. PubMed PMID: 23692637.

Chapter 2
Cell-Based Therapeutics in Stroke: An Industry Perspective

Martin M. Bednar

Stroke Background

An estimated 6.8 million Americans ≥20 years of age have had a stroke [1]. Each year, ~795,000 Americans experience a new or recurrent stroke. Approximately 610,000 of these are first attacks and 185,000 are recurrent attacks. In the USA, stroke ranks No. 4 among all causes of death and is the second leading cause of death worldwide, narrowly behind ischemic heart disease [2]. Of all the leading causes of death in both the USA and worldwide, stroke is unique for not having any specific therapy approved to improve outcome post insult, save for tissue plasminogen activator (t-PA) [3]. Although t-PA thrombolytic therapy was approved for acute stroke therapy nearly 20 years ago, its use worldwide remains extremely limited and the stroke field continues to search for neuronal protection and repair strategies.

Although the 1990s were labeled the "decade of the brain," significant nihilism crept into the field of stroke and was pervasive for approximately a decade from the late 1990s into the first decade of the twenty-first century [4]. During this period, all neuroprotective strategies examined in the clinic failed, despite careful review and recommendations from the Stroke Therapy Academic Industry Roundtable [5].

Perhaps not surprisingly, many major pharmaceutical companies began to de-emphasize neuroscience as a therapeutic area [6], although the most consistent failures have come in the attempts to treat acute neurologic disease such as stroke and traumatic brain injury (TBI). Inspired, transformational strategies were needed to reinvigorate the pursuit of medicines for one of the world's leading causes of death and disability.

A significant aspect of the de-emphasis of neuroscience drug development has been and continues to be the lack of clinical translatability of animal disease models. This is particularly evident in stroke where there have been more than 1000 pub-

M. M. Bednar (✉)
Pfizer Neuroscience Research Unit, 610 Main Street, 02139 Cambridge, MA, USA
e-mail: martin.m.bednar@pfizer.com

D. C. Hess (ed.), *Cell Therapy for Brain Injury*, DOI 10.1007/978-3-319-15063-5_2

lished preclinical studies, yet not a single neuroprotective agent has been approved for clinical use (considering t-PA a thrombolytic, not a neuroprotectant) [7]. When properly designed to recapitulate at least some aspects of the disease (e.g., timing of the therapeutic intervention, use of functional endpoints), animal models may be seen as useful screening tools—perhaps necessary, but certainly not sufficient for de-risking therapeutic approaches for central nervous system (CNS) diseases. However, the lack of translatability has resulted in a significant conundrum as clinical development, in general, requires massive resources (patients, professional personnel, and funding), limiting both the number of approaches that can be studied in the clinic and, as a result, the overall enthusiasm for the field. There are likely other contributors to the universal failure of neuroprotective therapeutics in both stroke and TBI, perhaps most notably both the presumed limited time window afforded for neuronal rescue and the consistent strategy of pursuing potential therapeutics which possess a single mechanism of action (MOA) to treat very complex diseases in an organ which has multiple redundant systems.

Next-Gen Stroke (CNS) Strategies

The solution for CNS drug development that has begun to emerge is twofold. While neither strategy specifically targets stroke or TBI, the field could be a major beneficiary. This "new path" arises both from a paradigm shift in drug development strategy and from major scientific advances in the field.

The first part of this disruptional thinking is to limit the emphasis on animal models. It is acknowledged that significant work has recently gone into improving preclinical stroke models, including the use of more relevant species, aged animals, animals with comorbid diseases, and chronic endpoints that may possibly reflect outcome more than acute measures of biochemistry and histopathology. However, improved guidelines for stroke research have not yet provided evidence of clinical translatability. Currently, the value of animal models in defining drug toxicity and determining initial exposure limits in early clinical trials remains, although animal models continue to have very limited utility in assessing drug efficacy for most human CNS diseases. Thus, a more rational strategy may be to advance therapeutics into the clinic as soon as they have demonstrated an appropriate safety profile in rigorous animal testing. The preclinical testing should define specific organ toxicity and the appropriate exposure/dose limits where the safety findings have occurred. These promising therapeutics are then taken into clinical testing. Small, focused clinical trials would rigorously study safety and determine relevant pharmacodynamic activity in order to provide safety profiling, risk mitigation, and greater confidence in rationale for the mechanism(s). This strategy is a rather bold but necessary move, shifting from animals to humans to understand the earliest signal of efficacy. For stroke, such early, surrogate clinical endpoints are in evolution but include neuroimaging (e.g., magnetic resonance imaging (MRI) measures of water content and directional flow; blood–brain barrier (BBB) integrity) and biochemical markers, either by employing cerebrospinal fluid (CSF) or blood-based markers and/or studying magnetic resonance spectroscopy which can combine both a biochemical signal with some indication of cell viability or activity. Although these surrogate markers lack full validation, they may be viewed as "fit for purpose" and a significant advance from relying on improvements in stroke animal models.

The second catalyst for the stroke field is the explosion of deep science on stem cells. Considering that human CNS stem cells were only identified ~15 years ago [8], the progress has been quite remarkable. Neurorestoration is an exciting alternative (and complementary) strategy to neuroprotection, removing the seemingly severe time constraints imposed by the latter. The emergence of the neurorestorative concept as applied to stroke, however, is not without its drawbacks. It seems appropriate to say that the technical capabilities of utilizing stem cells as a therapy for stroke have outstripped our understanding of how they may be efficacious in this disease setting.

Cell Therapy for Stroke: Considerations

There are multiple aspects of stem cell therapy to consider:

- Cell source. Broadly, three potential exogenous cell sources could conceivably be used for stroke therapy: embryonic, adult pluri- and multipotent stem cells, and induced pluripotent stem (iPS) cells. Additionally, although this chapter focuses on the strategy of exogenous stem cell therapy, there have also been considerations of encouraging a greater response from endogenous neural stem cells. For example, one possibility is to locally administer trophic factors that promote greater efficiency of endogenous neural stem cell proliferation, migration, differentiation, and/or survival [9].
- Cell type. If adult stem cells are considered the "gold standard," what cell type should be advanced? One could consider stem cells from a neural lineage, although non-CNS cells, especially autologous mesenchymal stem cells (MSCs), have many advantages (no/minimal immunogenicity and tumorigenic capability, no ethical issues) and also seem to improve outcome in animal models just as effectively as neural stem cells. Most of the clinical trials to date have used autologous mesenchymal stem cells derived from bone marrow [10], although adult stem cells from a wide variety of alternative sources have been used, including cells harvested from umbilical cord [11], olfactory ensheathing cells [12], adipose tissue [13], and placenta [14]. The data from animal models [15] suggest that mesenchymal stem cells are effective regardless of the route of administration. Mesenchymal stem cells also appear to modulate the local inflammatory response that contributes to the hostile environment for repair and recovery. Many of the clinical trials employing autologous mesenchymal stem cells have fulfilled the objectives of demonstrating safety and feasibility, although assessment of stroke outcome has not yet demonstrated clear evidence of efficacy, perhaps in part due

to the limited sample sizes. Autologous therapies pose challenges when it comes to large-scale manufacturing and time constraints for meaningful production, while allogeneic therapies run a higher risk of causing an immune response.

- Stem cell role (mechanism of action, MOA). There is no strong understanding of how stem cells actually would mediate neural recovery. One may think of the "3R's" when considering how stem cells may improve poststroke outcome: repair, replacement, and redirection (Fig. 2.1). Of course, these roles are not necessarily mutually exclusive:
  - Repair. In this context, there is consideration of immune response modulation, release of soluble trophic factors to create a more permissive environment either through local effects and/or niche upregulation, and instructing a specific endogenous stem cell fate. Facilitating a more permissive environment (reduced neuroinflammation, improved regional cerebral blood flow, either directly or through facilitating angiogenesis) should balance enhanced plasticity with the downregulation of inhibitory pathways that provide the CNS with the necessary feedback to maintain a homeostatic environment [16]. Mesenchymal stem cells have demonstrated experimental success in a number of pathologic scenarios. Such a broad protective/reparative effect suggests that these cells may be capable of releasing a diverse array of factors. Growth factors likely contribute to the beneficial effect [17], although much attention has recently focused on exosomes [18–20], a very heterogeneous group (both in size and content) of secreted lipid vesicles that may have therapeutic effects under a multitude of conditions. Although different exosomes may have competing actions, this is an area of medicine that provides both a rationale for the beneficial effect(s) of MSCs and is spawning a novel field of both diagnostics and therapeutics to improve the recognition of tissue injury and aid in customized tissue repair following various pathologies.
  - Replacement (including trans-differentiation). Generally, long-term functional engraftment/integration of exogenous stem cells at the site of pathology is much more the exception than the rule [12, 21, 22]. Nonetheless, even the integration of a small percentage of stem cells into the infarct area may result in a meaningful improvement. It is not clear if there is one or more critical variables exhibited by certain stroke patients that may facilitate a more permissive environment. Genetic polymorphisms, concomitant medications, or overall medical status may contribute in ways that are currently unclear. Additionally, the endogenous secretion of trophic factors directly within the stroke/penumbral region may be an important mechanism for neural repair.
  - Redirection (scaffold, bridge). This is a relatively new concept and is a hybrid between the repair and replacement mechanisms. The exogenous stem cells are necessary prerequisites at the stroke site, facilitating the directed migration of endogenous stem cells to the site of injury [23]. The exogenously placed stem cells have a limited presence at the stroke site, although it appears that the endogenously migrated stem cells may be able to integrate and improve the brain cytoarchitecture in this region.

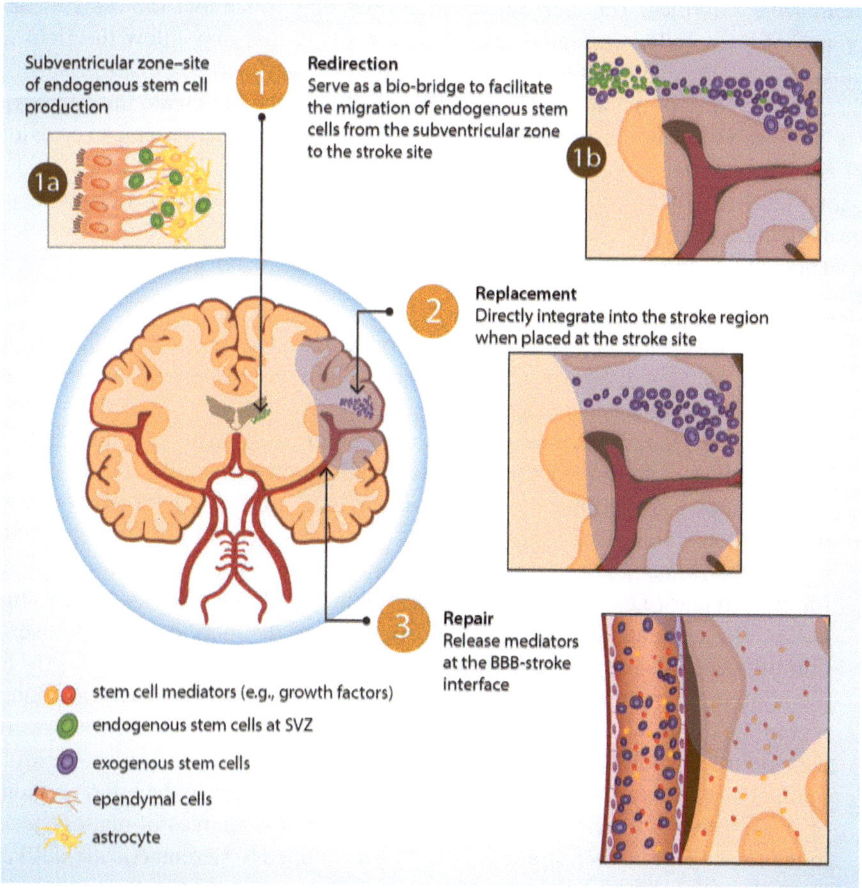

Fig. 2.1 Integrated approach to outcomes/clinical trials. The three general approaches by which stem cells could exert a beneficial effect following stroke (*1–3*). The direct placement of exogenous (autologous or allogeneic) within the stoke cavity could either serve to redirect endogenous stem cells from the subventricular zone (*SVZ*) to the infarcted region, facilitating their local engraftment and/or trophic effects or the transplanted exogenous cells could contribute more directly by replacing cells in the peri-infarct region and/or providing for a more favorable local environment. Stem cells delivered remotely (e.g., intravenously) likely exert their effect at the blood–brain barrier (*BBB*) interface through the activation of second messengers and/or the release of trophic factors, exosomes, etc. that facilitate a more permissive environment within the peri-infarct region

Even within each of the general strategies outlined above, there are multiple mechanisms by which a neurorestorative effect may be realized, just as there are multiple mechanisms that mediate the endogenous neuroplasticity that is constantly occurring. These include modulation of inhibitory circuits and facilitation of more permissive activities. The above discussion does emphasize that the "single" approach of stem cells is actually a pleiotropic strategy that may allow the field to move beyond the single MOA approach to address a very complex insult.

Certainly, any of the approaches mentioned would need to create the appropriate permissive environment that will allow for the differentiation and expression of any/all neuronal and glial types needed to reconstitute an effective and efficient network. This is irrespective of cellular augmentation by endogenous elements or through the delivery of exogenous stem cells (or both).

There are additional critical points for discussion when considering cell therapy for stroke:

- Stem cell route of administration. To a large extent, the route of administration will be dictated by the purported role for stem cells as a poststroke therapy. Currently, many contemporary stem cell studies for stroke utilize intravenous delivery which is seen as more of a reflection of pragmatism than an understanding of their MOA. The limited ability of stem cells to cross the BBB would not be an impediment to peripheral (intravenous or intra-arterial) administration [15] if the primary role of the stem cell was to provide second messenger signaling or growth factors that would then facilitate a more permissive environment for brain repair (e.g., shed exosomes have demonstrated the ability to cross the BBB). There are animal model data to suggest that the intravenous administration of stem cells may be as efficacious as administering stem cells directly to the injured region of the brain. A direct comparison of different delivery routes within the same animal model may be of great value and would not necessarily demand translatability to the clinical scenario. Alternatively, if the pluri- or multipotent stem cells are intended to serve as replacement cells in the infarct region, or if the intent is to serve as a biobridge from internal sources of neurogenesis to the stroke region, then strategies must be developed to stereotactically deliver the cells to the appropriate brain region.
- Stem cell timing of delivery. Ischemic brain injury is accompanied by a major inflammatory response which includes both CNS intrinsic (e.g., microglial upregulation) and extrinsic (e.g., leukocyte and cytokine upregulation) activation [24]. The inflammatory response can be seen as either well orchestrated or, alternatively, well-intentioned but suboptimally regulated. A critical question is when to intervene with a stem cell therapy. If the ongoing (at least initial) inflammatory response is beneficial for self-repair (e.g., removal of necrotic debris, facilitation of a more permissive environment), then the early downregulation may be counterproductive. The ultimate answer may only be understood through direct intervention and rigorous, methodical assessment of relevant clinical (and possibly imaging/biochemical) endpoints.
- Timing of primary endpoint measurement. Traditionally, key endpoints of neurologic outcome and function have been measured at 3 months post stroke. This was the time point used in the pivotal t-PA study and continues to be used nearly two decades later for a large number of putative neuroprotective and now, neurorestorative therapies. But is the best time point for assessing improvement? Neu-

rologic recovery continues for a year or longer post stroke. For therapies that are likely to provide cellular replacement and integration or otherwise improve plasticity, significant periods of time (weeks–months) may be needed to establish these new/repaired networks, in addition to the period of time for the affected individual to actually functionally improve. This may be especially relevant for the more eloquent domains such as speech. This time frame is likely longer than for a neuroprotective agent that focuses on salvaging ischemic neurons, yet very few studies perform randomized, double-blind placebo controlled trials for more than 3 months.

- Combination therapy. This paradigm is often favorably discussed but rarely initiated. This is likely for a number of reasons, including the logistics of understanding the optimal dose of each therapeutic alone and in combination and also ensuring that the animal toxicology package will support the clinical program. The plethora of permutations typically results in therapeutics being developed independently. That said, it is of interest that in a rat model of TBI, a combination of human umbilical cord blood cells and G-CSF, administered intravenously one week post insult, provided the most significant reduction in TBI-induced behavioral deficits when compared to either agent administered alone [25]. The behavioral improvement also complemented a histologic reduction in inflammation associated with improvement in neurogenesis. This combination strategy was recently studied in a clinical trial of chronic stroke patients, enrolled 6 months to 5 years post stroke [21]. In this study, subjects were given G-CSF for 5 days prior to intracerebral implantation of autologous peripheral blood stem cells ($CD34^+$). After a 12-month follow-up period, the subjects treated with the combination G-CSF and autologous $CD34^+$ stem cells demonstrated significant improvement in neurologic and functional scales as well as MRI evidence of structural improvement when compared to the control group. Although this was a small clinical study ($N=15$/group), the data are very encouraging.

Adding greatly to the complexity of combination therapy is a recent small study conducted in ten subjects 6 months to 20 years post stroke [12]. These subjects were treated with a combination of various stem cells including olfactory ensheathing cells, neural progenitor cells, umbilical cord mesenchymal cells and Schwann cells given through both systemic and local administration. Safety and efficacy were reported, although the extremely small sample size limits the strength of the conclusions and the multiple cell therapies and routes of administration will likely limit the use of this strategy, especially given the encouraging results with more straightforward clinical trial designs.

Noninvasive imaging will have an increasingly important role in the development of neurorestorative therapies in general, especially for stem cell and for gene therapies, where there is insufficient information on outcomes, in part due to the small sample sizes employed. In Parkinson's disease, neuroimaging is already part of the suite of modalities to understand outcomes [26]. Noninvasive imaging is a means by which promising mechanisms can be de-risked by seeking a surrogate (intermediate) biologic endpoint that can convey relevant pharmacodynamic activity and justify larger clinical studies. Examples include diffusion tensor imaging (DTI) to assess edema and fMRI to evaluate structural (brain, BBB) integrity. Neuroimaging may also be used to visualize the stem cells as well, determining their ultimate location and viability over time in a noninvasive way [27]. These data may then be related to either surrogate biochemical markers and/or to clinical endpoints.

There are additional critical factors that need to be addressed when one is considering if a potential therapeutic will improve stroke outcome, but many of these go beyond the breadth of a chapter devoted to cell therapies. One of the topics that will be critical to any potential therapy is the stroke type. This relates not only to location vis-à-vis infarcts involving gray matter or white matter (or specific fiber tract involvement) as these different anatomic locations most certainly will have unique strategies, but also to the individual patient deficit [28]. Employing endpoints for future stroke trials should consider a weighting based both on the patient-specific deficit at baseline (or at the time of enrollment in a clinical trial) and a potentially flexible timeline to primary endpoint analysis (based on specific deficits), possibly extending this timeframe to 6–12 months post initiation of therapy.

Summary

To summarize, our understanding of the complexities of stroke pathophysiology is very incomplete and this has contributed to the dearth of therapeutic options for this devastating disease. The examination of stem cell therapies for stroke is an emerging field which is also inadequately understood. However, this should not serve to dissuade investigations from the treatment of stroke and specifically stem cell therapy. Given the data to date, intravenous delivery of autologous stem cells provides the most pragmatic strategy, in regard to both the demonstrated safety and ease of administration but also having an efficacy profile that is at least comparable to the more aggressive routes of local stem cell administration. Mesenchymal stem cells have clearly garnered the greatest interest to date, although the optimal cell type to use awaits future investigative work. What is desperately needed for the evaluation of stroke therapies (stem cells and any other potential therapeutics) are both mechanistic and general endpoints that can be quantified in early phase clinical trials to de-risk these programs, providing confidence that they can demonstrate robust efficacy in larger and less controlled clinical trial scenarios. It is encouraging that cell therapy clinical trials for stroke and TBI have moved beyond the mindset of neuronal cell replacement to the use of other stem cell types (e.g., mesenchymal) that may facilitate endogenous repair and regeneration through the creation of a more permissive environment. Multifactorial approaches are clearly needed for these complex and devastating diseases and the future of stem cell therapies may assist in achieving this goal.

References

1. Go AS, Mozaffarian D, Roger VL, et al. Heart disease and stroke statistics- update. Circulation. 2014;129:e28–e292.
2. http://who.int/mediacentre/factsheets/fs310/en/. Accessed: 16 June 2014.
3. The National Institute of Neurological Disorders and Stroke rt-PA Stroke Study Group. Tissue plasminogen activator for acute ischemic stroke. N Eng J Med. 1995;333:1581–1587.
4. Hachinski V. Neurology: from nihilism to therapy. Arch Neurol. 2000;57:58.
5. Neuhaus AA, Rabie T, Sutherland BA, et al. Importance of preclinical research in the development of neuroprotective strategies for ischemic stroke. JAMA Neurol. 2014. doi:10.1001/jamaneurol.2013.6299. Accessed: 3 Mar 2014.
6. Miller G. Is pharma running out of brainy ideas? Science. 2010;329:502–504.
7. O'Collins VE, Macleod MR, Donnan GA, Horky LL, van der Worp BH, Howells DW. 1026 experimental treatments in acute stroke. Ann Neurol. 2006;59:467–477.
8. Eriksson PS, Perfilieva E, Björk-Eriksson T, et al. Neurogenesis in the adult human hippocampus. Nat Med. 1998;4:1313–7.
9. Hoag H. Brain food. Nature. 2014;510:S6–S7.
10. Miller RH, Bai L, Lennon DP, Caplan AI. The potential of mesenchymal stem cells for neural repair. Discov Med. 2010;9:236–42.
11. Guan Y-M, Zhu Y, Liu X-C, et al. Effect of human umbilical cord blood mesenchymal stem cell transplantation on neuronal metabolites in ischemic rabbits. BMC Neurosci. 2014;15:41–50.
12. Chen L, Xi H, Huang H, et al. Multiple cell transplantation based on an intraparenchymal approach for patients with chronic phase stroke. Cell Transplant. 2013;22 Suppl 1:s83–91.
13. Gutiérrez-Fernández M, Rodríguez-Frutos B, Otero-Ortega L, et al. Adipose tissue-derived stem cells in stroke treatment: from bench to bedside. Discov Med. 2013;16:37–43.
14. http://www.clinicaltrials.gov/ct2/show/NCT01310114?term=PDA+001&rank=2. Accessed: 16 June 2014.
15. Vu Q, Xie K, Eckert M, et al. Meta-analysis of preclinical studies of mesenchymal stromal cells for ischemic stroke. Neurology. 2014;82:1277–1286.
16. Bavelier D, Levi DM, Li RW, et al. Removing brakes on adult brain plasticity: from molecular to behavioral interventions. J Neurosci. 2010;30(45):14964–14971.
17. Shichinohe H, Ishihara T, Takahashi K. Bone marrow stromal cells rescue ischemic brain by trophic effects and phenotypic change toward neural cells. Neurorehabil Neural Repair. 2014. [Epub ahead of print] doi:10.1177/1545968314525856. Accessed: 14 Mar 2014.
18. Yeo RWY, Lai RC, Tan KH, at al. Exosome: a novel and safer therapeutic refinement of mesenchymal stem cell. Exosomes Microvesicles. 2013;1:1–12.
19. Zhang B, Yin Y, Lai RC, et al. Mesenchymal stem cells secrete immunologically active exosomes. St cells Dev. 2014;23:1233–1244.
20. Ibrahim AG-E, Cheng K, Marba′n E. Exosomes as critical agents of cardiac regeneration triggered by cell therapy. Stem Cell Reports. 2014;2:606–19.
21. Chen D-C, Lin S-Z, Fan J-R, et al. Intracerebral implantation of autologous peripheral blood stem cells in stroke patients: a randomized phase II study. Cell Transplant. 2014. doi:10.3727/096368914X678562.
22. Cramer S, Steinberg G. A novel phase 1/2A study of intraparenchymal transplantation of human modified bone marrow derived cells in patients with stable ischemic stroke. April 7, 2014, presentation, American Association of Neurological Surgeons annual meeting, San Francisco. "Transplantation cell therapy offers hope to stroke patients." Science Daily 2014. www.sciencedaily.com/releases/2014/04/140407153808.htm. Accessed: 7 Apr 2014.
23. Tajiri N, Kaneko Y, Shinozuka K, Ishikawa H, et al. Stem cell recruitment of newly formed host cells via a successful seduction? Filling the gap between neurogenic niche and injured brain site. PLOS ONE. 2013;8:e74857. Published Sept. 4, 2013.
24. Lemmens R, Steinberg GK. Stem cell therapy for acute cerebral injury: what do we know and what will the future bring? Curr Opin Neurol. 2013;26:617–625.

25. Acosta SA, Tajiri N, Shinozuka K, et al. Combination therapy of human umbilical cord blood cells and granulocyte colony stimulating factor reduces histopathological and motor impairments in an experimental model of chronic traumatic brain injury. PLoS ONE. 2014;9:e90953. doi:10.1371/journal.pone.0090953.
26. Hayashi T, Onoe H. Neuroimaging for optimization of stem cell therapy in Parkinson's disease. Expert Opin Biol Ther. 2013;13:1631–1638.
27. Ha BC, Jung J, Kwak BK. Susceptibility-weighted imaging for stem cell visualization in a rat photothrombotic cerebral infarction model. Acta Radiol. 2014 Feb 26. pii: 0284185114525605. [Epub ahead of print].
28. Cramer SC. Stratifying patients with stroke in trials that target brain repair stroke. Stroke. 2010;41:S114–S116.

Chapter 3
Autologous Mesenchymal Stem Cell Therapy in Patients with Stroke

Oh Young Bang

Limitations in Self-Repair and Current Stroke Therapy in Stroke Patients

Along with cancer and coronary heart disease, stroke is a leading cause of death, and the most common cause of physical disability in adults. The only specific therapies currently available for stroke are interventions to prevent inappropriate coagulation, surgical procedures to repair vascular abnormalities, and thrombolytic therapy. However, thrombolytic treatment can only be applied to certain patients, and various approaches to protect the brain from ischemic damage have met with limited success in clinical practice. Consequently, a large proportion of stroke survivors struggle with severe disabilities. Although rehabilitation therapy is important to maximize functional recovery in the early stages following stroke, no definitive treatment exists to restore lost brain function.

Cell therapy is an emerging paradigm in the stroke treatment field and, along with acute recanalization therapy and neuroprotective agents, is considered a potential regenerative strategy for patients with fixed neurologic deficits. Various cell types have been used to improve function and recovery after stroke, including embryonic stem cells (ESCs), immortalized pluripotent stem cells (iPSCs), neural stem/progenitor cells, mononuclear cells, cell lines, and nonneuronal adult stem cells, such as mesenchymal stem cells (MSCs). This chapter focuses on the clinical applications of adult stem cell therapy, in particular, the use of autologous MSCs in stroke patients.

O. Y. Bang (✉)
Department of Neurology, Samsung Medical Center, Sungkyunkwan University, 50 Irwon-dong, Gangnam-gu, 135–710 Seoul, South Korea
e-mail: ohyoung.bang@samsung.com

D. C. Hess (ed.), *Cell Therapy for Brain Injury,* DOI 10.1007/978-3-319-15063-5_3

MSCs as Candidate Cells for Cell Therapy in Stroke Patients

The ideal candidate cells for transplantation would: be autografted, i.e., easy to obtain and culture in order to produce sufficient cell dosages with no immune suppression; require minimal manipulation as per Food and Drug Administration (FDA) recommendations; and have appropriate stem cell characteristics, i.e., be self-renewing, noncarcinogenic cells that migrate to injured areas and undergo site-specific differentiation to an appropriate phenotype. MSCs fulfill all of these criteria. MSCs can migrate to injured brain regions (trophism) and self-renew with reportedly no carcinogenesis. Sufficient numbers of MSCs can be easily obtained under minimal culture expansion (as defined by a period of incubation not exceeding 60 days and number of culture passages not exceeding ten passages after colony formation). The International Cellular Medicine Society classified culture expanded autologous MSCs as a clinical cell line. However, they also stated that due to a lack of data regarding safe use, ESCs, iPSCs, allogeneic adult stem cells, and genetically modified stem cells were not ready for clinical translation (www.cellmedicinesociety.org).

Transplantation cell choice may depend on the mechanism that is deemed most beneficial. Even when the same cell types are transplanted, the beneficial action of transplanted cells may differ depending on disease conditions [1]. In stem cell therapy for stroke, most preclinical studies have emphasized the need to enhance self-repair systems rather than to replace cells, regardless of the type of cells used (MSC [2] and iPSC [3]). A recent study showed that although iPSC-derived neural stem cells (NSCs) induce neurogenesis, iPSC-derived NSCs enhance endogenous neurogenesis via trophic support, in a manner similar to adult nonneuronal stem cells (e.g., MSCs), not by cell replacement with exogenous iPSC-derived NSCs [3]. Thus, MSCs may be a good choice for stroke cell therapy because MSCs secrete a variety of bioactive substances, including neurotrophic factors, in the injured brain which may be associated with enhanced neurogenesis, angiogenesis, and attenuation of inflammation [4, 5].

MSCs can be obtained from various tissues, including bone marrow, adipose, liver, and umbilical cord blood. MSCs extracted from bone marrow and umbilical cord blood are most widely used as they can be separated through continuous culture. Extraction of MSCs from other tissue needs an additional collagenase digestion step, which can injure the MSCs. Adipose-derived MSCs can be considered a good alternative to bone-marrow-derived MSCs because the former have many advantages, in terms of proliferative capacity and secretome profile, compared to the latter [6, 7]. In addition, there have been recent advances in isolation and cultivation techniques, such as minimizing collagenase exposure and using a bone marrow filter for rapid purification and prevention of contamination [8].

Clinical Trials of MSC Therapy in Stroke Patients

After the first report of MSC therapy in stroke patients in 2005 [9], several recent clinical trials have used stem cells in stroke patients. Although all clinical trials in patients with stroke have shown that MSC therapy is feasible and safe, the results from these trials have raised important issues. Specifically, these trials varied in terms of patient characteristics, cell therapy timing, dose and type of cells delivered, and mode of treatment. In addition, many factors that could be critical for transplantation success, including the location/extent of lesions, were not adequately considered. Finally, the assessment of functional improvement, adverse effects, and pretreatment screening tests for safety were not standardized (Table 3.1).

We have reported the results of the stem cell Application Research and Trials In NeuroloGy (STARTING) trial, a randomized controlled trial of intravenous application of autologous MSCs, culture expanded with fetal bovine serum, in the subacute phase of stroke [10]. In this study, patients were randomly allocated to one of two groups, those who received intravenous autologous ex vivo cultured MSCs (n=16) or those in the control group who did not (n=36). Intravenous autologous MSC transplantation was found to be safe for stroke patients through a long-term period (~5 years), and may improve recovery, as evaluated via observer-blind evaluation of modified Rankin scores (mRSs) (Fig. 3.1). More recently, Honmou et al. described dramatic neurological improvements after administration of autologous MSCs, culture expanded with autologous serum, in 12 patients with chronic stroke [11]. However, the mechanisms by which transplanted MSCs result in functional benefits after stroke are unclear. Clinical trials have shown that functional improvement occurs shortly after cell therapy and diminishes with time, suggesting that the effects of MSCs are mediated via trophic support, rather than replacement of damaged cells [9].

A cost-effectiveness analysis showed that the societal value of stem cell therapy was US$ 166,500, and that therapy was cost-effective under a wide range of assumptions, such as the size of the effects, age, and functional status at discharge [12]. However, it should be noted that none of the studies aimed to evaluate the efficacy of MSC therapy in stroke patients. All of the studies aimed to assess the feasibility and safety of stem cell treatments, compared to conventional treatments, in patients with ischemic stroke. Most studies were small series, and several studies did not even include a control group. Moreover, assessments for functional improvement varied among studies. Therefore, it is too early to determine whether MSC therapy can improve functional outcomes in patients with stroke [13]. In the cardiology field, a recent meta-analysis concluded that transplantation of adult bone marrow cells improved left ventricular function, infarct size, and remodeling in patients with ischemic heart disease compared with standard therapies, and that these benefits persisted. A total of 50 studies (enrolling 2625 patients) were included in this analysis and patients received echocardiographic evaluations and long-term follow-ups in order for researchers to arrive at this conclusion [14].

Table 3.1 Clinical trials of mesenchymal stem cells or bone marrow mononuclear cells in stroke patients

Author, reference	Study design	Stroke	Cell dose	Manipulation	Mode of application	Efficacy	Adverse effects	Safety test
Autologous bone marrow mononuclear cells								
Suarez-Monteaqudo, 2009 [40]	No control group Treatment, $n=5$ 1 year f/u	Chronic Ischemic or ICH	N/A	Isolation using normal saline	Intracerebral	N/A	None	N/A
Battistella, 2011 [41]	No control group Treatment, $n=6$ 6 months f/u	Subacute MCA infarct	$1–5 \times 10^8$	Isolation using human albumin-containing normal saline	Intra-arterial	N/A	Seizure after 200 days	N/A
Savitz, 2011 [42]	No control group Treatment, $n=10$ 6 months f/u	Acute (24–72 h) Large MCA infarct	1×10^6 /kg		Intravenous	mRS 1 shift vs. historical control	None	Cell viability, MSC surface markers, bacteria, fungi, mycoplasma
Friedrich, 2012 [43]	No control group Treatment, $n=20$ 6 months f/u	Acute (3–7 days) Non-lacunar infarct	2.2×10^8		Intra-arterial	mRS 0–2 in 40 %	None	Cell viability
Li, 2013 [44]	Control, $n=40$ Treatment, $n=60$ 6 months f/u	Acute (5–9 days) ICH	1.33×10^{13}	Isolation using normal saline	Intracerebral	NIHSS and Barthel index improved	None	Cell viability

Table 3.1 (continued)

Author, reference	Study design	Stroke	Cell dose	Manipulation	Mode of application	Efficacy	Adverse effects	Safety test
Autologous bone-marrow-derived mesenchymal stem cells								
Bang, 2005 [9]	Control, $n=25$ Treatment, $n=5$ 1 year f/u	Subacute Large MCA infarct	1×10^8	Ex vivo culture expansion using fetal bovine serum	Intravenous	Barthel index improved at 3 months	None	Cell viability, MSC surface markers, bacteria, fungi, viral, mycoplasma
Lee, 2010 (STARTING trial) [10]	Control, $n=36$ Treatment, $n=16$ 5 years f/u	Subacute Large MCA infarct	1×10^8		Intravenous	mRS 0–3 increased in MSC but not in control group	None	
Honmou, 2011 [11]	No control group Treatment, $n=12$ 1 year f/u	Subacute to chronic variable	1×10^8	Ex vivo culture expansion using autologous serum	Intravenous	Improvement in daily rate of NIHSS changes	None	Cell viability, MSC surface markers, bacteria, syphilis, fungi, viral, mycoplasma, endotoxin
Bhasin, 2011 [45]	Control, $n=6$ Treatment, $n=6$ 24 weeks f/u	Chronic Ischemic or ICH	$5–6\times10^7$	Ex vivo culture expansion using serum-free media (Stem Pro)	Intravenous	Modest increase in Fugl-Meyer and modified Barthel index	None	Cell viability, mycoplasma, endotoxin

N/A not available, *f/u* follow-up, *ICH* intracerebral hemorrhage, *MCA* middle cerebral artery, *mRS* modified Rankin score, *MSC* mesenchymal stem cell, *NIHSS* National Institutes of Health Stroke Scale, *STARTING* stem cell Application Research and Trials In NeuroloGy

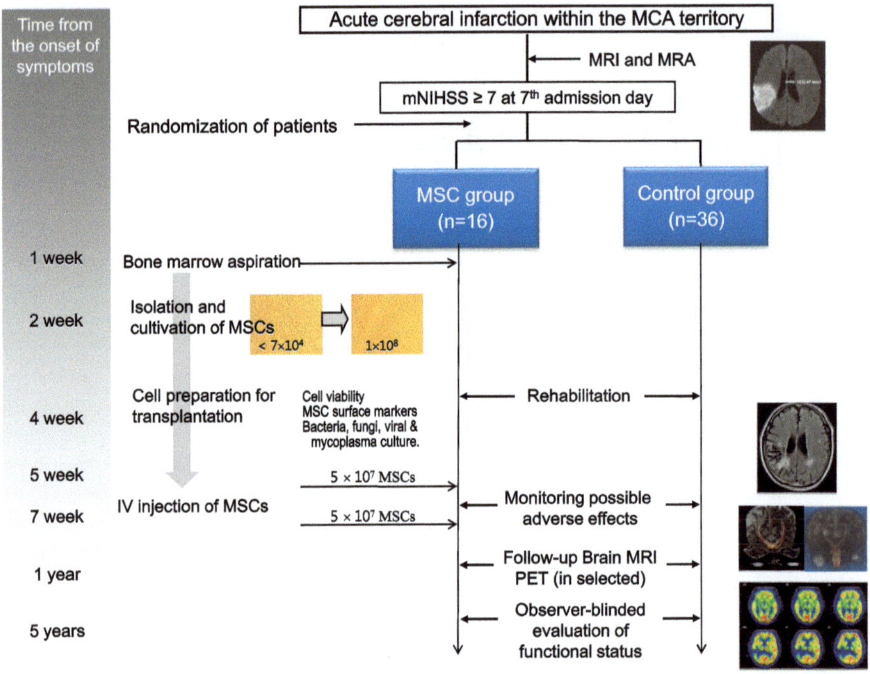

Fig. 3.1 Study protocol for the stem cell Application Research and Trials In NeuroloGy (STARTING) trial. *mNIHSS* modified National Institutes of Health Stroke Scale, *PET* positron emission tomography, *MRA* magnetic resonance angiography

Interestingly, a recent study showed that patients have unfounded expectations that stem cell therapy will improve function [15]. These fantasies may be a result of inappropriate media coverage and lack of information, and could be a hurdle in conducting randomized control trials of stem cell therapy in stroke patients.

In terms of safety, no adverse cell-related events have been reported in clinical adult stem cell trials in stroke patients. A recent meta-analysis of prospective clinical trials that used intravascular (intravenous or intra-arterial) MSC delivery, which included 1012 participants with ischemic stroke, Crohn's disease, cardiomyopathy, myocardial infarction, graft versus host disease, and healthy volunteers, showed that MSC therapy was not associated with any systemic complication, infection (except transient fever), malignancy, or death [16].

Considerations for Clinical Trials of MSC Therapy in Stroke Patients

For a stem cell therapy to be effective in augmenting recovery after stroke, the following criteria should be satisfied [17]: therapy should be both safe and effective; applicable to a broad spectrum of stroke patients, in terms of stroke subtypes (ischemic/hemorrhagic), time (acute/chronic), and vascular territories; and it should be inexpensive and cost-effective.

At this time, rigorous reasoning is required to replicate experimental results in stroke patients. The Stroke Treatment Academic Industry Roundtable (STAIR) recommendations were developed to improve the quality of preclinical and clinical research on neuroprotective and neurorestorative drugs, and to reduce the divide between them [18]. Based on these recommendations, studies should be randomized, controlled trials. After randomization, experimental procedures may not be blinded, as bone marrow sham aspiration should not be performed in control patients. A MSC dose equivalent to that used in animal studies (1×10^5–3×10^6 cells/rat), and based on mean body weight, should be used. Patient selection should be based on severity and location of lesions, and time of application. Although the optimal time window for administration is undecided, subacute periods of stroke should be the first candidate time period tested, i.e., after the acute period, when patients are likely to experience rapid worsening/improvement, and before the blood–brain barrier (BBB) closes and chemokine expression (i.e., stromal cell-derived factor 1α) in the infarcted brain disappears. As all preclinical studies were conducted in animal models of middle cerebral artery occlusion, patients with middle cerebral artery territory (or anterior circulation) stroke should be selected. Patients with moderate to severe neurological disabilities could be ideal candidates. Patients suffering from very severe strokes are likely to have poor outcomes regardless of intervention, whereas patients with minor strokes are not suitable for these potentially risky experimental treatments. In addition to the clinical outcomes measured, laboratory and neuroimaging findings should be used as surrogate markers of efficacy. Lastly, patients should be followed for more than 90 days. However, long-term monitoring (>6 months) is likely unnecessary because autologous MSCs are a clinical cell line and die within days/weeks of administration [19].

Recently, the Stem Cell Therapies as an Emerging Paradigm in Stroke (STEPS) committee suggested guidelines for both preclinical and clinical trials on stem cell therapies in stroke. The committee has suggested guidelines for bridging basic and clinical studies [20], early stage clinical trials [19], and phase II/III trials [21].

Perspectives in Stem Cell Therapy for Stroke

Although, to date, clinical trials of stem cell therapy have focused on the feasibility and safety of application of autologous MSCs or bone marrow cells in stroke patients, further trials to understand the therapeutic effects of stem cells are needed. Although the proportion of patients in the STARTING trial with a mRS of 0–3 significantly increased following MSC treatment, many patients in the MSC group remained significantly disabled [10].

The primary hurdles for current stem cell therapies include: the long culture period required to obtain sufficient stem cells, the relatively small number of MSCs that migrate into the injured brain, the death of stem cells within a toxic environment, limited trophic support by transplanted stem cells, and the use of xenogenic serum, with concomitant risk of transmission of prion diseases and zoonoses [22]. The following issues should be addressed in order to overcome these hurdles.

Allogenic Cells Allogenic MSCs could be a good alternative to autologous MSCs for a number of reasons. First, stroke commonly occurs in advanced age, often in conjunction with chronic illness, and MSCs from these patients can show reduced growth rates. In addition, although titers of hematopoietic bone marrow stem cells remain constant throughout life, the number of MSCs decline dramatically with age [23]. Second, the earlier the MSCs can be administered, the more the MSCs can migrate to the infarcted brain. The use of allogenic MSCs reduces the time required to obtain a sufficient number of cells. Finally, MSCs are both immunosuppressive and immunoprivileged, expressing little to no major histocompatibility class II or costimulatory molecules [24]. However, it has been reported that after contact with serum, allogenic MSCs can be injured by complement, and that viability of allogenic MSCs after infusion is greatly reduced compared to autologous MSCs [25].

Mode of Treatment The mode of application of MSCs may significantly influence the number of cells delivered to target regions, as well as the incidence of adverse effects. A major problem in introducing stem cells systemically is that cells may become trapped within organs that filter the bloodstream (first-pass effect). To avoid this, strategies to minimize lung adhesion and improve homing of systemically introduced cells are used, including the use of vasodilators, reduction in number of cells administered, or different routes of administration. For example, an intra-arterial approach can bypass pulmonary circulation; intra-arterial transplantations resulted in superior delivery and sustained presence of stem cells in the ischemic brain compared to intravenous infusions [26]. However, an arterial approach may cause arterial occlusion, resulting in stroke, and is reportedly not superior to intravenous approach for recovery after stroke [26, 27]. There have been relatively few studies directly comparing the efficacy of intravenous and intra-arterial delivery of MSCs. Although both intravenous and intra-arterial administration of MSC leads to functional recovery, the method of delivery may fundamentally alter their mechanism of action [28].

Extracellular Vesicles There may be problems associated with stem cell therapy including tumor formation, vascular occlusion causing infarcts, large cells resulting in limitations in crossing the BBB (especially during the chronic stage of stroke), and zoonosis by internalization of xenogenic serum (fetal bovine/calf serum) by MSCs during ex vivo culture expansion [22]. Stem cells secrete extracellular vesicles (EVs; e.g., microvesicles, exosomes) as well as soluble factors (e.g., trophic factors). Cell therapy using EVs derived from stem cells could avoid the cell-related problems described above, and could represent a new, clinically feasible and relatively safe paradigm. We have recently reported that strokes in humans trigger the mobilization of MSC-derived EVs [29]. The numbers of circulating MSC-derived microvesicles increased in patients with extensive ischemic stroke. However, the role of these microvesicles, and their biodistribution during stroke, is unclear. We and others have shown that intravenous administration of EVs derived from MSC culture media promote functional recovery and neurovascular plasticity after stroke in rats [30].

BBB Manipulation Molecules larger than 400 Da cannot pass through the BBB, which may affect the efficacy of cell therapy in stroke patients. Coadministration of stem cells and mannitol, an osmotic agent that shrinks and opens the BBB, may improve outcomes in stroke patients. In a preclinical study, BBB manipulation using intravenous mannitol prior to MSC treatment resulted in increases in trophic factors in the infarcted brain [31]. As mannitol is already widely used in clinical practice, BBB manipulation using mannitol should be considered in the future clinical trials.

Culture Expansion Conditions Current culture methods need to be improved. During a long period of ex vivo culture expansion, MSC characteristics may change significantly [5]. Several methods of culture expansion that improve the proliferation, survival, and trophic support of MSCs, and reduce senescence have been reported. First, hypoxic conditions (i.e., conditions similar to bone marrow) are beneficial to MSCs and may stimulate MSCs to exhibit adaptive responses. Preconditioning with hypoxia (0.1–2 % O_2) [32–36] reportedly increases anti-apoptotic gene expression, trophic factor release, ischemic tolerance, and C-X-C chemokine receptor type 4 expression. Second, treatment with trophic factors may alter MSC characteristics. We have shown that ex vivo trophic factor treatment during MSC cultivation enhances trophic support in the ischemic brain and further increases the production of trophic factors by MSCs, suggesting autocrine regulation of MSCs [37].

Finally, MSCs obtained at stroke onset may be optimal for use in cell therapy. MSCs derived from stroke patients may be better than MSCs from healthy donors. MSCs derived from rat stroke models exhibit increased trophic factor gene expression and enhanced restorative properties towards endogenous brain parenchymal cells compared to MSCs from normal rats [38]. Conversely, the characteristics of MSCs from stroke patients could change after the long process of stable culture expansion, and signals to MSCs in the blood may disappear at the time of cell administration [5]. Our data showed that culture expansion using ischemic serum could constitute a novel, feasible (using clinical grade cell lines as per the International Cellular Medicine Society), and effective preconditioning method for neurorestoration of stroke (unpublished data). Compared to MSC cultured with fetal bovine serum, MSCs preconditioned with autologous serum, obtained during the acute phase of stroke, exhibit higher cell proliferation rates and increased trophic factor release, and superior survival under ischemic brain conditions. Given these preclinical results, we have recently initiated STARTING-2, the first study that aims to evaluate the efficacy of MSC application in patients with ischemic stroke [39]. This clinical trial will determine the effectiveness and safety of autologous MSCs that are culture expanded in autologous ischemic serum and obtained as soon after stroke as possible. This trial (NCT01716481) is registered at http://clinicaltrials.gov.

Selection of Patients Selection of candidate patients for cell-based therapies should be optimized based on stroke severity, lesion location, and stroke chronicity. Due to the experimental nature of this treatment, clinical trials of cell-based therapies for stroke have studied patients with severe disabilities or chronic stroke, sometimes several years after stroke onset. However, it may be difficult to demonstrate therapeutic benefits in these cases. Preclinical studies in animal models of stroke have demonstrated the importance of neurogenesis; transplanted stem cells may enhance endogeneous neurogenesis in certain areas, including the subventricular zone [5, 18]. Our data from a clinical trial of MSC administration showed that patients with severe damage in preventricular areas, which limited endogeneous neurogenesis, had poor response to MSCs [27]. Therefore, it seems reasonable to preclude those patients from clinical trials of neurorestorative strategies, particularly for strategies that stimulate endogenous neurogenesis.

Ongoing Clinical Trials

At the time of writing, we are aware of at least nine active clinical trials using adult stem/progenitor cells to treat ischemic stroke (http://clinicaltrials.gov; Table 3.2). It should be noted that some of these trials are randomized controlled studies, aiming to test the efficacy of MSC therapy, and that two studies are testing the efficacy and safety of allogenic MSCs in stroke patients.

Clinical protocols must be established that consider recent advances in understanding the mechanisms by which stem cells aid recovery after stroke. In the STARTING-2 trial, we are incorporating ischemic preconditioning using ischemic serum, BBB manipulation, and strict selection of candidates in order to improve the therapeutic effects and safety of MSCs [39].

Conclusions

The therapeutic efficacy of stem cells can be improved. ESCs, and recently iPSCs, are not likely to be administered to stroke patients in the near future. Strategies that meet the FDA's regulations on stem cell use for clinical applications and enhance therapeutic efficacy require further preclinical and clinical trials (Fig. 3.2). These strategies should consider advances from bench (diverse cell sources and culture processes) to bedside (patient selection, BBB manipulation, and outcome measures).

Table 3.2 Ongoing clinical trials of mesenchymal stem cell therapy in adult stroke patients

Location	Diagnosis	*N*	Source	Route	Time after onset	Dosage	Measures of outcomes
Univ. Hospital Grenoble, France	Ischemic stroke	Treat = 30	Autologous BM-MSCs	Intravenous	< 6 weeks	N/A	1': Safety study of feasibility and tolerance 2': Clinical and functional effects
Chaitanya Hospital, India	Ischemic or hemorrhagic	Treat = 50	Autologous BM-MNCs	Intrathecal	N/A	1×10^8 cells	1': Improvement in power of body/face muscle 2': Improvement in walking/speech/vision
Imperial college, UK	Anterior circulation infarct	Treat = 10	Autologous CD34+stem cells	Intra-arterial	< 7 days	N/A	1': Safety 2': Improvements in mRS, NIHSS
Aldagen, USA	MCA infarct	Treat = 60 Control = 40	Autologous MSCs that express high levels of enzyme (ALD-401)	Carotid	13–19 days	N/A	1': Safety 2': Efficacy at 1 year (NIHSS, mRS, BI, EQ-5D)
Southern Medical Univ, China	MCA infarct	Treat = 60 Control = 30	Autologous BM-MSCs or EPCs	Intravenous	< 1 week	2.5×10^6 cells/kg, twice	1': Number of adverse events 2': Changes in mRS, BI
Wenzhou Medical Univ. China	Ischemic or hemorrhagic	Treat = 30	Autologous BM-MSCs	Intracerebral	3–60 months	$2–4 \times 10^6$ cells	1': NIHSS change from baseline at 12 months 2': Improvement of infarct size on MRI
Stemedica Cell Technologies, Inc	Chronic ischemic stroke	Treat = 33	Allogenic BM-MSCs; hypoxic growth conditions	Intravenous	> 6 months	$0.5–1.5 \times 10^6$ cells/kg	1': Safety 2': NIHSS, MMSE, BI, geriatric depression scale

Table 3.2 (continued)

Location	Diagnosis	*N*	Source	Route	Time after onset	Dosage	Measures of outcomes
Athersys, Inc, USA	MCA infarct	Treat = 72 Control = 64	Allogenic BM-adherent, multipotent adult progenitor cells (Multistem®)	Intravenous	24–36 h	4×10^8 or 12×10^8 cells	1': Dose determination, efficacy (90 day mRS) 2': Shift analysis, excellent outcome. Explorative: spleen and infarct size on CT/MRI
Samsung Medical Center, Korea (STARTING-2 trial)	MCA infarct	Treat = 40 Control = 20	Autologous BM-MSCs cultured with autologous serum obtained at acute phase	Intravenous	<90 days	1×10^6 cells/kg	1': Efficacy, mRS shift on 90th day 2': Changes in NIHSS; mRS, modified BI, motricity index, and Fugl-Meyer on 90th day

BM-MSC bone marrow mesenchymal stem cells, *MNC*s mononuclear cells, *EPC* endothelial progenitor cells, *N/A* not available, *NIHSS* National Institutes of Health Stroke Scale, *mRS* modified Rankin score, *BI* Barthel index, *MMSE* mini mental status examination, *EQ-5D* European quality of life five-dimensions questionnaire, *CT* computed tomography, *MRI* magnetic resonance imaging

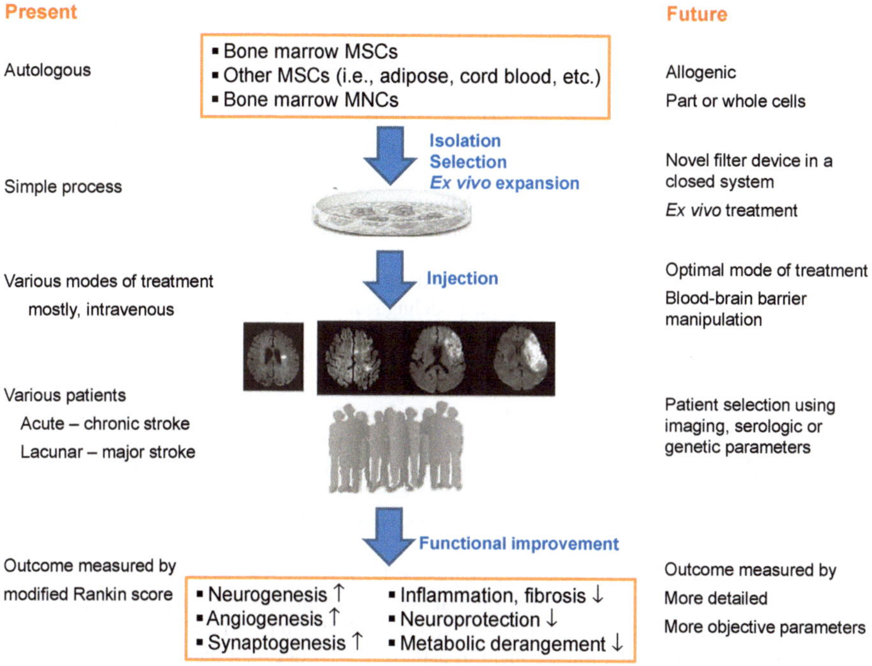

Fig. 3.2 Current status and perspectives on the application of adult mesenchymal stem cells for stroke. *MSC* mesenchymal stem cells, *MNC* mononuclear cells

References

1. Einstein O, Ben-Hur T. The changing face of neural stem cell therapy in neurologic diseases. Arch Neurol. 2008;65(4):452–6. Epub 2008/04/17.
2. Chopp M, Li Y. Treatment of neural injury with marrow stromal cells. Lancet Neurol. 2002;1(2):92–100. Epub 2003/07/10.
3. Chang DJ, Lee N, Park IH, Choi C, Jeon I, Kwon J, et al. Therapeutic potential of human induced pluripotent stem cells in experimental stroke. Cell Transplant. 2013;22(8):1427–40. Epub 2012/10/10.
4. Chen X, Li Y, Wang L, Katakowski M, Zhang L, Chen J, et al. Ischemic rat brain extracts induce human marrow stromal cell growth factor production. Neuropathology (Official Journal of the Japanese Society of Neuropathology). 2002;22(4):275–9. Epub 2003/02/05.
5. Li WY, Choi YJ, Lee PH, Huh K, Kang YM, Kim HS, et al. Mesenchymal stem cells for ischemic stroke: changes in effects after ex vivo culturing. Cell Transplant. 2008;17(9):1045–59. Epub 2009/01/31.
6. Nakanishi C, Nagaya N, Ohnishi S, Yamahara K, Takabatake S, Konno T, et al. Gene and protein expression analysis of mesenchymal stem cells derived from rat adipose tissue and bone marrow. Circ J (Official Journal of the Japanese Circulation Society). 2011;75(9):2260–8. Epub 2011/07/13.
7. Chen Y, Wang G, Zeng L. Adipose tissue or bone marrow, store for purchasing mesenchymal stem cells? Circ J (Official journal of the Japanese Circulation Society). 2011;75(9):2060–1. Epub 2011/08/06.

8. Chan TM, Harn HJ, Lin HP, Chou PW, Chen JY, Ho TJ, et al. Improved human mesenchymal stem cell isolation. Cell Transplant. 2014;23(4):399–406. Epub 2014/05/13.
9. Bang OY, Lee JS, Lee PH, Lee G. Autologous mesenchymal stem cell transplantation in stroke patients. Ann Neurol. 2005;57(6):874–82. Epub 2005/06/02.
10. Lee JS, Hong JM, Moon GJ, Lee PH, Ahn YH, Bang OY. A long-term follow-up study of intravenous autologous mesenchymal stem cell transplantation in patients with ischemic stroke. Stem Cells. 2010;28(6):1099–106. Epub 2010/05/28.
11. Honmou O, Houkin K, Matsunaga T, Niitsu Y, Ishiai S, Onodera R, et al. Intravenous administration of auto serum-expanded autologous mesenchymal stem cells in stroke. Brain. 2011;134(Pt 6):1790–807. Epub 2011/04/16.
12. Svensson J, Ghatnekar O, Lindgren A, Lindvall O, Norrving B, Persson U, et al. Societal value of stem cell therapy in stroke—a modeling study. Cerebrovasc Dis. 2012;33(6):532–9. Epub 2012/05/11.
13. Boncoraglio GB, Bersano A, Candelise L, Reynolds BA, Parati EA. Stem cell transplantation for ischemic stroke. Cochrane Database Syst Rev. 2010;(9):CD007231. Epub 2010/09/09.
14. Jeevanantham V, Butler M, Saad A, Abdel-Latif A, Zuba-Surma EK, Dawn B. Adult bone marrow cell therapy improves survival and induces long-term improvement in cardiac parameters: a systematic review and meta-analysis. Circulation. 2012;126(5):551–68. Epub 2012/06/26.
15. Kim YS, Chung DI, Choi H, Baek W, Kim HY, Heo SH, et al. Fantasies about stem cell therapy in chronic ischemic stroke patients. Stem Cells Dev. 2013;22(1):31–6. Epub 2012/07/13.
16. Lalu MM, McIntyre L, Pugliese C, Fergusson D, Winston BW, Marshall JC, et al. Safety of cell therapy with mesenchymal stromal cells (SafeCell): a systematic review and meta-analysis of clinical trials. PLoS One. 2012;7(10):e47559. Epub 2012/11/08.
17. Adams HP Jr, Nudo RJ. Management of patients with stroke: is it time to expand treatment options? Ann Neurol. 2013;74(1):4–10. Epub 2013/05/31.
18. Stroke Therapy Academic Industry Roundtable (STAIR). Recommendations for standards regarding preclinical neuroprotective and restorative drug development. Stroke. 1999;30(12):2752–8. Epub 1999/12/03.
19. Savitz SI, Chopp M, Deans R, Carmichael T, Phinney D, Wechsler L. Stem Cell Therapy as an Emerging Paradigm for Stroke (STEPS) II. Stroke. 2011;42(3):825–9. Epub 2011/01/29.
20. Stem Cell Therapies as an Emerging Paradigm in Stroke Participants. Stem Cell Therapies as an Emerging Paradigm in Stroke (STEPS): bridging basic and clinical science for cellular and neurogenic factor therapy in treating stroke. Stroke. 2009;40(2):510–5. Epub 2008/12/20.
21. Savitz SI, Cramer SC, Wechsler L. Stem cells as an emerging paradigm in stroke 3: enhancing the development of clinical trials. Stroke. 2014;45(2):634–9. Epub 2013/12/26.
22. Spees JL, Gregory CA, Singh H, Tucker HA, Peister A, Lynch PJ, et al. Internalized antigens must be removed to prepare hypoimmunogenic mesenchymal stem cells for cell and gene therapy. Mol Ther (The Journal of the American Society of Gene Therapy). 2004;9(5):747–56. Epub 2004/05/04.
23. Caplan AI. Why are MSCs therapeutic? New data: new insight. J Pathol. 2009;217(2):318–24. Epub 2008/11/22.
24. Trounson A, Thakar RG, Lomax G, Gibbons D. Clinical trials for stem cell therapies. BMC Med. 2011;9:52. Epub 2011/05/17.
25. Li Y, Lin F. Mesenchymal stem cells are injured by complement after their contact with serum. Blood. 2012;120(17):3436–43. Epub 2012/09/12.
26. Pendharkar AV, Chua JY, Andres RH, Wang N, Gaeta X, Wang H, et al. Biodistribution of neural stem cells after intravascular therapy for hypoxic-ischemia. Stroke. 2010;41(9):2064–70. Epub 2010/07/10.
27. Yang B, Migliati E, Parsha K, Schaar K, Xi X, Aronowski J, et al. Intra-arterial delivery is not superior to intravenous delivery of autologous bone marrow mononuclear cells in acute ischemic stroke. Stroke. 2013;44(12):3463–72. Epub 2013/10/12.
28. Eckert MA, Vu Q, Xie K, Yu J, Liao W, Cramer SC, et al. Evidence for high translational potential of mesenchymal stromal cell therapy to improve recovery from ischemic stroke. J Cereb Blood Flow Metab. 2013;33(9):1322–34. Epub 2013/06/13.

29. Kim SJ, Moon GJ, Cho YH, Kang HY, Hyung NK, Kim D, et al. Circulating mesenchymal stem cells microparticles in patients with cerebrovascular disease. PLoS One. 2012;7(5):e37036. Epub 2012/05/23.
30. Xin H, Li Y, Cui Y, Yang JJ, Zhang ZG, Chopp M. Systemic administration of exosomes released from mesenchymal stromal cells promote functional recovery and neurovascular plasticity after stroke in rats. J Cereb Blood Flow Metab. 2013;33(11):1711–5. Epub 2013/08/22.
31. Borlongan CV, Hadman M, Sanberg CD, Sanberg PR. Central nervous system entry of peripherally injected umbilical cord blood cells is not required for neuroprotection in stroke. Stroke. 2004;35(10):2385–9. Epub 2004/09/04.
32. Hu X, Yu SP, Fraser JL, Lu Z, Ogle ME, Wang JA, et al. Transplantation of hypoxia-preconditioned mesenchymal stem cells improves infarcted heart function via enhanced survival of implanted cells and angiogenesis. J Thorac Cardiovasc Surg. 2008;135(4):799–808. Epub 2008/04/01.
33. Liu H, Liu S, Li Y, Wang X, Xue W, Ge G, et al. The role of SDF-1-CXCR4/CXCR7 axis in the therapeutic effects of hypoxia-preconditioned mesenchymal stem cells for renal ischemia/reperfusion injury. PLoS One. 2012;7(4):e34608. Epub 2012/04/19.
34. Liu H, Xue W, Ge G, Luo X, Li Y, Xiang H, et al. Hypoxic preconditioning advances CXCR4 and CXCR7 expression by activating HIF-1alpha in MSCs. Biochem Biophys Res Commun. 2010;401(4):509–15. Epub 2010/09/28.
35. Pasha Z, Wang Y, Sheikh R, Zhang D, Zhao T, Ashraf M. Preconditioning enhances cell survival and differentiation of stem cells during transplantation in infarcted myocardium. Cardiovasc Res. 2008;77(1):134–42. Epub 2007/11/17.
36. Tang YL, Zhu W, Cheng M, Chen L, Zhang J, Sun T, et al. Hypoxic preconditioning enhances the benefit of cardiac progenitor cell therapy for treatment of myocardial infarction by inducing CXCR4 expression. Circ Res. 2009;104(10):1209–16. Epub 2009/05/02.
37. Choi YJ, Li WY, Moon GJ, Lee PH, Ahn YH, Lee G, et al. Enhancing trophic support of mesenchymal stem cells by ex vivo treatment with trophic factors. J Neurol Sci. 2010;298(1–2):28–34. Epub 2010/09/25.
38. Zacharek A, Shehadah A, Chen J, Cui X, Roberts C, Lu M, et al. Comparison of bone marrow stromal cells derived from stroke and normal rats for stroke treatment. Stroke. 2010;41(3):524–30. Epub 2010/01/09.
39. Kim SJ, Moon GJ, Chang WH, Kim YH, Bang OY. Intravenous transplantation of mesenchymal stem cells preconditioned with early phase stroke serum: current evidence and study protocol for a randomized trial. Trials. 2013;14:317. Epub 2013/10/03.
40. Suarez-Monteagudo C, Hernandez-Ramirez P, Alvarez-Gonzalez L, Garcia-Maeso I, de la Cuetara-Bernal K, Castillo-Diaz L, et al. Autologous bone marrow stem cell neurotransplantation in stroke patients. An open study. Restor Neurol Neurosci. 2009;27(3):151–61. Epub 2009/06/18.
41. Battistella V, de Freitas GR, da Fonseca LM, Mercante D, Gutfilen B, Goldenberg RC, et al. Safety of autologous bone marrow mononuclear cell transplantation in patients with nonacute ischemic stroke. Regen Med. 2011;6(1):45–52. Epub 2010/12/24.
42. Savitz SI, Misra V, Kasam M, Juneja H, Cox CS, Jr, Alderman S, et al. Intravenous autologous bone marrow mononuclear cells for ischemic stroke. Ann Neurol. 2011;70(1):59–69. Epub 2011/07/26.
43. Friedrich MA, Martins MP, Araujo MD, Klamt C, Vedolin L, Garicochea B, et al. Intra-arterial infusion of autologous bone marrow mononuclear cells in patients with moderate to severe middle cerebral artery acute ischemic stroke. Cell Transplant. 2012;21(Suppl 1):S13–21. Epub 2012/04/25.
44. Li ZM, Zhang ZT, Guo CJ, Geng FY, Qiang F, Wang LX. Autologous bone marrow mononuclear cell implantation for intracerebral hemorrhage-a prospective clinical observation. Clin Neurol Neurosurg. 2013;115(1):72–6. Epub 2012/06/05.
45. Bhasin A, Srivastava MV, Kumaran SS, Mohanty S, Bhatia R, Bose S, et al. Autologous mesenchymal stem cells in chronic stroke. Cerebrovasc Dis Extra. 2011;1(1):93–104. Epub 2011/01/01.

29. Kim SJ, Moon GJ, Cho YH, Kang HY, Hyung NK, Kim D, et al. Circulating mesenchymal stem cells microparticles in patients with cerebrovascular disease. PLoS One. 2012;7(5):e37036. Epub 2012/05/23.
30. Xin H, Li Y, Cui Y, Yang JJ, Zhang ZG, Chopp M. Systemic administration of exosomes released from mesenchymal stromal cells promote functional recovery and neurovascular plasticity after stroke in rats. J Cereb Blood Flow Metab. 2013;33(11):1711–5. Epub 2013/08/22.
31. [illegible]
32. Hu X, Yu SP, Fraser JL, Lu Z, Ogle ME, Wang JA, et al. Transplantation of hypoxia-preconditioned mesenchymal stem cells improves [illegible] via enhanced survival of [illegible] 2008;135(4):799–808. Epub 2008/04/01.
33. [illegible]
34. [illegible]
35. [illegible]
36. [illegible]
37. [illegible]
38. [illegible]
39. [illegible]
40. [illegible] with moderate to [illegible]
41. Li ZM, Zhang ZT, Guo CJ, Geng FY, Qiang F, Wang LX. Autologous bone marrow mononuclear cell implantation for [illegible]
42. [illegible]

Chapter 4
Treatment of Autologous Bone Marrow Mononuclear Cells for Acute and Subacute Stroke

Cell Therapy for Acute/Subacute Stroke

Yukiko Kasahara, Tomohiro Matsuyama and Akihiko Taguchi

Introduction

Stroke is the third leading cause of death in developed countries after heart disease and cancer [1], and the leading cause of disability worldwide. More than 50 % of stroke survivors are unable to completely recover, and 20 % of stroke patients require assistance with their daily activities [2]. Acute ischemic stroke is the most common type and has only thrombolysis as a therapeutic option [3]. Although thrombolytic therapy is effective for acute cerebral ischemia, it must be given within 4.5 h after stroke onset [4], and no definitive treatment exists after that period other than rehabilitation. Thus, development of novel therapies to regenerate neuronal function after stroke is eagerly awaited (Fig. 4.1). Recently, many studies have focused on cell-based therapies to repair the ischemic brain [5–9].

There are two main types of cells to enhance endogenous neurogenesis after stroke, i.e., mononuclear cells and mesenchymal stem cells. Although some clinical trials of mesenchymal stem cells have demonstrated safety, feasibility, and preliminary efficacy in stroke patients [10, 11], autologous mesenchymal stem cells require cell culture to obtain the required dose and cannot be administered in patients with acute stroke. In contrast, mononuclear cells can be prepared rapidly within a few hours and permit autologous administration, which avoids the problem of immunological rejection.

To develop novel therapies for patients after stroke, the therapeutic potential of bone marrow mononuclear cells has been investigated in experimental stroke

A. Taguchi (✉) · Y. Kasahara
Department of Regenerative Medicine Research, Institute of Biomedical Research and Innovation, Hyogo, Japan
e-mail: taguchi@fbri.org

T. Matsuyama
Institute for Advanced Medical Sciences, Hyogo College of Medicine, Hyogo, Japan

D. C. Hess (ed.), *Cell Therapy for Brain Injury*, DOI 10.1007/978-3-319-15063-5_4

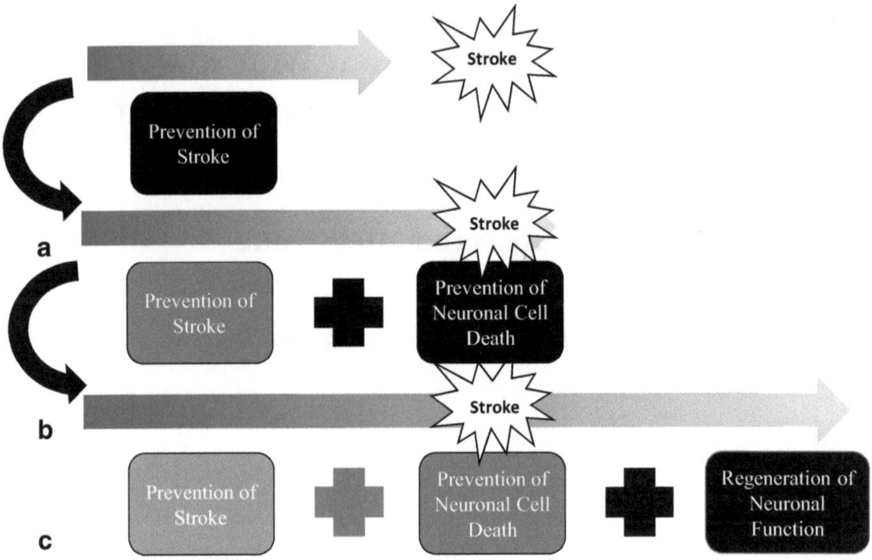

Fig. 4.1 Change in the strategy for stroke. **a** Treatment of cardiovascular factors and administration of antiplatelet/coagulant drugs significantly contribute to the prevention of stroke onset. **b** In addition to the prevention of stroke onset, development of thrombolysis and neurothrombectomy in the acute period enables prevention of neuronal cell death after cerebral vascular occlusion. **c** Furthermore, establishment of novel therapies that extend the therapeutic time window and broaden treatment options to regenerate neuronal function after stroke is eagerly awaited

models, followed by various clinical trials. This chapter summarizes the findings of recent basic science and clinical studies that have focused on regeneration of the injured brain using autologous bone marrow mononuclear cells in the acute/subacute stage of stroke.

Neuronal Regeneration After Cerebral Ischemia

Although it had been generally believed that the neuroregeneration in the adult mammalian brain does not occur until the mid-twentieth century, it became recognized that new neurons are continuously generated throughout life in the adult mammalian brain. Under normal, nondisease physiological conditions, neurogenesis is principally restricted to two regions, the subventricular zone (SVZ) of the lateral ventricles and the subgranular zone (SGZ) of the hippocampal dentate gyrus [12, 13], where unique niche architectures permit continuous neurogenesis [14, 15]. However, accumulating evidence has indicated the presence of neuronal stem cells in a variety of adult brain regions [16, 17]. Moreover, although it remains a matter of controversy as to whether neuronal stem cells in these regions are essentially similar to SVZ-derived neuronal stem cells, the presence of stroke-induced neuronal stem cells at the cerebral cortex has also been suggested in the adult murine brain [18]. Following an ischemic insult, the proliferation and/or dedifferentiation of endogenous neuronal stem cells is activated in various brain regions, including the SVZ, SGZ, striatum, and cerebral cortex [19] and these neuronal stem cells have been shown to migrate into the injured area, where neurogenesis does not occur under normal conditions [20–22]. Similar to the findings from the murine stroke model, we demonstrated the presence of stroke-induced neural stem/progenitor cells in patients, and that the peak of endogenous neurogenesis is approximately 1–2 weeks after a stroke [23]. These findings indicate the potential for a novel therapeutic strategy using injury-induced neurogenesis for functional recovery in patients with cerebral infarction.

However, the neurogenic response eventually yields only a very small number of mature neurons, as most of these stroke-induced neural stem/progenitor cells do not survive, nor do they contribute to functional recovery after stroke [20]. Thus, appropriate support for the survival of these stroke-induced neural stem/progenitor cells is essential for functional recovery after cerebral ischemia.

Angiogenesis for the Survival of Injury-Induced Neuronal Stem Cells

In the peri-infarction area, microvascular density decreases [24] and most injury-induced neural stem/progenitor cells are unable to survive there [20]. Angiogenesis after stroke had been investigated as the key element for the survival of injury-induced neural stem/progenitor cells and functional recovery after cortical infarction. Recent studies have indicated that there is a tight correlation between angiogenesis and neurogenesis under both physiological and pathological conditions in the adult brain. In the adult songbird, testosterone-induced angiogenesis leads to neuronal recruitment into the higher vocal center [25]. In the adult rat, endogenous neurogenesis and neovascularization occur in proximity to one another in the cortex following focal ischemia [26]. Moreover, angiogenesis and neurogenesis are regulated by an overlapping set of molecules—for example, sphongosine-1-phosphate plays a critical role in neurogenesis and angiogenesis during embryonic development [27]. The accumulating evidence indicates a close relationship between the vascular system and neurogenesis in the central nervous system, and recent studies that have explored therapeutic strategy have focused on promotion of neurogenesis in association with angiogenesis [6, 9]. Although the coupling and cross talk between endogenous neurogenesis and neovascularization in the cortex of the ischemic brain are still not fully understood, these findings clearly indicate that therapeutic angiogenesis could have a significant role in the functional recovery of stroke patients by enhancing neurogenesis in the poststroke brain.

Cell-Based Therapy to Enhance Neurogenesis in the Ischemic Brain

To achieve angiogenesis in ischemic tissue, an approach using bone-marrow-derived mononuclear cells, a rich cell source of both hematopoietic stem cells and endothelial stem/progenitor cells, has been proposed. Increasing evidence shows that endothelial stem/progenitor cells play an important role in maintaining vascular homeostasis and repair. Endothelial stem/progenitor cells have been shown to contribute to vascular homeostasis through differentiation to endothelial cells [28] and as a source of numerous growth and angiogenesis factors (e.g., vascular endothelial growth factor (VEGF), hepatocyte growth factor, and insulin-like growth factor I) [29]. Endothelial stem/progenitor cells, mainly obtained from bone marrow cells, have been shown to reduce ischemic damage and enhance functional recovery in experimental models, including limb [30–33], myocardium [34–37], and cerebral ischemia [38, 39] models. Based on these observations, various clinical trials using bone-marrow-derived endothelial stem/progenitor cells are ongoing, with promising results that show improvement of regional perfusion and function in ischemic tissues [40–42].

In addition, we observed that decreased levels of circulating immature bone-marrow-derived cells, such as endothelial stem/progenitor cells, are associated with impaired cerebrovascular function [39] and cognitive impairment [43, 44]. In contrast, high levels of bone-marrow-derived immature cells are associated with neovascularization of the ischemic brain [45].

Based on these observations, we investigated the effect of administrating bone-marrow-derived stem/progenitor cells on stroke using a highly reproducible murine model [46]. We found that transplantation of bone marrow mononuclear cells or hematopoietic stem cells after stroke induces neovascularization at the border of the ischemic zone followed by reconstruction of blood flow, that neovascularization is essential for the survival of neural stem cells in the cortex of ischemic brain, and that the support survival of neural stem cells contributes to functional outcomes improvement [7, 9, 47]. To link these basic findings to clinical trials, we investigated the appropriate cell numbers and optimal therapeutic time window for bone-marrow-derived bone marrow cell transplantation for stroke. We found that the required minimum number of bone-marrow-derived mononuclear cells was 1×10^6/kg of body weight and the therapeutic time window of administration of bone-marrow-derived mononuclear cells was revealed to be between day 2 and day 14 after stroke [47]. It is notable that this therapeutic time window overlaps with the peak in endogenous neurogenesis after stroke [23]. This positive effect of bone-marrow-derived mononuclear cells was negated by administration of an anti-angiogenesis reagent [7]. These findings suggest that therapeutic angiogenesis, achieved by administering bone marrow mononuclear cells, could be a novel therapeutic strategy for patients after stroke.

Although the mechanisms that link endothelial stem/progenitor cells, including bone marrow mononuclear cell transplantation, and angiogenesis is not fully understood, a recent study suggested that treatment with bone marrow mononuclear cells at the acute stroke stage increases cerebral blood flow (CBF) through endothelial nitric oxide synthase (eNOS) activation and NO production which leads to vasodilation, and subsequently promotes angiogenesis [48].

Based on these findings, clinical trials of the administration of autologous bone marrow mononuclear cells for patients in the acute/subacute stroke stage have been initiated in many institutes, including our hospital.

Clinical Trials Using Bone Marrow Mononuclear Cells in Patients After Stroke

We conducted a clinical trial to enhance neurogenesis and functional recovery through activating angiogenesis in patients with cerebral infarction. Our trial was an unblinded, uncontrolled phase 1/2a clinical trial aimed at investigating the feasibility and safety of autologous bone marrow mononuclear cell transplantation in subacute stroke patients. Major inclusion criteria were patients with cerebral embolism, a National Institute of Health Stroke Scale (NIHSS) score higher than nine on day 7 after stroke and the improvement in the NIHSS in the first 7 days after onset of less than six points. On day 7–10 after stroke, patients had 25 ml (low-dose group, $N=6$) or 50 ml (high-dose group, $N=6$) of bone marrow cell aspiration from the posterior iliac bone under local anesthesia. Autologous bone-marrow-derived mononuclear cells were purified by the density gradient method and administered intravenously on the same day as the aspiration. Primary outcome measures were: worsening NIHSS score (primary safety outcome measure) and change in the NIHSS score evaluated on day 7 after onset of stroke and day 30 after cell transplantation (primary efficacy outcome measure). We also evaluated the changes in regional cerebral blood and regional cerebral metabolic rate of oxygen consumption using steady-state ^{15}O positron emission tomography at 1 and 6 months after cell transplantation. The results of the study showed that administration of autologous bone marrow mononuclear cells in patients with severe stroke was both feasible and safe. Furthermore, the positive trends favoring neurologic recovery and improvement in CBF and metabolism in poststroke patients receiving cell therapy underscored the potential of this approach. Details of the results are now under submission. The clinical findings further support our hypothesis that bone marrow mononuclear cells transplantation after stroke improves CBF and neuronal activity that results in acceleration of functional recovery (Fig. 4.2). Similar clinical trials to ours, such as transplantation of autologous bone marrow mononuclear cells in stroke patients, are being carried out in other countries, including the USA, India, Brazil, and Spain with promising results [49–52]. Though the route of administration (intravenous or intra-arterial) and the stage of stroke (acute or subacute) vary, no side effects or safety problems with cell therapy have been reported. The current status of most of these ongoing clinical trials can be searched through http://clinicaltrials.gov/.

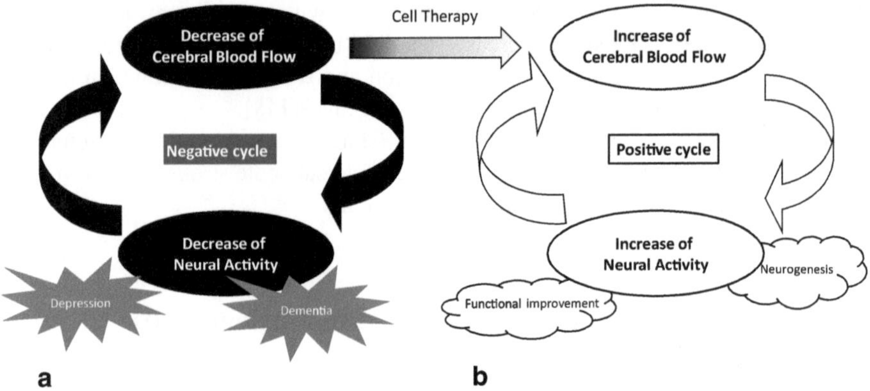

Fig. 4.2 Activation of microvasculature by cell therapy. **a** Cerebrovascular damage is closely related with cognitive impairment and depression with impaired neural activity and cerebral blood flow. **b** Cell therapy is expected to activate the cerebral microvasculature that enhances neurogenesis and functional recovery after stroke

Further Development of Cell Therapy for Patients After Stroke

Although several studies have indicated that bone-marrow-derived cells could be a source of endothelial cells [53, 54], a growing body of clinical and experimental evidence indicates that the number of injected cells reaching the brain parenchyma seem to be small, i.e., preclinical studies indicate that approximately 0.02–1 % of injected cells home into the brain [55–57]. Despite significant activation of angiogenesis by cell transplantation, the survival of transplanted cells has rarely been observed in experimental models; thus, the differentiation of the stem cells into endothelial cells in the ischemic brain may not play a critical role in angiogenesis after stroke. These findings suggest that cells do not need to remain in the brain to generate functional improvement. Therefore, many investigators have been focusing on where the cells go and what they do. Schwarting et al. suggested that homing of injected cells to spleen suppressed the infiltration of immune cells, such as T cells and monocytes, into the ischemic cerebral tissue, and consequently the infarct size was reduced [55]. Recent studies have reported that higher radioactive counts were observed in the lungs and spleen at 2 h post injection after technetium-99m labeled bone marrow mononuclear cell transplantation in animals and patients after stroke [51, 56]. The therapeutic effect of bone marrow mononuclear cells is achieved, we believe, by the activation of the systemic microvasculature as well as a local response. It is likely that multiple cytokines, growth factors, and cell adhesion molecules are involved, and the balance between these molecules may determine the fate of injured brain tissue.

In conclusion, the positive results of experimental stroke model and clinical trials indicate the potential of cell therapy for stroke patients, and larger scale, randomized controlled clinical trials are desirable in order to prove the efficacy and long-term safety of such treatment. Furthermore, elucidation of the therapeutic mechanism is one of the key elements in developing novel strategies to improve functional recovery in patients after stroke.

References

1. Pearson TA. Cardiovascular disease in developing countries: myths, realities, and opportunities. Cardiovasc Drugs Ther (sponsored by the International Society of Cardiovascular Pharmacotherapy). 1999;13(2):95–104.
2. Bonita R, Solomon N, Broad JB. Prevalence of stroke and stroke-related disability. Estimates from the Auckland stroke studies. Stroke J Cereb Circ. 1997;28(10):1898–902.
3. Adams HP Jr, del Zoppo G, Alberts MJ, et al. Guidelines for the early management of adults with ischemic stroke: a guideline from the American Heart Association/American Stroke Association Stroke Council, Clinical Cardiology Council, Cardiovascular Radiology and Intervention Council, and the Atherosclerotic Peripheral Vascular Disease and Quality of Care Outcomes in Research Interdisciplinary Working Groups: the American Academy of Neurology affirms the value of this guideline as an educational tool for neurologists. Stroke J Cereb Circ. 2007;38(5):1655–711. doi:10.1161/STROKEAHA.107.181486.
4. Xu ZP, Li HH, Li YH, Zhang Y, Wu Q, Lin L. Feasibility and outcomes of intravenous thrombolysis 3-4.5 hours after stroke in Chinese patients. J Clin Neurosci. 2014;21(5):822–6. doi:10.1016/j.jocn.2013.08.014.
5. Kim SS, Yoo SW, Park TS, et al. Neural induction with neurogenin1 increases the therapeutic effects of mesenchymal stem cells in the ischemic brain. Stem Cells. 2008;26(9):2217–28. doi:10.1634/stemcells.2008-0108.
6. Nakagomi N, Nakagomi T, Kubo S, et al. Endothelial cells support survival, proliferation, and neuronal differentiation of transplanted adult ischemia-induced neural stem/progenitor cells after cerebral infarction. Stem Cells. 2009a;27(9):2185–95.
7. Nakano-Doi A, Nakagomi T, Fujikawa M, et al. Bone marrow mononuclear cells promote proliferation of endogenous neural stem cells through vascular niches after cerebral infarction. Stem Cells. 2010;28(7):1292–302. doi:10.1002/stem.454.
8. Pendharkar AV, Chua JY, Andres RH, et al. Biodistribution of neural stem cells after intravascular therapy for hypoxic-ischemia. Stroke J Cereb Circ. 2010;41(9):2064–70. doi:10.1161/STROKEAHA.109.575993.
9. Taguchi A, Soma T, Tanaka H, et al. Administration of CD34 + cells after stroke enhances neurogenesis via angiogenesis in a mouse model. J Clin Invest. 2004b;114(3):330–8.
10. Eckert MA, Vu Q, Xie K, et al. Evidence for high translational potential of mesenchymal stromal cell therapy to improve recovery from ischemic stroke. J Cereb Blood Flow Metab. 2013;33(9):1322–34. doi:10.1038/jcbfm.2013.91.
11. Lalu MM, McIntyre L, Pugliese C, et al. Safety of cell therapy with mesenchymal stromal cells (SafeCell): a systematic review and meta-analysis of clinical trials. PloS ONE. 2012;7(10):e47559. doi:10.1371/journal.pone.0047559.
12. Lledo PM, Alonso M, Grubb MS. Adult neurogenesis and functional plasticity in neuronal circuits. Nat Rev Neurosci. 2006;7(3):179–93.
13. Ma DK, Bonaguidi MA, Ming GL, Song H. Adult neural stem cells in the mammalian central nervous system. Cell Res. 2009;19(6):672–82.
14. Alvarez-Buylla A, Garcia-Verdugo JM. Neurogenesis in adult subventricular zone. J Neurosci. 2002;22(3):629–34. doi:22/3/629 [pii].
15. Gage FH. Molecular and cellular mechanisms contributing to the regulation, proliferation and differentiation of neural stem cells in the adult dentate gyrus. Keio J Med. 2010;59(3):79–83.

16. Morshead CM, Reynolds BA, Craig CG, et al. Neural stem cells in the adult mammalian forebrain: a relatively quiescent subpopulation of subependymal cells. Neuron. 1994;13(5): 1071–82.
17. Yanamoto H, Miyamoto S, Tohnai N, et al. Induced spreading depression activates persistent neurogenesis in the subventricular zone, generating cells with markers for divided and early committed neurons in the caudate putamen and cortex. Stroke J Cereb Circ. 2005;36(7):1544–50. doi:10.1161/01.STR.0000169903.09253.c7.
18. Nakagomi T, Taguchi A, Fujimori Y, et al. Isolation and characterization of neural stem/progenitor cells from post-stroke cerebral cortex in mice. Eur J Neurosci. 2009b;29(9):1842–52. doi:10.1111/j.1460-9568.2009.06732.x.
19. Darsalia V, Heldmann U, Lindvall O, Kokaia Z. Stroke-induced neurogenesis in aged brain. Stroke. 2005;36(8):1790–5. doi:01.STR.0000173151.36031.be [pii]10.1161/01.STR.0000173151.36031.be.
20. Arvidsson A, Collin T, Kirik D, Kokaia Z, Lindvall O. Neuronal replacement from endogenous precursors in the adult brain after stroke. Nat Med. 2002;8(9):963–70. doi:10.1038/nm747.
21. Parent JM, Vexler ZS, Gong C, Derugin N, Ferriero DM. Rat forebrain neurogenesis and striatal neuron replacement after focal stroke. Ann Neurol. 2002;52(6):802–13. doi:10.1002/ana.10393.
22. Zhang R, Zhang Z, Wang L, et al. Activated neural stem cells contribute to stroke-induced neurogenesis and neuroblast migration toward the infarct boundary in adult rats. J Cereb Blood Flow Metab. 2004;24(4):441–8.
23. Nakayama D, Matsuyama T, Ishibashi-Ueda H, et al. Injury-induced neural stem/progenitor cells in post-stroke human cerebral cortex. Eur J Neurosci. 2010;31(1):90–8.
24. Taguchi A, Zhu P, Cao F, et al. Reduced ischemic brain injury by partial rejuvenation of bone marrow cells in aged rats. J Cereb Blood Flow Metab. 2011;31(3):855–67. doi:10.1038/jcbfm.2010.165.
25. Louissaint A Jr, Rao S, Leventhal C, Goldman SA. Coordinated interaction of neurogenesis and angiogenesis in the adult songbird brain. Neuron. 2002;34(6):945–60.
26. Shin HY, Kim JH, Phi JH, et al. Endogenous neurogenesis and neovascularization in the neocortex of the rat after focal cerebral ischemia. J Neurosci Res. 2008;86(2):356–67.
27. Mizugishi K, Yamashita T, Olivera A, Miller GF, Spiegel S, Proia RL. Essential role for sphingosine kinases in neural and vascular development. Mol Cell Biol. 2005;25(24):11113–21. doi:25/24/11113 [pii]10.1128/MCB.25.24.11113-11121.2005.
28. Asahara T, Murohara T, Sullivan A, et al. Isolation of putative progenitor endothelial cells for angiogenesis. Science. 1997;275(5302):964–7.
29. Majka M, Janowska-Wieczorek A, Ratajczak J, et al. Numerous growth factors, cytokines, and chemokines are secreted by human CD34(+) cells, myeloblasts, erythroblasts, and megakaryoblasts and regulate normal hematopoiesis in an autocrine/paracrine manner. Blood. 2001;97(10):3075–85.
30. Duong Van Huyen JP, Smadja DM, Bruneval P, et al. Bone marrow-derived mononuclear cell therapy induces distal angiogenesis after local injection in critical leg ischemia. Mod Pathol. 2008;21(7):837–46. doi:modpathol200848 [pii]10.1038/modpathol.2008.48.
31. Padilla L, Krotzsch E, De La Garza AS, et al. Bone marrow mononuclear cells stimulate angiogenesis when transplanted into surgically induced fibrocollagenous tunnels: results from a canine ischemic hindlimb model. Microsurgery. 2007;27(2):91–7. doi:10.1002/micr.20289.
32. Tachi Y, Fukui D, Wada Y, et al. Changes in angiogenesis-related factors in serum following autologous bone marrow cell implantation for severe limb ischemia. Expert Opin Biol Ther. 2008;8(6):705–12. doi:10.1517/14712598.8.6.70510.1517/14712598.8.6.705 [pii].
33. Talapkova R, Hudecek J, Sinak I, et al. [The salvage of ischaemic limb by therapeutical angiogenesis]. Vnitr Lek. 2009;55(3):179–83.
34. Tatsumi T, Matsubara H. Therapeutic angiogenesis for peripheral arterial disease and ischemic heart disease by autologous bone marrow cells implantation. Nihon Rinsho. 2006;64(11): 2126–34.

35. Tse HF, Siu CW, Zhu SG, et al. Paracrine effects of direct intramyocardial implantation of bone marrow derived cells to enhance neovascularization in chronic ischaemic myocardium. Eur J Heart Fail. 2007;9(8):747–53.
36. Yokokura Y, Hayashida N, Okazaki T, et al. Influence of angiogenesis by implantation of bone marrow mononuclear cells in the rat ischemic heart. Kurume Med J. 2007;54(3–4):77–84. doi:JST.JSTAGE/kurumemedj/54.77 [pii].
37. Zen K, Okigaki M, Hosokawa Y, et al. Myocardium-targeted delivery of endothelial progenitor cells by ultrasound-mediated microbubble destruction improves cardiac function via an angiogenic response. J Mol Cell Cardiol. 2006;40(6):799–809. doi:S0022-2828(06)00076-9 [pii]10.1016/j.yjmcc.2006.03.012.
38. Fan Y, Shen F, Frenzel T, et al. Endothelial progenitor cell transplantation improves long-term stroke outcome in mice. Ann Neurol. 2010;67(4):488–97. doi:10.1002/ana.21919.
39. Taguchi A, Matsuyama T, Moriwaki H, et al. Circulating CD34-positive cells provide an index of cerebrovascular function. Circulation. 2004a;109(24):2972–5. doi:10.1161/01.CIR.0000133311.25587.DE.
40. Hamano K, Nishida M, Hirata K, et al. Local implantation of autologous bone marrow cells for therapeutic angiogenesis in patients with ischemic heart disease: clinical trial and preliminary results. Jpn Circ J. 2001;65(9):845–7.
41. Taguchi A, Ohtani M, Soma T, Watanabe M, Kinosita N. Therapeutic angiogenesis by autologous bone-marrow transplantation in a general hospital setting. Eur J Vasc Endovasc Surg. 2003;25(3):276–8.
42. Tateishi-Yuyama E, Matsubara H, Murohara T, et al. Therapeutic angiogenesis for patients with limb ischaemia by autologous transplantation of bone-marrow cells: a pilot study and a randomised controlled trial. Lancet. 2002;360(9331):427–35. doi:10.1016/S0140-6736(02)09670-8.
43. Taguchi A, Matsuyama T, Nakagomi T, et al. Circulating CD34-positive cells provide a marker of vascular risk associated with cognitive impairment. J Cereb Blood Flow Metab. 2008;28(3):445–9. doi:10.1038/sj.jcbfm.9600541.
44. Taguchi A, Nakagomi N, Matsuyama T, et al. Circulating CD34-positive cells have prognostic value for neurologic function in patients with past cerebral infarction. J Cereb Blood Flow Metab. 2009;29(1):34–8. doi:10.1038/jcbfm.2008.92.
45. Yoshihara T, Taguchi A, Matsuyama T, et al. Increase in circulating CD34-positive cells in patients with angiographic evidence of moyamoya-like vessels. J Cereb Blood Flow Metab. 2008;28(6):1086–9.
46. Taguchi A, Kasahara Y, Nakagomi T, et al. A reproducible and simple model of permanent cerebral ischemia in CB-17 and SCID mice. J Exp Stroke Transl Med. 2010;3(1):28–33.
47. Uemura M, Kasahara Y, Nagatsuka K, Taguchi A. Cell-based therapy to promote angiogenesis in the brain following ischemic damage. Curr Vasc Pharmacol. 2012;10(3):285–8.
48. Fujita Y, Ihara M, Ushiki T, et al. Early protective effect of bone marrow mononuclear cells against ischemic white matter damage through augmentation of cerebral blood flow. Stroke J Cereb Circ. 2010;41(12):2938–43. doi:10.1161/STROKEAHA.110.596379.
49. Moniche F, Gonzalez A, Gonzalez-Marcos JR, et al. Intra-arterial bone marrow mononuclear cells in ischemic stroke: a pilot clinical trial. Stroke J Cereb Circ. 2012;43(8):2242–4. doi:10.1161/STROKEAHA.112.659409.
50. Prasad K, Mohanty S, Bhatia R, et al. Autologous intravenous bone marrow mononuclear cell therapy for patients with subacute ischaemic stroke: a pilot study. Indian J Med Res. 2012;136(2):221–8.
51. Rosado-de-Castro PH, Schmidt Fda R, Battistella V, et al. Biodistribution of bone marrow mononuclear cells after intra-arterial or intravenous transplantation in subacute stroke patients. Regen Med. 2013;8(2):145–55. doi:10.2217/rme.13.2.
52. Savitz SI, Misra V, Kasam M, et al. Intravenous autologous bone marrow mononuclear cells for ischemic stroke. Ann Neurol. 2011;70(1):59–69. doi:10.1002/ana.22458.

53. Hess DC, Hill WD, Martin-Studdard A, Carroll J, Brailer J, Carothers J. Bone marrow as a source of endothelial cells and NeuN-expressing cells after stroke. Stroke J Cereb Circ. 2002;33(5):1362–8.
54. Zhang ZG, Zhang L, Jiang Q, Chopp M. Bone marrow-derived endothelial progenitor cells participate in cerebral neovascularization after focal cerebral ischemia in the adult mouse. Circ Res. 2002;90(3):284–8.
55. Schwarting S, Litwak S, Hao W, Bahr M, Weise J, Neumann H. Hematopoietic stem cells reduce postischemic inflammation and ameliorate ischemic brain injury. Stroke J Cereb Circ. 2008;39(10):2867–75. doi:10.1161/STROKEAHA.108.513978.
56. Vasconcelos-dos-Santos A, Rosado-de-Castro PH, Lopes de Souza SA, et al. Intravenous and intra-arterial administration of bone marrow mononuclear cells after focal cerebral ischemia: is there a difference in biodistribution and efficacy? Stem cell research. 2012;9(1):1–8. doi:10.1016/j.scr.2012.02.002.
57. Willing AE, Lixian J, Milliken M, et al. Intravenous versus intrastriatal cord blood administration in a rodent model of stroke. J Neurosci Res. 2003;73(3):296–307. doi:10.1002/jnr.10659.

Chapter 5
Clinical Development of MultiStem® for Treatment of Injuries and Diseases of the Central Nervous System

Utilization of an Adult Stem Cell Product for Mitigating Adverse Aspects of Neuroinflammation

Robert W. Mays

Introduction

Because of the dynamic biology and potential for working through multiple mechanisms of action, there has been accumulating interest in developing stem cells as therapeutic agents for the treatment of central nervous system (CNS) injury and disease over the past decade [1–5]. Adult-derived stem cells have been isolated from multiple tissues by many laboratories [3, 4], and have been shown to demonstrate a number of favorable characteristics including genetic stability, extensive expansion capacity, and low immunogenicity profiles that support allogeneic utility [5]. Stem cells isolated from other developmental stages (embryonic, fetal, placental, amniotic fluid, etc.), through other distinct "stemness protocols" (induced pluripotency, genetic modification, somatic cell transfer, etc.) or via unique autologous cell isolation procedures or proprietary cell isolation devices, are also the focus of intense academic and industrial research, in an attempt to develop meaningful cell therapies for patients across a spectrum of disease and injury [6].

Regardless of methodology of cell derivation or source of the cell type under consideration, guidance provided by the Food and Drug Administration (FDA) make it clear that isolating, characterizing, and, when relevant, expanding the cell of interest, in a consistent, validated manner must be achieved prior to translating the therapy into patients [7]. Prior to submission of a Biologics License Application (BLA) following the completion of clinical trials for a potential new therapy, a potency assay, based on understanding the mechanism of action for how the cellular product is providing benefit, must be elaborated. And herein lies the challenge to scientists currently working to develop cellular therapeutics—how are cell therapies working? A vast literature demonstrating different cell types, isolated from different

R. W. Mays (✉)
Athersys, Inc., 3201 Carnegie Avenue, Cleveland, Ohio 44115, U.S.
e-mail: rwmays@athersys.com

D. C. Hess (ed.), *Cell Therapy for Brain Injury*, DOI 10.1007/978-3-319-15063-5_5

tissues, delivered via different routes, in different animal models, in different injury models, at different times, at different doses, almost universally demonstrate cell-mediated benefit—*but how and why*? If the culmination of these data is true and accepted at face value, is the basic biology underlying the mechanism of action for each cell type conserved? Is the mechanism for one cell type the same when administered through different routes? In different injury and disease models? At different times? Is it tissue replacement via the transplanted cells? Neoangiogenesis? Modulation of the immune system? All of the above? Something as yet undescribed?

Scientists at Athersys, in collaboration with leading academic researchers in the USA and Europe, have focused on studying multipotent adult progenitor cells (MAPC), and the clinical formulation of these cells known as MultiStem. Multiple publications describing the basic biology of these cells have been published. In parallel, experiments comparing and contrasting MAPC to other stem cell populations have been performed and reported. As the biology of MAPC was realized, preclinical studies in a variety of animal models of disease testing the safety and efficacy of the cells were performed. The composite results published led to a refined understanding of the multiple mechanisms through which MAPC may provide benefit when administered and subsequently, clinical translation and testing of the MultiStem product in humans. This chapter outlines the salient biology regarding MAPC, and a summary of the results highlighting the cell's effectiveness in animal models of CNS injury and disease. Results from those studies led to a conserved hypothesis for a mechanism of benefit for MAPC in acute CNS injury models, and helped define clinical endpoints being collected in the current ongoing phase I/II study enrolling patients having suffered an acute ischemic stroke treated with MultiStem (NCT01436487).

Properties of MAPC

MAPC are an adherent, population of adult stem cells, normally isolated from bone marrow, originally characterized, and described more than a decade ago [8]. Proprietary media formulations and growth conditions have been established and optimized to maintain the distinctive identity, long-term culture expansion, and potency of MAPC isolated from rodent [9] and human tissue [10, 11]. MAPC can be isolated from sources other than bone marrow [12], and have been reported to reconstitute tissues outside of the mesenchymal lineage including hematopoietic cells [13] and neuroectodermal cells [14], among others [8]. MAPC are 15–20 μm in diameter, distinctive from both bone marrow mononuclear cells (~5 μm) and traditional, mesenchymal stem cells (MSC; 30–35 μm) [15, 16]. Although numerous adult stem cell populations have been isolated and characterized over the past decades [3], MAPC have a unique secretome [17], as well as extended differentiation capability [18], transcriptome [18, 19], and microRNA (miRNA) profiles [10, 19] when compared to other adult adherent cells. Additional aspects of MAPC biology that have been reported and relevant to clinical development follow.

Immunomodulatory Properties

One of the most conserved and translationally important aspects of adult stromal cell biology is the ability to modulate the immune system. Scientific characterization of this class of cells in regard to immunomodulation has been extensive and thoroughly reviewed [20–23], a complete description of which is beyond the scope of this chapter. A summary of some of the immunomodulatory biology characteristics of these cells includes the ability to inhibit T cell activation, B cell activation, natural killer cell activation, dendritic cell maturation, as well as to differentially effect the secretion of a number of inflammatory cytokines.

MAPC are a member of this class of cells and, as such, display immunomodulatory mechanisms equivalent to those listed above, both *in vitro* and *in vivo* [20, 24–27], despite evidence demonstrating their distinctiveness from other cells of this class [18]. MAPC also express no detectable major histocompatibility class (MHC) II molecules on their cell surface, making them nonimmunogenic, with no activation of host T cells in in vitro or in vivo testing [24–27]. More recently, it was reported that the use of MAPC in an allogeneic, heterotypic rat heart transplant model leads to MAPC-mediated tolerance and immune acceptance of the heart, when transplanted sequentially into a third animal [16]. This tolerance was shown to be via MAPC-mediated effects on macrophage and T regulatory cell (Treg) populations carried forth in the heart into the recipient.

The nonimmunogenic, tolerizing, and immunomodulatory properties of MAPC led to the translation of the clinical cell product into a completed phase I study for prophylactic treatment of patients at risk for graft-versus-host disease (GvHD) [28], and ongoing clinical studies in a phase I study of solid organ transplant and a phase II study in treatment of ulcerative colitis. Continuing to study and understand the exact molecular mechanisms through which MAPC mediate the repertoire of immunomodulatory functions and applying this knowledge to clinical settings will be paramount in successfully moving these, as well as other cell therapies through the developmental pipeline for treatment of diseases where immune cell dysfunction underlies pathology [29].

Vasculogenic Properties

Numerous cell therapies, either allogeneic or autologous, including different adult stromal cell types, cardiac stem cells, and "cardiotrophic" stem cells, among others, have completed phase I studies in man and are into later-stage clinical evaluation for the treatment of acute myocardial infarction (AMI) or cardiac repair [30, 31]. Although given at different times, different doses, and via different routes across trials, these cell therapies have been generally safe and well tolerated, with a trend towards efficacy.

The evaluation of the MAPC clinical product, MultiStem, in a phase I open-label, dose escalation study of patients suffering AMI [32] was supported by preclinical safety [33] and efficacy data accessing the cells in rodent and pig models of AMI [34–36]. Data from these animal studies demonstrated that percutaneous administration of MAPC leads to a statistically significant improvement in cardiac function of cell-treated versus vehicle-treated animals.

In an attempt to determine the mechanism of benefit following cell administration observed in those studies, analyses of cardiac tissue from cell- versus vehicle-treated animals was completed. An increased neovascularization of the peri-infarct region of cell- versus vehicle-treated animals was observed [33, 34]. Subsequent studies in other models demonstrate that MAPC make relevant blood vessels *in vitro* [18, 37], more patent blood vessels when transplanted in normal mice [18], more effective revascularization in mice with peripheral limb ischemia injury [38], and increased neovascularization of the ischemic penumbra in mice following an ischemic stroke [39], when compared with equivalent dosages of other stromal cell populations.

In parallel, comparative analyses of MAPC with other adult cell types by both protein secretion and transcriptional profile were performed [17, 18]. Data from these studies were used to help identify three proteins differentially expressed and secreted by MAPC that initiate neo-angiogenesis and vessel formation [40]. The identification of these proteins, and bioactive doses required for vessel formation, has resulted in the development of a multiplex assay for surrogate potency of the cells for clinical manufacturing and lot release in treatment of cardiovascular injury.

Evaluation of MAPC Efficacy in Acute Models of CNS Injury

As a function of understanding the immunomodulatory properties of MAPC, scientists at Athersys initiated a series of collaborative experiments in animal models of acute CNS injury and disease with leading academic translational researchers. The hypothesis that an infusion of cells could downregulate the inflammatory response following administration while potentially providing trophic support for at-risk neurons and brain tissue after injury via vasculogenesis was attractive. The efficacy of MultiStem has subsequently been established in multiple CNS models including: hypoxic-ischemic (HI) injury, spinal cord injury (SCI), ischemic stroke, and traumatic brain injury (TBI), and results from these studies are summarized below.

HI Injury

The effect of MAPC on HI injury was assessed in neonatal rat studies performed in collaboration with Dr. Jim Carroll and Dr. David Hess at the Medical College of Georgia (now known as Georgia Regents University) [41, 42]. Data from these studies indicated that allogeneic or xenogeneic MAPC could be administered directly into the hippocampus 7 days after injury, or could be given intravenously (IV) 7 days after induction of injury, and that regardless of cell type or route, statistically significant improvements in locomotor behavior were maintained out to 4 weeks after a single-cell administration compared to vehicle-treated animals. Interestingly, animals that received cells IV had increased locomotor activity at 4 weeks when compared to animals that received the same dose of cells into the hippocampus, suggesting either enhanced benefit via the IV route or, alternatively, adverse aspects of a direct intraparenchymal administration.

Spinal Cord Injury

Previous work in a nontranslational dorsal column crush model of SCI, performed in collaboration with Dr. Jerry Silver and his colleagues at Case Western Reserve, demonstrated that administration of MAPC into the spinal cord immediately following injury results in a decrease in microglial/macrophage activation at the lesion over the first 7 days post injury [43]. There was also a concomitant decrease in axonal dieback and a statistically significant increase in axon regrowth towards the site of the lesion in cell- versus vehicle-treated rodents. Additional in vitro experiments and biochemistry supported the contention that the MAPC-mediated benefit observed in vivo was via a direct modulation of the macrophages from an M1 to an M2 phenotype, thereby decreasing macrophage-mediated dieback on the axons, which had been previously demonstrated to be the major initiator of long-range axonal retraction [44, 45].

To address the potential benefits of MAPC on SCI, additional studies were performed on a more relevant contusion model of injury. Pilot studies were performed to determine the optimal route of administration in the contusion injury model, and it was determined that IV administration was superior to direct injection into the cord following injury (DePaul et al., unpublished observations). Subsequently, we sought to determine the optimal window of administration and optimal single dose of the cells in the contusion model of SCI through a series of outcome assessments including the Basso–Beattie–Bresnahan locomotor scoring and catwalk regularity index, physiological readouts including measuring urination every 2 weeks using metabolic cages and endpoint urodynamic testing, tissue immunohistochemistry and biodistribution of labeled MAPC, and microarray analysis. The data demonstrate that a single IV dose of at least 4 million cells, administered 24 h after induction of the SCI, results in sustained and statistically significant locomotor benefit for at least 10 weeks after treatment [46]. Additionally, cell-treated animals urinated at a smaller bladder volume, had less residual volume, showed improved return to baseline pressure following a void, and had a decrease in bladder weight compared to vehicle-treated animals. Increasing the dose to 8 million cells showed no increase in additional locomotor recovery, but did improve bladder function [47].

Biodistribution analysis of labeled MAPC in collaboration with BioInVision indicated cells were found in the lungs, liver, and spleen at 24 h post IV administration. Normalizing cell counts to tissue weight showed a preferential homing to the spleen, with few cells found in the spinal column. Microarray analysis of the lesion, blood, and spleen suggests MAPC alters many injury-induced pathways including those involved in recruitment, activation, and migration of immune cells. In support of these data, we found a decrease of $ED1^{+}$macrophages by immune-histochemical staining at the lesion site 4 days post injury, when compared to untreated animals [47]. Taken as a whole, these data suggest that MAPC, when administered IV in an acute contusion model of SCI, are more likely to exert benefit through peripheral immune organ systems than via homing and direct interaction with the site of injury [46, 47].

Initial Ischemic Stroke Studies

While completing the neonatal HI injury studies being conducted at Georgia Regents University, adult rat ischemic stroke studies were also performed there to address relevant translational questions regarding the use of MAPC to treat ischemic injuries of the brain [48]. Data from these studies extended the understanding of MAPC-mediated benefit to include: (1) no immunosuppression was required when administering the cells through either intraparechymal or IV routes; (2) cell-mediated benefit was observed when given via both routes; however, higher doses were required, when given IV, to reach the same level of benefit observed at lower doses given directly into the brain (4 million IV = 400,000 direct injection); (3) the window for IV administration of cells can be extended out to 7 days for sustained significant benefits; however, earlier is better for minimizing the amount of brain tissue lost following onset of the stroke (i.e., 1 day > 2 days > 7 days). These results helped us address dose ranging, route of administration, and window of therapeutic benefit criteria for the Investigational New Drug (IND) submission for using the cells to treat ischemic stroke. However, an examination of the peri-infarct regions of the cell-treated stroke-injured brain revealed the presence of *very* few detectable MAPC, suggesting that the direct recovery observed was not likely the result of migration of MAPC to the site of injury and support of at-risk tissue through trophic factor secretion.

Traumatic Brain Injury

In an attempt to extend our understanding of how MAPC could be providing the neuroprotection observed in the initial stroke study [48], a collaboration with Dr. Charles Cox at the University of Texas, Houston (UT-Houston), was started to investigate the potential relationship between IV-infused MAPC and the spleen following acute injuries of the CNS. Previous publications from multiple laboratories had described the central role that the spleen has in exacerbating neuroinflammation following onset of an initial "inflammatory event," i.e., stroke, etc. [49–52].

The importance of the spleen as a target organ-mediating MAPC benefit was confirmed when cells were administered in the first 24 h following induction of TBI in rodents [53, 54]. After the injury, the blood–brain barrier (BBB) opens, contributing to edema and influx of inflammatory cells from the periphery to the CNS. Cells coming from the spleen are directly responsible for the loss of BBB integrity seen after TBI, as inducing injury in animals that had previously had their spleen removed results in little increase in BBB permeability [53]. IV infusion of MAPC blocks the spleen-mediated increased permeability of the BBB after injury; it is interesting to point out that labeled MAPC accumulate in the spleen in a dose-dependent way when IV infused after TBI injury [53]. The accumulation of MAPC into the spleen following IV administration within 24 h after acute CNS injury has now been confirmed in at least two other models in addition to TBI, ischemic stroke (Bang, unpublished results), and contusive SCI [46, 47].

The IV administration of MAPC within the first 24 h after TBI injury also results in other immune and behavioral changes. Cell-treated animals display an increase of the anti-inflammatory cytokines interleukin 4 (IL-4) and IL-10 emanating from the spleen with a simultaneous decrease in transcripts for the inflammatory cytokines IL-6 and interferon gamma when compared to TBI animals receiving vehicle [53]. There were statistically significant improvements in locomotor outcomes, and long-term spatial memory and memory retrieval (120 days post treatment) in MAPC-treated injured animals compared with vehicle-treated injured animals, and these improvements correlated with long-term diminution of microglial activation in the brains of the cell- versus vehicle-treated animals [55]. The compilation of these data lends strong support to the potential for moving MAPC into clinical development for treatment of TBI or concussive injuries.

Secondary Ischemic Stroke Injury Studies

In parallel to the TBI studies at UT-Houston, Athersys collaborated with Dr. Sean Savitz to confirm the efficacy of MAPC previously observed in rodent models of ischemic stroke [48]. A series of studies were designed to extend the translational information already established in ischemic stroke, coupled with the new observations relating to MAPC-mediated benefit via the spleen in models of TBI.

Experiments were performed in a new species of rat, via a new stroke injury protocol using a more robust series of outcome measures and confirmed that MAPC administration within 24 h of injury provided a statistically significant benefit [56]. There was also a statistically significant decrease in the inflammatory cytokines IL-6 and IL-1b when comparing cell-treated to vehicle-treated stroke animals along with a concomitant approximately sevenfold increase in IL-10 in cell-treated animals, 3 days after treatment (4 days after induction of injury). Finally, when the spleens of stroke animals given cells were compared with spleens from stroke animals given placebo, numerous statistically significant differences were observed including a preservation of splenic mass in the cell treatment, a decrease in TUNEL-positive apoptotic cells in the cell treatment, and upregulation of the transcript for IL-10 in the cell treatment [56]. These data led to an even greater focus and effort to understand the relationship between MAPC/spleen interactions following CNS injury on the molecular level.

To gain perspective, the collaborators isolated brain tissue and spleen tissue from stroke-injured rats that were treated 24 h after injury with cells or placebo. Tissue was isolated at both 3 days and 28 days after treatment from three different groups of animals: sham injured, stroke injured receiving placebo, and stroke injured receiving 4 million MAPC. Nucleic acid was isolated and microarray analyses were performed. Previously, microarray analyses of the brains of stroke-injured mice that received either a hippocampal injection of 100,000 human MSC or saline, 24 h after induction of the injury, had been reported [57]. This experiment yielded an enormous amount of data regarding MSC-mediated benefit in the brain following stroke. The authors of that study concluded the observed beneficial effects of the transplanted cells were largely explained by their modulation of inflammatory and immune responses, in the brain, by modulating activation of microglia and/or macrophages.

Although similar in many ways, the MAPC microarray stroke study differed most notably in the cell type being infused (MAPC vs. MSC), the species of investigation (rat vs. mouse), the route of administration (IV vs. direct hippocampal injection), and the number of cells being infused (4×10^6 vs. 1×10^5). Data generated in the MAPC/rat stroke study similarly suggested that transplanted cells were decreasing the inflammatory status of the stroke-injured brain, while simultaneously upregulating anti-inflammatory programs and signaling pathways in the brain despite IV administration (Hamilton, unpublished data). It is hypothesized that the observed inflammatory changes in the brain were taking place as a result of decreased migration and infiltration of peripheral immune cells to the site of injury, specifically from the spleen, as quantification of microarray markers for activated T cells and M1 macrophages were upregulated hundreds of fold in vehicle-treated stroke animals, and decreased almost back to the levels of sham in cell-treated stroke animals (Hamilton, unpublished data). Migration of a significant number of activated cells from the spleen to the peri-infarct region of the brain in the first days after stroke has previously been demonstrated [58]. This observed splenocyte migration correlates with a decrease in splenic mass similar to what has been observed by several groups [56, 58, 59]. Microarray analysis of spleen tissue after stroke suggests that cell-treated animals have a normal "healthy" splenic phenotype comparable to sham-injured animals (i.e., normal bioenergetics, cell division, oxidative phosphorylation pathway signatures, etc.), while vehicle-treated stroke-injured animals have an increase in proapoptotic genes, and immunosuppressive gene ontogeny signatures not observed in cell-treated injured animals (Hamilton, unpublished).

To directly test the importance of the spleen in MAPC-mediated benefit in stroke, a series of experiments were performed comparing injured animals that received MAPC to animals that had their spleens removed 14 days prior to stroke injury, and then received IV cell infusion [60]. The first observation was that the cell-mediated functional recovery observed in stroke animals with their spleens was completely lost in splenectomized cell-treated animals [60]. Interestingly, animals that had their spleens removed had a lower initial baseline deficit score, regardless of treatment group, when compared to animals with spleens, although no deficit improvement was observed over the 4-week testing period. Also, animals that had splenectomies prior to stroke injury showed a statistically decreased infarct volume compared to animals with spleens, as had previously been reported [51]. MAPC administration to stroke-injured animals, both with or without their spleens, resulted in a statistically significant approximately fourfold preservation of brain tissue 28 days after treatment. This is remarkable in light of the fact that no significant functional recovery was seen in cell-treated animals at 28 days, despite the preservation of brain tissue [60].

Finally, the serum cytokines previously characterized after stroke were comparatively analyzed in animals with and without spleens after stroke injury [60]. Three days after IV MAPC treatment, IL-6, IL-1b, and IL-10 displayed a different profile as a function of cell treatment and spleen status. The inflammatory cytokines IL-6 and IL-1b, which are both significantly downregulated by IV administration of MAPC following stroke, demonstrate different profiles in splenectomized animals. IL-6 levels, which are elevated three- to fourfold in normal stroke-injured animals *do not* increase in the 4 days after inducing stroke in animals without spleens. This suggests that the spleen is a major source of systemic IL-6 signaling in the subacute time frame after stroke. MAPC administration was able to significantly reduce IL-6 levels in the splenectomized animals, indicating that there is additional IL-6 released outside the spleen, and MAPC is able to dampen this secondary release. IL-1b modulation by MAPC, on the other hand, is lost in animals that have been splenectomized. IL-1b levels increase ~1.5-fold in the serum when comparing splenectomized to normal stroke-injured animals, and MAPC administration has no effect on this increase. Levels of the anti-inflammatory cytokine IL-10, which increased ~7x in MAPC-treated stroked animals with spleens, were completely abolished in animals without spleens [60]. These data combined with the observations testing MAPC in the other preclinical injury models discussed provided sufficient support for a novel conserved mechanism for cell-mediated benefit via modulation of the spleen, and to move into clinical testing for patients suffering an acute ischemic stroke.

Clinical Trial

As part of the translational development of cellular therapies, the FDA has provided guidance on several short- and long-term safety parameters that must be demonstrated prior to moving into the clinical setting [61]. To address this guidance, Athersys and collaborators have evaluated the safety of MAPC, and the clinical formulation of the cells known as MultiStem, in multiple experiments. Transplantation of the cells into multiple species and via multiple routes of administration has resulted in no evidence of infusional toxicity in both published and unpublished reports ([33]; Mays, unpublished results; Ting, unpublished results). Additional safety evaluation of MAPC has been performed in preclinical animal studies including: good laboratory practice (GLP) tumorigenicity studies in nonobese diabetic-severe combined immunodeficiency (NOD-SCID) and nude mice, GLP safety studies in stroke-injured rats, and long-term safety (> 1 year) studies in stroke-injured and HI-injured rats and rat pups, respectively (Mays, unpublished results), with no evidence of cell-mediated toxicity or safety issues. Immune sensitization analysis of the cells through multiple administrations into the same animal have been performed, and shown to have no reactivity as measured by respiration rates and lack of allogeneic antibody formation [33]. Gene expression, protein expression, and single nucleotide polymorphism analysis of clinical grade MultiStem cells derived from multiple production runs have shown no changes or significant variability [10], demonstrating the ability to repeatedly manufacture the same cell product.

This body of safety data, coupled with efficacy data presented in the section "Evaluation of MAPC Efficacy in Acute Models of CNS Injury," was submitted as part of IND #13852, MultiStem for treatment of patients suffering an acute ischemic stroke. The clinical trial, NCT01436487, is a double-blind, placebo-controlled, dose escalation study of ~136 patients having suffered an acute ischemic stroke. At the time of the writing of this chapter, the trial was open and actively enrolling patients at 33 clinical sites in the USA and 6 sites in the UK.

The specifics of this clinical trial, including design; patient population being evaluated as determined by inclusion and exclusion criteria; primary, secondary, and exploratory endpoints; and clinical procedures to be performed have previously been published [62]. Briefly, the trial is a phase I/II study designed to determine the safety and efficacy of MultiStem when administered into adult patients having suffered an ischemic stroke 24–48 h from the onset of symptoms. The trial consists of three cohorts. Cohort 1 is a low-dose treatment group (400 million cells) of eight patients randomized 6:2 (cell treatment: placebo). Cohort 2 is a high-dose treatment group (1.2 billion cells) of eight patients randomized 6:2 (cell treatment: placebo). Both of these cohorts have been completely enrolled, and evaluation of the safety data from these cohorts by an Independent Safety Committee (ISC), determined that the administration of the cells at both doses was safe and well tolerated, and that the trial should proceed to enrolling patients in cohort 3 at the highest well-tolerated dose of 1.2 billion cells. Cohort 3 consists of 120 patients randomized 1:1 (1.2 billion cell treatment: placebo).

The total duration for safety and efficacy evaluation is 12 months following infusion of the investigational product. Patients must be 18–83 years of age inclusive, with a diagnosis of cortical cerebral ischemic stroke. All animal models in which we tested the efficacy of the cells in stroke involved cortical involvement. An additional inclusion criterion of note was that patients must have a National Institute of Health Stroke Scale (NIHSS) score of 8–20, inclusive, defining a population of moderate to moderately severe injured patients. The patients must be stable, i.e., not to have shown a > 4 point change in NIHSS score during a 6-h window prior to treatment, as patients with either improving or worsening trajectories could confound interpretation of the data. Exclusion criteria of note include previous ipsilateral stroke lesion, and or brain injury that would complicate evaluation, and the patients could not have previously had their spleen surgically removed.

In addition to demonstrating safety of the cells in this population of stroke-injured patents, the primary efficacy endpoint is analysis of the MultiStem treatment group compared to subjects in the placebo treatment group using global stroke recovery at day 90 in modified intent-to-treat population, evaluating in accordance with Modified Rankin Score <=2, NIHSS improvement >=75 %, Barthel Index >=95. The basis for use of a global statistic is well documented [63], and is increasingly recognized as a better metric for examining "multidimensional" measurements of benefit, specifically in diseases of the CNS [64]. The use of a composite global recovery scale has previously been used in evaluating stroke therapeutics, including the only FDA-approved therapy tissue plasminogen activator (TPA) [65]. Other secondary and exploratory endpoints of note in the current MultiStem stroke trial include outcomes measures (NIHSS, Modified Rankin and Barthel Index) during the trial (days 1, 7, 30, 90, 365), stroke lesion volume as determined by magnetic resonance imaging (MRI) at baseline, day 30 and day 365, the measurement of potential biomarkers for cell-mediated effect such as IL-10, based on preclinical data, from baseline to day 2, 7, and 30, as well as measurement of various immune cell types present in the blood during the same time course. It is anticipated that enrollment will be completed and topline data for the primary efficacy endpoints to be reported in early 2015.

Summary and Potential of MAPC-Mediated Benefit

The reality of cellular therapies for treatment of human disease seems close to being realized after years of premature hype and promise. New sources for cells, and methodologies for isolating, deriving, and subsequently growing therapeutically relevant cell therapies are published or announced weekly, if not daily. Of note, there are approximately as many trials listing "cell therapy" as keywords on clinicaltrials.gov the past 2 years, as there were the 12 years previous. The final hurdle for the mainstream acceptance of cell therapies, however, will be definitive proof of efficacy in a phase II, and subsequently, a pivotal phase III study.

Over the past 8 years, collaborative research with leading academic translational scientists and physicians have demonstrated the efficacy of MAPC in diverse animal models of CNS injury, culminating in the submission of an IND for testing the cells in a currently enrolling phase I/II study treating patients suffering an acute ischemic stroke. The evidence supporting the use of the cells points to a mechanism wherein IV administration of MAPC during the first 24 h results in the accumulation of the cells in the spleen, in an injury- and dose-dependent way ([46, 47, 53]; Yang, unpublished; Busch, unpublished). The importance of the spleen in mitigating MAPC-mediated benefit is unquestioned; removing the spleen prior to inducing injury results in a loss of cell-mediated benefit in both ischemic stroke and TBI models [53, 60]. Splenectomized cell-treated stroke animals show the complete loss of serum IL-10 upregulation observed in cell-treated animals with spleens [60], and it is believed that this is a significant result pointing to the specific therapeutic biology MAPC confers when administered in the acute time frame following CNS injury. Recently, Melief and colleagues reported that multipotent stromal cells isolated from bone marrow specifically induce the upregulation of Tregs when incubated with peripheral blood mononuclear cells (PBMC) in in vitro assays [66]. The effect of the stromal cells on the PBMC required the presence of monocytes for induction of the Tregs, and was concomitant with a significant increase in measurable IL-10. The researchers then demonstrated that the major effect the multipotent stromal cells had on upregulating Tregs was via modulating the monocytes into M2 macrophages, an alternatively activated phenotype thought of as anti-inflammatory or reparative macrophages, which then secreted trophic factors, including CCL-18, leading to the direct upregulation of Tregs. Coculture of PBMC with MAPC results in detectable increases in CCL-18 *in vitro* (Stubblefield and Zilka, unpublished results).

Tregs have previously been identified as neuroprotective in the post-ischemic brain [67–69], although contradicting data for the efficacy of Tregs activity post-stroke exist [70, 71]. The data in the section "Evaluation of MAPC Efficacy in Acute Models of CNS Injury" support the contention that MAPC treatment can upregulate Treg levels after injury, and suggest that this upregulation correlates with improvement in cell versus untreated injured animals. It had previously been shown that coculturing MAPC with macrophages leads to a change in polarization from an M1 to an M2 phenotype *in vitro* [43], and that the IV administration of MAPC within the first 24 h after TBI results in a significant systemic upregulation of M2 macrophages versus M1 macrophages when compared to vehicle-treated animals for the first 5 days after treatment [54]. There is also a transient measurable upregulation of Tregs in these same MAPC-treated TBI animals, first at 24 h after treatment in the spleen, and then subsequently at 48 h in the blood [54]. There is also a statistically significant upregulation of Tregs in the blood of stroke-injured animals 72 h after treatment when compared to vehicle-treated animals [60]. Interestingly, in an allotypic heart transplant model, animals treated with MAPC showed an increase in Tregs in grafted heart tissue which was not observed in transplanted tissue maintained under cyclosporine administration [16]. The current MultiStem stroke clinical trial will evaluate Treg levels in all stroke patients at baseline, day 2, day 7, and day 30 after treatment, as an exploratory endpoint, and determine if there is a difference in Treg levels in cell-treated and placebo-treated patients, and if there is any correlation between Treg levels and clinical outcomes.

Other researchers are beginning to evaluate therapeutics previously tested in clinical trials as agents to increase Tregs and modulate spleen-mediated damage, as a means to potentially accelerate clinical development of agents for treatment of stroke. A publication by Na and colleagues described a series of in vivo experiments testing the efficacy of the superagonistic antibody CD28A, known to increase levels of Tregs, when given by intraperitoneal injection 3–6 h after inducing stroke in mice [72]. This antibody had previously been tested in a phase I study of immune dysfunction, but clinical development was stopped due to severe side effects [73]. The results testing the antibody in the stroked animals were compelling demonstrating an increase in brain localized Tregs, a significant increase in IL-10, significant increases in functional improvement, and significant decreases in brain tissue loss, when compared to saline. A decade of important research by Dr. Pennypacker and colleagues is culminating in the search for specific therapies and modulators to reverse the negative impact that the spleen has on exacerbating poststroke outcomes [74].

Is it possible that MAPC, or other cellular therapies, can simultaneously down-regulate or shut off the adverse, overt effects of the peripheral immune response to injury (i.e., migration of injury promoting cells to the site), while simultaneously initiating or enhancing repair? Is it even possible to uncouple those two parts of post-injury recovery? It is interesting to point out from the stroke studies reported in the section "Secondary Ischemic Stroke Injury Studies," when comparing the splenectomized rats that received MAPC to stroke animals with spleens that received cells, there was an approximately twofold increased preservation of brain tissue in splenectomized animals that got cells compared to stroked animals *with* spleens that got cells, yet the splenectomized cell-treated animals showed *no* locomotor or neurological improvement over placebo [60]. The splenectomized injured animals also had no upregulation of serum IL-10 levels. One explanation could be that stopping or eliminating immune cells from migrating to and infiltrating the brain limits additional damage, but that the upregulation of a separate IL-10-mediated repair and/or recovery process, possibly driven by Tregs, is necessary to initiate or allow for endogenous neuro-repair and behavior. Maybe it is impossible to restore lost neural function or improve deficit, for example in chronic injury models, until one first stops the inflammatory cascade, and initiates the repair program. Additional experiments testing adult stem cells in any number of models will be required before this speculation can be addressed.

In the "Introduction," the challenge of understanding the mechanism of benefit of cell-based therapies for researchers in this space was described, as was its importance to designing and effectively executing clinical trials. This will ultimately be the test for which the entire field will be judged, in the next several years. This chapter describes the sequential translational investigation of MAPC in a wide array of animal models of CNS injury leading to the design and execution of a phase I/II clinical trial testing a conserved hypothesis for cell-mediated benefit by mitigating the adverse effects the spleen and potential other peripheral immune organs present in the acute time frame. Clinical data in support of this hypothesis will have important implications for patients, patient advocates, clinicians, and researchers who believe in the future of cellular medicine.

Acknowledgments

RWM would like to acknowledge the hard work, intellectual involvement, and cooperative spirit of the academic colleagues with whom we collaborated in generating preclinical data and subsequently clinical aspirations cited in this chapter, specifically Dr. David Hess, Dr. Chuck Cox, Dr. Jerry Silver, Dr. Jim Carroll, and

Dr. Sean Savitz. RWM would like to recognize the ongoing efforts and daily inspiration of the Neuroscience team at Athersys, especially Dr. Sarah Busch, Dr. Samantha Stubblefield, Rochelle Cutrone, Marc Palmer, Sarah Zilka, and all others who contributed. The support of the senior management team at Athersys must be cited, especially Dr. Bob Deans and Dr. Gil van Bokkelen. Support for research performed and cited in this study came, in part, from the following grants to RWM: # 1U44NS077511-01 for developing MultiStem therapy for treatment of Traumatic Brain Injury; OTFBP Grant #10–833: Clinical Development of MultiStem for Treatment of Spinal Cord Injury; and #R42NS055606: Clinical development of an Adult Stem Cell Product for Treatment of Hypoxic Ischemic Injury. Finally, RWM would like to acknowledge the dedication of Dr. Jason Hamilton in realizing the potential of MAPC in the CNS, gone but not forgotten.

References

1. Boucherie C, Hermans E. Adult stem cell therapies for neurological disorders: benefits beyond neuronal replacement? J Neurosci Res. 2009;87:1509–21.
2. Kim S, de Vellis J. stem-cell-based cell therapy in neurological diseases: a review. J Neurosci Res. 2009;87:2183–200.
3. Mays RW, et al. Development of adult pluripotent stem cell therapies for ischemic injury and disease. Expert Opin Biol Ther. 2007;7:173–84.
4. Ting A, et al. Therapeutic pathways of adult stem cell repair. Crit Rev Oncol Hematol. 2008;65:81–93.
5. Gnecchi M, et al. Paracrine mechanisms in adult stem cell signaling and therapy. Circ Res. 2008;103:1204–19.
6. Dimmler S, et al. Translational strategies and challenges in regenerative medicine. Nature Med. 2014;20:814–20.
7. United States Department of Health and Human Services, Food and Drug Administration. Guidance for Industry Potency Tests for Cellular and Gene Therapy Products. http://www.fda.gov/BiologicsBloodVaccines/GuidanceComplianceRegulatoryInformation/Guidances/default.htm (2011).
8. Jiang Y, et al. Pluripotent nature of adult marrow derived mesenchymal stem cells. Nature. 2002;418:41–9.
9. Breyer A, et al. Multipotent adult progenitor cell isolation and culture procedures. Exp Hematol. 2006;34:1596–601.
10. Boozer S, et al. Global characterization and genomic stability of human multistem, a multipotent adult progenitor cell. J Stem Cells. 2009;4:17–28.
11. Roobrouck V, et al. Concise review: culture mediated changes in fate and/or potency of stem cells. Stem Cells. 2011;29:583–9.
12. Jiang Y, et al. Multipotent progenitor cells can be isolated from postnatal murine bone marrow, muscle, and brain. Exp Hemato. 2002;30:896–904.
13. Serafini M, et al. Hematopoietic reconstitution by multipotent adult progenitor cells: precursors to long-term hematopoietic stem cells. J Exp Med. 2007;204:129–39.
14. Jiang Y, et al. Neuroectodermal differentiation from mouse multipotent adult progenitor cells. PNAS. 2003;100 Suppl 1:11854–60.
15. Fischer U. Pulmonary passage is a major obstacle for intravenous stem cell delivery: the pulmonary first-pass effect. Stem Cells Dev. 2009;18:683–92.

16. Eggenhofer E, et al. Heart grafts tolerized through third-party multipotent adult progenitor cells can be re-transplanted to secondary hosts with no immunosuppression. Stem Cells Transl Med. 2013;2:595–606.
17. Burrows G, et al. Dissection of the human multipotent adult progenitor cell secretome by proteomic analysis. Stem Cells Transl Med. 2013;2:745–57.
18. Roobrouck V, et al. Differentiation potential of human postnatal mesenchymal stem cells, mesoangioblasts, and multipotent adult progenitor cells reflected in their transcriptome and partially influenced by the culture conditions. Stem Cells. 2011;29:871–82.
19. Aranda P, et al. Epigenetic signatures associated with different levels of differ-entiation potential in human stem cells. PLoS ONE. 2009;4:e7809. doi:10.1371/journal.pone.0007809.
20. Vaes B, et al. Application of multiStem allogeneic cells for immunomodulatory therapy: clinical progress and pre-clinical challenges in prophylaxis for graft versus host disease. Front Immuno. 2012;3:1–9.
21. Murphy M, et al. Mesenchymal stem cells: environmentally responsive therapeutics for regenerative medicine. Exp Mol Ned. 2013;45:e54. doi: 10.1038/emm.2013.94.
22. Stagg J, Galipeau J. Mechanisms of immune modulation by mesenchymal stromal cells and clinical translation. Curr Mol Med. 2013;13:856–67.
23. Gebler A, et al. The immunomodulatory capacity of mesenchymal stem cells. Trends Mol Med. 2012;18:128–34.
24. Reading J, et al. Clinical-grade multipotent adult progenitor cells durably control pathogenic T cell responses in human models of transplantation and autoimmunity. J Immunol. 2013;190:4542–52.
25. Jacobs S, et al. Human multipotent adult progenitor cells are non-immunogenic and exert potent immunomodulatory effects on alloreactive T cell responses. Cell Transpl. 2012;22:1915–28.
26. Kovacsovics-Bankowski M, et al. Clinical scale expanded adult pluripo tent stem cells prevent graft-versus-host disease. Cell Immunol. 2009;255:55–60.
27. Highfill S, et al. Multipotent adult progenitor cells can suppress graft-versus-host disease via prostaglandin E2 synthesis and only if localized to sites of allopriming. Blood. 2009;114:693–701.
28. Maziarz R, et al. Prophylaxis of acute GVHD using Multistem stromal cell therapy: preliminary results after administration of single or multiple doses in a phase 1 trial. Biol.Blood Marrow Transpl. 2012;18:S264–5.
29. Auletta J, et al. Regenerative stromal cell therapy in allogeneic hematopoietic stem cell transplantation: current impact and future directions. Biol Blood Marrow Transpl. 2010;16:891–906.
30. Behfar A, et al. Cell therapy for cardiac repair-lessons from clinical trials. Nature Rev Cardiol. 2014;11:232–46.
31. Telukuntla K, et al. The advancing field of cell-based therapy: insights and lessons from clinical trials. JAMA. 2013;10:e000338. doi: 10.1161/JAHA.113.000338.
32. Penn M, et al. Adventitial delivery of an allogeneic bone marrow-derived adherent stem cell in acute myocardial infarction: phase I clinical study. Circ Res. 2012;110:304–11.
33. Kovacsovics-Bankowski M, et al. Pre-clinical safety testing supporting clinical use of allogeneic multipotent adult progenitor cells. CytoTherapy. 2008;10:730–42.
34. Pelacho B, et al. Multipotent adult progenitor cell transplantation increases vascularity and improves left ventricular function after myocardial infarction. J Tiss Eng Reg Med. 2007;1:51–9.
35. Van't H. Direct delivery of syngeneic and allogeneic large-scale expanded multipotent adult progenitor cells improves cardiac function after myocardial infarct. CytoTherapy. 2007;9:477–87.
36. Medicetty S, et al. Percutaneous adventitial delivery of allogeneic bone marrow-derived stem cells via infarct-related artery improves long-term ventricular function in acute myocardial infarction. Cell Transpl. 2012;21:1109–20.

37. Aranguren X, et al. *In vitro* and *in vivo* arterial differentiation of human multipotent adult progenitor cells. Blood. 2007;109:2634–42.
38. Aranguren X, et al. Multipotent adult progenitor cells sustain function of ischemic limbs in mice. J Clin Invest. 2008;118:505–14.
39. Mora-Lee S, et al. Therapeutic effects of hMAPC and hMSC transplantation after stroke in mice. PLoSONE. 2012;7:e43683. doi:10.1371/journal.pone.0043683
40. Lehman N, et al. Development of a surrogate angiogenic potency assay for clinical-grade stem cell production. CytoTherapy. 2012;14:994–1004.
41. Yasuhara T, et al. Behavioral and histological characterization of intrahippocampal grafts of human bone marrow-derived multipotent progenitor cells in neonatal rats with hypoxic-ischemic injury. Cell transpl. 2006;15:231–8.
42. Yasuhara T, et al. Intravenous grafts recapitulate the neurorestoration afforded by intracerebrally delivered multipotent adult progenitor cells in neonatal hypoxic-ischemic rats. J Cereb Blood Flow Metab. 2008;28:1804–10.
43. Busch S, et al. Multipotent adult progenitor cells prevent macrophage-mediated axonal dieback and promote regrowth after spinal cord injury. J Neurosci. 2011;19:944–53.
44. Horn K, et al. Another barrier to regeneration in the CNS: activated macrophages induce extensive retraction of dystrophic axons through direct physical interactions. J Neurosci. 2008;28:9330–41.
45. Busch S, et al. Overcoming macrophage-mediated axonal dieback following CNS injury. J Neurosci. 2009;29:9967–76.
46. Busch S, et al. Optimizing administration of Multistem® for the treatment of acute contusion spinal cord injury. Paper presented at the American Society for Neuroscience, San Diego, California. 2013 Nov 9–13. 2013.
47. DePaul M, et al. Intravenous Multipotent Adult Progenitor Cell Treatment for Acute Spinal Cord Injury: Promoting Recovery Through Immune Modulation. Paper presented at the American Society for Neuroscience, Washington, D.C. 2014 Nov 15–19. 2014.
48. Mays RW, et al. Development of an allogeneic adherent stem cell therapy for treatment of ischemic stroke. J Exp Stroke Trans Med. 2010;3:34–46.
49. Walker, et al. Bone marrow-derived stromal cell therapy for traumatic brain injury is neuroprotective via stimulation of non-neurologic organ systems. Surgery. 2009;152:790–3.
50. Offner H, et al. Splenic atrophy in experimental stroke is accompanied by increased regulatory T cells and circulating macrophages. J Immunol. 2006;176:6523–31.
51. Ajmo C Jr, et al. The spleen contributes to stroke-induced neurodegeneration. J Neurosci Res. 2008;86:2227–34.
52. Offner H, et al. Effect of experimental stroke on peripheral immunity: CNS ischemia induces profound immunosuppression. Neurosci. 2009;158:1098–111.
53. Walker P, et al. Intravenous multipotent adult progenitor cell therapy for traumatic brain injury: preserving the blood brain barrier via an interaction with splenocytes. Exp Neurol. 2010;225:341–52.
54. Walker P, et al. Intravenous multipotent adult progenitor cell therapy after traumatic brain injury: modulation of the resident microglia population. J Neuroinflamm. 2012;28. doi:10.1186/1742-2094-9-228.
55. Bedi S, et al. Intravenous multipotent adult progenitor cell therapy attenuates activated microglial/macrophage response and improves spatial learning after traumatic brain injury. Stem Cells Transl Med. 2013;2:953–60.
56. Yang B, et al. human multipotential bone marrow stem cells exert immunomodulatory effects, prevent splenic contraction, and enhance functional recovery in a rodent model of ischemic stroke. Paper presented at the American Heart Association International Stroke Conference, Los Angeles, California. 2011 Feb 8–10. 2011.
57. Ohtaki H, et al. Stem/progenitor cells from bone marrow decrease neuronal death in global ischemia by modulation of inflammatory/immune responses. PNAS. 2008;105:14638–43.
58. Seifert H, et al. A transient decrease in spleen size following stroke corresponds to splenocyte release into systemic circulation. J Neuroimmune Pharmacol. 2012;7:1017–24.

59. Kim E, et al. Role of spleen-derived monocytes/macrophages in acute ischemic brain injury. J Cereb Blood Flow Metab. 2014;34:1411–9.
60. Yang B, et al. The spleen is a pivotal target of functional recovery after treatment with MultiStem for acute ischemic stroke. Paper presented at the American Heart Association International Stroke Conference, New Orleans, Louisiana. 2012 Jan 31-Feb 2. 2012.
61. Halme D, Kessler D. FDA regulation of stem-cell-based therapies. N Engl J Med. 2006;355:1730–5.
62. Hess D, et al. A double-blind placebo-controlled clinical evaluation of MultiStem for the treatment of ischemic stroke. Int J Stroke. 2014;9:381–6.
63. Lachin J. Applications of the wei-lachin multivariate one-sided test for multiple outcomes on possibly different scales. PLoS One. 2014;17:e108784. doi:10.1371/journal.pone.0108784. (eCollection 2014).
64. Rahlfs V, et al. The new trend in clinical research the multidimensional approach instead of testing individual endpoints. Pharma Med. 2012;3:160–5.
65. NINDS. Tissue plasminogen activator for acute ischemic stroke. N Engl J Med. 1995;333:1581–7.
66. Melief S, et al. Multipotent stromal cells induce human regulatory T cells through a novel pathway involving skewing of monocytes toward anti-inflammatory macrophages. Stem Cells. 2013;9:1980–91.
67. Liesz A, et al. Regulatory T cells are key cerebroprotective immunomodulators in acute experimental stroke. Nat Med. 2009;15:192–9.
68. Planas A, Chamorro A. Regulatory T cells protect the brain after stroke. Nat Med. 2009;15:138–9.
69. Li P, et al. Adoptive regulatory T-cell therapy protects against cerebral ischemia. Ann Neurol. 2013;74:458–71.
70. Schabitz WR. Regulatory T cells in ischemic stroke: helpful or hazardous? Stroke. 2013;44:e84. doi:10.1161/STROKEAHA.113.002228.
71. Xu X, et al. The paradox role of regulatory T cells in ischemic stroke. Sci.World J. 2013;174373. doi:10.1155/2013/174373
72. Na S, et al. (2014) Amplification of regulatory T Cells using a cd28 superagonist reduces brain damage after ischemic stroke in mice. Stroke: pii: STROKEAHA. 114.007756. [Epub ahead of print].
73. Suntharalingam G, et al. Cytokine storm in a phase 1 trial of the anti-CD28 monoclonal antibody TGN1412. N Engl J Med. 2006;355:1018–28.
74. Pennypacker K. Targeting the Peripheral immune response to stroke: role of the spleen. Trans Stroke Res. 2014;5:635–7.

59. Kim E, et al. Role of spleen-derived monocytes/macrophages in acute ischemic brain injury. J Cereb Blood Flow Metab. 2014;34:1411–9.
60. Yang B, et al. The spleen is a pivotal target of [illegible] recovery after treatment with MultiStem [illegible]. Paper presented at the American Heart Association International Stroke Conference [illegible] 2012 [illegible]
61. [illegible] FDA [illegible] of stem-cell-based therapies. N Engl J Med [illegible]
62. [illegible]
63. [illegible]
64. [illegible]
65. [illegible]
66. [illegible]
67. [illegible]
68. [illegible]
69. [illegible]
70. [illegible]
71. [illegible]
72. [illegible]
73. [illegible]
74. [illegible]

Chapter 6
Intra-arterial Approaches to Stem Cell Therapy for Ischemic Stroke

Vikram Jadhav, Pallab Bhattacharya and Dileep R. Yavagal

Introduction

Ischemic stroke occurs due to disruption of blood flow to an area of the brain resulting in tissue infarction and death of neurons and other brain cells. The loss of neurons in the infarcted core is irreversible, whereas the penumbra surrounding the infarcted core has cells that are functionally impaired but not dead. Acute ischemic stroke treatment (minutes to hours) is targeted to protect the penumbra. The only proven pharmacological treatment for acute ischemic stroke is clot thrombolysis using tissue plasminogen activator (tPA) delivered either by intravenous (IV) or intra-arterial route. However, it is subject to therapeutic time window of 4.5 h for IV delivery [1, 2]. Moreover, it is estimated that only 2–5 % of patients in the USA and 4 % in the UK receive tPA for acute ischemic stroke [3, 4]. Only 6 out of 1000 patients who receive IV-tPA do not have any disabilities, but most end up with

D. R. Yavagal (✉)
Neurology and Neurosurgery, University of Miami &Jackson Memorial Hospitals,
1120 NW 14th Street, 33136 Miami, FL, USA
e-mail: DYavagal@med.miami.edu

V. Jadhav
Neurology, Neurological Institute, University Hospitals Case Medical Center,
11100 Euclid Ave, Bolwell 5120, 44106 Cleveland, OH, USA
e-mail: vikram.jadhav@uhhospitals.org

P. Bhattacharya
Department of Neurology, Leonard M. Miller School of Medicine, TSL, Lab-230,
1420 NW 9th Avenue, 33136 Miami, FL, USA
e-mail: p.bhattacharya@med.miami.edu

D. C. Hess (ed.), *Cell Therapy for Brain Injury*, DOI 10.1007/978-3-319-15063-5_6

significant long-term disability. Endovascular acute ischemic stroke therapy when provided as standard of care may only reach 20% of ischemic stroke cases due to limited time window. Moreover, the recently completed Dutch clinical trial "Multicenter Randomized Clinical trial of Endovascular treatment for Acute ischemic stroke in the Netherlands" (MR CLEAN) showed that even with improved clinical outcomes after endovascular treatment, approximately 80% patients may not be able to return to their job and/or perform previous activities [5, 6].

Thrombolysis of the clot aims to limit the size of infarction. However, ischemic stroke also results in secondary brain injury (hours to days) due to multiple processes including but not limited to inflammation, apoptosis, oxidative stress, cytotoxic edema, loss of blood–brain barrier (BBB), and subsequent vasogenic edema. This is the phase of delayed neuronal death [7]. The preclinical pharmacological treatments target myriad processes including but not limited to neurogenesis, angiogenesis, and release of growth and trophic factors that can aid in promoting neuroplasticity. None of the preclinical therapies have managed to make their way into clinical practice. Presently, we do not have any Food and Drug Administration (FDA)-approved therapy for the prevention of secondary brain injury after ischemic stroke.

It is estimated that first time strokes affect 16 million individuals resulting in 6 million deaths [8]. If the patient survives, there is a huge burden of disease not only for the individual but also for the community due to the huge loss of man-hours and special rehabilitation needs for the patients [9]. Stem cell therapies have the potential to limit long-term disability and improve function by promoting repair and remodeling. Numerous preclinical studies and early clinical trials suggest that stem cell therapies may provide benefits in early as well chronic stages of ischemic stroke thus having a wide therapeutic window. In this chapter, we focus mainly on preclinical studies and clinical trials of intra-arterial (IA) stem cell therapies in ischemic stroke.

Preclinical Studies

Numerous of preclinical studies have been published in the past decade on the use of different types of stem cells for the treatment of ischemic stroke. Apart from establishing the efficacy of stem cells, these studies have been invaluable in understanding the likely signaling mechanisms and involved pathways, comparison of various routes and modes of stem cell delivery, optimal dosing of cells, and most importantly establishing preclinical safety. We have elucidated the key preclinical studies in Table 6.1 that demonstrated the safety and efficacy of IA stem cell therapy for ischemic stroke.

Table 6.1 Published and ongoing intra-arterial stem cell therapy preclinical studies

Study reference/ principal investigator	Study title	Route	Cell type/ source	Recipient species	Therapeutic window	Dose	Outcome measures	Follow-up/ outcome measurements	Results summary
[10]	Treatment of stroke in **rat** with intracarotid administration of marrow stromal **cells**	IA	MSC/ allogenic	Rat	1 day after ischemia	2×10^6	Neurological severity scores, Adhesive-removal test	At 1, 7, and 14 days after MCA	Homing of MSCs to ipsilateral MCA territory and functional improvement
[11]	Intracarotid transplantation of bone marrow stromal cells increases axon-myelin remodeling after stroke	IA	Rat MSC/ allogenic	Rat	After 24 h	2×10^6	Infarct volume, mNSS, adhesive removal	1, 7, 14, 21 and 28 days after MCA	Functional neurological improvement with treatment. Behavioral benefit is partly related to axon and myelin remodeling in the brain
[12]	Bone marrow stromal cells upregulate expression of bone morphogenetic proteins 2 and 4, gap junction protein connexin-43 and synaptophysin after stroke in rats	IA	Rat MSC/ allogenic	Rat	At 24 h after MCAo	2×10^6	Infarct volume, mNSS, adhesive removal	Baseline before MCA, one day after MCAo, imme-diately prior to BMSC administration and at 7, 14, 21, and at 28 days after MCAo	Significantly improved functional neurological recovery in rats at 14 days, 21 and 28 days after MCAo, without any significant decrease in lesion volume
[13]	Brain protection using autologous bone marrow cell, metalloproteinase inhibitors, and metabolic treatment in cerebral ischemia	IA	Rat MSC/ autologous	Rat	6 h after MCA	2×10^7	Infarct volume, rotarod	Rotarod at 2 hr after MCAo to 14 days	Combined treatment of BMCs with TIMPs and metabolic supplementation enhances neurogenesis and angiogenesis

Table 6.1 (continued)

Study reference/ principal investigator	Study title	Route	Cell type/ source	Recipient species	Therapeutic window	Dose	Outcome measures	Follow-up/ outcome measurements	Results summary
[14]	One-year follow-up after bone marrow stromal cell treatment in middle-aged female rats with stroke	IA	Rat MSC/ allogenic	Rat	At 1 day after stroke	$2 x 10^6$	Infarct volume, mNSS, adhesive removal	1 day, 2 weeks, 4 weeks, and then monthly	Improved neurological outcomes from 2 weeks through 1 year after BMSC injection
[15]	Intra-arterial transplantation of bone marrow mononuclear cells immediately after reperfusion decreases brain injury after focal ischemia in rats	IA/IV	BM-MNCs/ autologous	Rat	Immediately after reperfusion(90 min)	1×10^7	Infarct Volume, Motor function	24 h or 7 days after reperfusion IA superior to IV	Infarct volume decreased in the IA group, but not in the IV group
[16]	Intra-arterially delivered human umbilical cord blood-derived mesenchymal **stem cells** in **canine** cerebral ischemia	IA	MSC/ Xenogenic	Canine	1 day after ischemia	1×10^6	Neurological evaluation, MR imaging	1,7, 10, and 22 days until 4 weeks	Improved infarction volume (MRI) at one week and early recovery from neurological deficits
[17]	Functional recovery after hematic administration of allogenic mesenchymal stem cells in acute ischemic stroke in rats	IA/IV	Rat MSC/ allogenic	Rat	30 min after CC reperfusion	$2 x 10^6$	Infarct volume, MRI analysis of migration and implantation of stem cells	24 h and 14 days	IA equal to IV route in improving neurological recovery, and decreasing cerebral damage

Table 6.1 (continued)

Study reference/ principal investigator	Study title	Route	Cell type/ source	Recipient species	Therapeutic window	Dose	Outcome measures	Follow-up/ outcome measurements	Results summary
[18]	In vivo MR imaging of intra-arterially delivered magnetically labeled mesenchymal stem cells in a canine stroke model	IA	MSC/ autologous	Canine	At 1 week	3×10^6	MR imaging	In-vivo MRI images before cell grafting, one and 24 h after trans-plantation and weekly thereafter until 4 weeks	In vivo MR imaging is useful for tracking IA delivered MSCs after superparamagnetic iron oxide labeling
[19]	Intra-arterial cell transplantation provides timing-dependent cell distribution and functional recovery after stroke	IA	Human MSCs/ xenogenic	Rat	At 1, 4, or 7 days after MCA	$1 x 10^6$	Behaviour test, histology	0, 1, 4, 7, 14, and 21 days after stroke	Improved motor func-tion at day 1 and 4 but not day 7
[20]	Intracarotid transplantation of autologous adipose-derived mesenchymal stem cells significantly improves neurological deficits in rats after MCA	IA	Adipose-derived MSCs/ autologous	Rat	At day 3 after MCA	$2 x 10^6$	Behaviour test, adhesive-removal test	1, 7, 14, 21 and 28 days after MCAo	No significant differ-ence in infarct volume. Improved neurological and behavior scores 14 to day 28 after MCAo in treatment group
[21]	*Intra-arterial delivery* of human bone marrow mesenchymal *stem cells* is a safe and effective way to treat cerebral ischemia in rats	IA / IV	Human BM-derived MSCs/ xenogenic	Rat	At 24 h	3×10^6	Functional outcome, LDF, SPECT, PET	1, 3, 7, 14, and 28 days after transplantation	IA route promoted angiogenesis and improved functional recovery compared to IV

Table 6.1 (continued)

Study reference/ principal investigator	Study title	Route	Cell type/ source	Recipient species	Therapeutic window	Dose	Outcome measures	Follow-up/ outcome measurements	Results summary
[56]	Human bone marrow mesenchymal *stem*/stromal *cells* produce efficient localization in the brain and enhanced angiogenesis after *intra-arterial* delivery in *rats* with cerebral ischemia, but this is not translated to behavioral recovery	IA	Human BM-derived MSCs/ xenogenic	Rat	At day 2 or 7	1×10^6	Cylinder test, sticky label test	At 6, 21 and 42 days	No improvement in functional recovery despite effective initial homing to the ischemic hemisphere and enhanced angiogenesis
[22]	Efficacy and dose-dependent safety of intra-arterial delivery of mesenchymal stem cells in a rodent stroke model	IA	MSCs	Rat	1 and 24 h after reperfusion/MCA model	5×10^4 to 1×10^6	Infarct size and neurological evaluation	1 day through 28 days	Improved infarct size and neurological scores. Optimal dose calculated at 1×10^5

IA intra-arterial, *MSC* mesenchymal stem cell, *BM* bone marrow, *MNCs* mononuclear cells

Types of Stem Cells

Stem cells can be obtained from myriad sources. They can differentiate into diverse cell populations. Pluripotent stem cells, such as embryonic stem cells, are more potent with the ability to generate multiple cell types compared to multipotent stem cells that are derived from human tissue and differentiate into only mature cells [23]. However, the pluripotent stem cells have the drawback of tumorigenicity and forming potential teratomas [24]. Stem cells are also produced endogenously in response to tissue injury and can migrate to the ischemic lesions. Preclinical studies by various groups have shown that stem cells can act through multiple mechanisms including but not limited to neuroprotection, neurogenesis, angiogenesis, synaptogenesis, immunomodulation, release of growth, and trophic factors [25, 26].

Leong et al. in an extensive review of preclinical studies using stem cells in rodent models of ischemic stroke (most commonly used species) noted that among different types of stem cells, mesenchymal stem cells (MSCs, also called mesenchymal stem cell) and neuronal stem/progenitor cells reported most significant improvements in functional outcomes [27]. A recent meta-analysis of 46 preclinical studies of MSCs in ischemic stroke by Vu et al. suggested very promising results [28]. In their meta-analysis, they concluded that MSCs showed improved outcomes across various animal species, time of delivery, degree of immunogenicity, total dose/concentration, and in the presence of other comorbidities. Intracerebral (ICR) administration route was better than IA which in turn was better than IV route. ICR administration, however, is an invasive procedure compared to the alternative IA and IV modes of delivery. In our latest preclinical studies, we have shown the efficacy of IA delivery of MSCs and its superiority over the IV route in a rodent model of ischemic stroke [22]. The pros and cons of all routes of administration of stem cells are discussed in subsequent subsections. The meta-analysis by Vu et al. also suggested that MSCs used in primates showed most improvement in effect sizes compared to rodent and murine models [28]. The easy procurability and harvesting of MSCs, and improvement in outcomes after administration in various species and animal models of ischemic stroke, have been supported by various groups [29, 30]. MSCs can be obtained with relative ease; their procurement is not hampered by ethical considerations, and thus they have several advantages over other cell types.

There are various sources for MSCs, including but not limited to bone marrow (BM), umbilical cord blood cells (UCBCs), stem cells/progenitor cells derived from adipose tissue, stem cells derived from placenta [31], and human adult dental pulp [32]. UCBCs are pluripotent and are a source for MSCs. They are considered to have high regenerative potential and their naive immunologic phenotype makes them preferable for transplantation. Human adult dental pulp stem cells (DPSCs) that are derived from molar teeth are multipotent and have the capacity to differentiate into neurons. A preclinical study showed improved neurological outcomes with DPSCs in stroke animal model [32]. Olfactory ensheathing cells and endogenous stem cells from the subventricular zone (SVZ) and hippocampus are other potential therapies shown in preclinical studies for ischemic stroke [27, 33, 34].

Signaling Mechanisms

Complex signaling cascades and processes like vascular rolling, adhesion, endothelial transmigration, and migration through the extracellular space to the injury site are needed to recruit stem cells to the area of brain ischemia [35, 36]. Myriad chemical messengers are implicated in these processes. The cytokine stroma-derived factor (SDF)-1 from vascular endothelium and activated platelets acts on CXCR4 chemokine receptor on MSCs and influences the MSC mobilization and homing [37–41]. SDF-1 is regulated partially by the transcription modulator, hypoxia-inducible factor (HIF-1α) [42], matrix metalloproteinases (MMP), and nitric oxide (NO) [40, 41, 43–45]. Other chemokines like monocyte chemoattractant protein-1 (MCP-1) are also reported to have a role in the recruitment of MSCs [38]. Growth factors such as vascular endothelial growth factor (VEGF) and granulocyte colony-stimulating factor (G-CSF) are implicated in BM stem cell mobilization. G-CSF also promotes mobilization of hematopoietic stem cells (HSCs) and endothelial progenitor cells (EPCs) by decreasing SDF-1 expression in BM and CXCR4 expression on HSC [46–50].

Preclinical studies in stroke models suggest that MSCs themselves do not differentiate into neurons in vivo, however, promote differentiation of endogenous neural progenitor cells [27]. MSCs can secrete neurotrophic factors such as brain-derived neurotrophic factor (BDNF), basic fibroblast growth factor (bFGF), angiopoietin-2, transforming growth factor (TGF-B), and insulin-like growth factor (IGF-1). These factors promote the remodeling of the neurons, glia, and neurovascular cells, and enhance axonal sprouting and synaptogenesis [11]. MSCs also decrease neuronal apoptosis and activation of astrocytes to minimize scar formation, and promote angiogenesis by secreting VEGF, bFGF, nerve growth factor (NGF), and other angiogenic factors [13, 51]. Furthermore, MSCs also influence immunomodulation via TGF-B, NO and prostaglandin pathways, JAK-STAT signaling pathways and release of interleukin-6 [51]. Thus, stem cells could be considered to have pleiotropic effects by modulating numerous targets through multiple signaling cascades.

Radioimaging studies can be very helpful in tracking and localization of tracer-labeled stem cells which in turn can help in understanding homing mechanisms. Preclinical studies on rodent models [52, 53] as well as canine model [18] showed that bioluminescence imaging and superparamagnetic iron oxide labeling helped to track and localize IA-delivered stem cells to the ischemic areas after stroke. Other groups [16] also demonstrated the feasibility of magnetic resonance imaging (MRI) in canine model of ischemic stroke.

Therapeutic Window for Stem Cell Therapy in Ischemic Stroke

Therapies provided in the early hours after ischemic stroke are aimed at reducing the injury/limiting the infarct size. On the other hand, therapies started weeks to months after stroke are aimed at promoting remodeling and repair. MSCs and other stem cells may have potential for wide-ranging therapeutic time windows [27, 32,

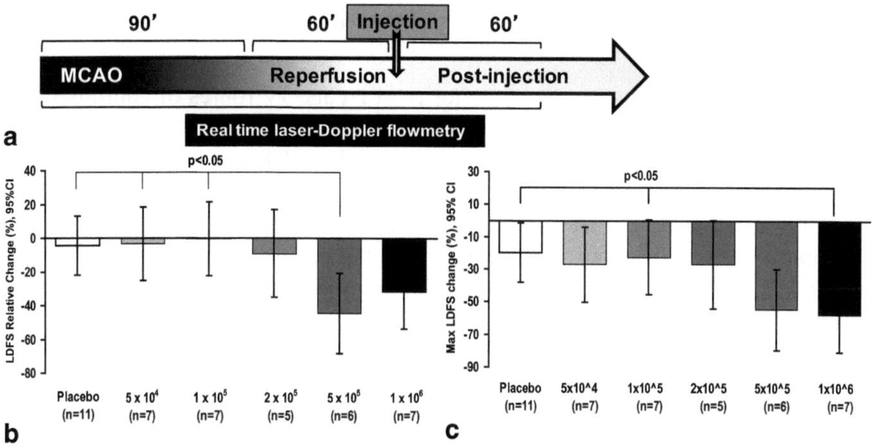

Fig. 6.1 Lower doses of IC MSCs mitigate adverse effect IC injection on MCA blood flow. **(a)** Experimental timeline showing rMCAo for 90 minutes followed by withdrawal of the suture to allow reperfusion. At 60 minutes of reperfusion, IC MSC or vehicle only injection was given, followed by LDF monitoring for 60 min. **(b)** Comparison of relative LDFS worsening from baseline to final recording, among de-escalating dose groups **(c)** The 1 x 10^5 dose and placebo has significantly less maximum LDF worsening as compared to the 1 x 10^6 dose. The comparisons in (b) and (c) were done using general linear modeling (GLM) to compare mean differences among groups. LDFS = Laser Doppler Flow Signal. IC=intracarotid. (Yavagal et al., 2014)

54]. A compromised BBB after ischemic stroke may allow better delivery of the MSCs to the target ischemic lesion. On the other hand, the BBB may be intact up to 24 h after stroke and may deter the engraftment of target tissue. Similarly, in the chronic phase, remnant glial scar tissue is likely to obstruct the passage of stem cells [27].

We showed that the therapeutic window lasts at least 24 h in our preclinical studies using a rodent model of ischemic stroke [22] (Fig. 6.1). One benefit of early infusion of stem cells is that the inflammatory chemokines that are released after ischemic stroke can help in attracting the stem cells. Some researchers suggested that the effect of MSCs may be reduced if provided 7 days after stroke [55]. However, there are other preclinical IA studies that have shown that stem cell treatment showed significant enhancement of functional recovery when given up to 7 days after ischemic stroke [19] and increased angiogenesis [56]. It is understandable that preclinical studies mostly focused on early therapeutic window in order to demonstrate better outcomes. Nonetheless, the preclinical studies demonstrate a wide therapeutic window for stem cells compared to existing standard of care therapies. On the other hand, some preclinical studies have unequivocally demonstrated that the beneficial effects of stem cells extend for a long duration from 4 months [32] up to a year [14] from the time of administration after ischemic stroke. This is very encouraging for clinical trials, if these time frames in rodent models were converted to human years.

Advantages and Disadvantages of Intra-arterial Route

Preclinical studies point out the superiority of IA and IV routes of administration over direct implantation with ICR and intracerebroventricular (ICV) modes of delivery. Firstly, IA and IV routes are less invasive than ICR and ICV routes. IA may be considered more invasive than IV, however compared to all other routes (ICR, ICV, IV, intrathecal), the IA route should be able to deliver significantly increased number of cells with more uniform cell distribution in the target ischemic brain tissue [11, 52]. The verdict is unclear with many preclinical studies showing diverse results.

Many groups have also shown that IA and IV are equally effective for improving neurological recovery, decreasing cerebral damage, and promoting protection mechanisms in ischemic stroke [57–59]. Vasconcelos-dos-Santos et al. used bone marrow mononuclear cells (BM-MNCs) via IA and IV routes, and Zhang et al. used human umbilical tissue-derived cells via IA, IV, ICR, ICV, and intrathecal routes. Both groups showed improved outcomes with all routes of delivery of stem cells. This was echoed by Savitz et al. who used autologous BM-MNCs and showed no difference in outcomes between IA and IV routes [58]. These studies suggest that IA delivery of BM-MNCs may not be superior to their IV delivery in ischemic stroke. Also, Gutierrez-Fernandez et al. used allogenic MSCs in a permanent middle cerebral artery occlusion (MCAo) rat model and showed no difference in ICR and IV routes of administration on the functional outcomes [17]. The permanent occlusion of the internal carotid artery (ICA)-MCA on the side of IA administration would block direct delivery of MSCs to the infarct area and may be the crucial reason for the absence of superior of IA over IV route in efficacy in this study. On the other hand, several preclinical direct comparison studies between IV and IA routes have shown that IA delivery provides higher concentration of cells to the target tissue than IV delivery [15, 21, 53, 60, 61]. BM-MNCs when delivered via IA route showed better results than IV route on the size of infarct volume, target cell delivery, and motor function after ischemic stroke [21, 61]. Other showed this in ischemic stroke using human MSCs in rodent model [60] and mouse neural stem cells in a mouse model [53]. Guzman et al. have shown using bioluminescence imaging that IA injection results in significantly higher and more sustained cell delivery to the brain [53]. Du et al. [21] in a direct comparison study showed greater functional recovery and increased angiogenesis with IA route compared to IV route using human BM-MSCs in a rat ischemic model. Lundberg et al. proved this concept in the traumatic brain injury (TBI) model. They showed in a rat TBI model, using various cell populations, human MSCs, human neural progenitor cells, and rat neural progenitor cells, that targeted IA route is more effective than IV administration [61]. The superiority of IA route over IV route may partly be due to circumvention of cell sequestration in the peripheral filtering organs such as lung capillaries encountered with the IV route [62]. We feel that there are multiple critical factors that can influence superiority of IA route over IV route including but not limited to homing of stem cells in response to chemoattractants, cell migration, level of arteriosclerosis and resulting stenosis, surviving arterial arborization, and disruption of BBB after stroke.

IA delivery of cells, however, does pose certain safety hurdles that need to be addressed fully prior to clinical translation. Microvascular occlusions leading to decreased cerebral blood perfusion have been reported in preclinical studies with IA stem cell delivery. Chua et al. reported that micro-occlusions can be prevented by preserving anterograde flow [63]. Whether related to microvascular occlusions or not, IA delivery of neural progenitor cells and MSCs has been reported to worsen ischemia and increased mortality in some preclinical studies [64–66]. We investigated this issue in our preclinical studies using multiple concentrations of MSCs delivered via IA route [22]. Dose de-escalation studies with real-time monitoring of MCA flow showed that de-escalation of IA MSC cells doses to 1×10^5 MSCs mitigated the decrease in MCA Doppler flow signal that was seen at higher doses (Figs. 6.2 and 6.3). This IA maximum tolerated dose (MTD) of 1×10^5 MSCs given at 24 h post rMCAO also resulted in significantly superior attenuation of neurological deficits and reduction in infarct size after ischemic stroke as compared to IA vehicle and IV MSC groups (Figs. 6.1 and 6.4). Moreover, delivering MSCs using IA route, we also showed that the therapeutic window extended to 24 h. In fact, MSCs delivered at 24 h after ischemic stroke showed greater beneficial effects as compared to MSCs administered at 1 h. The findings were very promising because the benefits of arterial delivery of MSCs translated into long-term neurological improvement as well as reduction in infarct size (Figs. 6.1 and 6.4). Considering all published preclinical studies till date, the IA route of stem cells delivery remains promising with a number of clinically attractive advantages over other modes of delivery.

Clinical Studies and Trials of Intra-arterial Delivery of Stem Cell Therapies in Ischemic Stroke

There has been a surge of early clinical trials of stem cells in acute ischemic stroke and interest has spiked across the globe in the last 5 years [34, 67]. It is of paramount importance to first establish clinical safety during the development of novel therapies. Rapid strides have been made over the past decade in establishing the clinical safety of stem cells in ischemic stroke patients. Kondziolka et al. first demonstrated the feasibility and safety of stem cells in ischemic stroke [68]. They directly implanted stem cells derived from a human teratocarcinoma cell line (NT2N, Layton BioScience Inc, Sunnyvale, CA) via stereotactic ICR delivery in ischemic stroke patients. In this phase I, non-randomized, observer-blinded study involving 12 patients with basal ganglia strokes and motor deficits, they showed that ICR transplantation of stem cells ($2–6 \times 10^6$) was feasible. All patients were immunosuppressed with cyclosporine A. Two isolated events of a generalized seizure in one patient and a distant brain stem stroke in another patient were thought to be unrelated to the stem cell injections. The 6- and 12-month follow-up positron emis-

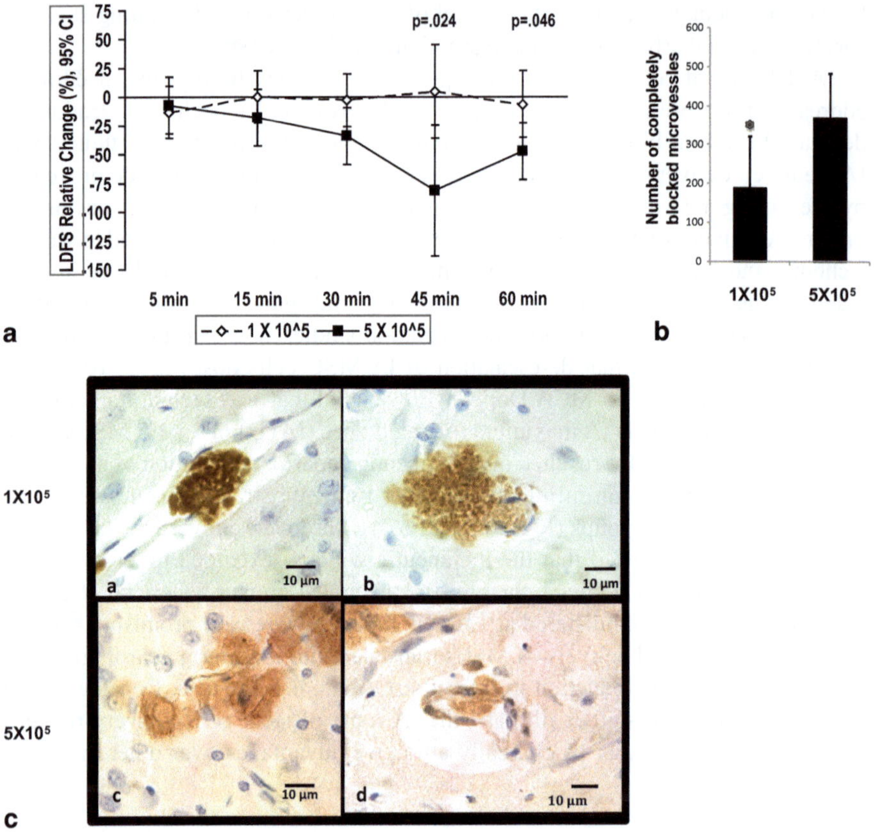

Fig. 6.2 Comparison of 1 x 10^5 to 5 x 10^5 dose-groups for LDFS change over time and microvascular occlusion. **(a)** IC MSC dose of 1 x 105 has a transient and less MCA LDF worsening during 60 minutes post injection as compared to 5 x 10^5 dose using mixed model. LDFS = Laser Doppler Flow Signal. **(b)** On comparing the total number of microvessels with complete occlusion among the two dose groups, there were a significantly lower number of complete occlusions in the lower dose group. mean ± SD, * $P< 0.05$, ANOVA. **(c)** Representative brain sections from IC MSC 1 x 10^5 dose group showing GFP + MSCs identified by 3, 3"-diaminobenzidine (DAB), showing localized complete filling of microvessels at 3-5 days post-injection, as well as MSCs just outside the vessel wall in brain parenchyma and high power field of representative brain sections from IC 5 x 10^5 dose group showing single MSC partly inside and partly outside microvessel wall IC=intracarotid. (Yavagal et al., 2014)

sion tomography (PET) scans showed improved fluorodeoxyglucose uptake at the implant site in seven and three patients, respectively [68, 69]. Postmortem analysis on one patient who died of myocardial infarction 27 months after transplantation showed the presence of neurons derived from the transplanted cells at the implantation site [70]. Kondziolka et al. followed this study with a phase II, randomized study in nine patients with ischemic and hemorrhagic stroke each [71]. They treated seven patients in each group and four patients served as controls. They reported

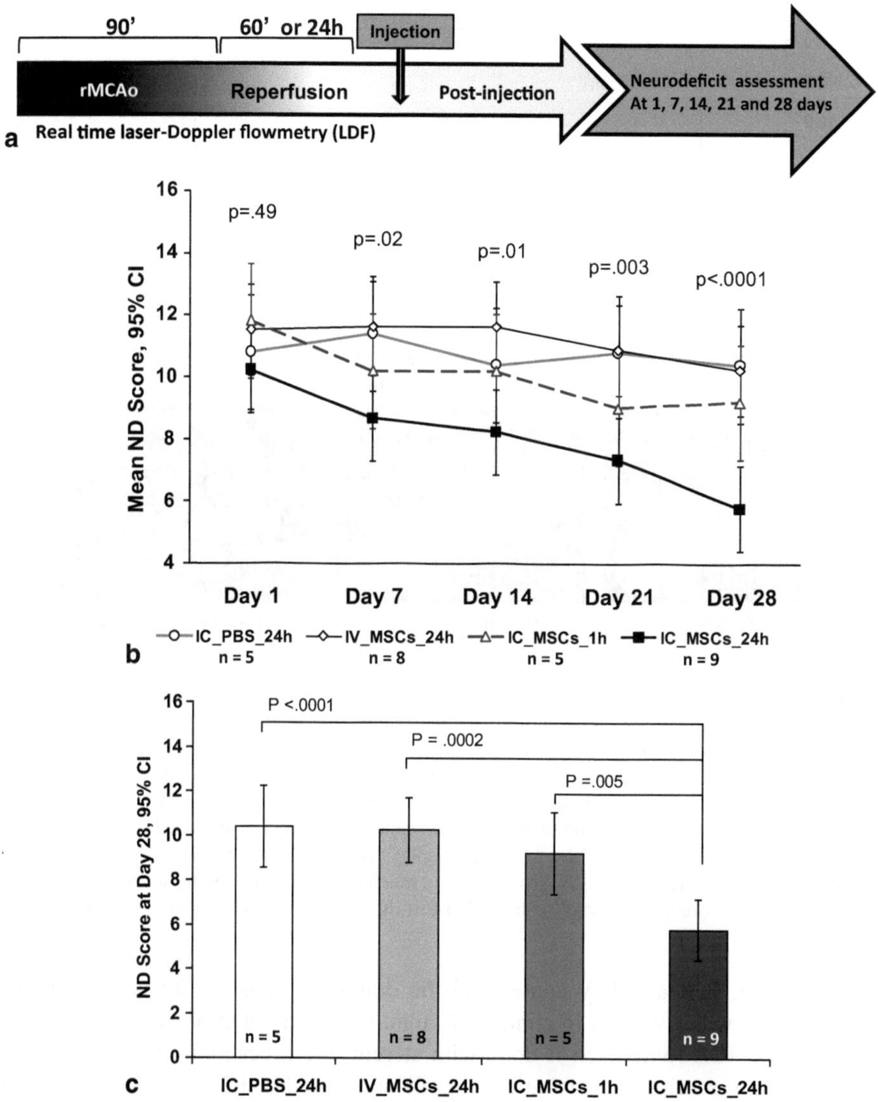

Fig. 6.3 Functional neurologic outcomes are superior in the group treated with 1 x 10^5 IC MSCs at 24 hours (24h IC MSC). **(a)** Experimental timeline of the efficacy study showing 90 min rMCAo followed by 24 hour reperfusion except in group receiving IC MSCs at 1 hour reperfusion followed by ND score assessment at 1,7,14,21 & 28 days. **(b)** On day 1 post rMCAo, the ND scores were not significantly different among groups. The ND score of the 1 x 10^5 group progressively decreased over time and at 28 days was significantly lower than the other groups. **(c)** The day 28 ND score was significantly lower in the 24h IC MSC group as compared to all the remaining groups. ND = Neurodeficit. IC=intracarotid. (Yavagal et al., 2014)

improved "Action Research Arm Test" gross hand-movement scores compared with the control and baseline values.

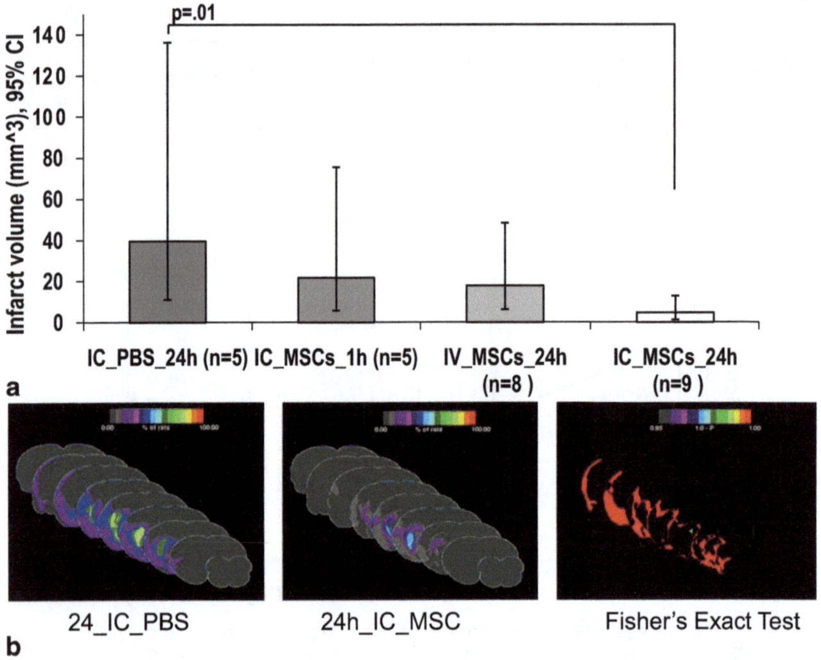

Fig. 6.4 Comparison of infarct volume among treatment groups. **(a)** Geometric mean infarct volumes are compared among groups after log transformation to achieve normal distribution. Only the 24h IC MSC group shows significantly reduced infarct volume as compared to the 24h IC PBS. **(b)** Frequency infarct map statistically comparing the location of the mean infarct volume in the 24h IC MSC and 24h IC PBS groups using color- coded representation of the percent of rats showing infarction in each brain region using "Fisher's exact test" with a color coded representation of the "p-value", the color bar in "1-p" format. Displays are in "Coronal presentation"; middle sections are selected. The 24h IC MSC group shows a much reduced infarction frequency, particularly surrounding the core as quantized by the Fisher test. IC=intracarotid. (Yavagal et al., 2014)

However, ICR and ICV routes had the drawback of being invasive. Furthermore, there was always a likelihood of nonuniform distribution within the target lesion with ICR route even if done with stereotactic precautions [72, 73]. Moreover, there was a risk of postsurgical complications such as hematomas, inadvertent lesions, and other complications including but not limited to seizures, syncope, and transient loss of motor function [71, 74]. The ICV route though less invasive than ICR implantation, also had the drawback of variable distribution of stem cells. Furthermore, some patients who underwent ICV delivery had fever and developed meningeal signs [75].

Various groups showed that peripheral/vascular delivery of stem cells was not only feasible and safe but also preliminarily beneficial. Bang et al. used the IV route and showed that 5 stroke patients receiving IV autologous MSCs (5×10^7) had better functional recovery at 1 year as compared to 25 matched controls, and did not suffer additional morbidity and mortality over a total follow-up period of 5 years [64, 76].

Currently, a multicenter phase II randomized, double-blind, placebo-controlled trial using MultiStem® is actively recruiting patients [77]. MultiStem® is an allogeneic cell therapy treatment with multipotent adherent BM cells, harvested from healthy donors, that has been shown to be safe in clinical trials of myocardial infarction and graft versus host disease, and preclinical models of stroke. The clinical trial is designed to deliver the stem cells within 48 h of anterior circulation ischemic stroke and evaluate long-term outcomes from 3 to 12 months. Other IV- and ICR-delivered stem cell studies have been extensively discussed in recent reviews [67, 78, 79].

In order to harness the multiple advantages of IA cell delivery in acute ischemic stroke seen in preclinical studies and the high clinical applicability of this route, several early clinical trials of IA delivery of stem cells have been conducted in the last 5 years. Mendonca et al. using BM-MNCs showed for the first time that IA mode of delivery was safe [80]. They injected 30×10^7 of autologous BM-MNC directly into the left MCA of a patient who had suffered an ischemic stroke yet had some preserved penumbra in the L-MCA territory. They injected the patient on day 5 after stroke; monitored, reported no microembolization or any electroencephalography (EEG) changes, and noted neurological improvement at 2-month follow-up. In a subsequent case, the same group also demonstrated that the autologous BM cells could be tracked by labeling with Tc-99 m-hexamethylpropylene amine oxime [81]. They reported that single-photon emission computed tomography (SPECT) images 8 h after cell transplantation showed the homing of 99 mTc hexamethylpropyleneamine oxime (HMPAO)-labeled cells mainly in the anterior division of left MCA territory, while the stroke was in the posterior MCA division, presumably because of the occlusion of the posterior branch. Recently, Barbosa da Fonseca et al. [34] used technetium-99 m-labeled aBM-MNC-SC in 12 patients with ischemic stroke and showed no difference in brain radioactive counts between IA and IV administration of same dose of cells. However, they reported higher radioactive counts in liver and spleen and lower counts in lungs in the IA group compared to the IV group. This partly supports the preclinical studies, which had showed advantages of IA over IV delivery by circumventing the systemic circulation and bypassing the sequestration in lungs [62]. Preclinical studies have shown that IA route improves stem cell delivery to the target lesion in the brain [15, 21, 53, 60, 61]. This was not seen in the data obtained by Barbosa Da Fonseca's group. Perhaps, the stem cells were also diverted via chemotactic signaling to other organs such as heart and kidneys which are affected in patients with ischemic strokes [82, 83]. More clinical studies with larger cohorts will be helpful to confirm the extrapolated information from preclinical studies that IA route delivers more stem cells to target stroke lesion than IV route. The same group [84] had also provided critical information that IA stem cell delivery was feasible in subacute stroke. They reported no adverse effects or any neurological worsening up to 6 months after aBM-MNC-SC delivery even if the stem cells were delivered via IA route 90 days after the ischemic stroke. This data along with numerous preclinical studies are helpful in acknowledging that IA stem cell therapy for ischemic stroke may hold an edge over current standard of care therapies in terms of an extended therapeutic window and having minimal side effects.

Friedrich et al. showed in a larger cohort of patients ($n=20$) that delivery of autologous BM mononuclear stem cells (aBM-MNCs) directly in the ipsilateral MCA was safe and could be helpful in delivering large number of stem cells (up to 6×10^8) without any adverse effects [85]. Eight of the 20 patients had improved outcomes defined by modified Rankin score (mRS) ≤ 2 at 3 months after ischemic stroke; however, they had no controls for comparison. Jiang et al. were the first to report safety of IA stem cell therapy in a patient with hemorrhagic stroke [86]. They used MSCs from umbilical cord source (Jiangsu Stem CellBank, Jiangsu, China) and noted improvements in this patient as well as three other patients with MCA ischemic stroke. They noted that mRS improved in two of the four patients they treated with stem cells. More recently, Banerjee et al. [87] showed that delivery of autologous, immunoselected $CD34^+$ stem/progenitor cells via ipsilesional MCA was safe in patients presenting within 7 days of ischemic stroke. They limited the study to patients with severe anterior circulation ischemic stroke with the National Institutes of Health Stroke Scale (NIHSS) of at least eight. Although, the recruitment was limited to 5 patients out of 82 eligible patients, the results were promising and showed improvement in functional outcomes up to 6 months of follow-up without any adverse effects.

The initial clinical studies were case reports and phase I, non-randomized, open-label studies. The first single-blind randomized clinical trial was reported from Spain. Moniche et al. [90] enrolled 20 consecutive MCA stroke patients (10 treatments and 10 controls) and injected the treatment group between 5 and 9 days with $\sim 1.6 \times 10^8$ aBM-MNCs. Though, they did not see any significant differences in neurological function at 6-month follow-up, they reported a trend to positive correlation between number of injected $CD34^+$ cells and Barthel index. But, most importantly, they reported that IA therapy was safe. Barring an isolated partial seizure in two patients in treatment group, there were no deaths, stroke recurrence, or any tumor formation during the follow-up period. The two patients were treated with antiepileptic medications and did not have any seizure recurrence. The same group has recently suggested that aBM-MNCs can induce changes in serum levels of granulocyte macrophage colony-stimulating factor (GM-CSF), platelet-derived growth factor (PDGF-BB), and MMP-2 up to 3 months after transplantation, which could be associated with better functional outcomes [88].

These initial studies have given impetus to IA and IV therapy for ischemic stroke and paved the way for many randomized clinical trials with larger patient population. The ongoing registered clinical trials for IA stem cell therapy are listed in Table 6.2 (source: www.clinicaltrials.gov) along with the related published reports. The first clinical trial for IA cell therapy for acute ischemic stroke, RECOVER-Stroke, was a phase II randomized clinical trial, evaluating the efficacy of ALD-401 cells (autologous BM cells, Aldagen Inc, Durham, NC). Subjects experiencing an ischemic stroke were to undergo either a BM or sham harvest on days 11–17 and then be dosed with ALD-401 or a sham procedure 13–19 days after the primary event. BM cells were processed, sorted, and formulated into a 3-mL suspension of ALD-401. Within 48 h of harvestation, subjects in the ALD-401 group had their processed BM cells (ALD-401) injected via intracarotid/MCA

Table 6.2 Published and ongoing intra-arterial stem cell therapy clinical studies and trials. (Sources: www.pubmed.gov and www.clinicaltrials.gov)

Study reference/country/study director	Phase/stage	Study design/title	Route	Cell type	Type of stroke	Age range (years)	Thera-peutic window (after stroke)	Number of patients (and controls)	Number of cells injected, volume, rate and duration	Follow-up
Registered clinical trials [Source: www.clinicaltrials.gov] and related published clinical studies (Source: www.pubmed.gov)										
NCT00473057 Andre C, De-Freitas GR, Mendez-Otero R, and Barbosa da Fonseca LM Brazil Universidade Federal do Rio de Janeiro	Phase I completed	Study of autologous stem cell transplantation for patients with ischemic stroke	IA/IV	aBM-SC	MCA ischemia	18–75 years	3–90 days	15 (IA) 5 (IV)	500×10^6	4 months
[80, 81] Brazil (Hospital Pró-Cardíaco, Universidade Federal do Rio de Janeiro e Instituto do Milênio de Bioengenharia Tecidual)	Case reports on Phase I non-randomized, open label	Safety and feasibility of autologous transplantation of bone marrow mononuclear cells for acute ischemic stroke	IA	aBM-MNC-SC	Acute MCA stroke	37 and 54 years	5 days and 9 days	2 patients	3×10^7 in each patient	2 months and 4 months
[84, 89] Rosado-de-Castro 2013 Brazil (Hospital Universitário Clementino Fraga Filho, Universidade Federal do Rio de Janeiro)	Phase I, non-randomized, open label	Safety of autologous bone marrow mononuclear cell transplantation in patients with nonacute ischemic stroke	IA	aBM-MNC-SC	MCA stroke	–	19–89 days (mean 64.5 days)	12 (no controls)	1×10^8 to 5×10^8(mean 3.1×10^8)	6 months
[85] Brazil (Hospital São Lucas, Porto Alegre)	Phase I, non-randomized, open label	Intra-arterial infusion of autologous bone marrow mononuclear cells in patients with moderate to severe middle cerebral artery acute ischemic stroke	IA	aBM-MNC-SC	MCA stroke	–	3–7days	20 (no controls	5.1×10^7 to 6×10^8(mean 2.2×10^8)	3–6 months

Table 6.2 (continued)

Study reference/country/study director	Phase/stage	Study design/title	Route	Cell type	Type of stroke	Age range (years)	Thera-peutic window (after stroke)	Number of patients (and controls)	Number of cells injected, volume, rate and duration	Follow-up
NCT02178657 Moniche F et al. Spain (Hos-pitales Universitarios Virgen del Rocío)	II not recruiting	Intra-arterial autologous bone-marrow mononuclear cells infusion for acute ischemic stroke	IA	aBM-SC	Acute isch-emic stroke	18—80 years	1–7 days	76	2×10^{6} cells/kg	6 months—2 years
[90] Spain (Hospitales Universitarios Virgen del Rocío)	Phase I non-randomiz, single-blind	Intra-arterial bone marrow mononuclear cells in isch-emic stroke: a pilot clinical trial.	IA	aBM-MNC-SC	MCA stroke	–	5–9d (mean 6.4 d)	10 (10 controls)	mean 1.59×10^{8}	6 months
China Nanjing Univer-sity School of Medicine (Funded by National Science Foundation of China [86]	Phase I, non-randomized, open label	Feasibility of delivering mesenchymal stem cells via catheter to the proximal end of the lesion artery in patients with stroke in the territory of the middle cerebral artery	IA	Allo-genic UC-MSC	MCA isch-emic stroke (3) + Hemor-rhagci stroke (1)	40–50 years	11–50 days (mean 25.5)	4 (no controls)	2×10^{7}	6 months
NCT00535197 Habib N et al. United Kingdom Imperial College Health-care NHS Trust, London	I/II active, recruiting	Autologous bone marrow stem cells in ischemic stroke	IA	aBM-SC	MCA isch-emic stroke	30–80 years	7 days	10 patients	1×10^{8}	6 months

Table 6.2 (continued)

Study reference/country/study director	Phase/stage	Study design/title	Route	Cell type	Type of stroke	Age range (years)	Therapeutic window (after stroke)	Number of patients (and controls)	Number of cells injected, volume, rate and duration	Follow-up
[87] United Kingdom Imperial College Healthcare NHS Trust, London	Phase I, non-randomized, open label	Intra-Arterial immunoselected CD34+ stem cells for acute ischemic stroke	IA	Autologous, immunoselected CD34+ stem/progenitor cells	Anterior circulation stroke	–	7 days	5 (no controls)	–	6 months
NCT00761982 Spain (Hospital Universitario Central de Asturias)	I/II completed	Autologous bone marrow stem cells in middle cerebral artery acute stroke treatment	IA	aBM-SC	Acute ischemic stroke	18–80 years	5–9 days	20	Not specified	1, 3 and 6 months
NCT01273337 Hinson et al. USA (Aldagen Inc, Durham, NC, now Cytomedix Inc Gaithersburg MD)	II active, not recruiting	Study of ALD-401 Via intracarotid infusion in ischemic stroke subjects	IA	ALD-401 derived from aBM	MCA ischemia	30–83 years	13–19 days	~100	3 mL autologous BM suspension	3 months -1 year[a]
NCT01453829 Morales V et al. Mexico (Instituto de Medicina Regenerativa, S.A. de C.V)	I/II not recruiting	Study to assess the safety and effects of autologous adipose-derived stromal cells in patients after stroke	IV/IA	Autologous adipose-derived MSCs	Ischemic/hemorrhagic stroke	18–80 years	Not specified	10 (no controls)	Not specified	1 week -6 months

aPB autologous peripheral blood, *aMSC* autologous mesenchymal stem cell, *aBM* autologous bone marrow, *UC* umblical cord, *SC* stem cell, *IV* intravenous, *IA* intra-arterial, *ICR* intracerebral

[a]NCT01273337 clinical trial had a press release on 5 May, 2014, as discussed in the text of the accompanying chapter

infusion, while control subjects had a sham infusion. ALD-401 had previously been shown to have beneficial outcomes in preclinical studies. Clinical and functional outcomes were to be monitored at 3, 6, and 12 months. Preliminary results at 3-month follow-up showed no difference in the functional outcomes between the treatment and sham groups ("Cytomedix Announces Results of RECOVER-Stroke Phase 2 [91] Study," Press release: May 2, 2014). Nonetheless, there were no serious adverse events attributable to the use of ALD-401, demonstrating good tolerability and safety. The 6- and 12-month follow-up results are eagerly awaited by the scientific community.

In summary, early clinical trials of IA stem cell therapy for ischemic stroke have shown the safety of this approach without any feared adverse effects such as tumorigenicity, microembolization, and stroke expansion. They have set the stage for the next phase of clinical trials in this emerging therapeutic area. Careful designing of the next round of studies including using the most promising type of stem cells and determining the optimal time range of administration of stem cells and optimal cell dosing will be critical in translating the tremendous promise of the IA cell therapy in ischemic stroke.

References

1. Hacke W, Donnan G, Fieschi C, Kaste M, Kummer R von, Broderick JP, Brott T, Frankel M, Grotta JC, et al. Association of outcome with early stroke treatment: pooled analysis of ATLANTIS, ECASS, and NINDS rt-PA stroke trials. The Lancet. 2004;363:768–74.
2. Lansberg MG, Schrooten M, Bluhmki E, et al. Treatment time-specific number needed to treat estimates for tissue plasminogen activator therapy in acute stroke based on shifts over the entire range of the modified Rankin Scale. Stroke. 2009;40:2079–84.
3. Kleindorfer D, Lindsell CJ, Brass L, Koroshetz W, Broderick JP. National US estimates of recombinant tissue plasminogen activator use: ICD-9 codes substantially underestimate. Stroke. 2008;39:924–8.
4. Macrae IM. Preclinical stroke research—advantages and disadvantages of the most common rodent models of focal ischaemia. Br J Pharmacol. 2011;164:1062–78.
5. Berkhemer OA, Dippel DW and MR CLEAN Investigators. A randomized trial of intra-arterial treatment for acute ischemic stroke. N Engl J Med. 2015 Jan 1;372(1):11–20. doi: 10.1056/NEJMoa1411587. Epub 2014 Dec 17.
6. Dippel D, et al. 9th World Stroke Congress: Session: Main Theme 43: Late Breaking News. No abstract number. Presented October 25, 2014.
7. Dirnagl U, Iadecola C, Moskowitz MA. Pathobiology of ischaemic stroke: an integrated view. Trends Neurosci. 1999 Sep;22(9):391–7. (Review).
8. Strong K, Mathers C, Bonita R. Preventing stroke: saving lives around the world. Lancet Neurol. 2007;6:182–7.
9. Taylor TN, Davis PH, Torner JC, et al. Lifetime cost of stroke in the United States. Stroke. 1996;27:1459–66.
10. Li Y, Chen J, Wang L, Lu M, Chopp M. Treatment of stroke in rat with intracarotid administration of marrow stromal cells. Neurology. 2001;56:1666–72.
11. Shen LH, Li Y, Chen J, Zhang J, Vanguri P, Borneman J, Chopp M. Intracarotid transplantation of bone marrow stromal cells increases axon-myelin remodeling after stroke. Neuroscience. 2006;137:393–9.

12. Zhang ZG, Chopp M. Neurorestorative therapies for stroke: underlying mechanisms and translation to the clinic. Lancet Neurol. 2009;8:491–500.
13. Baker AH, Sica V, Work LM, et al. Brain protection using autologous bone marrow cell metalloproteinase inhibitors, and metabolic treatment in cerebral ischemia. Proc Natl Acad Sci USA. 2007;104:3597–602.
14. Shen LH, Li Y, Chen J, et al. One-year follow-up after bone marrow stromal cell treatment in middle-aged female rats with stroke. Stroke. 2007;38:2150–56.
15. Kamiya N, Ueda M, Igarashi H, et al. Intra-arterial trans- plantation of bone marrow mononuclear cells immediately after reperfusion decreases brain injury after focal ischemia in rats. Life Sci. 2008;83:433–7.
16. Chung DJ, Choi CB, Lee SH, Kang EH, Lee JH, Hwang SH, Han H, Lee JH, Choe BY, Lee SY, Kim HY. Intraarterially delivered human umbilical cord blood-derived mesenchymal stem cells in canine cerebral ischemia. J Neurosci Res. 2009 Dec;87(16):3554–67.
17. Gutierrez-Fernandez M, Rodriguez-Frutos B, Alvarez-Grech J, et al. Functional recoveryafter hematic administration of allogenic mesenchymal stem cells in acute ischemic stroke inrats. Neuroscience. 2011;175:394–405.
18. Lu SS, Liu S, Zu QQ, Xu XQ, Yu J, Wang JW, Zhang Y, Shi HB. In vivo MR imaging of intraarterially delivered magnetically labeled mesenchymal stem cells in a canine stroke model. PLoS ONE. 2013;8(2):e54963.
19. Ishizaka S, Horie N, Satoh K, Fukuda Y, Nishida N, Nagata I. Intra-arterial cell transplantation provides timing-dependent cell distribution and functional recovery after stroke. Stroke. 2013;44:720–6.
20. Jiang W, Liang G, Li X, Li Z, Gao X, Feng S, et al. Intracarotid transplantation of autologous adipose-derived mesenchymal stem cells significantly improves neurological deficits in rats after MCAo. J Mater Sci Mater Med. 2014;25:1357–66.
21. Du S, Guan J, Mao G, Liu Y, Ma S, Bao X, Gao J, Feng M, Li G, Ma W, Yang Y, Zhao RC, Wang R. Intra-arterial delivery of human bone marrow mesenchymal stem cells is a safe and effective way to treat cerebral ischemia in rats. Cell Transplant. 2014 Nov 4;23(suppl 1):73–82.
22. Yavagal DR, Lin B, Raval AP, Garza PS, Dong C, et al. Efficacy and dose-dependent safety of intra-arterial delivery of mesenchymal stem cells in a rodent stroke model. PLoS ONE. 2014;9(5):e93735.
23. Weissman IL, Anderson DJ, Gage F. Stem and progenitor cells: origins, phenotypes, lineage commitments, and transdifferentiations. Annu Rev Cell Dev Biol. 2001;17:387–403. (Review).
24. Thomson JA, Itskovitz-Eldor J, Shapiro SS, Waknitz MA, Swiergiel JJ, Marshall VS, Jones JM. Embryonic stem cell lines derived from human blastocysts. Science. 1998 Nov 6;282(5391):1145–7.
25. Burns TC, Steinberg GK. Stem cells and stroke: opportunities, challenges and strategies. Expert Opin Biol Ther. 2011 Apr;11(4):447–61. doi: 10.1517/14712598.2011.552883. Epub 2011 Feb 16. Review.
26. Zhang C, Li Y, Chen J, et al. Bone marrow stromal cells upregulate expression of bone-morphogenetic proteins 2 and 4, gap junction protein connexin-43 and synaptophysin after stroke _ in rats. Neuroscience. 2006;141:687–95.
27. Leong WK, Lewis MD, Koblar SA. Concise review: Preclinical studies on human cell-based therapy in rodent ischemic stroke models: where are we now after a decade? Stem Cells. 2013;31(6):1040–3.
28. Vu Q, Xie K, Eckert M, Zhao W, Cramer SC. Meta-analysis of preclinical studies of mesenchymal stromal cells for ischemic stroke. Neurology. 2014 Mar 7;82(14):1277–86.
29. Eckert MA, Vu Q, Xie K, Yu J, Liao W, Cramer SC, Zhao W. Evidence for high translational potential of mesenchymal stromal cell therapy to improve recovery from ischemic stroke. J Cereb Blood Flow Metab. 2013 Sep;33(9):1322–34.
30. Gutiérrez-Fernández M, Rodríguez-Frutos B, Ramos-Cejudo J, Teresa Vallejo-CremadesM, Fuentes B, Cerdán S, Díez-Tejedor E. Effects of intravenous administration of allogenic

bone marrow- and adipose tissue-derived mesenchymal stem cells on functional recovery and brain repair markers in experimental ischemic stroke. Stem Cell Res Ther. 2013;4(1):11.
31. Kranz A, Wagner DC, Kamprad M, et al. Transplantation of placenta-derived mesenchymal stromal cells upon experimental stroke in rats. Brain Res. 2010;1315:128–36.
32. Leong WK, Henshall TL, Arthur A, Kremer KL, Lewis MD, Helps SC, Field J, Hamilton-Bruce MA, Warming S, Manavis J, Vink R, Gronthos S, Koblar SA. Human adult dental pulp stem cells enhance poststroke functional recovery through non-neural replacement mechanisms. Stem Cells Transl Med. 2012 Mar;1(3):177–87.
33. Burns TC, Verfaillie CM, Low WC. Stem cells for ischemic brain injury: a critical review. J Comp Neurol. 2009;515(1):125–44.
34. Rosado-de-Castro PH, Schmidt Fda R, Battistella V, Lopes de Souza SA, Gutfilen B, Goldenberg RC, Kasai-Brunswick TH, Vairo L, Silva RM, Wajnberg E, Alvarenga Americano do Brasil PE, Gasparetto EL, Maiolino A, Alves-Leon SV, Andre C, Mendez-Otero R, Rodriguez de Freitas G, Barbosa da Fonseca LM. Biodistribution of bone marrow mononuclear cells after intra-arterial or intravenous transplantation in subacute stroke patients. Regen Med. 2013a;8(2):145–55. doi:10.2217/rme.13.2.
35. Chen FM, Wu LA, Zhang M, Zhang R, Sun HH. Homing of endogenous stem/progenitor cells for *in situ* tissue regeneration: promises, strategies, and translational perspectives. Biomaterials. 2011;32(12):3189–209.
36. Chavakis E, Urbich C, Dimmeler S. Homing and engraftment of progenitor cells: a prerequisite for cell therapy. J Mol Cell Cardiol. 2008;45(4):514–22.
37. Abbott JD, Huang Y, Liu D, Hickey R, Krause DS, Giordano FJ. Stromal cell-derived factor-1α plays a critical role in stem cell recruitment to the heart after myocardial infarction but is not sufficient to induce homing in the absence of injury. Circulation. 2004;110(21):3300–5.
38. Belema-Bedada F, Uchida S, Martire A, Kostin S, Braun T. Efficient homing of multipotent adult mesenchymal stem cells depends on FROUNT-mediated clustering of CCR2. Cell Stem Cell. 2008;2(6):566–75.
39. Ceradini DJ, Kulkarni AR, Callaghan MJ, et al. Progenitor cell trafficking is regulated by hypoxic gradients through HIF-1 induction of SDF-1. Nat Med. 2004;10(8):858–864. (One of the first reports that recruitment of CXCR4-positive progenitor cells to regenerating tissues is mediated by hypoxic gradients via HIF-1-induced expression of SDF–1).
40. Hiasa K, Ishibashi M, Ohtani K, et al. Gene transfer of stromal cell-derived factor-1α enhances ischemic vasculogenesis and angiogenesis via vascular endothelial growth factor/endothelial nitric oxide synthase-related pathway: next-generation chemokine therapy for therapeutic neovascularization. Circulation. 2004;109(20):2454–61.
41. Kim SJ, Moon GJ, Cho YH, Kang HY, Hyung NK, Kim D, Lee JH, Nam JY, Bang OY. Circulating mesenchymal stem cells microparticles in patients with cerebrovascular disease. PLoS ONE. 2012;7(5):e37036.
42. Youn SW, Lee SW, Lee J, et al. COMP-Ang1 stimulates HIF-1α-mediated SDF-1 overexpression and recovers ischemic injury through BM-derived progenitor cell recruitment. Blood. 2011;117(16):4376–86.
43. Heissig B, Hattori K, Dias S, et al. Recruitment of stem and progenitor cells from the bone marrow niche requires MMP-9 mediated release of kit-ligand. Cell. 2002;109(5):625–37.
44. Kaminski A, Ma N, Donndorf P, et al. Endothelial NOS is required for SDF-1α/CXCR4-mediated peripheral endothelial adhesion of c-kit+ bone marrow stem cells. Lab Invest. 2008;88(1):58–69.
45. Li N, Lu X, Zhao X, et al. Endothelial nitric oxide synthase promotes bone marrow stromal cell migration to the ischemic myocardium via upregulation of stromal cell-derived factor-1alpha. Stem Cells. 2009;27(4):961–70.
46. Hattori K, Dias S, Heissig B, et al. Vascular endothelial growth factor and angiopoietin-1 stimulate postnatal hematopoiesis by recruitment of vasculogenic and hematopoietic stem cells. J Exp Med. 2001;193(9):1005–14.

47. Hopkins SP, Bulgrin JP, Sims RL, Bowman B, Donovan DL, Schmidt SP. Controlled delivery of vascular endothelial growth factor promotes neovascularization and maintains limb function in a rabbit model of ischemia. J Vasc Surg. 1998;27(5):886–4. (discussion 895).
48. Levesque JP, Hendy J, Takamatsu Y, Simmons PJ, Bendall LJ. Disruption of the CXCR4/CXCL12 chemotactic interaction during hematopoietic stem cell mobilization induced by GCSF or cyclophosphamide. J Clin Invest. 2003;111(2):187–96.
49. Pitchford SC, Furze RC, Jones CP, Wengner AM, Rankin SM. Differential mobilization of subsets of progenitor cells from the bone marrow. Cell Stem Cell. 2009;4(1):62–72.
50. Guzman R, Uchida N, Bliss TM, He D, Christopherson KK, Stellwagen D, Capela A, Greve J, Malenka RC, et al. Long-term monitoring of transplanted human neural stem cells in developmental and pathological contexts with MRI. Proc Natl Acad Sci. 2007;104:10211–16. (USA).
51. McGuckin CP, Jurga M, Miller AM, Sarnowska A, Wiedner M, Boyle NT, Lynch MA, Jablonska A, Drela K, Lukomska B, Domanska-Janik K, Kenner L, Moriggl R, Degoul O, Perruisseau-Carrier C, Forraz N. Ischemic brain injury: a consortium analysis of key factors involved in mesenchymal stem cell-mediated inflammatory reduction. Arch Biochem Biophys. 2013 Jun;534(1-2):88–97.
52. Wu X, Wang K, Cui L, et al. Effects of granulocyte-colony stimulating factor on the repair of balloon-injured arteries. Pathology. 2008;40(5):513–9.
53. Pendharkar AV, Chua JY, Andres RH, Wang N, Gaeta X, Wang H, De A, Choi R, Chen S, Rutt BK, Gambhir SS, Guzman R. Biodistribution of neural stem cells after intravascular therapy for hypoxic-ischemia. Stroke. 2010 Sep;41(9):2064–70.
54. Sinden JD, Muir KW. Stem cells in stroke treatment: the promise and the challenges. Int J Stroke. 2012;7:426–34.
55. Yang M, Wei X, Li J, et al. Changes in host blood factors and brain glia accompanying the functional recovery after systemic administration of bone marrow stem cells in ischemic stroke rats. Cell Transplant. 2010;19:1073–84.
56. Mitkari B, Nitzsche F, Kerkelä E, Kuptsova K, Huttunen J, Nystedt J, Korhonen M, Jolkkonen J. Human bone marrow mesenchymal stem/stromal cells produce efficient localization in the brain and enhanced angiogenesis after intra-arterial delivery in rats with cerebral ischemia, but this is not translated to behavioral recovery. Behav Brain Res. 2014;259:50–9.
57. Vasconcelos-dos-Santos A, Rosado-de-Castro PH, Lopes deSSA, da Costa SJ, Ramos AB, Rodriguez deFG, Barbosa daFLM, Gutfilen B, Mendez-Otero R. Intravenous and intra-arterial administration of bone marrow mononuclear cells after focal cerebral ischemia: Is there a difference in biodistribution and efficacy? Stem Cell Res. 2012 Jul;9(1):1–8. doi: 10.1016/j.scr.2012.02.002.
58. Yang B, Migliati E, Parsha K, Schaar K, Xi X, Aronowski J, Savitz SI. Intra-arterial delivery is not superior to intravenous delivery of autologous bone marrow mononuclear cells in acute ischemic stroke. Stroke. 2013 Dec;44(12):3463–72.
59. Zhang L, Li Y, Romanko M, Kramer BC, Gosiewska A, Chopp M, Hong K. Different routes of administration of human umbilical tissue-derived cells improve functional recovery in the rat after focal cerebral ischemia. Brain Res 2012; 1489:104–2.
60. Byun JS, Kwak BK, Kim JK, Jung J, Ha BC, Park S. Engraftment of human mesenchymal stem cells in a rat photothrombotic cerebral infarction model : comparison of intra-arterial and intravenous infusion using MRI and histological analysis. J Korean Neurosurg Soc. 2013;54(6):467–76.
61. Lundberg J, Sodersten E, Sundstrom E, et al. Targeted intra-arterial trans- plantation of stem cells to the injured CNS is more effective than intravenous administration: engraftment is dependent on cell type and adhesion molecule expression. Cell Transplant. 2012;21:333–43.
62. Fischer UM, Harting MT, Jimenez F, et al. Pulmonary passage is a major obstacle for intravenous stem cell delivery: The pulmonary first-pass effect. Stem Cells Dev. 2009;18:683–92.
63. Chua JY, Pendharkar AV, Wang N, et al. Intra-arterial injection of neural stem cells using a microneedle tech- nique does not cause microembolic strokes. J Cereb Blood Flow Metab. 2011;31:1263–71.

64. Lee JS, et al. A long-term follow-up study of intravenous autologous mesenchymal stem cell transplantation in patients with ischemic stroke. Stem Cells. 2010;28:1099–106.
65. Li L, Jiang Q, Ding G, Zhang L, Zhang ZG, Li Q, Panda S, Lu M, Ewing JR, Chopp M. Effects of administration route on migration and distribution of neural progenitor cells transplanted into rats with focal cerebral ischemia, an MRI study. J Cereb Blood Flow Metab. 2010;30:653–62.
66. Walczak P, Zhang J, Gilad AA, et al. Dual-modality monitoring of targeted intraarterial delivery of mesenchymal stem cells after transient ischemia. Stroke. 2008;39:1569–74.
67. Misra V, Lal A, El Khoury R, Chen PR, Savitz SI. Intra-arterial delivery of cell therapies for stroke. Stem Cells Dev. 2012;21(7):1007–15.
68. Kondziolka D, Wechsler L, Goldstein S, Meltzer C, Thulborn KR, Gebel J, Jannetta P, DeCesare S, Elder EM, McGrogan M, Reitman MA, Bynum L. Transplantation of cultured human neuronal cells for patients with stroke. Neurology. 2000;55(4):565–9.
69. Meltzer CC, Kondziolka D, Villemagne VL, Wechsler L, Goldstein S, Thulborn KR, Gebel J, Elder EM, DeCesare S, Jacobs A. Serial [18F] fluorodeoxyglucose positron emission tomography after human neuronal implantation for stroke. Neurosurgery. 2001;49:586–91. (discussion 591–582).
70. Nelson PT, Kondziolka D, Wechsler L, et al. Clonal human (hNT) neuron grafts for stroke therapy: neuropathology in a patient 27 months after implantation. Am J Pathol. 2002;160:1201–6.
71. Kondziolka D, Steinberg GK, Wechsler L, et al. Neu- rotransplantation for patients with subcortical motor stroke: a phase 2 randomized trial. J Neurosurg. 2005;103:38–45.
72. Jiang Q, Zhang ZG, Ding GL, et al. Investigation of neural progenitor cell induced angiogenesis after embolic stroke in rat using MRI. Neuroimage. 2005;28:698–707.
73. Modo M, Stroemer RP, Tang E, Patel S, Hodges H. Effects of implantation site of stem cell grafts on behavioral recovery from stroke damage. Stroke. 2002;33:2270–8.
74. Savitz SI, Dinsmore J, Wu J, Henderson GV, Stieg P, Ca- plan LR. Neurotransplantation of fetal porcine cells in patients with basal ganglia infarcts: a preliminary safety and feasibility study. Cerebrovasc Dis. 2005;20:101–7.
75. Rabinovich SS, Seledtsov VI, Banul NV, et al. Cell therapy of brain stroke. Bull Exp Biol Med. 2005;139:126–8.
76. Bang OY, et al. Autologous mesenchymal stem cell transplantation in stroke patients. Ann Neurol. 2005;57:874–2
77. Hess DC, Sila CA, Furlan AJ, Wechsler LR, Switzer JA, Mays RW. A double-blind placebo-controlled clinical evaluation of multistem for the treatment of ischemic stroke. Int J Stroke. 2014 April;9:381–386.
78. Banerjee S, Williamson DA, Habib N, Chataway J. The potential benefit of stem cell therapy after stroke: an update. Vasc Health Risk Manag. 2012;8:569–80.
79. Rosado-de-Castro PH, Pimentel-Coelho PM, da Fonseca LM, de Freitas GR, Mendez-Otero R. The rise of cell therapy trials for stroke: review of published and registered studies. Stem Cells Dev. 2013b;22(15):2095–111.
80. Mendonca ML, Freitas GR, Silva SA, et al. Safety of intra-arterial autologous bone marrow mononuclear cell trans- plantation for acute ischemic stroke. Arq Bras Cardiol. 2006;86:52–5.
81. Correa PL, Mesquita CT, Felix RM, Azevedo JC, Barbirato GB, Falcão CH, Gonzalez C, Mendonça ML, Manfrim A, de Freitas G, Oliveira CC, Silva D, Avila D, Borojevic R, Alves S, Oliveira AC Jr, Dohmann HF. Assessment of intra-arterial injected autologous bone marrow mononuclear cell distribution by radioactive labeling in acute ischemic stroke. Clin Nucl Med. 2007 Nov;32(11):839–41.
82. Kraitchman DL, Tatsumi M, Gilson WD, Ishimori T, Kedziorek D, Walczak P, Segars WP, Chen HH, Fritzges D, Izbudak I, Young RG, Marcelino M, Pittenger MF, Solaiyappan M, Boston RC, Tsui BM, Wahl RL, Bulte JW. Dynamic imaging of allogeneic mesenchymal stem cells trafficking to myocardial infarction. Circulation. 2005;112(10):1451–61.
83. Hauger O, Frost EE, van Heeswijk R, et al. MR evaluation of the glomerular homing of magnetically labeled mesenchymal stem cells in a rat model of nephropathy. Radiology 2006;238:200–10.

84. Battistella V, de Freitas GR, da Fonseca LM, Mercante D, Gutfilen B, Goldenberg RC, Dias JV, Kasai-Brunswick TH, Wajnberg E, Rosado-de-Castro PH, Alves-Leon SV, Mendez-Otero R, Andre C. Safety of autologous bone marrow mononuclear cell transplantation in patients with nonacute ischemic stroke. Regen Med. 2011 Jan;6(1):45–52. doi:10.2217/rme.10.97.
85. Friedrich MA, Martins MP, Araújo MD, Klamt C, Vedolin L, Garicochea B, Raupp EF, Sartori EAmmarJ, Machado DC, Costa JC, Nogueira RG, Rosado-de-Castro PH, Mendez-Otero R, Freitas GR. Intra-arterial infusion of autologous bone marrow mononuclear cells in patients with moderate to severe middle cerebral artery acute ischemic stroke. Cell Transplant. 2012;21(Suppl 1):S13–21. doi:10.3727/096368912X612512.
86. Jiang Y, Zhu W, Zhu J, Wu L, Xu G, Liu X. Feasibility of delivering mesenchymal stem cells via catheter to the proximal end of the lesion artery in patients with stroke in the territory of the middle cerebral artery. Cell Transplant. 2013;22(12):2291–8.
87. Banerjee S, Bentley P, Hamady M, Marley S, Davis J, et al. Intra-Arterial Immunoselected CD34+ Stem Cells for Acute Ischemic Stroke. Stem Cells Trans Med. 2014;3:1–9.
88. Moniche F, Montaner J, Gonzalez-Marcos JR, Carmona M, Piñero P, Espigado I, Cayuela A, Escudero I, de la Torre-LavianaFJ, Boada C, Rosell A, Mayol A, Jimenez MD, Gil-Peralta A, Gonzalez A. Intra-arterial bone marrow mononuclear cell (BM-MNC) transplantation correlates with GM-CSF, PDGF-BB and MMP-2 serum levels in stroke patients: Results from a clinical trial. Cell Transplant. 2014;23:57–64.
89. Barbosa da Fonseca LM, Battistella V, de Freitas GR, Gutfilen B, Dos Santos Goldenberg RC, Maiolino A, Wajnberg E, Rosado de Castro PH, Mendez-Otero R, Andre C. Early tissue distribution of bone marrow mononuclear cells after intra-arterial delivery in a patient with chronic stroke. Circulation. 2009;120(6):539–41. doi:10.1161/CIRCULATIONAHA.109.863084.
90. Moniche F, Gonzalez A, Gonzalez-Marcos JR, et al. Intra-arterial bone marrow & mononuclear cells in ischemic stroke: a pilot clinical trial. Stroke. 2012;43:2242–4.
91. Cytomedix Announces Results of RECOVER-Stroke Phase 2 Study (May 5, 2014). http://www.irdirect.net/pr/release/id/607621. Accessed 17 Nov 2014.

Chapter 7
Neural Stem Cells in Stroke: Intracerebral Approaches

Nathan C. Manley, Ricardo L. Azevedo-Pereira, Tonya M. Bliss and Gary K. Steinberg

Applying Stem Cell Therapy to Stroke: Early Development as a Cell-Replacement Strategy

Stem cell therapy for stroke was originally conceived as a cell-replacement strategy to repair circuits damaged after stroke and, as such, intracerebral (IC) transplantation was considered to be the preferred method of delivery in order to precisely target the cells to the area of injury. This strategy was based on pioneering Stem cell transplantation studies in experimental models of Parkinson's disease where IC delivery of fetal mesencephalic tissue containing a high proportion of dopaminergic neurons could engraft and induce functional recovery in hemiparkinsonian rats [1, 2]. Additional preclinical studies in the Parkinson's disease field paved the way for the initial IC stem cell clinical trials [3, 4], which led to the first clinical demonstration of neurological recovery associated with IC stem cell treatment [5, 6]. These early discoveries and successes achieved for Parkinson's disease provided much hope and guidance for parallel efforts in the stroke field. However, the complex pathophysiology associated with stroke injury also presented several unique challenges.

In Parkinson's disease, progressive loss of dopaminergic neurons in the basal ganglia defined a specific cell type needed for replacement. In stroke, however, cell

G. K. Steinberg (✉) · N. C. Manley · R. L. Azevedo-Pereira · T. M. Bliss
Department of Neurosurgery and Stanford Stroke Center, Stanford University School of Medicine, MSLS, 1201 Welch Rd, Stanford, CA, USA
e-mail: gsteinberg@stanford.edu

N. C. Manley
Asterias Biotherapeutics, Inc., 230 Constitution Drive, Menlo Park, CA, USA
e-mail: nmanley@asteriasbio.com

R. L. Azevedo-Pereira
e-mail: razevedo@stanford.edu

T. M. Bliss
e-mail: tbliss1@stanford.edu

D. C. Hess (ed.), *Cell Therapy for Brain Injury*, DOI 10.1007/978-3-319-15063-5_7

replacement was a more ambitious goal given that associated brain damage could cause indiscriminate loss of all cell types within the lesion site, including neurons, astrocytes, oligodendrocytes, as well as the surrounding vascular system [7]. To address this, initial cell therapies for stroke sought to identify appropriate sources of multipotent cells, among which neural stem cells (NSCs), largely defined by their capacity to generate neurons, astrocytes, and oligodendrocytes, were viewed as ideal candidates. Pioneering stroke studies used rat fetal brain tissue as a source of NSCs and involved IC delivery of partially dissociated tissue chunks into adult rats subjected to hypoxia–ischemia. This approach reduced dystrophy and death of host cortical neurons, and demonstrated that the cells within the engrafted fetal tissue could both adopt neural and glial morphologies and form physical associations with surrounding host neurons [8–10]. Additional studies indicated that graft survival was improved when delivered to the stroke penumbra relative to the core [11], and that fetal grafts could be used to increase local neurotransmitter levels and improve functional recovery [12]. These initial preclinical studies provided key proof-of-concept support for applying cell therapy to stroke, and further suggested that the therapeutic action of transplanted cells might extend beyond cell replacement. Despite these promising results, both the ethical concerns and limited availability associated with human fetal brain tissue limited its clinical applicability and prompted efforts to identify alternative stem cell sources as well as methods to expand stem cell populations in vitro.

Different Types of Cultured Neural Stem Cells Can Promote Stroke Recovery

Immortalized Cell Lines

Concomitant to the first stroke studies with transplanted fetal tissue, experiments with the teratocarcinoma-derived cell line NT2 showed that these immortalized cells could differentiate into postmitotic, neuron-like cells upon stimulation with retinoic acid [13]. The use of immortalized cells provided a way to address the scale of production needed for clinical use, but also raised the concern that cells derived from an immortalized line might revert back to a transformed state after transplantation. Subsequent studies with the NT2-derived cells showed that IC delivery into stroke-injured rats resulted in increased functional recovery and, critically, showed no signs of teratoma or ectopic tissue formation [14]. Based on these promising results, NT2 cells became the first reported cell therapy clinical trial for stroke, and consisted of IC delivery of NT2 cells into patients after subcortical stroke [15]. Cell survival in the brain was even demonstrated 27 months after the transplantation, in one patient who died of a myocardial infarct [16]. Based on an acceptable safety profile and suggestion of efficacy during phase I clinical testing [15, 17], IC delivery of NT2 cells advanced to a phase II randomized controlled trial, with positive improvements observed in some patients and minimal adverse reactions [18]. Additionally, conditionally immortalized cell lines have also promoted recovery in experimental stroke, including the tamoxifen-dependent, c-myc-transformed cell line CTX030. A phase I study in chronic stroke patients using IC delivery of these cells was recently completed with promising early results, and a phase II clinical stroke study has been launched (discussed in "Clinical Testing of IC-based Cell Therapies for Stroke").

Fetal-Derived Neural Stem Cells

Beginning in the 1990s, methods for in vitro culturing and expansion of adult-derived NSCs were established [19], as were techniques to culture both fetal- and embryonic-derived cells that could be used to generate NSCs in vitro and ultimately differentiate into mature neurons and/or glia [20–27]. Subsequently, many of these NSC products were tested by several groups, including our laboratory, for their relative safety and efficacy in preclinical stroke models. In agreement with the initial stroke studies involving transplanted fetal brain tissue [8–10], human fetal-derived, cultured NSCs delivered intracerebrally in rodent stroke models have been shown to survive, migrate towards the stroke lesion, differentiate into neurons and glia [28], and improve functional recovery (Fig. 7.1) [29–31]. Although IC delivery of fetal-derived NSCs has not yet been used to treat stroke clinically, results of a recent phase I trial for Batten's disease indicate the general safety and feasibility of this approach [32]. Of note, the clinical trial for Batten's disease utilized fetal-derived NSCs very similar to those shown by our group to improve functional recovery in stroke-injured rats [30], thus indicating that this NSC type may be applicable to both indications.

Embryonic-Derived Neural Stem Cells

Following the establishment of cultured human embryonic stem cell lines [26], several groups, including ours, have developed methods to generate NSCs from human or primate embryonic stem cells and have demonstrated their ability to enhance functional recovery following IC administration in preclinical models of stroke [21, 33–37]. Interestingly, in a study comparing IC delivery of embryonic- and adult-derived NSCs into stroke-injured rats, embryonic-derived NSCs exhibited substantially higher survival that was correlated with a decreased host immune response [38]. These promising preclinical results, along with the capacity for potentially limitless scale-up of the starting embryonic stem cell population, make embryonic-derived NSCs an attractive cell therapy candidate for stroke, and continued efforts towards clinical application are underway (Steinberg/Carmichael: California Institute for Regenerative Medicine (CIRM) Disease Team Grant DR1 01480) [39, 40].

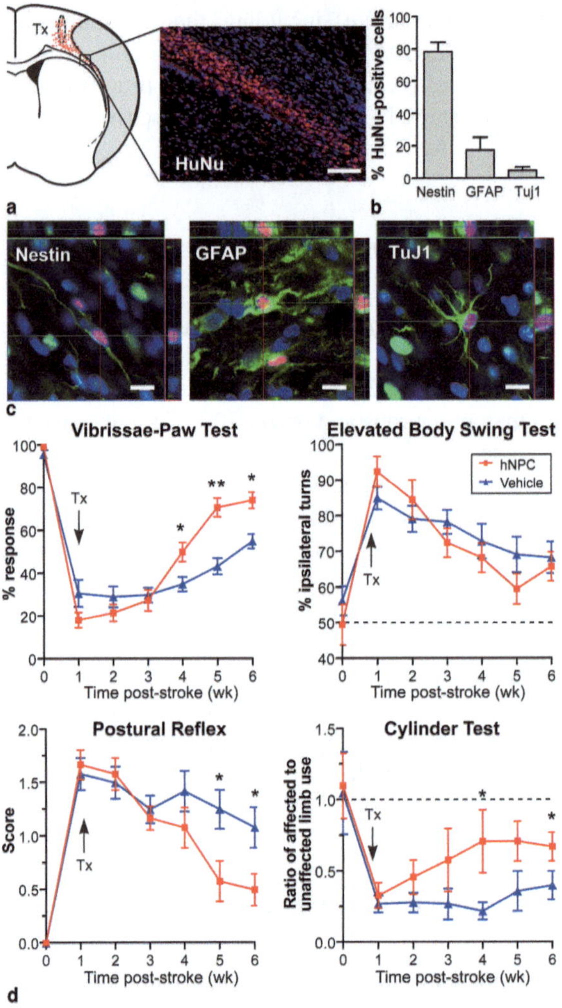

Fig. 7.1 Differentiation fate of transplanted human fetal-derived NSCs and their effects on behavioral recovery after stroke. **a** Human fetal-derived NSCs (human nuclear antigen (*HuNu*)-positive cells, *red*) survive and migrate towards the stroke lesion at 5 weeks post-IC transplantation in athymic nude rats subjected to cortical ischemic stroke. Representative photomicrograph of human fetal-derived NSCs in the peri-infarct region. *Gray-shaded area* indicates the lesion. **b** Differentiation profile of human fetal-derived NSCs at 5 weeks post transplant. **c** Confocal images of differentiation markers colocalizing with HuNu in the peri-infarct area. **d** Rats treated with human fetal-derived NSCs have significantly improved functional recovery compared to vehicle-treated controls in three out of four behavior tests; $^{*}P<0.05$, $^{**}P<0.01$; $n=12$ per group, except cylinder test, $n=6$. *Scale bars:* A=50 µm, C=10 µm. *GFAP* glial fibrillary acidic protein, *TuJ1* neuronal class III β-tubulin, *Tx* transplantation. (This figure was reprinted with the publisher's permission from [29])

Induced Pluripotent Stem Cells

Recently, excitement has increased regarding the possibility of using adult-derived NSC therapies to treat stroke due to the emergence and continued development of induced pluripotent stem cell (iPSC) technologies [41, 42] as well as methods to differentiate iPSCs into NSCs [43]. Preclinical stroke studies have confirmed the ability of iPSC-derived NSCs to enhance stroke recovery following IC delivery [44–47], including in neonatal stroke [48], and have further shown that iPSC-derived NSCs can differentiate into multiple types of functionally active neurons following IC delivery into the stroke-injured brain [45, 46]. Similar to embryonic-derived NSCs, the potential to form tumors or ectopic tissue by iPSC-derived NSCs remains a concern, prompting development of methods to generate induced NSCs (iNSCs) or induced neuronal (iN) cells directly from somatic cells without initial reversion to a pluripotent state [49, 50]. Although iNSCs have not yet been tested for safety/efficacy in preclinical stroke models, the prospects of patient specificity and decreased tumorigenicity make the clinical attraction of this approach clear.

Overall, NSCs from multiple sources have been shown to improve recovery in preclinical stroke models, with specific benefits and drawbacks associated with each NSC type. In order to maximize the potential benefit of NSC-based approaches, it is important to understand the underlying mechanism(s) by which NSCs improve stroke outcome. While there are still many remaining questions regarding the therapeutic actions of transplanted NSCs, it is becoming increasingly clear that NSCs have the capacity to impact multiple pathways involved in brain repair and regeneration after stroke.

Mechanisms of Action for Transplanted NSCs in Stroke

Cell Replacement Versus Paracrine Signaling

While the initial notion was that transplanted stem cells would repair stroke-damaged circuits by replacing lost neurons, it soon became apparent that this was not the primary mechanism driving stem cell-induced recovery. Early evidence that alternative mechanisms of action were involved included the observations that NSC-associated functional improvements occurred prior to any detectable integration of the transplanted cells, and for those studies that reported some positive integration, the number of observed synapses between transplanted NSCs and host neurons was small, with a large proportion of engrafted cells remaining in an immature, precursor state [51, 52]. Based on these observations, several nonintegrating mechanisms of NSC efficacy were investigated. These focused largely on the idea that transplanted NSCs, through secretion of paracrine factors, enhance endogenous brain repair pathways that are activated after stroke. Significant data support this mechanism of action, yet a growing number of studies indicate that transplanted NSCs can also integrate into the host brain, thus renewing the idea that cell replacement, or at least host integration, may also be important.

Stimulation of Endogenous Brain Repair by NSC Paracrine Signaling

Trophic Support

NSCs have been shown to express trophic factors (proteins and other molecules) that promote survival, proliferation, and differentiation of host brain cells after stroke, and this has led to the general hypothesis of NSC-mediated recovery through trophic support [53]. In vitro profiling of NSCs has shown that they secrete such key factors as brain-derived neurotrophic factor (BDNF), glial-derived neurotrophic factor (GDNF), vascular endothelial growth factor (VEGF), and several others [29, 54–56]. However, relatively little is known about which trophic factors mediate efficacy in vivo, due in part to the need for improved methods to profile transplanted NSC expression/secretion in vivo, as well as the need for additional studies that selectively upregulate or suppress trophic factors within transplanted NSCs. For example, IC delivery of VEGF-transfected NSCs into stroke-injured rats resulted in higher levels of vascular repair and functional recovery compared to unmodified NSCs [57]. Further, a recent study by our group showed that in stroke-injured rats, the recovery enhancement associated with IC delivery of human NSCs could be suppressed by systemic infusion of bevacizumab (Avastin), an antibody specific for human VEGF [30], thus enabling us to tease apart the role of the NSC-secreted VEGF from that of endogenous rodent VEGF. Results of this study further indicated that recovery enhancement by NSCs coincided with increased neovascularization and blood–brain barrier integrity, along with decreased microglial/immune cell activation within the peri-infarct region, and that both of these changes could be suppressed by cotreatment with Avastin. These results highlight the concept that NSC-associated functional improvements are unlikely to occur via a single mechanism and instead may involve simultaneous modulation of multiple endogenous repair pathways.

Vascular Repair

Regeneration of the vascular system by localized activation of angiogenesis is one form of endogenous repair that is correlated with improved stroke outcome [58, 59] and has the potential for modulation by NSCs and other cell therapies. Different types of NSCs have been shown to express trophic factors that can stimulate neovascularization (angiogenesis or vasculogenesis), including VEGF, epidermal growth factor (EGF), insulin-like growth factor 1 (IGF1), and basic fibroblast growth factor (bFGF) [30, 60]. IC delivery of NSCs into stroke-injured rats was correlated with increased angiogenesis or neovascularization [30, 61], and poststroke stimulation of neovascularization by transplanted human fetal-derived NSCs could be suppressed by selective blockade of human VEGF [30]. Further evidence of VEGF as a key mediator of these effects was shown by intravenous coadministration of NSCs and recombinant VEGF [62] or IC delivery of VEGF-transfected NSCs [57] into stroke-injured rats; both approaches led to increased vascular repair and functional recovery beyond that of NSCs alone or unmodified NSCs. The trophic factors placental growth factor and angiopoietin-1 have been identified as candidate mediators of neovascularization by bone marrow-derived stromal cells [63–65], but their potential involvement in NSC-based therapies has not yet been elucidated.

Immunomodulation

The immune response after stroke involves both an acute phase that is associated with injury progression and a delayed phase that can aid subsequent repair [66]. Preclinical studies suggest that IC delivery of NSCs can modulate this response in favor of repair by reducing the number of activated microglia and macrophages that infiltrate surviving peri-infarct tissue [28, 30, 67]. Similarly, systemic delivery of NSCs has been shown to reduce infiltration of activated monocytes into the brain of stroke-injured rodents, and this was correlated with reduced levels of proinflammatory cytokines in the spleen and brain, including tumor necrosis factor, interferon gamma, and interleukin-6 [68, 69]. Additional in vitro studies have shown that NSCs can suppress T cell activation and dendritic cell maturation in a coculture setting, and that these effects are mediated in part by NSC secretion of such immunomodulatory factors as nitric oxide and prostaglandin E2 [70–72].

Brain Plasticity

Activation of brain plasticity after stroke in the form of outgrowth and remapping of neuronal projections is correlated with improved behavioral outcome, and thought to be a driving factor behind spontaneous recovery after stroke [73–75]. Based on studies from our group, IC delivery of human fetal-derived NSCs augmented several aspects of brain plasticity in stroke-injured adult rats, including remapping and increased axonal sprouting of contralesional neurons into the ischemic hemisphere and through the cortical thalamic and cortical spinal tracts, increased dendritic branching in both cortical hemispheres, and increased axonal transport [29, 30]. Similarly, IC delivery of human embryonic-derived NSCs enhanced axonal sprouting of contralesional neurons in a rodent model of neonatal hypoxia–ischemia [76]. Moreover, the plasticity-enhancing effects of fetal-derived NSCs could be recapitulated in vitro with a primary neuron coculture system, and the use of neutralizing antibodies implicated SLIT and thrombospondin 1 and 2 as NSC-secreted factors potentially driving these effects [29]. In addition, IC-transplanted human embryonic-derived NSCs were shown to promote recruitment and/or proliferation of endogenous oligodendrocytes in the ischemic hemisphere [77], as well as differentiate into oligodendrocytes themselves [33, 77, 78], thus demonstrating the potential to support or even directly participate in the myelination of newly formed circuits.

Neurogenesis

In response to a stroke injury, endogenous NSCs residing in the subventricular zone of the lateral ventricles can undergo proliferation and migrate to the lesion's boundary, where a small proportion become mature neurons and integrate into surviving circuits [79–81]. In rats subject to striatal ischemic stroke, subsequent IC delivery of human fetal-derived NSCs was correlated with increased migration and survival of immature neurons in the infarct area [67]. However, given that the majority of recruited cells either die or remain in an immature state upon arrival to the infarct site [79, 82], the exact mechanisms by which injury- and cell therapy-induced neurogenesis might lead to improved stroke recovery still need to be resolved and may be more akin to various types of therapeutic paracrine signaling attributed to transplanted NSCs.

Integration into the Host Brain

Although an increasing number of preclinical studies suggest that transplanted NSCs improve stroke outcome by stimulating endogenous repair through paracrine actions, there are a growing number of reports that transplanted NSCs can functionally integrate into surviving host circuits, raising the question regarding the significance of this integration. The degree to which transplanted NSCs differentiate into neurons varies by study and NSC type, and includes ranges of 34–46 % for human fetal-derived NSCs [28, 31, 83], 40–66 % for human iPSC-derived NSCs [46, 84], and 30 % for human embryonic-derived NSCs [85]. Evidence of integration by transplanted NSCs includes the expression of synaptic proteins [86], identification of synapse formation with neighboring host neurons by electron microscopy [31, 77], and detection of voltage-gated sodium currents by electrophysiology [46, 77, 85]. While it is unlikely that integration into host circuits by transplanted NSCs can account for the enhanced stroke recovery in these studies, given the delayed timing relative to stroke recovery onset, and low frequency of integration, a remaining question is whether integrated NSCs might instead play a role in the maintenance or progression of long-term functional recovery. And if so, what role would the integrated cells play? The original idea of simple cell replacement and replacement of lost circuits is unlikely, given the low numbers of integrating neurons compared to the vast number of neurons lost after stroke. A more attractive hypothesis is that the integrating neurons influence the excitability of remaining circuits and in doing so affect brain plasticity, which is a major component driving stroke recovery.

Cell integration applies not only to neurons but also to astrocytes and oligodendrocytes that can be derived from NSCs. Astrocytes play a major role in brain plasticity from coordinating brain activity with blood supply and energy, maintaining the blood–brain barrier, to regulating synapse formation and activity [87–90]. While transplanted NSCs have been shown to mature into astrocytes in the stroke-injured brain [29, 77], the relevance of this to stroke recovery requires further investigation. In addition, stroke injury often includes significant damage to white matter, indicating that strategies to replace lost oligodendrocytes and support remyelination could be effective. Remyelination by human NSCs has been demonstrated in the injured spinal cord of rats and mice [23, 91], and some studies have shown oligodendrocyte differentiation by transplanted NSCs in the ischemic brain [21, 33, 77, 78].

Since its initial conception as a cell-replacement strategy, IC transplantation of NSCs is now thought to enhance brain repair and stroke recovery via multiple mechanisms. In the early-phase posttransplantation, it is likely that NSC paracrine signaling plays a critical role in the activation of endogenous repair pathways, such as plasticity, angiogenesis, immunomodulation, and neurogenesis. Subsequent to this, the formation of physical connections between transplanted cells and surrounding host cells may facilitate additional recovery and/or the maintenance of enhanced recovery. Additional preclinical studies that track long-term recovery after stroke will be needed to increase our understanding of this. Nonetheless, an ever-growing list of preclinical studies has shown that IC delivery of NSCs can stimulate brain repair and improve stroke outcome and has allowed advancement into clinical trials.

Advancing to the Clinic: Progress and Remaining Challenges

Clinical Testing of IC-based Cell Therapies for Stroke

To date, very few stroke clinical trials involving IC delivery of NSCs or other human cell types have been approved (Fig. 7.2, Table 7.1), but those that have advanced generally produced encouraging results. IC delivery of NT2-derived neural cells became the first cell therapy to be applied to stroke, with results of the phase I and phase II trials demonstrating the general safety of the IC delivery method and indicating some potential for functional improvement with NT2-derived cells [15, 17, 18].

More recently, a conditionally immortalized NSC line developed by Reneuron, CTX0E03, was approved for clinical testing in the UK following promising preclinical safety and efficacy data [92]. The phase I open-label, single-site, dose-ascending trial (PISCES trial) consisted of IC delivery of CTX0E03 into male patients with ischemic stroke 6–60 months after injury at doses of 2, 5, 10, or 20 million cells. Results of this trial recently were presented at the European Stroke Conference, and indicated that as of January 1, 2014, no immunological or cell-related adverse

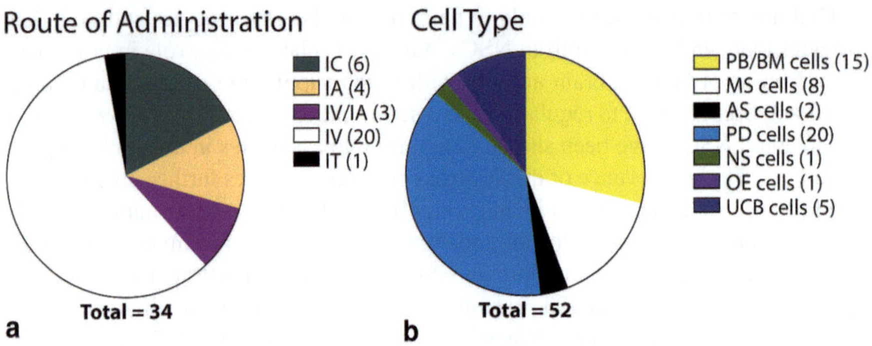

Fig. 7.2 Administration route and cell types used in stroke clinical trials. **a** Administration routes used in stroke clinical trials. **b** Cell types used in stroke clinical trials. *IC* intracerebral, *IA* intra-arterial, *IV* intravenous, *IT* intrathecal, *PB/BM* cell peripheral blood- or bone marrow-derived cells, *MS* cells mesenchymal stromal cells, *AS* cells adipose-derived stromal cells, *PD* cells placenta-derived cells, *NS* cells neural stem cells, *OE* cells olfactory ensheathing cells, *UCB* cells umbilical cord blood-derived cells

Table 7.1 Registered stroke clinical trials with IC administration of cells

ID	Study phase	Cell type	Administration time post stroke	Cell dose (million)	Recruitment status
NCT02117635	Phase 2	Human fetal cortical cells (CTX0E03 DP)	>4 weeks	20	Recruiting
NCT01151124	Phase 1	Human fetal cortical cells (CTX0E03 DP)	6–60 months	2; 5; 10; 20	Ongoing, not recruiting
NCT01714167	Phase 1	Autologous bone marrow mesenchymal stem cells	3–60 months	2–4	Recruiting
NCT00950521	Phase 2	Autologous peripheral blood CD34 stem cells	6–60 months	2–8	Completed
NCT01287936	Phase 1 Phase 2	Modified mesenchymal stromal cells (SB623)	6–60 months	2.5, 5, 10	Ongoing, not recruiting
NCT01327768	Phase 1	Olfactory ensheathing cells	6–60 months	2–8	Unknown

events were observed in any of the 11 patients treated with CTX0E03. Further, CTX0E03-treated patients exhibited functional improvements at 1 month and 1 year postimplantation [93]. Based on these encouraging results, a phase II trial was recently approved to include up to 41 patients treated with CTX0E03 8–12 weeks post stroke. It is expected that this earlier treatment timing better reflects the optimal therapeutic window of CTX0E03, which will be assessed by multiple functional outcome measures up to 6 months posttreatment (www.reneuron.com).

While CTX0E03 is the only IC-administered NSC product currently in clinical testing for stroke, IC delivery of mesenchymal stromal cells (MSCs) is also being clinically tested. SanBio completed the first North American stroke clinical trial to utilize IC delivery of MSCs, specifically testing the human MSC line, SB623. Following confirmation of the safety and efficacy of SB623 in preclinical stroke [94], SanBio initiated a phase II/phase IIa clinical trial testing SB623 delivered IC 6–60 months after ischemic stroke at doses of 2.5, 5, and 10 million cells. Results of this trial recently were presented at the International Stroke Conference and Western Neurosurgical Society, and indicated that as of September 2014, no cell-related adverse events were observed in any of the 18 patients treated with SB623. Three measures of efficacy (NIHSS, ESS, Fugl-Meyer) all demonstrated a statistically significant improvement at 6 months after treatment that continued during follow up to a year. Two patients showed remarkable improvement in their motor (2) and language function (1) within 24 h of surgery, effects which have been sustained during follow-up of 32 and 16 months [95, 96]. Although the mechanisms underlying the functional improvements noted in this trial are not fully defined, these results suggest that the IC cell-transplant surgery can have early beneficial effects on chronic stroke outcome that may persist in the long term.

Remaining Challenges for Clinical Translation

Scaling Up to Meet Clinical Demand

As we continue to explore the potential of NSCs and IC administration as therapeutic strategies for stroke, several challenges remain. First, to meet the demand of the increasingly high number of stroke cases that occur each year (estimated 795,000 new or recurrent stroke cases annually in the USA [97]), improved methods for large-scale production of clinical-grade NSCs must be established. While embryonic-derived NSCs or immortalized cell lines may be the best option for this, given the virtually unlimited scalability of the starting material, they may also pose the greatest safety risk due to the potential for undifferentiated cells to persist in the final product and undergo tumorigenesis or ectopic tissue formation in vivo, particularly when transplanted into the stroke-injured brain [98, 99]. Similarly, though iPSC technologies may ultimately allow reasonable scalability of patient-specific NSCs from a more ethically acceptable source, they will also require strategies to address any undifferentiated cells that remain in the final product [98, 100], as well as the carryover of aberrant genetic or epigenetic changes that can accumulate in aged somatic cells. Alternatively, fetal-derived NSCs pose less tumorigenic risk, but their scalability is more limited, both by the relative availability of appropriate fetal starting material and the extent to which intermediate precursors or the final NSC product can be expanded in vitro. Because extended time in culture can alter a cell's phenotype and even lead to chromosomal abnormalities, any efforts to scale up NSC production will require careful monitoring of karyotypic stability, as well as the specific molecular signatures that determine an NSC product's clinical release criteria (California Institute for Regenerative Medicine (CIRM) Disease Team Grant DR1 01480) [40].

Defining Clinical Release Criteria

A second major challenge faced by NSC and non-NSC cell therapies is the need to adequately characterize the final cell product, including quantitative measures of cell identity, purity, and any potential impurities (California Institute for Regenerative Medicine (CIRM) Disease Team Grant DR1 01480) [40]. Together, these measures define key elements of the release criteria that determine the consistency and clinical eligibility of each manufactured lot for a given cell-therapy product. Establishing appropriate quantitative measures of cell identity and purity remains difficult for NSCs, given that surface proteins and intracellular markers typically associated with NSCs either are expressed only transiently during differentiation or are also expressed by other cell types. Use of multiple identity/purity markers can help overcome these limitations, but also can increase the risk of a failed manufacturing run if any of the predetermined release criteria are not met. Furthermore, to ensure safety of an NSC product, its release criteria must also include appropriate impurity markers indicative of unwanted cell types, with acceptable impurity levels ideally based on preclinical studies where the known impurity is spiked into the NSC product at increasing levels and tested for its relative impact on safety. In the case of embryonic- and iPSC-derived NSCs, residual pluripotent cells represent the most concerning type of impurity given their known tumorigenic potential [98, 99], and pluripotency markers such as Oct4, Tra-1-60, and SSEA-4 often are used to detect these cells.

In addition, a detailed understanding of the production/derivation process is necessary to identify any other unwanted cell types that might arise and persist in the final product. An example of this was seen with the first ever clinical trial to test a human embryonic-derived cell therapy, which consisted of embryonic-derived oligodendrocyte progenitor cells (OPCs) developed by Geron Corporation as a direct intraparenchymal spinal treatment for patients with subacute thoracic spinal cord injury. Prior to its initiation, the trial was put on temporary hold after preclinical safety studies indicated the presence of microscopic epithelial cysts in some rats treated with the OPCs [101], thus indicating that this type of NSC product may require detection or removal of epithelial impurities as part of the manufacturing process. Based on a favorable safety profile with the five patients treated to date, it was recently announced that this OPC product, since acquired by Asterias Biotherapeutics, Inc., will advance into a phase I/phase IIa dose escalation study for patients with subacute cervical spinal cord injury (www.asteriasbiotherapeutics.com).

In Vivo Functional Profiling of Transplanted NSCs

Based on guidelines provided by the US Food and Drug Administration (FDA), the development of candidate NSC products must include extensive preclinical testing to sufficiently characterize key functional properties posttransplantation, including their proliferative capacity, differentiation potential, and putative mechanism(s) of action [40, 102]. The proliferative capacity of an NSC product for stroke is an important aspect of its safety profile, as unrestricted growth could lead to transplanted cells spreading beyond the intended graft site or even tumorigenesis. However, most preclinical stroke studies with NSCs indicate very low posttransplant survival and proliferation [21, 31, 51, 52]. Efforts to enhance this, such as hypoxia preconditioning [37, 103] or coadministration of NSCs with some kind of structural support matrix [104, 105], may improve efficacy of NSCs, particularly for large stroke lesions, but whether this will also increase the tumorigenic risk remains to be determined.

As discussed in "Mechanisms of Action for Transplanted NSCs in Stroke," maximizing the efficacy of NSCs in stroke requires an in-depth understanding of how transplanted NSCs act in the host brain to promote recovery, including their capacity to survive, differentiate, integrate into the host brain, and stimulate endogenous repair through paracrine signaling. For example, the in vivo secretion profile of transplanted NSCs remains largely unknown, but such knowledge could guide development of optimized NSCs that express higher levels of active paracrine factors or enrichment strategies used to purify active subpopulation(s) from an NSC product. It is widely accepted that NSCs are heterogeneous, and this may be advantageous especially if multiple cell types (e.g., neurons, astrocytes, and oligodendrocytes) are required for efficacy. However, if only a certain subpopulation of cells is required (e.g., VEGF-expressing cells or neuronal precursors) then enrichment for this active suppopulation could increase the "effective dose" of the cells.

An understanding of the mechanisms by which transplanted NSCs modulate brain repair and promote stroke recovery will also help optimize the location of transplantation and define the optimal temporal therapeutic window post stroke. Furthermore, understanding what elements of endogenous repair are affected by transplanted NSCs offers the potential for surrogate markers of clinical efficacy that could be included in clinical trial design. For example, if NSCs are expected to augment brain plasticity, treatment might be coupled with noninvasive monitoring of host fiber tract integrity using diffusion tensor magnetic resonance imaging. Finally, an understanding of in vivo activity can lead to the development of in vitro potency assays that can be used to screen new NSC products and assess lot-to-lot consistency during NSC manufacturing, both of which would facilitate the translation of NSC therapy to the clinic.

Additional Considerations for IC delivery of NSCs after Stroke

In addition to the challenges associated with developing a safe and effective NSC product, there are also questions regarding the relative safety of IC administration and the use of immunosuppressant drugs in stroke-injured patients. Although the majority of preclinical studies have assessed IC delivery of NSCs during the first week post stroke (corresponding to the acute/sub-acute phase), this timing likely poses the greatest risk for surgical complications during IC administration, including increased hemorrhage or neuroinflammation. In addition, the acute/subacute phase post stroke can involve significant endogenous immunosuppression and increased risk of sepsis [7], thus presenting an additional challenge for allogenic NSC therapies. To address these concerns, future studies should focus on IC delivery of NSCs during the chronic phase after stroke, as well as improving methods for generating NSCs with reduced immunogenicity or autologous NSCs, including induced reprogramming technologies. Progress in these realms as well as those discussed above will be critical to advancing new NSC therapies to the clinic, which will determine, ultimately, whether IC delivery of NSCs can successfully reverse the functional impairments caused by stroke.

Acknowledgments This chapter was supported in part by funding from California Institute for Regenerative Medicine (CIRM) grant DR1-01480, Bernard and Ronni Lacroute, the William Randolph Hearst Foundation, and Russell and Elizabeth Siegelman (to GKS). Figure 7.1 was reprinted with the publisher's permission from Andres, R. H., N. Horie, et al. (2011) ([29]).“Human neural stem cells enhance structural plasticity and axonal transport in the ischaemic brain.” Brain 134(Pt 6): 1777–1789. We thank Cindy Samos for assistance with the manuscript.

References

1. Olson L, Seiger A, et al. Intraocular transplantation in rodents: a detailed account of the procedure and examples of its use in neurobiology with special reference to brain tissue grafting. Adv Cell Neurobiol. 1983;4:407–42.
2. Perlow MJ, Freed WJ, et al. Brain grafts reduce motor abnormalities produced by destruction of nigrostriatal dopamine system. Science. 1979;204(4393):643–7.
3. Lindvall O, Rehncrona S, et al. Human fetal dopamine neurons grafted into the striatum in two patients with severe Parkinson's disease. A detailed account of methodology and a 6-month follow-up. Arch Neurol. 1989;46(6):615–31.
4. Madrazo I, Leon V, et al. Transplantation of fetal substantia nigra and adrenal medulla to the caudate nucleus in two patients with Parkinson's disease. N Engl J Med. 1988;318(1):51.
5. Lindvall O, Brundin P, et al. Grafts of fetal dopamine neurons survive and improve motor function in Parkinson's disease. Science. 1990;247(4942):574–7.
6. Peschanski M, Defer G, et al. Bilateral motor improvement and alteration of L-dopa effect in two patients with Parkinson's disease following intrastriatal transplantation of foetal ventral mesencephalon. Brain. 1994;117(3):487–99.
7. Dirnagl U, Iadecola C, et al. Pathobiology of ischaemic stroke: an integrated view. Trends Neurosci. 1999;22(9):391–7.
8. Polezhaev LV, Alexandrova MA. Transplantation of embryonic brain tissue into the brain of adult rats after hypoxic hypoxia. J Hirnforsch. 1984;25(1):99–106.

9. Polezhaev LV, Alexandrova MA, et al. Morphological, biochemical and physiological changes in brain nervous tissue of adult intact and hypoxia-subjected rats after transplantation of embryonic nervous tissue. J Hirnforsch. 1985;26(3):281–9.
10. Polezhaev LV, Alexandrova MA, et al. Normalization of dystrophic brain cortex neurons after hypoxia and transplantation of embryonic nervous tissue in rats. J Hirnforsch. 1986;27(5):501–13.
11. Hadani M, Freeman T, et al. Fetal cortical cells survive in focal cerebral infarct after permanent occlusion of the middle cerebral artery in adult rats. J Neurotrauma. 1992;9(2):107–12.
12. Aihara N, Mizukawa K, et al. Striatal grafts in infarct striatopallidum increase GABA release, reorganize GABAA receptor and improve water-maze learning in the rat. Brain Res Bull. 1994;33(5):483–8.
13. Andrews PW. Retinoic acid induces neuronal differentiation of a cloned human embryonal carcinoma cell line in vitro. Dev Biol. 1984;103(2):285–93.
14. Borlongan CV, Tajima Y, et al. Cerebral ischemia and CNS transplantation: differential effects of grafted fetal rat striatal cells and human neurons derived from a clonal cell line. Neuroreport. 1998;9(16):3703–9.
15. Kondziolka D, Wechsler L, et al. Transplantation of cultured human neuronal cells for patients with stroke. Neurology. 2000;55(4):565–9.
16. Nelson PT, Kondziolka D, et al. Clonal human (hNT) neuron grafts for stroke therapy: neuropathology in a patient 27 months after implantation. Am J Pathol. 2002;160(4):1201–6.
17. Stilley CS, Ryan CM, et al. Changes in cognitive function after neuronal cell transplantation for basal ganglia stroke. Neurology. 2004;63(7):1320–2.
18. Kondziolka D, Steinberg GK, et al. Neurotransplantation for patients with subcortical motor stroke: a phase 2 randomized trial. J Neurosurg. 2005;103(1):38–45.
19. Reynolds BA, Weiss S. Generation of neurons and astrocytes from isolated cells of the adult mammalian central nervous system. Science. 1992;255(5052):1707–10.
20. Aubry L, Bugi A, et al. Striatal progenitors derived from human ES cells mature into DARPP32 neurons in vitro and in quinolinic acid-lesioned rats. Proc Natl Acad Sci U S A. 2008;105(43):16707–12.
21. Daadi MM, Maag AL, et al. Adherent self-renewable human embryonic stem-cell-derivedd neural stem cell line: functional engraftment in experimental stroke model. PLoS ONE. 2008;3(2):e1644.
22. Li XJ, Du ZW, et al. Specification of motoneurons from human embryonic stem cells. Nat Biotechnol. 2005;23(2):215–21.
23. Nistor GI, Totoiu MO, et al. Human embryonic stem cells differentiate into oligodendrocytes in high purity and myelinate after spinal cord transplantation. Glia. 2005;49(3):385–96.
24. Perrier AL, Tabar V, et al. Derivation of midbrain dopamine neurons from human embryonic stem cells. Proc Natl Acad Sci U S A. 2004;101(34):12543–8.
25. Svendsen CN, ter Borg MG, et al. A new method for the rapid and long term growth of human neural precursor cells. J Neurosci Methods. 1998;85(2):141–52.
26. Thomson JA, Itskovitz-Eldor J, et al. Embryonic stem cell lines derived from human blastocysts. Science. 1998;282(5391):1145–7.
27. Zhang SC, Wernig M, et al. In vitro differentiation of transplantable neural precursors from human embryonic stem cells. Nat Biotechnol. 2001;19(12):1129–33.
28. Kelly S, Bliss TM, et al. Transplanted human fetal neural stem cells survive, migrate, and differentiate in ischemic rat cerebral cortex. Proc Natl Acad Sci U S A. 2004;101(32):11839–44.
29. Andres RH, Horie N, et al. Human neural stem cells enhance structural plasticity and axonal transport in the ischaemic brain. Brain. 2011;134(6):1777–89.
30. Horie N, Pereira MP, et al. Transplanted stem-cell-secreted vascular endothelial growth factor effects poststroke recovery, inflammation, and vascular repair. Stem Cells. 2011;29(2):274–85.
31. Ishibashi S, Sakaguchi M, et al. Human neural stem/progenitor cells, expanded in long-term neurosphere culture, promote functional recovery after focal ischemia in Mongolian gerbils. J Neurosci Res. 2004;78(2):215–23.

32. Selden NR, Al-Uzri A, et al. Central nervous system stem cell transplantation for children with neuronal ceroid lipofuscinosis. J Neurosurg Pediatr. 2013;11(6):643–52.
33. Hicks AU, Lappalainen RS, et al. Transplantation of human embryonic stem-cell-derived neural precursor cells and enriched environment after cortical stroke in rats: cell survival and functional recovery. Eur J Neurosci. 2009;29(3):562–74.
34. Ikeda R, Kurokawa MS, et al. Transplantation of neural cells derived from retinoic acid-treated cynomolgus monkey embryonic stem cells successfully improved motor function of hemiplegic mice with experimental brain injury. Neurobiol Dis. 2005;20(1):38–48.
35. Jin K, Mao X, et al. Delayed transplantation of human neural precursor cells improves outcome from focal cerebral ischemia in aged rats. Aging Cell. 2010;9(6):1076–83.
36. Jin K, Xie L, et al. Effect of human neural precursor cell transplantation on endogenous neurogenesis after focal cerebral ischemia in the rat. Brain Res. 2011;1374:56–62.
37. Theus MH, Wei L, et al. In vitro hypoxic preconditioning of embryonic stem cells as a strategy of promoting cell survival and functional benefits after transplantation into the ischemic rat brain. Exp Neurol. 2008;210(2):656–70.
38. Takahashi K, Yasuhara T, et al. Embryonic neural stem cells transplanted in middle cerebral artery occlusion model of rats demonstrated potent therapeutic effects, compared to adult neural stem cells. Brain Res. 2008;1234:172–82.
39. Daadi MM, Steinberg GK. Manufacturing neurons from human embryonic stem cells: biological and regulatory aspects to develop a safe cellular product for stroke cell therapy. Regen Med. 2009;4(2):251–63.
40. Steinberg GK, Carmichael ST, et al. Embryonic-derived neural stem cells for treatment of motor sequelae following sub-cortical stroke, California Institute for Regenerative Medicine (CIRM) DR1-01480. 2010.
41. Takahashi K, Yamanaka S. Induction of pluripotent stem cells from mouse embryonic and adult fibroblast cultures by defined factors. Cell. 2006;126(4):663–76.
42. Takahashi K, Tanabe K, et al. Induction of pluripotent stem cells from adult human fibroblasts by defined factors. Cell. 2007;131(5):861–72.
43. Sareen D, Gowing G, et al. Human induced pluripotent stem cells are a novel source of neural progenitor cells (iNPCs) that migrate and integrate in the rodent spinal cord. J Comp Neurol. 2014;522(12):2707–28.
44. Chang DJ, Lee N, et al. Therapeutic potential of human induced pluripotent stem cells in experimental stroke. Cell Transplant. 2013;22(8):1427–40.
45. Gomi M, Takagi Y, et al. Functional recovery of the murine brain ischemia model using human induced pluripotent stem-cell-derived telencephalic progenitors. Brain Res. 2012;1459:52–60.
46. Oki K, Tatarishvili J, et al. Human-induced pluripotent stem cells form functional neurons and improve recovery after grafting in stroke-damaged brain. Stem Cells. 2012;30(6): 1120–33.
47. Polentes J, Jendelova P, et al. Human induced pluripotent stem cells improve stroke outcome and reduce secondary degeneration in the recipient brain. Cell Transplant. 2012;21(12): 2587–602.
48. Chau M, Deveau TC, et al. iPS cell transplantation increases regeneration and functional recovery after ischemic stroke in neonatal rats. Stem Cells. 2014;32(12):3075–87.
49. Pang ZP, Yang N, et al. Induction of human neuronal cells by defined transcription factors. Nature. 2011;476(7359):220–3.
50. Yamashita T, Abe K. Direct reprogrammed neuronal cells as a novel resource for cell transplantation therapy. Cell Transplant. 2014;23(4–5):435–9.
51. Englund U, Bjorklund A, et al. Grafted neural stem cells develop into functional pyramidal neurons and integrate into host cortical circuitry. Proc Natl Acad Sci U S A. 2002;99(26):17089–94.
52. Song HJ, Stevens CF, et al. Neural stem cells from adult hippocampus develop essential properties of functional CNS neurons. Nat Neurosci. 2002;5(5):438–45.

53. Guzman R. Cellular stroke therapy: from cell replacement to trophic support. Expert Rev Cardiovasc Ther. 2009;7(10):1187–90.
54. Lu P, Jones LL, et al. Neural stem cells constitutively secrete neurotrophic factors and promote extensive host axonal growth after spinal cord injury. Exp Neurol. 2003;181(2):115–29.
55. Yan J, Welsh AM, et al. Differentiation and tropic/trophic effects of exogenous neural precursors in the adult spinal cord. J Comp Neurol. 2004;480(1):101–14.
56. Zhang YW, Denham J, et al. Oligodendrocyte progenitor cells derived from human embryonic stem cells express neurotrophic factors. Stem Cells Dev. 2006;15(6):943–52.
57. Zhu W, Mao Y, et al. Transplantation of vascular endothelial growth factor-transfected neural stem cells into the rat brain provides neuroprotection after transient focal cerebral ischemia. Neurosurgery. 2005;57(2):325–33 (discussion 325–333).
58. Arai K, Jin G, et al. Brain angiogenesis in developmental and pathological processes: neurovascular injury and angiogenic recovery after stroke. FEBS J. 2009;276(17):4644–52.
59. Krupinski J, Kaluza J, et al. Role of angiogenesis in patients with cerebral ischemic stroke. Stroke. 1994;25(9):1794–8.
60. Hicks C, Stevanato L, et al. In vivo and in vitro characterization of the angiogenic effect of CTX0E03 human neural stem cells. Cell Transplant. 2013;22(9):1541–52.
61. Jiang Q, Zhang ZG, et al. Investigation of neural progenitor cell induced angiogenesis after embolic stroke in rat using MRI. Neuroimage. 2005;28(3):698–707.
62. Chu K, Park KI, et al. Combined treatment of vascular endothelial growth factor and human neural stem cells in experimental focal cerebral ischemia. Neurosci Res. 2005;53(4):384–90.
63. Liu H, Honmou O, et al. Neuroprotection by PlGF gene-modified human mesenchymal stem cells after cerebral ischaemia. Brain. 2006;129(10):2734–45.
64. Onda T, Honmou O, et al. Therapeutic benefits by human mesenchymal stem cells (hMSCs) and Ang-1 gene-modified hMSCs after cerebral ischemia. J Cereb Blood Flow Metab. 2008;28(2):329–40.
65. Toyama K, Honmou O, et al. Therapeutic benefits of angiogenetic gene-modified human mesenchymal stem cells after cerebral ischemia. Exp Neurol. 2009;216(1):47–55.
66. Lo EH. A new penumbra: transitioning from injury into repair after stroke. Nat Med. 2008;14(5):497–500.
67. Mine Y, Tatarishvili J, et al. Grafted human neural stem cells enhance several steps of endogenous neurogenesis and improve behavioral recovery after middle cerebral artery occlusion in rats. Neurobiol Dis. 2013;52:191–203.
68. Bacigaluppi M, Pluchino S, et al. Delayed post-ischaemic neuroprotection following systemic neural stem cell transplantation involves multiple mechanisms. Brain. 2009;132(8):2239–51.
69. Lee ST, Chu K, et al. Anti-inflammatory mechanism of intravascular neural stem cell transplantation in haemorrhagic stroke. Brain. 2008;131(3):616–29.
70. Pluchino S, Gritti A, et al. Human neural stem cells ameliorate autoimmune encephalomyelitis in non-human primates. Ann Neurol. 2009;66(3):343–54.
71. Shahbazi M, Kwang TW, et al. Inhibitory effects of neural stem-cell-derived from human embryonic stem cells on differentiation and function of monocyte-derived dendritic cells. J Neurol Sci. 2013;330(1–2):85–93.
72. Wang L, Shi J, et al. Neural stem/progenitor cells modulate immune responses by suppressing T lymphocytes with nitric oxide and prostaglandin E2. Exp Neurol. 2009;216(1):177–83.
73. Benowitz LI, Carmichael ST. Promoting axonal rewiring to improve outcome after stroke. Neurobiol Dis. 2010;37(2):259–66.
74. Carmichael ST. Cellular and molecular mechanisms of neural repair after stroke: making waves. Ann Neurol. 2006;59(5):735–42.
75. Liu Z, Zhang RL, et al. Remodeling of the corticospinal innervation and spontaneous behavioral recovery after ischemic stroke in adult mice. Stroke. 2009;40(7):2546–51.
76. Daadi MM, Davis AS, et al. Human neural stem cell grafts modify microglial response and enhance axonal sprouting in neonatal hypoxic-ischemic brain injury. Stroke. 2010;41(3):516–23.

77. Daadi MM, Li Z, et al. Molecular and magnetic resonance imaging of human embryonic stem-cell-derived neural stem cell grafts in ischemic rat brain. Mol Ther. 2009;17(7): 1282–91.
78. Takagi Y, Nishimura M, et al. Survival and differentiation of neural progenitor cells derived from embryonic stem cells and transplanted into ischemic brain. J Neurosurg. 2005;103(2):304–10.
79. Arvidsson A, Collin T, et al. Neuronal replacement from endogenous precursors in the adult brain after stroke. Nat Med. 2002;8(9):963–70.
80. Jin K, Minami M, et al. Neurogenesis in dentate subgranular zone and rostral subventricular zone after focal cerebral ischemia in the rat. Proc Natl Acad Sci U S A. 2001;98(8): 4710–5.
81. Kernie SG, Parent JM. Forebrain neurogenesis after focal Ischemic and traumatic brain injury. Neurobiol Dis. 2010;37(2):267–74.
82. Zhang RL, Chopp M, et al. Patterns and dynamics of subventricular zone neuroblast migration in the ischemic striatum of the adult mouse. J Cereb Blood Flow Metab. 2009;29(7):1240–50.
83. Darsalia V, Kallur T, et al. Survival, migration and neuronal differentiation of human fetal striatal and cortical neural stem cells grafted in stroke-damaged rat striatum. Eur J Neurosci. 2007;26(3):605–14.
84. Jensen MB, Yan H, et al. Survival and differentiation of transplanted neural stem-cell-derived from human induced pluripotent stem cells in a rat stroke model. J Stroke Cerebrovasc Dis. 2013;22(4):304–8.
85. Buhnemann C, Scholz A, et al. Neuronal differentiation of transplanted embryonic stem-cell-derived precursors in stroke lesions of adult rats. Brain. 2006;129(12):3238–48.
86. Toda H, Takahashi J, et al. Grafting neural stem cells improved the impaired spatial recognition in ischemic rats. Neurosci Lett. 2001;316(1):9–12.
87. Allen NJ, Barres BA. Signaling between glia and neurons: focus on synaptic plasticity. Curr Opin Neurobiol. 2005;15(5):542–8.
88. Chen Y, Swanson RA. Astrocytes and brain injury. J Cereb Blood Flow Metab. 2003;23(2):137–49.
89. Lok J, Gupta P, et al. Cell-cell signaling in the neurovascular unit. Neurochem Res. 2007;32(12):2032–45.
90. Panickar KS, Norenberg MD. Astrocytes in cerebral ischemic injury: morphological and general considerations. Glia. 2005;50(4):287–98.
91. Cummings BJ, Uchida N, et al. Human neural stem cells differentiate and promote locomotor recovery in spinal cord-injured mice. Proc Natl Acad Sci U S A. 2005;102(39):14069–74.
92. Stroemer P, Hope A, et al. Development of a human neural stem cell line for use in recovery from disability after stroke. Front Biosci. 2008;13:2290–2.
93. Kalladka D, Sinden J, et al. PISCES—a phase I trial of CTX0E03 human neural stem cells in ischaemic stroke: interim results. European Stroke Conference. Nice, France. 2014.
94. Yasuhara T, Matsukawa N, et al. Notch-induced rat and human bone marrow stromal cell grafts reduce ischemic cell loss and ameliorate behavioral deficits in chronic stroke animals. Stem Cells Dev. 2009;18(10):1501–14.
95. Steinberg GK, Kondziolka D, et al. A novel phase 1/2A study of intraparenchymal transplantation of human modified bone marrow derived cells in patients with stable ischemic stroke. Int Stroke Conf San Diego CA. 2014a;45:A149.
96. Steinberg GK, Kondziolka D, et al. A novel phase 1/2A study of intraparenchymal transplantation of human modified bone marrow derived cells in patients with stable ischemic stroke. Annu Meet West Neurosurg Soc Progr Sun Valley Idaho. 2014b;60:30.
97. Go AS, Mozaffarian D, et al. Heart disease and stroke statistics-2014 update: a report from the American Heart Association. Circulation. 2014;129(3):e28–e92.
98. Ben-David U, Benvenisty N. The tumorigenicity of human embryonic and induced pluripotent stem cells. Nat Rev Cancer. 2011;11(4):268–77.
99. Seminatore C, Polentes J, et al. The postischemic environment differentially impacts teratoma or tumor formation after transplantation of human embryonic stem-cell-derived neural progenitors. Stroke. 2010;41(1):153–9.

100. Kawai H, Yamashita T, et al. Tridermal tumorigenesis of induced pluripotent stem cells transplanted in ischemic brain. J Cereb Blood Flow Metab. 2010;30(8):1487–93.
101. Chapman AR, Scala CC. Evaluating the first-in-human clinical trial of a human embryonic stem-cell-derived therapy. Kennedy Inst Ethics J. 2012;22(3):243–61.
102. Bailey AM, Mendicino M, et al. An FDA perspective on preclinical development of cell-based regenerative medicine products. Nat Biotechnol. 2014;32(8):721–3.
103. Sart S, Ma T, et al. Preconditioning stem cells for in vivo delivery. Biores Open Access 2014;3(4):137–149.
104. Bible E, Qutachi O, et al. Neo-vascularization of the stroke cavity by implantation of human neural stem cells on VEGF-releasing PLGA microparticles. Biomaterials. 2012;33(30):7435–46.
105. Wong FS, Chan BP, et al. Carriers in cell-based therapies for neurological disorders. Int J Mol Sci. 2014;15(6):10669–723.

Chapter 8
The CTX Human Neural Stem Cell Line and the PISCES Stroke Trial

Keith W. Muir and John D. Sinden

Background

Neural stem cells (NSCs) are multipotent cells derived from the developing or adult brain that have the capacity to undergo repeated cell division and to make identical copies of themselves under specific culture conditions. Depending on their environment, cells will undergo differentiation to neurons, astrocytes and oligodendrocytes in variable proportions, after initial growth arrest [1]. Most adult stem cells are inefficient at generating functional differentiated cells in vivo; however, they are able to provide paracrine, trophic and immune system support for endogenous repair mechanisms, stimulating angiogenesis and neurogenesis that may be more pertinent to their regenerative properties than cell replacement [2]. Cells are attracted to regions of damage or inflammation by chemokine or other immune signalling, and the existence of this "niche" appears key to adult stem cell survival in recipients. Clinical trials in stroke have targeted predominantly acute and subacute patients, when potential mechanisms of cell therapy effects are likely to include alteration of the brain environment, and influence over some mechanisms of secondary injury or early regeneration such as apoptosis, reactive angiogenesis, inflammatory cell infiltration and immunological response, all of which are relevant to known actions of NSCs [3–5]. In contrast, stem cells administered in the chronic stages (weeks, months or years) after stroke enter a very different cellular niche, and credible mechanisms of action at these late stages are likely to be dependent on neurogenesis, angiogenesis, growth factor secretion, cell differentiation and engraftment [6–8]. These mechanisms have been studied overwhelmingly in animal models of stroke that are likely to offer some, but not all insights into the clinical situation.

K. W. Muir (✉)
Institute of Neuroscience and Psychology, University of Glasgow,
Southern General Hospital, Glasgow G51 4TF, Scotland, UK
e-mail: Keith.Muir@glasgow.ac.uk

J. D. Sinden
ReNeuron Ltd, Surrey Research Park, Guildford GU2 7AF, Surrey, UK

D. C. Hess (ed.), *Cell Therapy for Brain Injury*, DOI 10.1007/978-3-319-15063-5_8

Previous Clinical Studies in Stroke

Implantation of cells from a range of sources including NSCs derived from neural tissues or differentiated from pluripotent cells such as embryonic stem cells or induced pluripotent stem cells, as well as cord blood-derived cells, mesenchymal stem cells, peripheral blood mononuclear cells (MNCs), adipose tissue-derived cells and cells derived from menstrual blood have promoted recovery of neurological function in animal models of stroke [4, 6, 9, 10]. Selected cells from bone marrow, identified based on surface markers such as CD34+, aldehyde dehydrogenase enzyme activity (ALDH+) or genetically modified Notch-1 activated mesenchymal stromal cells (MSCs), have all shown promise in preclinical studies and some are being developed as proprietary cell therapy products. The heterogeneity of cell sources and the very limited degree (or indeed complete absence) of cell survival and engraftment for many cell types [11] suggests strongly that cell replacement is not the main mechanism of therapeutic benefit, and has prompted the search for potential alternative mechanisms of action.

In distinction to autologous cell approaches that take cells from bone marrow, adipose tissue or similar sources, with uncertain yield, and inevitable delays from sampling to dosing if culture expansion is undertaken, an allogeneic cell product carries the hypothetical advantages of immediate availability, suitability for all patients and predictable dosing. Significant technical challenges include achievement of manufacturing standards, characterising potency and quality assurance to guarantee a consistent and stable therapeutic product.

Neural Stem Cells

Mode of Administration

A number of previous investigations have explored safety and efficacy of cerebral implantation of cellular therapies in both animal models and small clinical trials [12–15]. Studies in healthy animals indicated that brain implantation of human neuron-like teratocarcinoma-derived (hNT) cells appears to be safe [16] and formed the basis for clinical studies of these cells in patients with chronic stroke. Safety and feasibility of the stereotaxic implantation of two or six million hNT cell doses (60 and 180 μL, respectively) was initially demonstrated in 12 patients [14]. There were no reports of cell-related adverse events. Postmortem examination of a patient, who died of an unrelated illness, identified implanted cells at the injection site 27 months after the implantation but found no signs of inflammation, neoplasia or infection [17]. In the initial open label study, 50 % of participants had evidence of neurological improvement by using summated scores on the European Stroke Scale over a time period of 6 months. In a subsequent phase 2 randomised clinical trial, also involving the stereotaxic implantation of five or ten million hNT cells in 250 μL total volume, there were no safety issues in 25 patients with subcortical stroke resulting in motor deficits [15]. Some improvement in the Barthel Index (BI) of Activities of Daily Living (ADL) was reported by some patients; however, no significant improvement was found in motor function.

Development of CTX, a Human Neural Stem Cell Line

Somatic or lineage-restricted stem cells may be derived from human fetal tissue, but their therapeutic use is limited by tissue availability and variable purity or quality of cell product. Several approaches for the development of NSCs as therapeutics have been adopted. HuCNS-SC (developed by Stem Cells Inc. Palo Alto CA, USA) provides an example of a neural stem product which has been successfully expanded from isolated fetal somatic stem cells in culture [18]. A phase 1 clinical trial of intracerebral implants of HuCNS-SC in fatal paediatric Batten disease has been completed, with no treatment-related safety concerns reported, and trials are ongoing in Pelizaeus–Merzbacher disease, spinal cord injury and age-related macular degeneration (www.stemcellsinc.com).

In order to overcome the problem of cell senescence of expanded cell populations from fetal stem cells in culture and through a manufacturing process, immortalizing genes can be inserted into NSCs. This approach was taken by ReNeuron in the development of CTX, a human fetal-derived neural stem cell line, genetically modified by insertion of a single copy of the c-myc gene fused with a modified oestrogen receptor [19, 20]. The c-MycER^{TAM} protein enables cell cloning and increased cell proliferation as well as extending stable growth of the cells by upregulation of telomerase activity. The activity of the fusion protein is regulated by the addition of 4OHT to the culture media, enabling translocation of the fusion protein to the cell nucleus. The history of the development of ReNeuron's c-MycER^{TAM} neural stem cell lines has been outlined previously [21]. CTX was identified following in vitro and in vivo screening of a range of cell lines as having potential to differentiate into relevant cell lineages. The c-mycER^{TAM} technology allows CTX cells to undergo stable large-scale expansion and manufacture under current good manufacturing practice (cGMP) conditions in much the same way as a conventional biopharmaceutical agent, and that can be stored frozen in liquid nitrogen. The resultant CTX "Drug Product" (CTX-DP) is a manufactured cell line composed of CTX cells at a passage of ≤ 37 following conditional immortalization with c-mycER^{TAM}technology [19]. The active DP is a suspension of living cells formulated in sterile buffer suitable for intracranial administration using stereotaxic techniques.

Cell banking of the CTX cell line was undertaken early in the development program to ensure that all pivotal preclinical safety and efficacy studies would be conducted using the same DP material, which will in due course be used for any subsequent clinical trials and any authorised or marketed product. CTX will not need to be re-derived, as sufficient vials are available at every level of the manufacturing process to enable essentially limitless manufacture. Robotic automation of the CTX manufacturing process has also been demonstrated [22], further validating the potential of this cell line to be efficiently and safely scaled at a reasonable cost of goods.

Non-clinical Studies

Efficacy testing for cell therapies, as for other therapies being developed in ischaemic stroke, generally relies on rodent models of focal ischaemia. Middle cerebral artery occlusion (MCAo) in the rat is the most widely used and best characterised animal model of ischaemic stroke, albeit one that has been challenged on the grounds that no therapeutic strategy has yet been translated successfully to the clinic from this model system (although some successful clinical therapies have also demonstrated efficacy when evaluated in animal models, providing some reverse translational validation) [23–25]. Animals with MCAo are affected in the same anatomical location (i.e. basal ganglia and sensorimotor cortex) as humans with an ischaemic MCA territory stroke, producing the same core functional deficits such as unilateral paralysis, sensory dysfunction and visuospatial neglect. Rodent MCAo models of stroke do not generally attempt to reproduce the heterogeneity of human stroke, but rather offer a model system that recapitulates certain key aspects of stroke pathophysiology in order to allow early investigation of potential therapeutic agents with minimal sample size, reproducible in different laboratories, and thus affording sufficient evidence of potential efficacy to justify translation to the clinical environment. MCAo initially produces severe gross neurological dysfunction; most early dysfunctions resolve spontaneously within days to weeks [26]. However, spontaneous early improvement confounds attempts to measure treatment-related recovery (the clinical correlate of which represents a substantial hurdle for phase II trials). Certain behavioural tests demonstrate persistent dysfunction that does not spontaneously resolve, including bilateral asymmetry, the Morris water maze and rotameter testing [27]. A battery of sensorimotor and cognitive tests to assess the degree of injury to a specific area of a brain linked to the treatment [28] has been utilised and reasonably well characterised in rodent MCAo models.

Autologous stem cells isolated from bone marrow have, to date, been subjected to very little preclinical safety testing because of decades of clinical experience in bone marrow transplantation. Bone marrow-derived stem cells scaled up as an allogeneic cell product, for example as MSCs, require safety testing, but because these cells typically show little or no engraftment, the focus has been on acute toxicity rather than long-term tumorigenicity. In respect of engrafting stem cells, in particular those derived from pluripotent cells whose normal fate following implantation is to form teratomas, preclinical tumorigenicity is a major concern and requires extensive long-term testing, with high associated costs and time commitment.

Translation to clinical investigation of any novel stem cell product, including conditionally immortal hNSCs, requires extensive supporting data characterising the cells and offering quality assurance regarding cells, their manipulation and their manufacturing and a series of studies demonstrating short and long-term toxicology and safety.

Immediate impacts on the immune system can be assessed in standard toxicology study designs using functional observational batteries, haematology and necroscopy endpoints. Characterisation of migration and engraftment is important in both safety and efficacy assessment. Full in vivo biodistribution analysis, using a time course study of validated stem cell markers, involves a combination of immunohistochemistry, in situ hybridization or quantitative PCR for specific features of the stem cells or their tissues of origin, or of inserted labels specific to the stem cells. These studies require pilot feasibility and validation studies and suitable controls to eliminate false positives.

Long-term safety evaluation requires investigation of tumorigenic potential in immunodeficient strains of mice using large numbers of both sexes. The duration of studies will vary depending on the survival time of the cells, which can range from about 3 months for non-engrafting cell types to 12 months or longer for cells that survive well in vivo.

Additionally, because of the presence of the c-myc gene and its retroviral insertion, further studies of CTX were undertaken to demonstrate c-mycER downregulation and epigenetic silencing [29] as well as further cell-bank testing for retrovirus to eliminate any risk of infection transmission.

Following intracerebral implantation of CTX in MCAo rats, non-human primates and non-obese diabetic/severe combined immunodeficiency (NOD SCID) mice, general safety was assessed by a functional observation battery of tests for up to 6 months. No CTX-related adverse events were reported in any of these studies. Both the cell dose and volume of implant were well tolerated in all studies conducted.

Studies in vitro have confirmed that re-exposure of growth arrested/differentiated cells to 4-OHT does not return differentiated CTX cells to a proliferative state. Similarly, exposure of CTX cells to endogenous steroid hormones or to the drug tamoxifen as might occur in vivo does not activate the c-mycER^{TAM} technology in cells, leading to inappropriate cell proliferation. In addition, long-term (6 months) treatment of animals with tamoxifen had no impact on CTX cell survival and proliferation and there were no CTX, plus or minus tamoxifen, related reports of tumour formation. Studies in a range of tumorigenesis models in mice and rats have shown no incidence of CTX-related tumour pathologies, including in animals where CTX cells survived for extended periods of time. These data support the view that CTX does not present a tumour risk following implantation into the brain.

CTX Efficacy Studies

A validated MCAo rat model of ischaemic stroke was used for non-clinical studies of CTX [30–32]. Rats were transplanted 3–4 weeks post infarct, permitting prior recovery from acute phase neurological dysfunction and establishment of steady-state sensorimotor deficits.

Two studies using MCAo stroked rats have demonstrated long-term improvements in sensorimotor function following intracerebral CTX implantation. In the first study, CTX cells from early stage cell banks were implanted 3–4 weeks after MCAo. Animals were immunosuppressed using methylprednisolone and cyclosporine A. Transplantation of CTX cells to the striatum in this model of stroke caused statistically significant improvements in both sensorimotor function and gross motor asymmetry at 6–12 weeks post grafting. In addition, cell migration and long-term survival in vivo were not associated with significant cell proliferation [19].

A second study using CTX delivered adjacent to the infarcted region in rats with stable neurological deficits after MCAo [8] demonstrated a cell dose–response effect. Again, animals were immunosuppressed using methylprednisolone and cyclosporine A, albeit only for the first 2 weeks.

Statistically significant dose-related recovery in sensorimotor function deficits (bilateral asymmetry test in the mid- and high-dose groups and rotameter test after amphetamine exposure in the high-dose group) was found in the CTX cell implanted groups compared to the vehicle group. In-life functional improvements correlated with cell dose although these improvements did not correlate with survival of CTX cells measured at postmortem. There was differentiation of CTX cells into oligodendroglial (8 %) and endothelial phenotypes (6 %). MCAo-induced reduction of neurogenesis in the subventricular zone (SVZ) was partially restored to that observed in controls without MCA occlusion. No adverse CTX cell-related effects were observed during in-life observations or on tissue histology. These effects were seen at 3 months post implantation [8].

In a further study, intraparenchymal implantation of CTX cells in the rat MCAo model improved sensorimotor dysfunctions (bilateral asymmetry test) and motor deficits (foot-fault test, rotameter). Importantly, analyses based on lesion topology (striatal versus striatal plus cortical damage) revealed a more significant improvement in animals with a stroke confined to the striatum. No improvement in learning and memory (Morris water maze) was evident. In contrast to intraparenchymal implantation, intracerebroventricular implantation of cells did not result in any improvement. MRI-measured lesion, striatal and cortical volumes were unchanged in treated animals compared to those with stroke that received an intraparenchymal injection of suspension vehicle. Grafted cells only survived after intraparenchymal injection with a striatal plus cortical topology resulting in better graft survival (16,026 cells) than in animals with smaller striatal lesions (2374 cells). Almost 20 % of cells differentiated into glial fibrillary acidic protein positive (GFAP+) astrocytes, but <2 % turned into FOX3+ neurons. These results indicate that CTX cell implants are associated with robust recovery of behavioural dysfunction over a 3-month time frame and that this effect is specific to their site of implantation. Lesion topology is potentially an important factor in recovery, with a stroke confined to the striatum showing a better outcome compared to a larger area of damage [33]. Mechanistically, given the lack of correlation between CTX survival and functional recovery, there is limited evidence for like-for-like cell replacement as a major promoter of functional recovery. Instead, like other recent reports using neural stem cell implants in stroke brain, CTX implants act to promote endogenous repair mechanisms, including new blood vessel formation in ischaemic rat brain areas [34] and sustained increased striatal neurogenesis after MCAo stroke [35].

Translation into Clinical Trials: The PISCES I Study

The use of CTX DP, identical to that used in preclinical studies, provides assurance in some key aspects of translation to clinical trials. Pro-angiogenic and immunomodulatory characteristics of CTX are considered to represent the likely mechanism of action in the rodent ischaemic tissue environment [10, 29, 34],and are known potential clinical targets. Other aspects of translation remain more speculative: these include the appropriate dose scaling; equivalence of "chronic" stroke in a rodent and "chronic" stroke in human subjects; the role of immunosuppression; the mitigating effects of comorbidities (including drug treatments); and the confounding effects of concomitant physical therapies and environmental factors on recovery.

In addition, the early stages of clinical research for advanced therapy medicinal products are tightly regulated, and the populations deemed suitable for *first-in-man* studies may be atypical of the ultimate therapeutic target population, notably with respect to age and chronicity of stroke. The protocol for the first Pilot Investigation of Human Neural Stem Cells in Stroke (PISCES) trial attempted to balance the available preclinical data with regulatory requirements.

PISCES (trial registration NCT 01151124; EUDRACT number 2008-000696-19) was an open label, single site, ascending dose study in male patients aged 60 years or over with chronic, stable disability as defined by neurological impairment (National Institutes of Health Stroke Scale (NIHSS) score ≥ 6), and disability (modified Rankin Scale (mRS) > 2) after ischaemic stroke that had occurred 6–60 months previously. Sequential groups received a single implantation of 2, 5, 10 and 20 million cells by stereotaxic injection to the putamen ipsilateral to the stroke. The primary endpoint was safety, including adverse events and neurological change. Secondary functional endpoints included clinical scales of neurological impairment, ADL, limb spasticity, disability and health-related quality of life (NIHSS, mRS, Barthel Index (BI), Ashworth Scale, Euro-QoL) as well as brain imaging.

Detailed inclusion and exclusion criteria are listed in Table 8.1. The rationale for protocol criteria are discussed in depth.

Stereotaxic Intracerebral Implantation

The safety and feasibility of stereotaxic implantation of two or six million hNT cell doses (60 and 180 μL, respectively) was initially demonstrated by Kondziolka and colleagues [14] and confirmed in the subsequent randomised trial, involving the stereotaxic implantation of five or ten million hNT cells in 250 μL total volume, in 25 patients with subcortical stroke [15]. Intracerebral delivery has also been described

Table 8.1 Major inclusion and exclusion criteria for the PISCES I trial

Inclusion criteria
Males aged $\geq$60 years
Unilateral ischaemic stroke affecting subcortical white matter and/or basal ganglia (with or without cortical involvement) 6 months to 5 years prior to entry into the trial, with persistent hemiparesis
NIHSS score a minimum of 6, including $\geq$2 for motor arm and leg sub-scores
Stable neurological deficit, defined as change in NIHSS total score of two points or less over 2 months prior to cell implantation
mRS 2–4
Fitness to undergo neurosurgical procedures under general anaesthesia
Able to participate in the trial and willing to comply with all procedures either alone or with the aid of a responsible caregiver
Capable of providing informed consent
Minimum infarct diameter of 1 cm on MRI
Exclusion criteria
Structural cerebral vascular disease of size and location likely to require surgical intervention or increase the risk of stereotaxic stem cell implantation
Any unstable medical condition or any medical condition with an expected survival time 12 months, e.g. malignancy, uncontrolled diabetes mellitus (HbA1c 8 %)
Any disorders that could interfere with participation in the trial, including progressive neuro-degenerative disease such as Alzheimer's disease, severe Parkinson's disease or Huntington's disease, chronic alcohol or drug abuse, untreated major depression, schizophrenia
Planned or recent major surgery within previous 30 days; e.g. cardiac or carotid surgery
Previous allogeneic stem cell, tissue, organ or bone marrow transplant
Cognitive impairment with MMSE < 24
History of epilepsy
Blood coagulation disorder
Requirement for warfarin or anticoagulant treatment that cannot be interrupted
Received an unlicensed pharmaceutical product as part of a clinical trial within previous 3 months
Previous enrolment in this trial
Any condition contra-indicating MRI
Contra-indications to surgery found on the screening MRI scan:
Multifocal cerebral microbleeds suggestive of cerebral amyloid angiopathy
Cerebral aneurysms at potential risk of rupture
Arteriovenous malformations
Presence of antibodies to CTX HLA antigens
Coagulation test results that preclude surgery taking place
Current use of psychostimulant medications including; amphetamine, methylphenidate, sinemet, amantadine or bromocriptine
Current use of intermittent botulinum toxin therapy, phenol or other antispasticity medications (antispasticity medications are acceptable if taken regularly for at least a month)
Current requirement for tamoxifen or similar analogues

NIHSS National Institutes of Health Stroke Scale, *MRI* magnetic resonance imaging, *MMSE* mini mental state examination, *HLA* human leucocyte antigen

in stroke for fetal porcine cells, and autologous bone marrow-derived cells [36, 37]. Trials of primary fetal cell implantation for Huntington's or Parkinson's diseases include cell injections targeted to the basal ganglia of 0.5 up to 10 million cells in volumes of 10–200 μL.

Stereotaxic intraparenchymal delivery offers the advantage that a large and controlled number of cells are delivered adjacent to the site of ischaemic tissue damage, and circumvents the blood–brain barrier. Delivery by intraparenchymal injection imposes anatomical restrictions since some sites cannot be safely implanted (e.g. the brainstem), and carries a well-characterised risk of intracranial bleeding and seizures in the range of 1–2 % [38] and 2.4 % [39] respectively. In addition, there are procedural risks associated with general anaesthesia, hospitalisation and neurosurgery, all in a predominantly elderly population likely to be taking antithrombotic or anticoagulant treatment for stroke prevention.

Dose Selection

Allometric scaling suggests that a dose equivalent to the efficacious dose in rats is approximately 20 million cells in humans. While non-clinical studies have not indicated the dose ceiling above which there is no further increase in efficacy, the volume of material that can be injected into the brain suggested that the highest practical dose of CTX in man for stroke is 20 million cells.

A conservative approach for the starting dose and dose escalation was used, as appropriate for a first-in-man trial. Previously, cell dosing in the brain for treatment of neurological disease has employed small doses at multiple sites in the brain. A major consideration that is specific to intracerebral delivery is the volume of injection since this alone may lead to adverse events through compression of brain parenchyma. The maximum implantation volumes delivered into the brains across three animal species (NOD-SCID mouse, rat and cynomolgus monkey) in the non-clinical development of CTX were compared to the proposed first-in-man volume in order to evaluate the safety margin on volume of delivery in humans. In the rat studies the volume of the implantation was ~ 1/111th of the rat brain volume. The volume of the proposed first-in-man dose is 40 μL, equivalent approximately to 1/32500th of the volume of the human brain. This represents a large safety margin of 300x in the proposed clinical trial compared to the rat studies (83x and 667x compared to cynomolgous monkeys and NOD SCID mice, respectively). In the two clinical trials of the hNT cell line, doses of 2–10 million cells were used, delivered by multiple needle trajectories and multiple injection boluses of 10–20 μL each at points along each needle pass. The initial dose volume in PISCES (40 μL in three patients) was lower than the starting dose volume of 60 μL in four patients in the first-in-man hNT trial.

Since the concentration of CTX drug product is fixed at 50,000 cells/μL, a dose of two million cells requires 40 μL and a dose of five million cells 100 μL total volume. A single needle pass was used for the 2 and 5 million cell doses, with a second needle pass in close proximity to the first for the 10 million cell dose (total volume 200 μL) and four needle passes (total volume 400 μL) for the 20 million cell dose. Doses were allocated serially so that successive groups received 2, 5, 10 and 20 million cells. Only one patient was treated at a time.

Safety Review

Screening was conducted over approximately 8 weeks before surgery to ensure that eligible patients had stable functional deficits as measured by the National Institutes of Health Stroke Scale (NIHSS), a measurement of neurological deficit that is widely used to characterise the severity of stroke in the acute phase. NIHSS score predicts survival and functional recovery [40–42], and deterioration by 4 or more points on the total NIHSS score is a widely used threshold to identify clinically significant deterioration. NIHSS was therefore used to monitor clinical status before and after implantation.

In addition to standard biochemistry and haematology panels, participants were screened for the presence of specific antibodies to CTX HLA antigens before and at multiple time points after cell implantation.

Structural magnetic resonance imaging (MRI) scans were obtained pre-implantation and at 3, 12 and 24 months to specifically seek evidence of haemorrhage, new infarction, inflammation or tumour growth at the site of cell implantation.

All implanted patients were identified to a national registry that provides lifelong surveillance for events such as cancer and death, so long as they remain resident in the same country. Prospective consent for postmortem examination of brain tissue was also sought at trial entry.

An independent data and safety monitoring board (DSMB) reviewed all clinical, imaging and safety data throughout the study. A decision to continue dosing at each dose level followed satisfactory review of the 28-day safety data for the first patient at that dose level; and to increase the dose to the next level following satisfactory review of the 3-month safety data for the last patient in the lower dose group.

Rationale for Open Label Design

As a safety trial, a control group was not included. Inclusion of a small number of non-operated control patients would be highly unlikely to be informative given the heterogeneity of stroke patients, would significantly slow recruitment, and randomisation to non-implantation may be unacceptable to patients [43]. Logically, the appropriate control group would undergo all procedures (including general anaesthesia and craniotomy) except cell implantation, but the invasive nature of this treatment in a population with potential morbidity from such procedures, was not felt to be justified in a phase 1 study.

Rationale for Exclusion of Women

As a first-in-human trial and with no reproductive toxicology studies conducted with CTX cells to date, only male stroke patients were included. In addition, there is theoretical concern that if tamoxifen is required at a later stage, it might "switch" the CTX cells on again causing them to proliferate.

Rationale for Implanting CTX Cells into the Putamen

Implantation of CTX cells within nerve cell clusters near but not directly within the lesion was planned, consistent with intrastriatal implantation in preclinical models. Ischaemic lesions from stroke occurring within the internal capsule can interrupt large numbers and types of nerve fibres, including cerebral cortical efferents to the basal ganglia, thalamus, brainstem and spinal cord. In addition, afferent fibres projecting from the thalamus to the cortex can be interrupted. There is considerable variability between patients in the number and type of fibre affected, depending upon the size and specific location of the stroke within the white matter. It is considered hazardous to inject directly into the white matter lesion, as pressure resulting from the injection could cause further injury of the already damaged tissue. In seeking a neuronal cluster near most white matter ischaemic lesions, the putamen was selected as the best target, being large and easily accessible with stereotaxic approaches, and in close proximity to the site of many middle cerebral territory strokes.

Rationale for Inclusion of Patients 6 Months to 5 Years Postischemic Stroke

The timing for cell implantation in PISCES was selected to ensure the recruitment of patients with stable neurological and functional deficits. Studies have found consistently that functional recovery from disabling stroke reaches a maximal level by 6 months [44–48], although the great majority of such studies must be qualified by recognising that both speed and completeness of apparent recovery are dictated by the chosen measure, which may be insensitive to deficits that are nonetheless significantly disabling; the BI has well described ceiling effects, for example [49]. In addition, the setting of previously reported studies may introduce bias, as the majority were conducted in specialist rehabilitation services, referral or eligibility for which may be dictated in part by perceived rehabilitation potential. In the Copenhagen Stroke Study [44, 45], recovery of neurological function was complete within 12.5 weeks from stroke onset in 95 % of patients and time course of recovery was clearly related to initial stroke severity. The best recovery of ADL occurred within 8.5 weeks for mild stroke, within 13 weeks for moderate stroke, within 17 weeks for severe stroke and within 20 weeks for very severe stroke, with no further significant improvement after this point. In another study, half of the patients with disabling ischaemic stroke recovered within 18 months although recovery was greatest in the first 6 months [50]. A minimum delay of 6 months therefore ensures that spontaneous recovery of function is highly unlikely, and that participants are likely to be medically stable.

Rationale for Not Including Immunosuppressive Therapy

Transplantation of tissue or cell-based products is usually combined with immunosuppression to guard against rejection. In respect of CTX, non-clinical studies have shown that cell survival and associated beneficial effects are not influenced by treatment of the animals with immunosuppressive drugs. Furthermore, in vitro studies for MHC-DR and MHC-ABC showed that expression of MHC class I and class II protein levels was low in CTX cells, indicating that the potential for rejection was low. Infections commonly complicate the poststroke period, especially pneumonia and urinary tract infections, and are strongly associated with poor outcome [51]. A prolonged period of immunodepression is recognised after stroke [52] and is likely to increase vulnerability. Immunosuppressive therapy was therefore considered to have significant risks, without evidence of either need or avoidance of complications.

Study Procedures

Cell Preparation and Manufacture

CTX was manufactured according to cGMP, including testing for sterility, mycoplasma and endotoxin and determination of cell number and viability.

Cell Implantation

CTX cells were implanted under general anaesthesia by a neurosurgeon experienced in stereotaxic intracranial implantation. Stem cell delivery was performed using the technique used successfully in two previous clinical trials [53]. A burr-hole craniotomy was fashioned at an appropriate point on the patient's skull and the cells implanted using an implantation cannula of the same design as the Pittsburgh cell delivery cannula connected to a Hamilton syringe, as originally described by Kondziolka and colleagues [53]. All injections were made at a rate of 5 μL/min, with each deposit of one million cells (20 μL volume of cell suspension). The number of deposits was two for the lowest dose and thereafter increased to five for all subsequent higher doses. The number of trajectories required for dosing increased from one (2 and 5 million cell dose) to 2 (10 million cell dose) to 4 (20 million cell dose). Total procedure time increased accordingly.

Secondary Endpoints for Efficacy and Biological Effects

Secondary endpoints included the assessment of clinical outcomes post implantation at 12 and 24 months using the Barthel Index, a measure of performance of ten

ADL of which eight evaluate the level of dependence [54, 55] and mini mental state examination (MMSE) and at 12, 24 and 120 months using the modified Rankin Scale (mRS), an overall measure of disability and handicap and clinician-reported measure of function after stroke, divided into seven categories/steps from asymptomatic to dead [56, 57]. In addition, assessments were made of health-related quality of life using the EuroQoL (EQ-5D) instrument at 12 months, a summated score derived from the modified Ashworth Scale, a widely used method for assessing muscle spasticity, in each muscle group of the affected upper and lower limbs [58], and the Star Cancellation test, a screening tool for unilateral spatial neglect in patients with stroke [59].

Since clinical measures cannot offer a reliable index of activity with small patient numbers and marked heterogeneity, brain-imaging studies were used to explore potential indices of biological activity. These included a functional MRI motor activation task, single voxel spectroscopy, resting state BOLD MRI, and diffusion tensor imaging (DTI) pre-cell implantation and again at 1 and 12 months post implantation.

Rehabilitation Therapy

Rehabilitation therapies are a complex set of interventions targeting multiple aspects of neurological and physical function, with the aim of encouraging neural repair, and restoring function through re-training or adaptation to disability, together with avoidance of maladaptive physical strategies and secondary complications that may compromise function. While general principles are supported by evidence from clinical trials (e.g. a dose–response relationship between physical therapy and recovery of motor function), the complexity and individual approach involved in rehabilitation therapy represents a major challenge for trials of restorative therapy to even record, let alone to control for.

No formal rehabilitation programme was undertaken as part of the study, and the trial selection criteria for inclusion of chronic but stable patients meant that ongoing participation in any rehabilitation therapy was unlikely, although not specifically prohibited. The trial recorded participation in any rehabilitation therapies throughout follow-up.

Future Directions

Observations from the first PISCES trial support further studies, which will begin to explore efficacy. Initial experience permits some relaxation of recruitment criteria relating to initially cautious safety assessment, such that the lower age limit for the next stage of trials has been lowered to 40 years, and women are also included. Immunosuppression is again not planned. Earlier administration of cells than in PISCES will be undertaken since the mechanistically relevant recovery processes are most prevalent in the initial few weeks after stroke, which is also most consistent with the data from preclinical studies. For intraparenchymal administration at late time points after stroke, the blood–brain barrier might prevent systemic immune reactions and rejection, but this may not be so if subacute administration is undertaken, when the blood–brain barrier is compromised in the early stages of the ischaemic injury.

Determination of efficacy ultimately necessitates robust controlled trials, and presents a number of challenges. Appropriate controls for a randomised trial are likely to be a compromise in some respects. Historical controls are not valid, and concurrent controls who undergo only conventional medical and rehabilitation therapies (especially if deemed ineligible for cell therapy, as some trials have proposed) are also sub-optimal, since not exposed to the major confounders of invasive procedures and the potent placebo effect of "stem cell therapy" [60]. While essential scientifically, sham neurosurgery is of uncertain acceptability to patients, and the need for sham neurosurgery has been challenged by patient groups [43]. Patient acceptability of a proposed trial is essential to ensure recruitment, and rigid adherence to conventional parallel-group randomised controlled trial designs is probably impossible at early stages of efficacy testing since patients who accept the risks of surgery and experimental treatment are frequently unwilling to consider the possibility that they will be randomly allocated to a placebo control group. The reward–risk balance for neurosurgical administration will also mandate the use of clinically (not just statistically) significant endpoints in a relatively small number of patients which poses a major challenge itself in early clinical development.

Randomisation of participants to deferred or immediate therapy, as previously employed in a trial of ablative neurosurgery or deep-brain stimulation in Parkinson's disease (PD Surg) [61], may offer a suitable design compromise for stroke, although the difference between an acute condition with anticipated trajectory of early recovery (with many relevant mechanisms operating at these early stages) and a chronic degenerative process like Parkinson's disease may raise ethical questions around the biological equivalence of early and late interventions.

The ability to adjust for concomitant rehabilitation therapy in both active and control arms represents a major difficulty in trial design, as discussed earlier. It is also possible that rehabilitation training is a necessary facilitatory factor for regenerative effects of cell therapies to be seen, although animal models of non-specific physical therapies also suggest that negative interactions are also possible. Implementation of standardised therapy as part of a clinical trial is difficult when dealing with what is traditionally a highly individualised and time-limited intervention that varies widely across healthcare systems.

Patient-selection criteria are likely to be driven largely by choice of efficacy endpoints: a motor endpoint dictates selection of patients with a relevant motor deficit. As well as the prevalence of motor deficits after stroke, both functional scales and the natural history of recovery are better characterised than other neurological deficits such as speech or visuospatial neglect, and visual field problems are confounded by the prevalence of a diverse range of pathologies affecting vision that are prevalent in the stroke population.

Optimal timing for intervention remains a balance of practicality and interpretation of data largely derived from animal models on mechanisms of injury and recovery over time. The mode of cell delivery, delays due to requirements for manufacture or ex-vivo culture, and appropriate consent procedures, are all considerations. Intraparenchymal brain delivery entails patient preparation and adequate time to ensure clinical stability for anaesthesia. On the background of expected recovery, patient perspectives on acceptability of invasive intervention are likely to shift in the early weeks after stroke, presenting a further issue with recruitment. Assumptions about the mechanisms that pertain at different time points after stroke, and the mechanisms of cell action, are likely to be an over-simplification.

Phase 2 trials therefore face a challenging environment where studies will inevitably be small due to cell production constraints, and patient selection critical in ensuring a credible "responder" population is recruited at a reasonable rate. Careful short- and long-term safety review remains a requirement.

References

1. Gage FH. Mammalian neural stem cells. Science. 2000;287(5457):1433–8.
2. Horie N, Pereira MP, Niizuma K, et al. Transplanted stem-cell-secreted vascular endothelial growth factor effects poststroke recovery, inflammation, and vascular repair. Stem Cells. 2011;29(2):274–85.
3. Chopp M, Li Y, Zhang ZG. Mechanisms underlying improved recovery of neurological function after stroke in the rodent after treatment with neurorestorative cell-based therapies. Stroke. 2009;40(3 Suppl):S143–5.
4. Andres RH, Choi R, Steinberg GK, Guzman R. Potential of adult neural stem cells in stroke therapy. Regen Med. 2008;3(6):893–905.
5. Bliss TM, Andres RH, Steinberg GK. Optimizing the success of cell transplantation therapy for stroke. Neurobiol Dis. 2010;37(2):275–83.
6. Bliss T, Guzman R, Daadi M, Steinberg GK. Cell transplantation therapy for stroke. Stroke. 2007;38(2 Suppl):817–26.
7. Modo M, Stroemer RP, Tang E, Patel S, Hodges H. Effects of implantation site of stem cell grafts on behavioral recovery from stroke damage. Stroke. 2002;33(9):2270–8.
8. Stroemer P, Patel S, Hope A, Oliveira C, Pollock K, Sinden J. The neural stem cell line CTX0E03 promotes behavioral recovery and endogenous neurogenesis after experimental stroke in a dose-dependent fashion. Neurorehabil Neural Repair. 2009;23(9):895–909.
9. Hicks A, Schallert T, Jolkkonen J. Cell-based therapies and functional outcome in experimental stroke. Cell Stem Cell. 2009;5(2):139–40.
10. Miljan EA, Sinden JD. Stem cell treatment of ischemic brain injury. Curr Opin Mol Ther. 2009;11(4):394–403.
11. Pendharkar AV, Chua JY, Andres RH, et al. Biodistribution of neural stem cells after intravascular therapy for hypoxic-ischemia. Stroke. 2010;41(9):2064–70.
12. Sinden JD, Muir KW. Stem cells in stroke treatment: the promise and the challenges. Int J Stroke. 2012;7(5):426–34.
13. Sinden JD, Vishnubhatla I, Muir KW. Prospects for stem-cell-derived therapy in stroke. Prog Brain Res. 2012;201:119–67.
14. Kondziolka D, Wechsler L, Goldstein S, et al. Transplantation of cultured human neuronal cells for patients with stroke. Neurology. 2000;55(4):565–9.

15. Kondziolka D, Steinberg GK, Wechsler L, et al. Neurotransplantation for patients with subcortical motor stroke: a phase 2 randomized trial. J Neurosurg. 2005;103(1):38–45.
16. Kleppner SR, Robinson KA, Trojanowski JQ, Lee VM. Transplanted human neurons derived from a teratocarcinoma cell line (NTera-2) mature, integrate, and survive for over 1 year in the nude mouse brain. J Comp Neurol. 1995;357(4):618–32.
17. Nelson PT, Kondziolka D, Wechsler L, et al. Clonal human (hNT) neuron grafts for stroke therapy: neuropathology in a patient 27 months after implantation. Am J Pathol. 2002;160(4):1201–6.
18. Taupin P. HuCNS-SC (StemCells). Curr Opin Mol Ther. 2006;8(2):156–63.
19. Pollock K, Stroemer P, Patel S, et al. A conditionally immortal clonal stem cell line from human cortical neuroepithelium for the treatment of ischemic stroke. Exp Neurol. 2006;199(1):143–55.
20. Littlewood TD, Hancock DC, Danielian PS, Parker MG, Evan GI. A modified oestrogen receptor ligand-binding domain as an improved switch for the regulation of heterologous proteins. Nucleic Acids Res. 1995;23(10):1686–90.
21. Hodges H, Pollock K, Stroemer P, et al. Making stem cell lines suitable for transplantation. Cell Transpl. 2007;16(2):101–15.
22. Thomas RJ, Hope AD, Hourd P, et al. Automated, serum-free production of CTX0E03: a therapeutic clinical grade human neural stem cell line. Biotechnol Lett. 2009;31(8):1167–72.
23. Dirnagl U. Bench to bedside: the quest for quality in experimental stroke research. J Cereb Blood Flow Metab. 2006;26(12):1465–78.
24. Macleod MR, O'Collins T, Howells DW, Donnan GA. Pooling of animal experimental data reveals influence of study design and publication bias. Stroke. 2004;35(5):1203–8.
25. Sena E, van der Worp HB, Howells D, Macleod M. How can we improve the pre-clinical development of drugs for stroke? Trends Neurosci. 2007;30(9):433–9.
26. Markgraf CG, Green EJ, Watson B, et al. Recovery of sensorimotor function after distal middle cerebral artery photothrombotic occlusion in rats. Stroke. 1994;25(1):153–9.
27. Modo M, Stroemer RP, Tang E, Veizovic T, Sowniski P, Hodges H. Neurological sequelae and long-term behavioural assessment of rats with transient middle cerebral artery occlusion. J Neurosci Methods. 2000;104(1):99–109.
28. Schallert T. Behavioral tests for preclinical intervention assessment. Neuro Rx. 2006;3(4):497–504.
29. Stevanato L, Corteling RL, Stroemer P, et al. c-MycERTAM transgene silencing in a genetically modified human neural stem cell line implanted into MCAo rodent brain. BMC Neurosci. 2009;10:86.
30. Laing RJ, Jakubowski J, Laing RW. Middle cerebral artery occlusion without craniectomy in rats. Which method works best? Stroke. 1993;24(2):294–7 (discussion 7–8).
31. Virley D. Choice, methodology, and characterization of focal ischemic stroke models: the search for clinical relevance. Methods Mol Med. 2005;104:19–48.
32. Longa EZ, Weinstein PR, Carlson S, Cummins R. Reversible middle cerebral artery occlusion without craniectomy in rats. Stroke. 1989;20:84–91.
33. Smith EJ, Stroemer RP, Gorenkova N, et al. Implantation site and lesion topology determine efficacy of a human neural stem cell line in a rat model of chronic stroke. Stem Cells. 2012;30(4):785–96.
34. Hicks C, Stevanato L, Stroemer RP, Tang E, Richardson S, Sinden JD. In vivo and in vitro characterization of the angiogenic effect of CTX0E03 human neural stem cells. Cell Transplant. 2013;22(9):1541–52.
35. Hassani Z, O'Reilly J, Pearse Y, et al. Human neural progenitor cell engraftment increases neurogenesis and microglial recruitment in the brain of rats with stroke. PloS ONE. 2012;7(11):e50444.
36. Savitz SI, Dinsmore J, Wu J, Henderson GV, Stieg P, Caplan LR. Neurotransplantation of fetal porcine cells in patients with basal ganglia infarcts: a preliminary safety and feasibility study. Cerebrovasc Dis. 2005;20(2):101–7.

37. Suarez-Monteagudo C, Hernandez-Ramirez P, Alvarez-Gonzalez L, et al. Autologous bone marrow stem cell neurotransplantation in stroke patients. An open study. Restor Neurol Neurosci. 2009;27(3):151–61.
38. Muir KW, Sinden J, Miljan E, Dunn L. Intracranial delivery of stem cells. Transl Stroke Res. 2011;2(3):266–71.
39. Coley E, Farhadi R, Lewis S, Whittle IR. The incidence of seizures following deep brain stimulating electrode implantation for movement disorders, pain and psychiatric conditions. Br J Neurosurg. 2009;23(2):179–83.
40. Brott T, Adams HP, Olinger CP, et al. Measurements of acute cerebral infarction: a clinical examination scale. Stroke. 1989;20(7):864–70.
41. Lyden P, Brott T, Tilley B, et al. Improved reliability of the NIH stroke scale using video training. Stroke. 1994;25:2220–6.
42. Lyden PD, Lu M, Levine SR, Brott TG, Broderick J. A modified national institutes of health stroke scale for use in stroke clinical trials: preliminary reliability and validity. Stroke. 2001;32(6):1310–7.
43. Cohen PD, Isaacs T, Willocks P, et al. Sham neurosurgical procedures: the patients' perspective. Lancet Neurol. 2012;11(12):1022.
44. Jorgensen HS, Nakayama H, Raaschou HO, ViveLarsen J, Stoier M, Olsen TS. Outcome and time course of recovery in stroke. Part II: Time course of recovery. The Copenhagen Stroke Study. Arch Phys Med Rehabil. 1995;76:406–12.
45. Jorgensen HS, Nakayama H, Raaschou HO, ViveLarsen J, Stoier M, Olsen TS. Outcome and time course of recovery in stroke. Part I: Outcome. The Copenhagen stroke study. Arch Phys Med Rehabil. 1995;76:399–405.
46. Jorgensen HS, Reith J, Nakayama H, Kammersgaard LP, Raaschou HO, Olsen TS. What determines good recovery in patients with the most severe strokes? The Copenhagen Stroke Study. Stroke. 1999;30(10):2008–12.
47. Patel AT, Duncan PW, Lai SM, Studenski S. The relation between impairments and functional outcomes poststroke. Arch Phys Med Rehabil. 2000;81(10):1357–63.
48. Hankey GJ, Spiesser J, Hakimi Z, Bego G, Carita P, Gabriel S. Rate, degree, and predictors of recovery from disability following ischemic stroke. Neurology. 2007;68(19):1583–7.
49. Muir KW, Weir CJ, Murray GD, Povey C, Lees KR. Comparison of neurological scales and scoring systems for acute stroke prognosis. Stroke. 1996;27:1817–20.
50. Hankey GJ, Spiesser J, Hakimi Z, Carita P, Gabriel S. Time frame and predictors of recovery from disability following recurrent ischemic stroke. Neurology. 2007;68(3):202–5.
51. Aslanyan S, Weir CJ, Diener HC, Kaste M, Lees KR. Pneumonia and urinary tract infection after acute ischaemic stroke: a tertiary analysis of the GAIN International trial. Eur J Neurol. 2004;11(1):49–53.
52. Dirnagl U, Klehmet J, Braun JS, et al. Stroke-induced immunodepression: experimental evidence and clinical relevance. Stroke. 2007;38(2):770–3.
53. Kondziolka D, Steinberg GK, Cullen SB, McGrogan M. Evaluation of surgical techniques for neuronal cell transplantation used in patients with stroke. Cell Transpl. 2004;13(7–8):749–54.
54. Mahoney FI, Barthel DW. Functional evaluation: the Barthel index. Maryland State Med J. 1965;14:61–5.
55. van der Putten JJ Hobart JC Freeman JA Thompson AJ. Measuring change in disability after inpatient rehabilitation: comparison of the responsiveness of the barthel index and the functional independence measure. J Neurol Neurosurg Psychiatry. 1999;66(4):480–4.
56. Rankin J. Cerebral vascular accidents in patients over the age of 60. 2: Prognosis. Scottish Med J. 1957;2:200–15.
57. van Swieten JC Koudstaal PJ Visser MC Schouten HJ van Gijn J. Interobserver agreement for the assessment of handicap in stroke patients. Stroke. 1988;19:604–7.
58. Bohannon RW, Smith MB. Interrater reliability of a modified Ashworth scale of muscle spasticity. Phys Ther. 1987;67(2):206–7.

59. Wilson B, Cockburn J, Halligan P. Development of a behavioral test of visuospatial neglect. Arch Phys Med Rehabil. 1987;68(2):98–102.
60. Diederich NJ, Goetz CG. The placebo treatments in neurosciences: new insights from clinical and neuroimaging studies. Neurology. 2008;71(9):677–84.
61. Williams A, Gill S, Varma T, et al. Deep brain stimulation plus best medical therapy versus best medical therapy alone for advanced Parkinson's disease (PD SURG trial): a randomised, open-label trial. Lancet Neurol. 2010;9(6):581–91.

Chapter 9
Induced Pluripotent Stem Cells as a Cell-Based Therapeutic in Stroke

David C. Hess, Nasir Fakhri and Franklin D. West

The discovery of "induced pluripotent stem cells" (iPSC) represents a major scientific breakthrough. Yamanaka first reported in 2006 that mouse fibroblasts could be reprogrammed into embryonic-like stem cells with four virally carried genes Oct3/4, Sox2, c-Myc, and Klf4, and termed these cells "induced pluripotent stem cells" [1]. In 2007, he reported that human fibroblasts could be similarly reprogrammed into human iPSC with the same four factors [2]. The factors became known as "Yamanaka factors" (Fig. 9.1). James Thomson reported similar reprogramming of fibroblasts into iPSC, using four factors, two different from the Yamanaka factors Oct4, Lin28, Nanog, and SOX2 [3]. Reprogramming of amphibian cells had been accomplished decades earlier. In the early 1960s, John Gurdon showed that nuclei from frog somatic cells when introduced into enucleated oocytes could be reprogrammed and give rise to tadpoles [4, 5]. Gurdon and Yamanaka shared the Nobel Prize in Physiology or Medicine in 2012 for their discoveries of reprogramming of somatic cells.

The iPSC and somatic cell reprogramming field continues to advance rapidly. Further work has established that fewer reprogramming factors are needed for cells more undifferentiated than fibroblasts. iPSC can be reprogrammed from neural stem cells (NSC) with two factors, Oct4 and Klf4, or even one factor, Oct4 [6]. Fibroblasts can also be directly reprogrammed into NSC, termed induced neural stem cells (iNSC), without going through the iPSC stage [7]. Central to reprogramming, the master pluripotency gene, *Oct4,* appears to serve as "the gatekeeper into a reprogramming expressway" [8].

D. C. Hess (✉) · N. Fakhri
Department of Neurology, Medical College of Georgia, Georgia Regent's University, 30912 Augusta, Georgia
e-mail: dhess@gru.edu

F. D. West
Regenerative Bioscience Center, University of Georgia, Athens, Georgia

Department of Animal and Dairy Science, University of Georgia, Athens, Georgia

D. C. Hess (ed.), *Cell Therapy for Brain Injury,* DOI 10.1007/978-3-319-15063-5_9

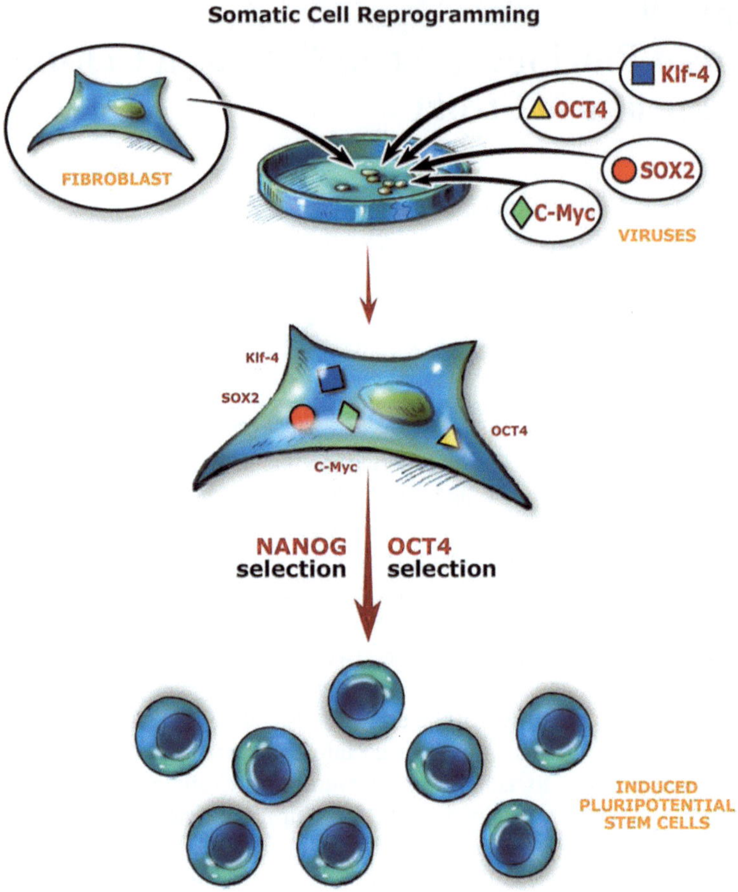

Fig. 9.1 Reprogramming fibroblasts

The major advantages of iPSC cells are their similar developmental plasticity to embryonic stem cells (ESC), yet iPSC lack the ethical concerns of ESC as they do not require destruction of a human embryo. iPSC can be "patient specific" and can be derived from patients with specific diseases. This allows them to have applications in modeling "diseases in a dish" and screening potential disease treatments. For example, iPSC can be isolated from the skin of patients with degenerative neurological diseases such as amyotrophic lateral sclerosis (ALS) or Parkinson's disease and then differentiated into the cell type of interest (e.g., motor or dopaminergic neurons) [9, 10]. This allows the human disease process to be studied on a developmental level, enabling the elucidation of the cellular and molecular mechanisms of the disease. iPSC also are of interest to pharmaceutical companies for testing of drugs in differentiated human cells. Often, drugs are screened on immortalized cell lines, cells that are not the exact target, or of a different species. All of these are confounding variables that limit the predictability of drug screens.

In addition, many cell types can only be obtained in small quantities such as mature neurons that no longer divide. Therefore, it makes it challenging to have sufficient cell numbers to perform large-scale screens. iPSC-derived cells overcome all of these limitations. Human iPSC are immortal and can be rapidly expanded to large cell numbers and differentiated into specific cell types of interest [11, 12]. These cells can then be utilized in high-throughput, high-content screens that provide the rapid and compressive data sets needed to determine safety and efficacy of potential drugs.

What has generated the most excitement for regenerative medicine is the potential of iPSC as a form of cell therapy and "cell replacement" for a variety of disabling diseases. Cell therapy with iPSC has been shown to be effective in preclinical murine models of sickle cell anemia [13], hemophilia A [14], and Parkinson's disease [15] holding out hope for effective treatments for these conditions. In this chapter, we focus on the use iPSC in cell therapy and transplantation after stroke.

Cell transplantation in stroke presents a challenge for iPSC therapy as multiple cell types need to be repaired and regenerated after ischemia—endothelial cells, pericytes, astrocytes, oligodendrocytes, and neurons. Therefore, unlike Parkinson's disease where the focus is on transplantation of dopaminergic cells into the nigrostriatum, stroke will require either undifferentiated iPSC or multiple more differentiated cells such as vascular and neuronal progenitors. On the other hand, one of the advantages of transplantation into stroke is that the host tissue is not undergoing ongoing degeneration. We know from the history of cell transplantation in Parkinson's disease, that the graft (fetal dopaminergic cells) undergoes degeneration with protein aggregation of alpha-synuclein deposits identical to Lewy bodies [16]. This would not be an issue with a "one-time" injury and "static" process such as stroke. In addition, development of iPSC therapies for stroke would not require genetic manipulation to repair mutated genes, such as the case of Huntington's disease, which would be technically challenging and make Food and Drug Administration (FDA) approval for therapy difficult. Stroke is a promising target for iPSC therapy, being a condition with limited treatment options, and the potential to treat a large number of patients.

iPSC Therapy in Preclinical Stroke Models

We are in the early stages of evaluating iPSC therapy in stroke and relatively few studies have been done (Table 9.1). Early efforts at iPSC transplantation for stroke involved direct injection of undifferentiated iPSC into the region of the infarct. Transplantation of mouse undifferentiated iPSC into the lesioned striatum and cortex 24 h after transient (30 min) middle cerebral artery occlusion (MCAo) in mice treated with cyclosporine resulted in teratoma formation by 28 days and worse functional outcome [17, 18]. The tumors were much larger in the post-ischemic brain than in the unlesioned cortex in sham-operated mice. Conversely,

Table 9.1 Reprogramming factors

Fibroblasts to iPSC	
	Oct4, SOX2, c-myc, Klf4 "Yamanaka factors" [2]
	Oct4, SOX2 Nanog, LIN28 [3]
NSC to iPSC	
	Oct4; Oct4, Klf4 [6]
Fibroblasts directly to NSC (iNSC)	
	Brn4/Pou3f4, Sox2, Klf4, c-Myc, plus E47/Tcf3 [7]

iPSC induced pluripotent stem cells, *NSC* neural stem cells

Jiang et al. transplanted human undifferentiated iPSC into the area of the striatum after MCA stroke in female rats [19]. Seven days after stroke, they treated animals with iPSC, fibroblast (cell control), or phosphate-buffered saline (PBS) control. They found iPSC treatment resulted in an improvement in functional outcome as early as 4 days post transplant and continuing through 16 days post transplant. iPSC transplantation also reduced infarct size. The iPSC had migrated from the injection site along the corpus callosum toward the infarct, and they showed differentiation into "neural cells." However, they only showed staining for the astrocyte marker glial fibrillary acidic protein (GFAP) and the progenitor marker nestin, which is not specific for neurons. Surprisingly, they found no tumors or teratomas in their animals, which were all immunosuppressed with corticosteroids. Similarly, Chen et al. transplanted iPSC with fibrin glue (FG) cells into the subdural space of rats in the area near the infarct after MCAo [20]. The iPSC-FG cells reduced pro-inflammatory and increased anti-inflammatory cytokines in the brain, reduced infarct size, and improved functional outcome. There was no evidence of teratoma formation. However, when iPSC without FG were transplanted into the brain before the cerebral infarct, they improved functional outcome but all animals at 1 month showed evidence of teratoma formation. The subdural delivery method avoided potential injury of the brain parenchyma during injection of the cells and the subdural transplantation method with iPSC-FG reduced teratoma formation. It is unclear what factors limited the potential of iPSCs to form tumors in these studies, yet it would be of significant interest as future studies of iPSC-derived therapies progress. There is precedent for limited tumor formation upon transplantation of iPSCs into ischemic tissue. In the heart, mouse iPSC engrafted and improved functional outcome (echocardiography) after myocardial infarction in the immunocompetent host without forming teratomas. However, teratomas did form when these cells were transplanted in immunodeficient hosts, and this was associated with worsening cardiac function [21]. Immunocompetence in the host recipients ensured controlled iPSC engraftment with tissue integration without perturbing electrical homeostasis and without teratoma formation.

Clearly, transplantation of undifferentiated iPSC is highly unlikely to be a viable option as a cell therapy due to the high levels of tumorigenicity associated with iPSC and the significant risk of tumor formation in patients. On the other hand, terminally differentiated neurons do not engraft and survive after transplantation so that an "intermediate" stage of differentiation is required. Therefore, most transplantation strategies of iPSC-derived cells in stroke have employed cells developmentally restricted to the neural pathway, alternatively termed "neural progenitor cells" (NPC), "neural stem cells" (NSC) or "neuroepithelial-like cells" (NES). These cells give rise to neurons and glial cells.

Treatment of Stroke with iPSC-Derived Neural Cells Leads to Improved Recovery

Chang et al. were among the first to demonstrate that transplantation of iPSC–NPC after ischemic stroke can promote functional recovery. They transplanted human iPSC–NPC into the contralateral striatum 7 days after an MCAo in a rat suture occlusion model. They tracked these cells with MRI and found they migrated toward the peri-infarct area on the contralateral side [22]. The migrating cells expressed C-X-C chemokine receptor type 4 (CXCR4), a ligand for stromal cell-derived factor 1 (SDF-1) and this SDF-1 CXCR4 axis is known to play a key role in stem cell migration. The iPSC-NPC differentiated into multiple types of mature neurons, astrocytes, and oligodendrocytes and integrated into damaged tissues. In addition, cell transplantation led to reduced inflammation and gliosis and improved functional outcomes on a battery of tests. Transplantation of iPSC-NPC also led to increased proliferation and migration of endogenous NSCs derived from the subventricular zone and led to an overall increase in neurogenesis in the ischemic brain. No evidence of tumor formation was found.

There is a publication bias for "positive" preclinical studies in experimental stroke and for other disease states, while negative studies are underreported in the literature [23, 24]. A "negative study" with iPSC–NSC was reported by Jensen et al. [25]. They transplanted 250,000 iPSC–NSC or PBS (group size of ten) into rats 7 days post MCAo. Grafting was observed in eight of ten animals. The cells displayed a neuronal phenotype and unbiased stereology of the tissue showed double the number of injected cells with many of the cells being positive for the proliferation marker Ki67. Although there was no tumor formation, there was graft overgrowth in some of the animals. They found no functional improvement out to 20 days using a battery of tests that included the elevated body swing test, adhesive removal, and the cylinder test and found no reduction in infarct size. These negative results may be related to the differentiation stage of NSC transplanted. Given the high proliferation of the NSC in vivo, the NSC may have been "too undifferentiated". Previous studies with NSC derived from ESC showed that graft overgrowth in the host was related to the in vitro differentiation state and proliferative rate in the transplanted cells and not the ischemic environment of the host tissues [26].

Oki et al. and the Lund group reported their work with iPSC differentiated in vitro into long-term neuroepithelial-like stem cells (lt-NES) [27]. They transplanted the lt-NES cells into C57 BL mice striatum 1 week after a 30 min MCAo. The mice were treated with cyclosporine to reduce graft rejection. At 10 weeks, 7 of 12 mice showed engraftment of the cells with only 10% of cells surviving engraftment.

Within 1 week after transplantation, mice receiving the lt-NES performed better on the staircase test than mice receiving vehicle injection but they found no difference in the corridor test. They also found no association with long-term engraftment of cells and performance on this test. This lack of association and the early improvement at 1 week made it unlikely that the improved functional performance was related to neuronal cell replacement and more likely due to the production of regenerative or neuroprotective factors. This was supported by the observed increase in vascular endothelial growth factor (VEGF) a known stimulator of angiogenesis, neurogenesis, neuroprotection, and glial growth [28] near the lt-NES transplant. However, they were not able to detect any increased blood vessel growth indicating the VEGF effect was not via increased angiogenesis, but through an alternative mechanism. They also demonstrated through flurogold studies that lt-NES-derived neuronal cells were capable of producing long axonal extensions from the striatum into the globus pallidus suggesting that these cells are capable of complex integrative activity.

Oki et al. also transplanted NES into the striatum of nude rats 48 h after 30 min MCAo and found that about 50 % of the engrafted cells survived at 4 months although not every rat showed engraftment. The majority of the cells (72 %) at 4 months expressed neuronal nuclei (NeuN), indicating a mature neuronal phenotype, while only a minority (6 %) were positive for the astrocyte marker GFAP. They also transplanted lt-NES into the cortex of 10 nude rats 48 h after they underwent distal MCAo. Seven survived and two were sacrificed at 2 months and five at 4 months. Graft survival was about 80 % at 2 months and 60 % at 4 months. The majority of cells (77 %) expressed NeuN at 4 months but about 5 % expressed dopamine- and cAMP-regulated neuronal phosphoprotein (DARPP)-32, a marker of striatal neurons, and similar to the percentage found after striatal transplantation. This finding that the site where the cells were transplanted (e.g., cortex vs. striatum) did not influence their final fate led the authors to conclude that the in vitro differentiation of cells was a more important determinant of fate than the in vivo environment in which they were transplanted. Perhaps the most significant finding of this study was that electrophysiological recordings of transplanted lt-NES in brain slices showed functional neuronal activity. In two of the ten engrafted cells, excitatory α-amino-3-hydroxy-5-methyl-4-isoxazolepropionic acid (AMPA) receptor-mediated currents could be evoked by stimulating a cortical region remote from the transplant. These data suggest that iPSC-derived NES were capable of successfully integrating into the brain synaptic circuitry.

While lt-NES improved functional outcome and showed engrafted and evidence of integration into functional circuitry, there remained a question of whether more differentiated neurons such as progenitors differentiated into cortical neurons in vitro might be a more optimal cell replacement. To test this hypothesis, a distal MCAo in nude rats was used to produce a mostly cortical infarct. Forty-eight hours after stroke, either "fated cells" differentiated into cortical neurons in vitro or "nonfated" lt-NES were transplanted into the cortex. The fated cells had a lower proliferation rate but more efficient conversion into neurons with greater pyramidal cell morphology. Fated cells also possessed a layered cortical structure and survived long term (2 months). Both the fated and nonfated neurons improved bilateral function in the stepping test compared to vehicle-treated controls. Electrophysiological recording in brain slices at 20–25 weeks post transplantation showed that both fated and nonfated cells showed functional properties of mature neurons and were able to fire action potentials induced by either depolarizing current injection or spontaneously. Moreover, following electrical stimulation of the intact cortex adjacent to the transplant of fated cells, a monosynaptic-evoked response was recorded in the fated cells suggesting functional synapses.

A major concern of iPSC is that the reprogramming genes used to generate these cells are often integrated into the genome and many of these genes are tumorigenic. To overcome this limitation, Mohammad et al. used vector-free and transgene-free human iPSC and iPSC–NPC differentiated from these iPSC [29]. These cells were transplanted into mice 7 days after they underwent distal MCAo with bilateral transient common carotid artery (CCA) occlusion resulting in a barrel cortex infarction. iPSC–NPC improved recovery of sensorimotor function as measured by the adhesive removal test and led to increased brain-derived neurotrophic factor (BDNF; a factor known to have neuroprotective effects in stroke and stimulates survival, growth, and synaptogenesis) levels. Intrinsic optical signals (IOS) evoked by whisker stimulation at the barrel cortex were assessed 30 days after transplantation and indicated that transplanted iPSC–NPC restored neurovascular coupling. No tumor formation was found up to 12 months. Liu et al. showed similar results with transplanted neural precursor cells (NPreC) derived from a "novel mouse iPSC" line reprogrammed from mouse fibroblasts using a virus and oncogene-free method with two factors under conditions of hypoxia [30]. These iPSC–NPreCs after transplantation into a mouse stroke model differentiated into neurons and astrocytes and improved functional outcome. These results demonstrate that safe and efficacious cells can be derived utilizing a nonviral and nonintegrating iPSC reprogramming approach.

Since stroke generally occurs in older patients, it is important to test cell therapies in aged animals. Tatarishvili et al. from the Lund group transplanted iPSC–NES in aged rats (24 months) 48 h after distal MCA occlusion [31]. They found that 49% of the injected cells engrafted at 8 weeks and differentiated into mature gamma γ-aminobutyric acid (GABAergic) neurons. The iPSC–NES improved functional outcome as measured by the cylinder test at 4 and 7 weeks and reduced microglial activation.

Timing of Transplantation

One of the important issues in transplant of iPS–NP is the optimal time to transplant in relation to the stroke event. Previous work with NSC (derived from fetal cells) showed that transplantation in the striatum after MCA stroke at 48 h was superior to transplantation at 6 weeks in terms of graft survival [32]. This was thought to be related to the more established and vigorous immune response with microglial activation at 6 weeks. While the 48-h time point was more effective, this is not a "clinically practical" time to transplant in humans. Logistically, it would be difficult to prepare the patient or the cells in such a brief period. Studies of iPSC with transplantation times of 7 days have shown engraftment and improvement in functional outcome indicating that this time point of transplantation is effective [27, 29]. Even 7 days would be logistically challenging for the production of therapeutic cells. Even the most rapid protocols for iPSC production require weeks [33]. Then iPSC must be expanded and differentiated, which again requires weeks before cells reach an NSC fate. Ultimately, these cells must still undergo rigorous testing of plasticity, homogeneity, and karyotype, under good manufacturing practice (GMP) conditions to ensure a safe product.

Cell Dose

There is paucity of data on optimized cell dose, and there are no dose response studies of iPSC transplantation in stroke. Most of the doses of iPSC–NSC have been extrapolated from dose response studies done with NSC derived from fetal cells in the rat. Darsila et al. transplanted 300,000 ($2 \times 150{,}000$), 750,000, and 1.5 million NSCs after stroke in the rat and found that doses above 300,000 were not associated with any increased graft survival suggesting that 300,000 cells were the maximum number in the rat [32]. As can be seen in Table 9.2, the dose range for transplantation in rodents ranges from 2×10^5 to 1×10^6. There are no dose response studies in large animals, and it is not clear if the number of cells in the rat can be multiplied by the ratio of human to rat brain size to find an appropriate dose in man.

Scaffolds/Matrices

The use of biodegradable scaffolds to serve as "biobridges" helps support the survival and function of the graft after intracerebral transplantation of NSC [34]. Ji et al. used Matrigel scaffolding to transplant NPC at a "late" time point (3 weeks) after distal MCAo–NPC grown in and transplanted with Matrigel showed improved engraftment, survival, and function of the NSC compared to NSC transplanted without Matrigel. The NSC with the Matrigel also reduced the size of the infarct cavity [35]. However, Matrigel is derived from Engelbreth–Holm–Swarm (EHS) mouse sarcoma cells, which will likely limit the ability of this matrix to be approved by the FDA. However, there are a host of other matrices including natural and synthetic systems that can be used to deliver cells and provide structural support, such as hydrogels, or protection from the host immune system through encapsulation [36, 37]. These matrices often possess unique properties that allow them to respond to changes in the environment due to temperature, pH, and other physiologically relevant parameters [38]. Matrices can be used to also transport and release other

Table 9.2 Preclinical studies of iPSC in rodent stroke models

Study	Model	Species/gender	Cell type and dose (immunosuppression)	Time of transplant	Engraftment	Functional outcome	Other
Kawai et al. 2010 [17]	tMCAo—30 min	C57mice/male	Mouse iPSC; 5×10^5	24 h post	Teratomas	Worse at 28 days	
Jiang et al. 2011 [19]	tMCAo—70 min	SD rats/females	Human iPSC; 8×10^5 (4 injections ipsilateral and contralateral; medrone)	7 days post	GFAP and nestin-positive cells that migrated	Improved 16 days post transplant	
Chen et al. 2010 [20]	tMCAo—1 h	Evans Long rats 8 weeks/male	Mouse iPSC; 1×10^6 with/without fibrin glue (FG)	Pre with direct iPS or iPS, iPS–FG 1 h post subdural	Teratomas with direct injection; no teratomas with iPS–FG	Reduction of infarct size with both direct iPS pre-stroke and post stroke with subdural iPS–FG; Improved function with pre-stroke direct injection at 1 week and at 4 weeks with iPS–FG	iPS–GF decreased pro-inflammatory and increased anti-inflammatory cytokines
Chang et al. 2013 [22]	tMCAo—90 min	SD rats/males	Human iPS–NPC; 2×10^5 cyclosporine	7 days post; contralateral striatum to infarct	NeuN and MAP2 expressing cells and some with spiny projections	Improvement in battery of functional tests beginning at 1 week and at 8 weeks	MRI tracking showed migration to contralateral infarct along corpus callosum
Jensen et al. 2013 [25]	tMCAo—30 min	Wistar rats/males	Human iPS-NSC; 2.5×10^5 total in five injections; Cyclosporine	7 days post	8/10 showed engraftment; Neuronal differentiation with beta III tubulin and MAPs in all, no teratomas	No improvement in functional outcome out to 5 weeks	Graft overgrowth with twice number of cells as injected

Table 9.2 (continued)

Study	Model	Species/gender	Cell type and dose (immunosuppression)	Time of transplant	Engraftment	Functional outcome	Other
Oki et al. 2012 [27]	MCAo—30 min	C57 BL mice; males	Human iPS–lt-NES 3×10^5; Cyclosporine	7 days in the striatum	At 10 weeks, 7/12 mice engrafted; 10% engraftment in those 7 animals; most differentiated to neurons	At 8 weeks, improvement in staircase test but not corridor test; no correlation between functional outcome and engraftment	Increased VEGF in brain with iPS–lt-NES but no increase in microvessels
Oki et al. 2012 [27]	Distal MCA and bilateral CCA occlusion—30 min; and tMCAo	Nude rats/males	Human iPS–lt-NES; No immunosuppressive drugs	48 h in striatum or cortex	Long term (4 mos) engraftment		Brain slices showed electrophysiological evidence of neurons and functional integration
Tornero et al. 2013 [57]	Distal MCA or tMCAo—30 min	SG or nude rats/male	Human iPS–lt-NES (nonfated) or cortically (fated); 3×10^5 total divided in two injection sites; Cyclosporine in SG only	48 h in cortex	Fated cells higher differentiation into cortical neurons; survival at 5 months and no teratomas	Improvement in stepping test at 8 weeks with both nonfated and fated cells	Fated cells showed electrophysiological characteristics of functional neurons at 5 mos
Mohammad et al. 2013 [29]	MCA occlusion, bilateral 7 min CCA occlusion. “barrel cortex” model	C57 mice 8–10 weeks;	Human iPS (vector, transgene-free)–NPC; 4×10^5 at 2 injection sites (core and penumbra)	7 days	Differentiated into neurons by 28 days No teratomas out to 12 months	Improvement in adhesive removal test out to 28 days	Improved neurovascular coupling in barrel cortex; increased BDNF in brain

Table 9.2 (continued)

Study	Model	Species/gender	Cell type and dose (immunosuppression)	Time of transplant	Engraftment	Functional outcome	Other
Yuan et al. 2013 [58]	tMCAo—2 h	Sprague Dawley rats/male	Human iPS–NSC 1×10^6; Cyclosporine	Immediately upon reperfusion; intrastriatal	Differentiated into B tubulin expressing neurons and astrocytes No teratomas	Improvement at 3 weeks in beam walking, Morris water maze, rope grabbing	
Tatarishvili et al. 2014 [31]	Distal MCAo	Rats (aged 24 mos)	Human iPs–lt-NES	48 h Intracortical	49 % cell engraftment at 8 weeks and markers of mature GABAric neuron	Improved performance in cylinder test at 4 and 7 weeks	Reduced microglial activation
Liu et al. 2014 [30]	Rt MCA and CCA occlusion for 120 min	Mouse	Mouse (viral and oncogene-free, grown under hypoxia) iPS-derived NPC; 1×10^6 in 3 injections	At time of MCAo	Differentiated into neurons and glia	Improved functional outcome at 14 and 21 days by beam walking, rotarod, activity	

MCAo middle cerebral artery occlusion, *tMCA* temporary middle cerebral artery, *CCA* common carotid artery, *NSC* neural stem cells, *NPC* Neural progenitor cells, *lt-NES* long-term neuroepithelial stem cells, *iPSC* induced pluripotent stem cells, *iPS* induced pluripotent stem, *GFAP* glial fibrillary acidic protein, *SD* Sprague Dawley, *MRI* magnetic resonance imaging, *BDNF* brain-derived neurotrophic factor bioactive compounds to help in tissue regeneration as well as transport cells. Biodegradable scaffolds may be important in late transplantation times after stroke (e.g., 3 months) when a cystic cavity has begun to form.

Preclinical Experimental Stroke Models

With the many failed clinical trials in stroke and other neurological diseases, there has been a movement and major "push" to improve the "quality" of preclinical studies and to apply "clinical trial standards." Important issues are randomization, blinding of the observers and investigators, an accounting of all the animals randomized, and sample size estimation [39]. Many preclinical studies that have formed the foundation for clinical trials in the past have fallen short of these recommendations. Moreover, most preclinical studies are done in young healthy adult male rodents, and there is a need to test new agents in females and animals with comorbidities such as age, hypertension, and diabetes [40]. While Stroke Therapy Academic Industry Roundtable criteria for preclinical evaluation were designed for acute neuroprotective agents, many of these criteria are relevant for stroke recovery and cell therapy. Table 9.3 shows the STAIR preclinical criteria and the present iPSC stroke studies. There has also been a concern that due to the lissencephalic architecture and the small proportion of white matter in the rodent brain, cell therapies should be tested in large animal models with brain size and architecture closer to humans. To date, all of the published preclinical studies of iPSC have been performed in rodents and most in young healthy adult males.

In order to both avoid xenograft issues and test in a large animal model that better models human stroke in terms of brain size and proportion of white matter, we are now proceeding to test porcine neural progenitors (iNPs) derived from porcine iPSC in a porcine model of stroke. This large animal work is a necessary step for translating a stroke therapy of iPS-derived neural progenitor cells to the bedside.

Nonviral Methods for iPS Cells

The original Yamanaka method relied upon insertion of transgenes by retroviral vectors into the host genome. This increased the risk of activation of proto-oncogenes in the host genome. Moreover, c-MYC, one of the inserted transgenes is "tumorigenic" [41]. These viral and genomic integrative methods of producing iPSC are associated with risk of tumors and will need to be avoided for future cell transplantation for regenerative medicine applications.

A number of methods have been used to avoid genomic integration. These can be divided into five approaches: (1) use of viruses that do not integrate such as adenovirus [42] or Sendai [43], (2) DNA in the form of episomal plasmids, PiggyBac transposons [44, 45], (3) RNA [46], (4) proteins [47, 48], and (5) small molecules

Table 9.3 iPSC transplantation in stroke and STAIR criteria for preclinical testing

STAIR Criterion	Description	Met?
Laboratory	Focal model tested in two or more laboratories	Met
Animal species	Focal model tested in two or more species	Only rodent
Health of animals	Focal model tested in old/diseased animals (diabetes, HTN)	Met with aged animals
Sex of animals	Focal model tested in males and females	One study in females
Reperfusion	Temporary and permanent models	Met
Time window	Administered at least 1 h after occlusion	Met
Dose response	At least 2 doses tested	Met
Route of delivery	Feasible model of delivery	Met
End point	Behavioral and histological outcomes measured	Met
Long-term effect	Outcome measured at 4 or more weeks	Met

HTN hypertension, *STAIR* Stroke Therapy Academic Industry Roundtable

[49] (see Table 9.4). However, the concern with all these methods is their lower efficiency compared to the efficiency of iPSC reprogramming with the retroviral method. For example, the initial plasmid approach used by Yamanaka's group was 1000 times less efficient than the retroviral method [41, 50]. This is compounded by the fact that generation of iPSC from human fibroblasts is ten times less efficient than from mouse fibroblasts. However, the nonviral reprogramming methods have improved in terms of efficiency, and many of these reprogramming methods are now commercial kits allowing them to be used by laboratories throughout the world, allowing a dissemination and "democratization" of iPSC technology. A systematic evaluation and comparison of these nonintegrative reprogramming methods in xeno-free conditions was conducted in both a fibroblast cell line and primary human fibroblasts [51]. In the fibroblast cell line, the messenger RNA (mRNA) method showed the highest reprogramming efficiency (1.89 % of starting cells) 20-fold higher than retroviral and plasmid methods. However, in reprogramming of primary human fibroblasts, the episomal plasmid method of Yamanaka [52] was

Table 9.4 Reprogramming and generation of iPSC without genomic integration

Nonintegrating viruses
Adenovirus [42]
Sendai [43]
DNA
Episomal plasmid [52]
PiggyBac transposon [45, 59]
RNA
Synthetic modified mRNA [46]
Proteins [47, 48]
Small Molecules
Seven small-molecule compounds [49]

mRNA messenger RNA

fourfold higher than the retroviral method and more than 50-fold higher than the mRNA method. There was no difference in the quality of the iPSC cells between the methods as measured using a reference map and "scorecard" that allows a "quick" characterization of iPSC by comparing the gene expression pattern of key pluripotency and germ lineage markers relative to a reference standard that consists of nine different human ES and iPSC lines [53]. This episomal plasmid method appeared to be the method most conformable to a GMP process and the generation of patient-specific iPSC line.

Autologous or Allogeneic iPSC?

One of the major concerns of present intracerebral cell transplantation approaches is that the transplanted cells are allografts, with concerns about limited long-term engraftment and the requirement for long-term immunosuppression. Long-term engraftment in the brain in human allografts may be limited by host rejection of the graft. One of the major advantages of iPSC-based therapy is that graft rejection can be avoided by using "personalized" autologous cells for transplant. Studies of syngeneic iPSC in mice differentiated into tissue-specific cells did not generate an immune response in the recipient suggesting that transplantation of cells differentiated from autologous iPSC will not elicit an immune response and will not require immunosuppression [54]. While an autologous therapy is not ideal for neurodegenerative diseases such as Parkinson's disease or ALS, where the cells may possess genetic defects, it is ideal for stroke where neurons of the individuals are not genetically defective but "bystanders" to an external vascular process. However, a "personalized" autologous approach has disadvantages. An autologous approach is difficult to "scale" and will likely be expensive. It is unlikely that a clinical "autologous" GMP process would be financially viable. The other major limitation would be the time needed to generate the iPSC and the iPSC-derived NPC from the individual stroke patient. This will likely take weeks to months even with advances in iPSC technology. If the optimal time window to transplant is within the first 2 weeks, this would not permit enough time to generate the iPS–NP. However, this personalized approach could be viable if preclinical studies indicated that iPSC–NP transplantation was effective "months" after the stroke. In this case, a time window of 1–6 months could be accommodated (see Table 9.5).

An alternative approach to the expense and complicated logistics of a "personalized" approach would be the establishment of an HLA haplotype-matched iPSC bank or library. A call for international collaboration to establish a global iPSC library has been sounded [55]. HLA matching would be analogous to organ and hematopoietic stem cell (HSC) HLA matching and banking. While iPSC transplantation would not involve contaminating T cells capable of "graft-versus-host" disease, allogeneic iPSC would likely generate an immune response, and therefore HLA matching is needed to reduce graft rejection [55]. It is estimated that 140 HLA homogeneous donors could produce iPSC lines that would match 90 % of the Japanese

Table 9.5 Advantages and disadvantages with autologous and allogeneic iPSC approaches

Autologous cells	Allogeneic cells
No rejection	Possibility of rejection
No need for immunosuppression	May require long-term immunosuppression
Require time to expand	"Off the shelf", ready to administer in shorter time
Difficult to scale	Scalable
Expensive	Less expensive after initial biobanking
Require local GMP facility	Need for HLA-matched library and biobanking

iPSC induced pluripotent stem cells, *HLA* human leukocyte antigen

population but this would require screening of 160,000 Japanese volunteers [52]. Gourraud et al. used a probabilistic model and showed that a bank of 100 iPSc cell lines with the most frequent HLA in each population would cover 78 % of European Americans, 63 % of Asians, 52 % of Hispanics, and 45 % of Americans [56]. However, this would require extensive screening of populations including over 25,000 European Americans and 100,000 African Americans. Turner et al. demonstrated that 150 HLA-matched homozygous donors could match 93 % of the UK population and might require minimal immunosuppression [55]. In this proposal, instead of extensive screening for donors, consultation and cooperation with HSC registries with donors who have already volunteered to donate HSC would be conducted, and donors would be approached for informed consent to donate their IPSC to the bank. This would require a skin biopsy or blood donation to isolate and generate the iPSCs. For regenerative medicine with allogeneic iPSC to move forward, a global, international effort will be required.

Summary

iPSC-based therapy has great potential in stroke and, unlike other cell therapy approaches, provides the opportunity to replace damaged cells. While "rebuilding" the damaged brain is a daunting task, there is evidence that iPSC can integrate as functional neurons and send axons toward targets. Moreover, iPSC–NPC are likely to work by "trophic" and perhaps even immunomodulatory effects on the brain. While other routes of delivery such as the intravenous have not been explored in stroke, direct stereotactic transplantation into the brain is presently the most attractive route. More work needs to be done on optimizing the types of iPSC-neural progenitors to transplant, the dose of cells, the timing, and targets of transplantation (penumbra or outside the area of infarct and allow migration to the infarct). There has been progress in viral-free and nongenomic integrative methods of reprogramming of cells. There will need to be major efforts at developing a clinical GMP product and, for reasons of scalability, allogeneic cells will be more feasible but this will require establishment of iPSC libraries and banks. However, with advances in iPSC technology such as direct fibroblast to iNP reprogramming and with the possibility of “later” transplantation times (out to 3 months) with biodegradable matrices, autologous transplantation may be a viable option in the future, allowing transplantation without the need for long-term immunosuppressive drugs.

References

1. Takahashi K, Yamanaka S. Induction of pluripotent stem cells from mouse embryonic and adult fibroblast cultures by defined factors. Cell. 2006;126(4):663–76.
2. Takahashi K, et al. Induction of pluripotent stem cells from adult human fibroblasts by defined factors. Cell. 2007;131(5):861–72.
3. Yu J, et al. Induced pluripotent stem cell lines derived from human somatic cells. Science. 2007;318(5858):1917–20.
4. Gurdon JB. The developmental capacity of nuclei taken from intestinal epithelium cells of feeding tadpoles. J Embryol Exp Morphol. 1962;10:622–40.
5. Gurdon JB. Adult frogs derived from the nuclei of single somatic cells. Dev Biol. 1962;4:256–73.
6. Kim JB, et al. Generation of induced pluripotent stem cells from neural stem cells. Nat Protoc. 2009;4(10):1464–70.
7. Han DW, et al. Direct reprogramming of fibroblasts into neural stem cells by defined factors. Cell Stem Cell. 2012;10(4):465–72.
8. Sterneckert J, Hoing S, Scholer HR. Concise review: Oct4 and more: the reprogramming expressway. Stem Cells. 2012;30(1):15–21.
9. Dimos JT, et al. Induced pluripotent stem cells generated from patients with ALS can be differentiated into motor neurons. Science. 2008;321(5893):1218–21.
10. Qiang L, Fujita R, Abeliovich A. Remodeling neurodegeneration: somatic cell reprogramming-based models of adult neurological disorders. Neuron. 2013;78(6):957–69.
11. Schondorf DC, et al. iPSC-derived neurons from GBA1-associated Parkinson’s disease patients show autophagic defects and impaired calcium homeostasis. Nat Commun. 2014;5:4028.
12. Juopperi TA, et al. Astrocytes generated from patient induced pluripotent stem cells recapitulate features of Huntington’s disease patient cells. Mol Brain. 2012;5:17.
13. Hanna J, et al. Treatment of sickle cell anemia mouse model with iPS cells generated from autologous skin. Science. 2007;318(5858):1920–3.
14. Xu D, et al. Phenotypic correction of murine hemophilia A using an iPS cell-based therapy. Proc Natl Acad Sci U S A. 2009;106(3):808–13.
15. Wernig M, et al. Neurons derived from reprogrammed fibroblasts functionally integrate into the fetal brain and improve symptoms of rats with Parkinson’s disease. Proc Natl Acad Sci U S A. 2008;105(15):5856–61.
16. Petit GH, Olsson TT, Brundin P. The future of cell therapies and brain repair: Parkinson’s disease leads the way. Neuropathol Appl Neurobiol. 2014;40(1):60–70.
17. Kawai H, et al. Tridermal tumorigenesis of induced pluripotent stem cells transplanted in ischemic brain. J Cereb Blood Flow Metab. 2010;30(8):1487–93.
18. Yamashita T, et al. Tumorigenic development of induced pluripotent stem cells in ischemic mouse brain. Cell Transplant. 2011;20(6):883–91.
19. Jiang M, et al. Induction of pluripotent stem cells transplantation therapy for ischemic stroke. Mol Cell Biochem. 2011;354(1–2):67–75.
20. Chen SJ, et al. Functional improvement of focal cerebral ischemia injury by subdural transplantation of induced pluripotent stem cells with fibrin glue. Stem Cells Dev. 2010;19(11):1757–67.
21. Nelson TJ, et al. Repair of acute myocardial infarction by human stemness factors induced pluripotent stem cells. Circulation. 2009;120(5):408–16.

22. Chang DJ, et al. Therapeutic potential of human induced pluripotent stem cells in experimental stroke. Cell Transplant. 2013;22(8):1427–40.
23. Dirnagl U. Bench to bedside: the quest for quality in experimental stroke research. J Cereb Blood Flow Metab. 2006;26(12):1465–78.
24. Bath PM, Macleod MR, Green AR. Emulating multicentre clinical stroke trials: a new paradigm for studying novel interventions in experimental models of stroke. Int J Stroke. 2009;4(6):471–9.
25. Jensen MB, et al. Survival and differentiation of transplanted neural stem cells derived from human induced pluripotent stem cells in a rat stroke model. J Stroke Cerebrovasc Dis. 2013;22(4):304–8.
26. Seminatore C, et al. The postischemic environment differentially impacts teratoma or tumor formation after transplantation of human embryonic stem-cell-derived neural progenitors. Stroke. 2010;41(1):153–9.
27. Oki K, et al. Human-induced pluripotent stem cells form functional neurons and improve recovery after grafting in stroke-damaged brain. Stem Cells. 2012;30(6):1120–33.
28. Rosenstein JM, Krum JM, Ruhrberg C. VEGF in the nervous system. Organogenesis. 2010;6(2):107–14.
29. Mohamad O, et al. Vector-free and transgene-free human iPS cells differentiate into functional neurons and enhance functional recovery after ischemic stroke in mice. PLoS ONE. 2013;8(5):e64160.
30. Liu SP, et al. Mouse-induced pluripotent stem cells generated under hypoxic conditions in the absence of viral infection and oncogenic factors and used for ischemic stroke therapy. Stem Cells Dev. 2014;23(4):421–33.
31. Tatarishvili J, et al. Human induced pluripotent stem cells improve recovery in stroke-injured aged rats. Restor Neurol Neurosci. 2014;32(4):547–58.
32. Darsalia V, et al. Cell number and timing of transplantation determine survival of human neural stem cell grafts in stroke-damaged rat brain. J Cereb Blood Flow Metab. 2011;31(1):235–42.
33. Durruthy-Durruthy J, et al. Rapid and efficient conversion of integration-free human induced pluripotent stem cells to GMP-grade culture conditions. PLoS ONE. 2014;9(4):e94231.
34. Park KI, Teng YD, Snyder EY. The injured brain interacts reciprocally with neural stem cells supported by scaffolds to reconstitute lost tissue. Nat Biotechnol. 2002;20(11):1111–7.
35. Jin K, et al. Transplantation of human neural precursor cells in Matrigel scaffolding improves outcome from focal cerebral ischemia after delayed postischemic treatment in rats. J Cereb Blood Flow Metab. 2010;30(3):534–44.
36. Zhong J, et al. Hydrogel matrix to support stem cell survival after brain transplantation in stroke. Neurorehabil Neural Repair. 2010;24(7):636–44.
37. Hoban DB, et al. The reduction in immunogenicity of neurotrophin overexpressing stem cells after intra-striatal transplantation by encapsulation in an in situ gelling collagen hydrogel. Biomaterials. 2013;34(37):9420–9.
38. Park JS, et al. Multi-lineage differentiation of hMSCs encapsulated in thermo-reversible hydrogel using a co-culture system with differentiated cells. Biomaterials. 2010;31(28):7275–87.
39. Landis SC, et al. A call for transparent reporting to optimize the predictive value of preclinical research. Nature. 2012;490(7419):187–91.
40. Fisher M, et al. Update of the stroke therapy academic industry roundtable preclinical recommendations. Stroke. 2009;40(6):2244–50.
41. Okita K, Yamanaka S. Induced pluripotent stem cells: opportunities and challenges. Philos Trans R Soc Lond B Biol Sci. 2011;366(1575):2198–207.
42. Zhou W, Freed CR. Adenoviral gene delivery can reprogram human fibroblasts to induced pluripotent stem cells. Stem Cells. 2009;27(11):2667–74.
43. Ban H, et al. Efficient generation of transgene-free human induced pluripotent stem cells (iPSCs) by temperature-sensitive Sendai virus vectors. Proc Natl Acad Sci U S A. 2011;108(34):14234–9.
44. Woltjen K, et al. Transgene-free production of pluripotent stem cells using piggyBac transposons. Methods Mol Biol. 2011;767:87–103.

45. Woltjen K, et al. piggyBac transposition reprograms fibroblasts to induced pluripotent stem cells. Nature. 2009;458(7239):766–70.
46. Warren L, et al. Highly efficient reprogramming to pluripotency and directed differentiation of human cells with synthetic modified mRNA. Cell Stem Cell. 2010;7(5):618–30.
47. Kim D, et al. Generation of human induced pluripotent stem cells by direct delivery of reprogramming proteins. Cell Stem Cell. 2009;4(6):472–6.
48. Zhou H, et al. Generation of induced pluripotent stem cells using recombinant proteins. Cell Stem Cell. 2009;4(5):381–4.
49. Hou P, et al. Pluripotent stem cells induced from mouse somatic cells by small-molecule compounds. Science. 2013;341(6146):651–4.
50. Okita K, et al. Generation of mouse induced pluripotent stem cells without viral vectors. Science. 2008;322(5903):949–53.
51. Goh PA, et al. A systematic evaluation of integration free reprogramming methods for deriving clinically relevant patient specific induced pluripotent stem (iPS) cells. PLoS ONE. 2013;8(11):e81622.
52. Okita K, et al. A more efficient method to generate integration-free human iPS cells. Nat Methods. 2011;8(5):409–12.
53. Bock C, et al. Reference Maps of human ES and iPS cell variation enable high-throughput characterization of pluripotent cell lines. Cell. 2011;144(3):439–52.
54. Guha P, et al. Lack of immune response to differentiated cells derived from syngeneic induced pluripotent stem cells. Cell Stem Cell. 2013;12(4):407–12.
55. Turner M, et al. Toward the development of a global induced pluripotent stem cell library. Cell Stem Cell. 2013;13(4):382–4.
56. Gourraud PA, et al. The role of human leukocyte antigen matching in the development of multiethnic "haplobank" of induced pluripotent stem cell lines. Stem Cells. 2012;30(2):180–6.
57. Tornero D, et al. Human induced pluripotent stem-cell-derived cortical neurons integrate in stroke-injured cortex and improve functional recovery. Brain. 2013;136(12):3561–77.
58. Yuan T, et al. Human induced pluripotent stem-cell-derived neural stem cells survive, migrate, differentiate, and improve neurologic function in a rat model of middle cerebral artery occlusion. Stem Cell Res Ther. 2013;4(3):73.
59. Kaji K, et al. Virus-free induction of pluripotency and subsequent excision of reprogramming factors. Nature. 2009;458(7239):771–5.

Chapter 10
Induced Pluripotent Stem-Cell-Derived Neural Cell Types in Treatment of Stroke

Vivian W. Lau, Simon R. Platt, Steven L. Stice and Franklin D. West

Introduction

Every year, approximately 800,000 individuals in the USA alone suffer a stroke, making stroke the leading cause of long-term disability and the fourth leading cause of death, and adding to the millions of stroke victims and families that care for them [1–4]. Despite considerable efforts to develop pharmacological treatments and devices, developed approaches are grossly inadequate. These treatments are predicated on limiting damage that occurs during an ischemic event, yet none of them enable large-scale tissue regeneration. The promise of stem cell therapies is the potential to replace ablated cells and damaged tissue, to form new functional neural networks that make appropriate connections and lead to the restoration of sensory, motor, and cognitive function in patients. The regeneration and replacement of lost tissue and improvements in functional deficits will enable the many stroke victims to return to a more productive lifestyle and relieve the family burden

V. W. Lau (✉) · S. R. Platt · S. L. Stice · F. D. West
Regenerative Bioscience Center, University of Georgia, 425 River Rd, 30602 Athens, GA, USA
e-mail: vwlau@uga.edu

V. W. Lau · S. R. Platt
Department of Small Animal Medicine and Surgery, University of Georgia, 501 D. Brooks Drive, 30602 Athens, GA, USA

S. L. Stice · F. D. West
Department of Animal and Dairy Science, University of Georgia, College of Veterinary Medicine, 501 D. W. Brooks Drive, 30602 Athens, GA, USA
e-mail: sstice@uga.edu

S. L. Stice
e-mail: srplatt@uga.edu

F. D. West
e-mail: westf@uga.edu

D. C. Hess (ed.), *Cell Therapy for Brain Injury,* DOI 10.1007/978-3-319-15063-5_10

of long-term care. Adult and embryonic stem cells (ESCs) and neural stem cells derived from ESCs have all been of keen interest to the stroke field. However, adult stem cells pose inherent difficulties, including isolation and expansion for some therapies, while ESCs have been mired in controversy since they were first isolated [5]. Some patients and practitioners may object to the use of ESCs and seek alternatives, despite publications demonstrating that viable embryos do not need to be used or destroyed in order to isolate ESCs [6].

In parallel, a new type of pluripotent stem cell has been generated—induced pluripotent stem cells (iPSCs). Although relatively new, iPSCs are believed to harbor all the same beneficial properties as ESCs, with both being pluripotent stem cells capable of forming any cell type in the body. It is a common belief that iPSCs will eventually be derived from the patient's own somatic cells so that immunological rejection associated with transplantation of any foreign cells or tissues may be averted. iPSCs are highly plastic and can be easily differentiated into neural stem cells (NSCs) that can be expanded to someday provide the volume of cells needed for therapeutic applications. Data in rodent stroke models have been very positive with induced pluripotent stem-cell-derived neural stem cell (iNSC) transplantations leading to functional recovery and decreased infarct sizes [7–10]. However, many challenges and questions remain before iNSC cell therapies can be deemed a safe and effective treatment in human patients.

iNSCs have the potential to transform the way researchers and physicians approach stroke treatments; transitioning from a paradigm of merely limiting further ischemic injury to one where lost tissue can be regenerated. For well over 50 years, tremendous effort has been committed to producing stroke therapies that limit the extent of injury through pharmaceutical and mechanical means with limited success. These approaches lead to recanalization of occluded vessels to restore blood flow to ischemic tissues or function as neuroprotectants that reduce cytotoxicity from inflammatory responses, damaging free radicals, or similar elements [11–14]. These efforts have had limited success with tissue plasminogen activator (tPA) being the only Food and Drug Administration (FDA)-approved pharmacological treatment in addition to a handful of FDA-approved clot-retrieval devices [14, 15]. These approaches are effective yet suffer from significant shortcomings. Only about 5 % of ischemic stroke patients receive tPA due to its restrictive 4.5 h window of use. The mechanical embolus removal in cerebral ischemia (MERCI) system (an FDA-approved clot-retrieval device) can be used in patients up to 8 h post stroke, but often fails to restore blood flow in ~50 % of occluded vessels [14, 15]. Neither of these clot-removal approaches can be utilized to treat patients that have suffered a hemorrhagic stroke, thereby excluding approximately 15 % of the stroke patient population [16]. A host of other neuroprotective treatments reducing secondary injury caused by inflammatory and immune responses have been developed yet have never made it beyond clinical trials (reviewed in [17, 18]). Even assuming that thrombolytic, neuroprotective, or similar approaches were 100 % effective, these treatments only prevent further damage, but have little regenerative capabilities. Therefore, the tissue damage caused by the initial ischemic event remains unchanged beyond normal healing. An assessment of the litany of failed treatments by the Stem Cell Emerging

Paradigm in Stroke Consortium meetings (STEPS I, II, and III), modeled on the stroke therapy academic industry roundtable (STAIR) model where leaders from academia, industry, and the FDA and National Institute of Neurological Disorders and Stroke (NINDS) participate, resulted in publications identifying several major factors needed to improve the development of stroke treatments. One of the major conclusions was the need for a regenerative cell therapy that will not only protect cells from ischemic injury but also replace lost and damaged tissues [19, 20]. This has resulted in a growing interest in potentially restorative treatments centered on stem cell therapies.

Recent studies have demonstrated that iNSCs may serve as an excellent regenerative therapy with a dual function: (1) acting as a cell-replacement therapy and as (2) a producer of regenerative paracrine factors (e.g., vascular endothelial growth factor, VEGF) that enhance endogenous tissue regeneration in rodent stroke models [7–10, 21–23] (Fig. 10.1). Transplanted cells migrate to the site of injury, differentiate, and functionally integrate forming new electrically active neural networks leading to improvement in neurological scores and motor function. These exciting and encouraging results have spurned considerable interest in the stroke community as a step forward in personalized regenerative medicine. In this chapter, we examine the development of iPSCs and derived NSCs, the current state of the art and areas of emphasis for improved translation to human medicine.

Development of Induced Pluripotent Stem Cell Technology

iPSCs are a recent discovery where mature somatic cells can be reprogrammed into pluripotent stem cells capable of differentiating into any cell type in the body through the over-expression of defined genes [24, 25]. The development of iPSC reprogramming technology resides at the convergence point of the fields of cellular reprogramming and ESCs where the conceptual framework to understand the genetic, epigenetic, and functional pluripotency networks were pioneered [26–32]. Based on prior knowledge, Yamanaka's research team hypothesized that "the factors that play important roles in the maintenance of ES cell identity also play pivotal roles in the induction of pluripotency in somatic cells" [25]. In the mouse, embryonic fibroblasts were retrovirally transduced with 24 pluripotency-associated genes resulting in the formation of nine colonies exhibiting ESC character with cells growing in colonies and displaying a rounded morphology, large nucleoli, and high nucleus-to-cytoplasm ratio. They went on to demonstrate that only four critical factors (Pou5f1 (also known as Oct3/4), Sox2, c-Myc and Klf4) were necessary to achieve complete reprogramming of embryonic and adult fibroblast cells. iPSCs demonstrated morphology, immunoreactivity, global gene expression, and epigenetic status indicative of a pluripotent state similar to ESCs. Functional tests of plasticity demonstrated that iPSCs were capable of forming embryoid bodies (EBs; Fig. 10.2a) in vitro and teratomas in vivo consisting of all three germ layers, ectoderm, endoderm, and mesoderm. iPSCs ultimately passed the most stringent of tests and were found capable

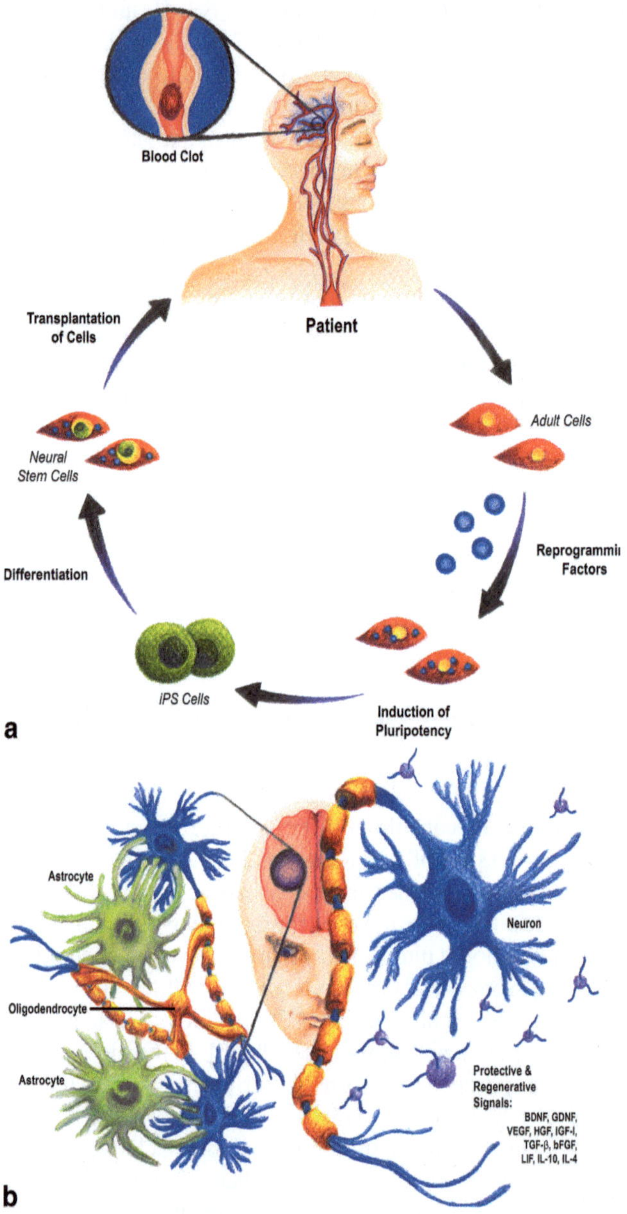

Fig. 10.1 iNSC functioning as a cell-replacement therapy and as a producer of regenerative therapeutics for stroke patients. **a** A patient who has an ischemic or hemorrhagic stroke (ischemic stroke shown) experiences significant brain tissue damage and loss. iNSCs could be generated from the patient's own body by collecting adult somatic cells and reprogramming these cells using pluripotency transcription factors into induced pluripotent stem cells (*iPSCs*). iPSCs could then be differentiated into iNSCs and transplanted back into the patient where they would differentiate into neurons and glia that functionally integrate into the site of injury. **b** Transplanted iNSCs and differentiated cells have been shown to produce and may generate other regenerative and protective signaling factors including vascular endothelial growth factor (*VEGF*) and interleukin-10 (*IL-10*). (Illustration by Leah K. Schultz)

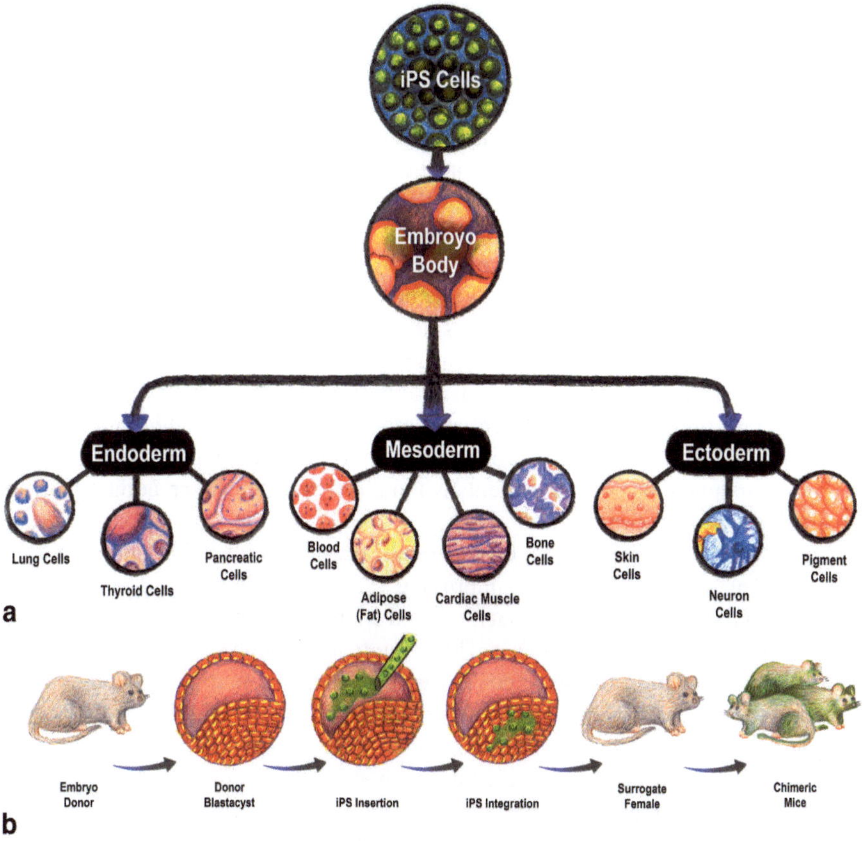

Fig. 10.2 Induced pluripotent stem cells (*iPSCs*) are capable of differentiating into any cell type in the body in vitro and in vivo. **a** To test iPSCs for their ability to form cell types of all three germ layers, a defining characteristic of iPSCs, EB differentiation is commonly used. iPSCs are induced to form large masses of cells reminiscent of developing embryos that induces spontaneous cell signaling that leads to the formation of endoderm, mesoderm, and ectoderm. Cell types representative of these lineages are commonly confirmed by immunocytochemistry of specific cell type marker expression. For example, cells can be immunostained for the neuron marker microtubule-associated protein 2 (*MAP2*) to identify cells of the ectoderm lineage. **b** To more stringently test the functional capacity of iPSCs, chimera formation is commonly performed. Embryos are collected from donor animals and are injected with iPSCs. iPSCs integrate and are transferred to a surrogate female. As the embryos develop, the integrated iPSCs are incorporated into tissues throughout the animal's body. Chimeric offspring are then composed of cells from the donor embryo and the inserted iPSCs. (Illustration by Leah K. Schultz)

of incorporating into all tissues of chimeric mice including the germline (Fig. 10.2b) and were successful in the tetraploid complementation pluripotency assay—a test where all cells of the embryo proper are derived solely from transplanted iPSCs [25, 33–35]. The significant value of this iPSC technology for basic mouse genetics was soon recognized. However, of perhaps even greater interest was their obvious potential in human medicine for cell-replacement therapy. In 2007, the Yamanaka lab was successful in deriving the first human iPSCs using the same reprogramming genes that were successful in the generation of mouse iPSCs, thus opening the door a bit wider for personalized medicine [24].

Intuitively, patients treated with their own iPSCs would be less immunogenic than those treated with allogeneic iPSCs (iPSCs derived from other patients) or ESCs; therefore, autologous iPSCs are thought to be similar to autologous human adult stem cell therapies used today in the clinic. However, there is still debate as to their immunogenicity. An early publication demonstrated that mouse iPSCs transplanted into a syngenic recipient animal, an animal that is genetically identical and transplant compatible to the mouse from which the iPSCs were derived, resulted in a T-cell-dependent immune response [36]. The researchers attributed this response to aberrant gene expression resulting from the reprogramming process. In contrast, recent studies showed little or no evidence of increased T cell proliferation or integration, antigen-specific secondary immune activity, or graft rejection in response to undifferentiated or differentiated mouse iPSCs transplanted into syngenic animals [37, 38]. These studies support the premise that autologous iPSCs may be safely transplanted into human patients without rejection; however, additional studies are needed to confirm these findings.

iPSC technology has made considerable advancements with alternative reprogramming strategies aimed at improving safety and efficiency. The initial Yamanaka lab reprogramming approach utilized spontaneous retrovirus integration of known oncogenes, including c-Myc. This random integration approach raised major concerns that insertion of genes could lead to insertional mutagenesis in addition to spontaneous reactivation of the c-Myc oncogene, which could potentially lead to tumor formation in human patients. However, recent advancements have led to novel nonintegrating approaches including minicircle DNA, modified mRNAs, and protein strategies to generate iPSCs without the need for permanent incorporation of reprogramming genes or the use of viral techniques [39–41] (Table 10.1). Efforts have also led to combinations of reprogramming factors that do not require the use of c-Myc [42]. These advances significantly improved many of the initial safety concerns that limited the potential of iPSC technology.

Differentiation of Induced Pluripotent Stem Cells into Therapeutic Cells for Stroke Treatment

Stroke results in the active recruitment of endogenous NSCs in the brain leading to proliferation and migration of NSCs from the subventricular zone to the ischemic region [43–46]. This natural regenerative cell response is insufficient, however, to restore most stroke patients to their normal pre-stroke function [45, 46]. iNSCs can act as a supplemental cell source to increase the number of NSCs and the regenerative capabilities of the stroked brain. It is preferential to differentiate iPSCs into iNSCs as the direct transplantation of undifferentiated cells is likely to lead to tumor formation. Previous studies by Kawai et al. and Chen et al. showed that transplantation of undifferentiated stem cells into middle cerebral artery occlusion (MCAO)

Table 10.1 Methods for reprogramming somatic cells to iPSCs

Vector type	Cell type	Advantage	Disadvantage	References
Integrating	Retroviral	Efficient, highly successful with numerous cells types	Genomic integration, incomplete proviral silencing and slow kinetics, formation of large numbers of partially reprogrammed colonies	[24, 25, 117, 118]
	Lentiviral	Efficient and transduces dividing and nondividing cells, highly successful with numerous cells types	Genomic integration and incomplete proviral silencing, formation of large numbers of partially reprogrammed colonies	[119–122]
Excisable	Transposon	Efficient and integrated regions can be removed	Screening of excised lines is labor intensive	[123]
	LoxP-flanked lentiviral	Efficient and integrated regions can be removed	Exogenous genes are removed, but loxP sites are retained in the genome	[124]
Nonintegrating	Adenoviral	No genomic integration	Low efficiency	[125]
	Plasmid	Occasional genomic integration	Low efficiency and occasional vector genomic integration	[126, 127]
DNA free	Protein	No genomic integration, direct delivery of transcription factors and no DNA-related complications	Low efficiency, short half-life, and requirement for large quantities of pure proteins	[128, 129]
	Modified mRNA	No genomic integration, faster reprogramming kinetics, controllable and high efficiency	Labor Intensive	[130]

DNA deoxyribonucleic acid, *mRNA* messenger ribonucleic acid stroke models led to the development of large tumors containing cells of the ectoderm, endoderm, and mesoderm lineages [47, 48]. The differentiation of iPSCs into iNSCs has been successfully achieved using a number of different protocols originally developed for hESCs [7–10]. Oki et al. utilized a previously developed ESC approach where iPSCs were detached and grown in suspension as EBs to enhance spontaneous differentiation [9, 49]. To better direct these cells down the neural lineage, EBs were then cultured in chemically defined neural medium composed of DMEM/F12, supplemented with insulin, transferrin, progesterone, putrescine, sodium selenite, and heparin in the presence of FGF-2. Plated EBs flattened and formed small, elongated cells that generated rosette structures resembling the early neural tube. In addition to the typical neural stem cell markers SOX2 and Nestin, neural rosette cells expressed the rosette-associated transcription factors DACH1 and PLZF with apical expression of ZO-1. Neural rosettes were isolated and ultimately lost the rosette morphology and further developed into NSCs. However, these iNSCs are capable of long-term expansion, while maintaining SOX2 and Nestin expression [9]. Yuan et al. used a similar EB approach, where EBs were formed and plated but were also exposed to retinoic acid (RA) leading to the formation of rosettes [10]. Upon removal of RA, neural rosette cells detached, continued to grow in suspension and formed neural spheres. These spheres were then plated on poly-ornithine and laminin-coated dishes in serum-free media with derived cells being a homogeneous population of NSCs. Other groups have used similar systems with variations including the addition of unique growth factors, inhibitors, supporting stromal cells (e.g., PA6) and changes in timing of differentiation steps [7, 8]. Despite the variability in protocols, iNSCs are SOX1 and Nestin positive and should be capable of differentiating into multiple lineages of neurons and glia.

Intuitively it may seem that iNSCs would be the best cell type for transplantation to regenerate lost and damaged tissue. However, the plasticity of iNSCs is such that they may differentiate into any neural cell type and may differentiate into cells that are regionally incorrect. Therefore, it is of potential value to generate iPSC-derived progenitors that are regionalized. A recent report described the derivation of telencephalic progenitors, which may be valuable for treating stroke regionalized to the forebrain [21]. iPSCs were differentiated using a serum-free EB approach. Telencephalic progenitors expressed the pallial telencephalic marker PAX6 and the telencephalic marker BF1 in addition to the neural stem cell markers SOX1 and Nestin. Tornero et al. recently generated cortical neuron progenitors for the treatment of stroke noting that "Clinical and imaging data showing the distribution of ischemic cell loss underlying the most severe symptoms in stroke patients indicate that cell replacement approaches should focus on the reconstruction of damaged cortex" [23]. To produce cortically fated cells, Tornero et al. differentiated iPSCs in the presence of Wnt3A, BMP4, and cyclopamine. These cortical progenitors expressed the cortex-specific neuronal marker TBR1 and cortex markers CTIP2 and CDP (markers associated with the deeper and superficial cortex layers respectively). hESCs and iPSCs have been found to be capable of differentiating into numerous specialized neural cell types making the potential cell type options and combinations for therapeutic use numerous. The ability to transplant multiple combinations of various neural cell types to match regional-specific areas of the brain is intriguing yet adds an additional layer of complexity that will take significant consideration.

Direct Reprogramming of Fibroblasts into NSCs

Two major limitations of transplanting iNSCs into stroke patients are (1) the potential of transplanting a contaminating iPSC subpopulation that spontaneously develops into a tumor and (2) the somewhat lengthy time period it takes to generate iNSCs. Typically, it can take months to generate and sufficiently characterize iNSCs with the need to first isolate and expand the somatic cells, then reprogram the cells into iPSCs, differentiate these cells into iNSCs and then perform the necessary quality control tests on these cells prior to transplantation (e.g., cellular phenotyping, functionality assessments, karyotype analysis). A recent breakthrough in reprogramming has led to the development of technologies where somatic cells can be directly reprogrammed into neurons and NSCs without a pluripotent stem cell intermediate [50–53]. Direct neural stem cell reprogramming has been accomplished with various combinations of reprogramming factors (Table 10.2). Ring et al. was the first to show that both mouse and human fibroblasts could be reprogrammed into iNSCs with simple culture manipulations and the overexpression of the single reprogramming gene SOX2 [53]. Human cells formed clusters of SOX2 and Nestin positive cells 5 days after SOX2 retroviral transduction. These cells then underwent multiple rounds of neurosphere culture and could be maintained under standard NSC conditions. Human iNSCs were capable of differentiation into TUJ1/MAP2+ neurons, glial fibrillary acidic protein (GFAP)+ astrocytes and O4/OLIG2+ oligodendrocytes. Mouse cells showed similar developmental plasticity and upon further differentiation were proven to be functionally active. Neurons derived from mouse iNSCs formed synapses marked by synapsin with patch-clamp recordings showing functional membrane properties and activity. Neither human nor mouse cells formed tumors upon transplantation into noninjured animals. Direct iNSC reprogramming provides a rapid and safe reprogramming approach with the only major limitation being the need for viral delivery and integration of reprogramming factors. Yet, building upon nonviral and nongenomic DNA-integrating approaches created for generating iPSCs, it is very likely that similar approaches can be developed for direct reprogramming of somatic cells into iNSC.

iNSC Transplantation into Stroke Models Leads To Promising Yet Mixed Success

iNSCs have been transplanted into mouse and rat MCAO reperfusion models with cells showing promising results [7–10, 21, 22]. However, it is difficult to compare outcomes and efficacy across studies as transplantation parameters were variable

Table 10.2 Direct reprogramming into neural progenitors

	Factors	Mode of reprogramming	Lineages	Species	Reference
1	ASCL1 & BRN2 & MYT1 L or ZIC1	Lentivirus	Neuron	Mouse	[50]
2	ASCL1 & BRN2 & MYT1 L & NEUROD1	Lentivirus	Neuron	Human	[51]
3	miR-124 and BRN2 and MYT1L	Lentivirus	Neuron	Human	[131]
4	SOX2 & FOXG1 & BRN2	Lentivirus	Neurons, astrocytes and oligodendrocytes	Mouse	[52]
5	SOX2 & BRN4, KLF4, C-MYC (4 Factor) or with E47 (5 Factor)	Retrovirus	Neurons, astrocytes and oligodendrocytes (low)	Mouse	[132]
6	OCT4 & SOX2 & KLF4 & c-MYC + Significant media manipulations	Lentivirus (pre-transduced dox controlled TEFS)	Neurons and astrocytes	Mouse	[133]
7	ASCL1 & BRN2 & MYT1 L	Lentivirus	Neurons	Mouse	[134]
8	ASCL1 & NGN2 & HES1 & ID1& PAX6 & BRN2 & SOX2 & C-MYC & KLF4	Retrovirus	Neurons, astrocytes and oligodendrocytes	Mouse	[135]
9	SOX2 & KLF4 & C-MYC & highly regulated OCT4	Retrovirus	Neurons, astrocytes and oligodendrocytes	Mouse	[136]
10	SOX2	Retrovirus	Neurons, astrocytes and oligodendrocytes	Mouse and human	[53]

with transplant cell numbers ranging from 100,000 to 1,000,000, and timing of transplantation ranging from immediately post-reperfusion to 7 days later. The site of injection was also variable with cells being injected proximal to the lesion or in the contralateral hemisphere to the site of injury. These differences may also account for the significant amount of variability with respect to results. In general, 200,000–250,000 cells were injected 7 days later, avoiding the extreme levels of cytotoxicity immediately after stroke, into the ipsilateral lobe of stroked animals. Transplants showed survival in most studies, but the exact cell number is questionable with one study estimating 10% cell survival [9]. Transplanted cells regularly showed differentiation into neurons, astrocytes, and oligodendrocytes with quantitative data showing higher levels of neuron differentiation than glia [9, 22]. iNSC-derived neurons showed specialization with cells differentiating into dopaminergic and gabaminergic neurons [7–9]. Functionally, whole-cell patch-clamp recordings of brain slices from iNSCs-treated mice at 5 months showed that the majority of iNSC-derived neurons tested were able to produce action potentials in response to depolarizing current and were sensitive to the voltage-gated Na+ channel blocker tetrodotoxin (TTX) and the voltage-gated K+ channel blocker tetraethylammonium (TEA) [9]. iNSC-derived neurons were also sensitive to type-A γ-aminobutyric acid ($GABA_A$) receptor and glutamate receptor antagonists. These results and additional findings showed that iNSC-derived neurons were capable of receiving synaptic input from host neurons and functionally integrating into the neural circuitry [9].

The effect of iNSCs on endogenous tissue was inconsistent between studies. iNSCs had a protective and regenerative effect on host tissues, likely caused by paracrine signaling with the release of factors such as VEGF, as demonstrated in the Cheng et al. study [7]. They found that cell transplantation resulted in a 36 and 11% reduction in Iba-1+ and ED1+ immune cells respectively—cells that are often associated with increased cytotoxicity. At week 8, they demonstrated a 55% reduction in gliosis and a 17% reduction in apoptosis. However, Oki et al. found no significant effect on immune cells (Iba-1 or ED1) or gliosis [9]. Similar studies showed no significant difference in stroke volume suggesting a minimal neuroprotective effect [9, 22].

Functional assessments again showed mixed results across studies, yet were promising. Studies showed improvements in modified neurological scores, rotarod, stepping, and staircase assessments in animals treated with iNSCs relative to non-treated controls [7–9, 21]. Animals showed mixed results with the tape removal test and failed to show significant improvement over control in the corridor, elevated body swing, and cylinder tests [9, 22]. Interestingly, the study that demonstrated the most significant functional improvement also showed the largest decrease in immune cell number, apoptosis, and gliosis [9]. This suggests a strong correlation between tissue-level improvements and positive functional outcomes.

iNSCs in Stroke May Confer Neuroprotection and Enhance Neuroplasticity and Angiogenesis Through Trophic Factor Effects

Transplantation of human iNSCs has been associated with improved functional recovery and a reduction in secondary neural degeneration in various models of ischemic stroke [7–9, 21, 23]. Most of the beneficial effects of iNSCs are observed shortly after transplantation and appear to be independent of iNSC survival suggesting that beneficial effects of iNSCs are not all attributable to cell replacement [8, 9, 21]. While the exact mechanisms through which iNSCs are able to contribute to neural recovery are not well understood, proposed mechanisms include secretion of angiogenic factors such as VEGF, neurotrophic factors such as brain-derived neurotrophic factor (BDNF) and glial cell-line-derived neurotrophic factor (GDNF), and downregulation of inflammatory mediators such as interleukin-6 (IL-6), IL-1β, and tumor necrosis factor alpha (TNF-α) [8, 9, 54]. These neuroprotective trophic factors may be secreted by the stem cells themselves or act through stimulation of endogenous protective pathways decreasing inflammation, promoting neural regeneration, angiogenesis, plasticity, and recruitment of axons from ipsilateral and contralateral hemispheres [8, 9, 55].

Routes of Cell Delivery

A variety of approaches to delivering therapeutic cells to sites of neural injury have been described [56–58]. These include intraparenchymal, intravascular, intracisternal, and intracerebroventricular injections (Fig. 10.3). To date, delivery of iNSCs for the treatment of ischemic stroke in animal models has been limited to intraparenchymal injection through transcranial approaches. Intraparenchymal approaches have also been used in several human clinical trials involving the administration of fetal porcine cells and cultured human neuronal cells to patients suffering from chronic stroke injuries [59, 60]. It is important to note, however, that intraparenchymal injections are by no means the only method of delivering cell therapy to sites of neural injury. To appreciate alternative delivery options, it is necessary to explore methods utilized with other stem cell therapies (e.g., mesenchymal stem cells, embryonic stem cell-derived NPCs, and umbilical cord-derived cells). There are pros and cons to each delivery method, which will be described in more detail in the sections below.

Intraparenchymal

Intraparenchymal injections are the most commonly reported approach. This may be due to the advantages conferred by this method such as site specificity, guaranteed cell delivery to the site of injury, and direct penetration through the blood–brain

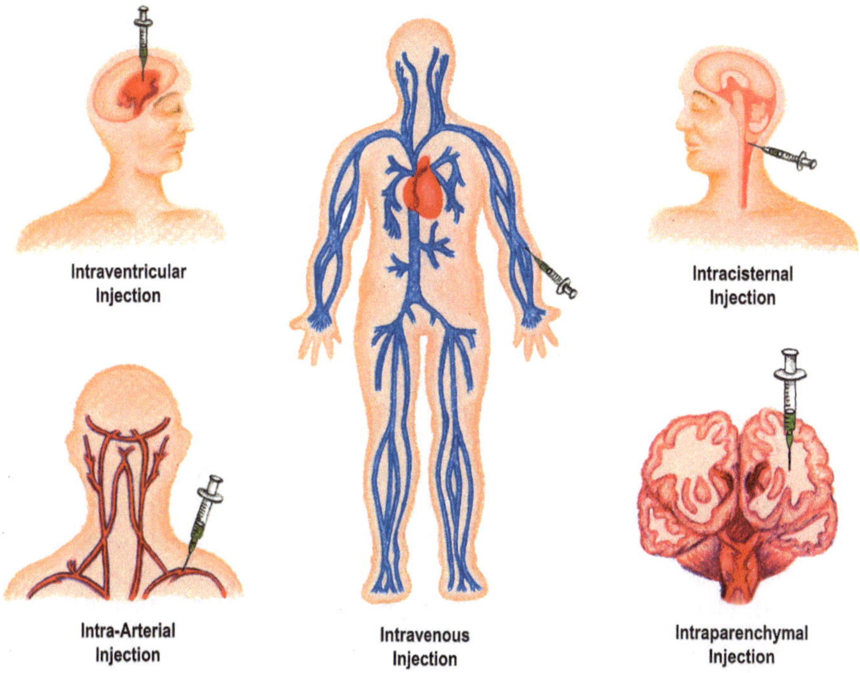

Fig. 10.3 Routes of iNSC transplantation. iNSCs can be transplanted utilizing a number of approaches: intravenous, intra-arterial, intracisternal, intraparenchymal, and intraventricular. Intravenous routes are the least invasive and least technically challenging approach with cells being injected into a peripheral vein of the patient. Successfully transplanted cell numbers are generally low with IV injections as many cells are lost to the pulmonary first-pass effect. Intra-arterial injections of iNSCs provide superior cell delivery but are associated with increased morbidity from thrombosis and hemorrhage. Intracisternal injections are moderately invasive with cells being injected into one of the subarachnoid cisterns (injection into cisterna magna shown in figure). Intraparenchymal and intraventricular injections allow more direct cell delivery but are relatively more invasive and require a transcranial approach with injection directly into the brain matter or lateral ventricles respectively. (Illustration by Leah K. Schultz)

barrier [58]. Unfortunately, intraparenchymal injections typically require more invasive approaches to the site of injury through burr hole craniectomies. Intraparenchymal injections may also result in more clustered, uneven distributions of cells within injured tissue relative to other cell delivery techniques such as intra-arterial (IA) injections that can accomplish a diffuse, extensive spread of cells throughout an injured region [57]. The location of the injection is dependent on the specific injury with a large proportion of MCAO models targeting injections at the site most consistently associated with infarction—the striatum [8, 9, 21]. With advanced imaging techniques such as magnetic resonance imaging (MRI), it is also possible to target injections into the peri-infarct tissue rather than into the infarct core, which may allow for improved cell survival and engraftment [61]. Implantation into either the ipsilateral or contralateral hemisphere to the injury has resulted in beneficial effects, with evidence that cells are able to migrate across midline from the contralateral hemisphere towards the site of injury [7, 62]. Injected neural progenitor cells display a predilection for injured tissues—a trait described as pathotropism [63, 64]. Ischemic injured brain tissue can secrete a variety of signals such as stromal derived factor-1 (SDF-1) and monocyte chemotactic factor 1 (MCP-1), which attract cells, including iNSCs, carrying the receptors CXCR4, CXCR7, and CCR2 [65–67]. This pathotropism will likely enhance the ability of cells to treat ischemic tissue through trophic factor signaling and improve engraftment of cells.

Intravenous

Intravenous (IV) injections are another popular route as they are generally less invasive and pose less of a technical challenge. The number of cells that need to be administered is normally greatly increased from what is permissible via intraparenchymal delivery [57]. In general, IV cell delivery results in the reduced cell engraftment and is associated with cell-uptake by systemic, nontarget organs with many of these cells being trapped in the lungs and liver [68]. The common occurrence of cells being trapped in the small vasculature of lungs is commonly referred to as the pulmonary first-pass effect [69]. Nonetheless, studies with NPCs (not of human-induced pluripotent stem cell origin) have shown that IV administration of cells in ischemic neural injury models can result in a reduction of ischemia-associated learning dysfunction, even when administration was delayed beyond the typical acute therapeutic window [70]. It is believed that these benefits arise from the production of anti-inflammatory and regenerative factors that have a systemic effect including the injured brain. In rare instances, IV delivery of cells has resulted in detectable cell engraftment within the brain [56, 71]. The ability of cells to travel from the vasculature into the brain may reflect the permeability of a compromised blood–brain barrier at the site of injury.

Intra-arterial

A means of avoiding the pulmonary first-pass effect is IA delivery of cells. This method is generally more invasive with higher patient risk for morbidity (due to hemorrhage and thrombosis) and mortality than intravenous approaches [57]. While riskier, cells administered IA have demonstrated increased migration, dissemination, and transplantation success than cells administered IV or intrathecally (IT) [57]. In a study where human mesenchymal stem cells (MSCs) were delivered IA in an MCAO rodent model, it was demonstrated that the location of transplanted cells was dependent on the timing of cell delivery [72]. Cells delivered 1 day post injury were distributed to the peri-infarct region and core of the stroke. Cells delivered on day 4 post injury demonstrated only a peri-infarct distribution. No functional improvements and only very few cells were successfully delivered when injections were administered 7 days post stroke. This would imply a limitation in the timeframe in which IA treatments are effective, although this remains to be shown with iNSCs. Timing of delivery IA may also have an impact on the phenotypic fate of transplanted NSCs with cells transplanted in the first 24 h expressing significantly more GFAP and cells transplanted at 7 and 14 days expressing more βIII-tubulin, indicating astrocyte and neuron differentiation respectively [73].

Intracerebroventricular and Intracisternal

Intracerebroventricular or intrathecal injections have also been reported as a means of delivering cells to the ischemic brain [56, 57]. These are generally associated with less patient risk than IA and intraparenchymal approaches and permit the injection of higher cell numbers. Following intra-ventricular injections, the cells are able to adhere to the walls of the ventricles and migrate through the ependymal lining into the damaged tissues, especially through the lateral versus medial walls of the ventricles [56, 57]. The exact mechanism through which the cells traverse through the ependymal lining is unknown but theories include transport through macrophage-associated regional specializations termed "fractones" [56, 74].

With intrathecal (IT) injections, cells are delivered into the cisterna magna. Again, larger cell numbers can be delivered than with intraparenchymal approaches. Unfortunately, given the flow of cerebrospinal fluid, cells can be lost to other parts of the central nervous system (CNS). In comparisons between IA, IV, and IT in rodent models, IT injections were more effective at delivering cells than IV injections, but IA was considered the superior delivery method in terms of total number of NPCs successfully delivered to the targeted tissue and the achievement of a diffuse, widespread distribution of cells within the injury [57, 75].

Delivery with an Extracellular Matrix

Survival rates of engrafted cells, regardless of the method of cell delivery, are typically low with less than half of injected cells surviving for any period of time ([76]; reviewed in [58]). One method of increasing cell survivability in the cytotoxic acute ischemic injury environment is through implantation of iNSCs with supportive extracellular matrices (ECMs). ECMs can be derived from natural materials like collagen, polyglycosaminoglycans, and ornithine/laminin or from synthetic polymers and hydrogels including polyglycolic acid (PGA), polyethylene glycol (PEG), and poly-L-lactic acid (PLLA) [77, 78]. These materials can be transformed into scaffolds with a variety of shapes and sizes with various porosities and stiffnesses to promote engraftment and recovery. In areas of severe or cystic tissue loss, as is seen in ischemic stroke, biomaterial scaffolds can act as bridging substrates to allow cell attachment and engraftment [79]. There is also suggestion that biomaterial scaffolds can play a part in inhibiting glial scar formation in some neural injury models [80]. Encapsulating scaffolds can act as barriers for grafted cells protecting from host immune rejection, while permitting signaling factors to diffuse between the graft and the injured environment [81]. As neural injuries often possess irregular boundaries, malleable and liquid substrates such as injectable hydrogels and microspheres have been particular targets of investigation [79, 82, 83]. In addition to acting as structural support, scaffolds can also be engineered to contain various growth factors, peptides, and chemical signals such as heparin and hyaluronan to promote microenvironments conducive to graft survival [78, 83–86]. Multiple studies demonstrated that ECM or NSC alone did not improve sensory motor function recovery nor decreased infarct size after focal cerebral ischemia in rodents. However, when ECM and NSCs were combined, there was a significant improvement in both functional and anatomical outcomes [82].

Despite their anticipated benefits, biomaterials can also present unique challenges including inhibition of neurite outgrowth by the scaffold [79, 87–89] and variable matrix degradation times [90]. In some cases, the scaffold may interfere with graft cell differentiation and integration [91]. For synthetic polymers, there is particular concern about harmful degradation by-products that can increase local acidity, inflammation, and tissue damage [92]. Immunoreactivity and tumorgenicity of biomaterials are also of concern, especially with undefined natural materials harvested from plant and animal sources [77]. In addition to the materials in its composition, the macro-architecture of the implants appears to play a role in the host-immune response [93]. With some scaffolds, fibrous tissue buildup and foreign-body reactions around the implant can also cause interference with tissue integration, angiogenesis, and trophic factor diffusion to and from the grafted cells ([93]; reviewed by [92]). Nonetheless, ECMs offer an exciting and viable option for improving the success of iNSC transplantation in the ischemic-stroke environment.

Cell Dosage

Currently, there are no clear guidelines on how to determine the therapeutic number of cells to be transplanted for any cell therapy to achieve optimum treatment of stroke. Albeit there are some key factors that are likely to be critical in the development of guidelines for therapeutic dose. These potential factors include:

1. Severity, localization and type of stroke injury
2. Whether the therapeutic is acting through paracrine signaling as a producer of neuroprotectants, regenerative factors or as a replacement therapy
3. Comorbidities such as hypertension and diabetes
4. Patient age, sex, and size
5. Delivery mechanism

When delivering cells through the vasculature, cistern, or ventricle, it is possible to administer higher cell numbers with some rodent IV dosages approximating 5×10^6 cells [56, 94, 95]. The beneficial effects may also be dose-dependent as shown in a rat ischemic stroke study involving IV administration of bone marrow stromal cells

(BMSCs); rats receiving higher cell numbers displayed better outcomes [96]. In rodent models of ischemic stroke, cell numbers for intraparenchymal implantation have ranged anywhere from 5000 to 1.5 million [97]. When translating this to human patients, consideration should be given to the significant size disparity between rodents and humans. A cells-per-body-mass dosage can be extrapolated from rodent studies but may not be the best method for determining an optimal dose in human patients similar to pharmacokinetic studies where differences in species metabolism and physiology contribute significantly to appropriate dose scaling [98, 99]. Some clinical trials have adopted this approach by calculating the effective IV dose of BMSCs in rodents and determining an equivalent dosage in humans as about 1×10^8 cells/patient [100]. In one study looking at intraparenchymal cell delivery in a rodent model, the injection of higher numbers of cells resulted in higher total number of cells surviving [97]. Cell survivability on a percentage basis, however, was actually higher when lower numbers of cells were injected, suggesting an optimal threshold for cell numbers to be engrafted. It is thought that beyond this threshold, cell survivability decreases due to limited supply of local nutrients.

Timing

A potential benefit of cell therapy is that it offers a broader therapeutic time window than current FDA-approved therapies like tPA, which require administration in the hyperacute phase (<6 h) from the time of injury [101]. The precise optimal therapeutic time window for stem cell treatment of stroke is still unclear and likely varies between stroke conditions. One factor that should be considered is the route of cell delivery. For routes of administration like intravascular injection that rely on a compromised blood–brain barrier and inflammatory signaling for cells to home to the site of injury, therapy within the acute period post stroke may be more relevant as reviewed in Bliss et al. 2010 [101]. For IA routes, the timing of the injection can affect the distribution, survival, and the phenotypic fates of the injected cells [73].

There are also concerns for cell survivability with transplantation during the acute stroke phase due to the cytotoxic environment, which suggests that the subacute or chronic injury periods may be more optimum transplant points [101]. This is highly dependent, however, on the anticipated primary effect of the transplanted cells. In some cases where the primary effect is through neuroprotection via trophic effects rather than cell differentiation and replacement, transplantation during the acute post-stroke period may be optimal. Whereas if a cell is predicted to have an anti-inflammatory or neuroplasticity effect, it is perhaps more relevant to transplant cells during the subacute stroke phase [101]. Earlier intraparenchymal injection times are also supported by reports where the beneficial effects of intraparenchymal injections were independent of cell survival [8, 9, 56]. One thought is that earlier intracerebral injection times may improve cell survivability as the microglial response has not yet had a chance to establish itself [97]. Rosenblum et al. [73] compared injection of neural progenitor cells (NPCs) at various time points in a hypoxia–ischemia rodent model and demonstrated that intra-striatal injections 3 days post injury yielded the highest cell engraftment as compared to injections administered at 6 and 24 h and 7 and 14 days. Current clinical trials have surveyed the effects of BMSCs and Human NT2N neurons (derived from the NTera2 teratocarcinoma cell line) on stroke injuries in the late subacute period (4–5 weeks post stroke) and chronic periods, respectively [100, 102]. While some patients appeared to benefit from the treatment, the benefits were not considered to be statistically significant [102]. The exact mechanisms of action of cell therapy at these later treatment points have not been specified. To date, studies investigating the optimal therapeutic time frame for administration of iNSCs have not been investigated.

Future Directions

While there have been successful transplantations of iNSCs into rodent models of ischemic stroke, there are still many unanswered questions regarding the specifics of cell dosage, use of ECM, transplant location, optimal timing for transplantation, or the best vehicle and approach for cell delivery. The lack of consensus in a relatively homogeneous model species like the rat or mouse suggests that optimal cell transplantation conditions are likely varied and case dependent.

Long-term studies on the safety and efficacy of iNSCs have yet to be completed in non rodent models. As outlined by the STAIR [103] and STEPS II [104] meetings, it is vital that successful rodent therapies be confirmed in other animal models of stroke prior to advancing to human clinical trials. Several large-animal models of ischemic stroke have been developed and iNSC transplantation studies in these species are eagerly awaited [105–107]. Due to similarities between humans and primates, the primate model would seem to be a natural fit for studies of iNSC treatments. However, the cost, specialized facilities, regulatory burden, and ethical issues associated with primate models make alternative large-animal models such as sheep and pigs more attractive in some respects. The pig stroke model offers a significant advantage over rodent models as pigs have much larger gyrencephalic brains with gray–white matter composition more similar to humans [108, 109]. Utilization of animal models with similar white matter composition is of significant importance as white matter injuries uniquely contribute to clinical deficits in stroke patients and it will be important to determine if iNSC treatment will be able to appropriately differentiate and integrate in both gray- and white-matter compartments [110, 111]. Both the human and pig brain is composed of > 65 % white matter, while white matter in the rodent brain is < 10 %, making the pig a potentially excellent surrogate [107, 110, 112–115]). Moreover, human and pig brains are both gyrencephalic, while the rodent brain is lissencephalic, a key architectural difference that has a direct correlation with brain connectivity and complexity [107, 112, 113]. Brain size is also another major variable when considering a cell therapy. The human brain is approximately 650 times the size of the average rodent brain, while only being 7.5 times the size of the pig brain—a size comparable to typical nonhuman primate models [116]. Size affects the number of cells to be transplanted, the sites of injection, the ability of the graft to be vascularized, and the distances axons must travel to form connections. To achieve maximum clinical translatability, using animal models as similar to humans as possible will be of critical importance in testing additional factors affecting iNSC therapy efficacy and safety.

iNSC treatment of stroke in rodent models have led to justified enthusiasm with cells showing long-term integration and functionality with treated animals showing improvement in functional deficiencies [7–10]. In the light of these initial successes, and with an eye towards clinical applications, additional studies are now needed to assess basic questions such as cell dosage, treatment window, and route of delivery in suitable large animal models. These studies should be performed as randomized double-blinded trials to prevent any unintended bias from researchers, under the most stringent testing conditions possible. Utilizing strict testing protocols, regenerative iNSC therapy will hopefully move from promise and potential to a realized clinical therapy that will help millions of stroke victims lead more normal and productive lives.

References

1. Young JA, Tolentino M. Stroke evaluation and treatment. Top Stroke Rehabil. 2009;16(6):389–410.
2. Towfighi A, Ovbiagele B, Saver JL. Therapeutic milestone: stroke declines from the second to the third leading organ- and disease-specific cause of death in the United States. Stroke. 2010;41(3):499–503.
3. Hess DC, Borlongan CV. Cell-based therapy in ischemic stroke. Expert Rev Neurother. 2008;8(8):1193–201.
4. Madri JA. Modeling the neurovascular niche: implications for recovery from CNS injury. J Physiol Pharmacol. 2009;60 Suppl 4:95–104.
5. Wright LS, Prowse KR, Wallace K, Linskens MH, Svendsen CN. Human progenitor cells isolated from the developing cortex undergo decreased neurogenesis and eventual senescence following expansion in vitro. Exp Cell Res. 2006;312(11):2107–20.
6. Klimanskaya I, Chung Y, Becker S, Lu SJ, Lanza R. Human embryonic stem cell lines derived from single blastomeres. Nature. 2006;444(7118):481–5.
7. Chang DJ, Lee N, Park IH, Choi C, Jeon I, Kwon J, Oh SH, Shin DA, Do JT, Lee DR, Lee H, Moon H, Hong KS, Daley GQ, Song J. Therapeutic potential of human induced pluripotent stem cells in experimental stroke. Cell Transplant. 2013;22(8):1427–40.
8. Polentes J, Jendelova P, Cailleret M, Braun H, Romanyuk N, Tropel P, Brenot M, Itier V, Seminatore C, Baldauf K, Turnovcova K, Jirak D, Teletin M, Come J, Tournois J, Reymann K, Sykova E, Viville S, Onteniente B. Human induced pluripotent stem cells improve stroke outcome and reduce secondary degeneration in the recipient brain. Cell Transplant. 2012;21(12):2587–602.
9. Oki K, Tatarishvili J, Woods J, Koch P, Wattananit S, Mine Y, Monni E, Prietro DT, Ahlenius H, Ladewig J, Brustle O, Lindvall O, Kokaia Z. Human induced pluripotent stem cells form functional neurons and improve recovery after grafting in stroke-damaged brain. Stem Cells. 2012;30(6):1120–33.
10. Yuan T, Liao W, Feng NH, Lou YL, Niu X, Zhang AJ, Wang Y, Deng ZF. Human induced pluripotent stem-cell-derived neural stem cells survive, migrate, differentiate, and improve

neurological function in a rat model of middle cerebral artery occlusion. Stem Cell Res Ther. 2013;4(3):73.
11. Kirmani JF, Alkawi A, Panezai S, Gizzi M. Advances in thrombolytics for treatment of acute ischemic stroke. Neurology. 2012; 79(13 Suppl 1):S119–25.
12. Sarraj A, Grotta JC. Stroke: new horizons in treatment. Lancet Neurol. 2014;13(1):2–3.
13. Sutherland BA, Minnerup J, Balami JS, Arba F, Buchan AM, Kleinschnitz C. Neuroprotection for ischaemic stroke: translation from the bench to the bedside. Int J Stroke. 2012;7(5):407–18.
14. Turner RC, Dodson SC, Rosen CL, Huber JD. The science of cerebral ischemia and the quest for neuroprotection: navigating past failure to future success. J Neurosurg. 2013;118(5):1072–85.
15. Hassan AE, Aman MM, Chauhdry SA, Grigoryan M, Tekle WG, Rodriguez GJ, Qureshi AI. Value of other endovascular techniques among patients with MERCI device failure during the treatment of acute ischemic stroke: what to do when MERCI fails? J Vasc Interv Neurol. 2013;5(2):9–13.
16. Sahni R, Weinberger J. Management of intracerebral hemorrhage. Vasc Health Risk Manag. 2007;3(5):701–9.
17. Cheng YD, Al-Khoury L, Zivin JA. Neuroprotection for ischemic stroke: two decades of success and failure. NeuroRx. 2004;1(1):36–45.
18. Savitz SI, Fisher M. Future of neuroprotection for acute stroke: in the aftermath of the SAINT trials. Ann Neurol. 2007;61(5):396–402.
19. Stem Cell Therapies as an Emerging Paradigm in Stroke Participants. Stem cell therapies as an emerging paradigm in stroke (STEPS): bridging basic and clinical science for cellular and neurogenic factor therapy in treating stroke. Stroke. 2009;40(2):510–5.
20. Savitz SI, Chopp M, Deans R, Carmichael ST, Phinney D, Wechsler L. Stem cell therapy as an emerging paradigm for stroke (STEPS) II. Stroke. 2011;42(3):825–9.
21. Gomi M, Takagi Y, Morizane A, Doi D, Nishimura M, Miyamoto S, Takahashi J. Functional recovery of the murine brain ischemia model using human induced pluripotent stem-cell-derived telencephalic progenitors. Brain Res. 2012;1459:52–60.
22. Jensen MB, Yan H, Krishnaney-Davison R, Al Sawaf A, Zhang SC. Survival and differentiation of transplanted neural stem-cell-derived from human induced pluripotent stem cells in a rat stroke model. J Stroke Cerebrovasc Dis. 2013;22(4):304–8.
23. Tornero D, Wattananit S, Gronning Madsen M, Koch P, Wood J, Tatarishvili J, Mine Y, Ge R, Monni E, Devaraju K, Hevner RF, Brustle O, Lindvall O, Kokaia Z. Human induced pluripotent stem-cell-derived cortical neurons integrate in stroke-injured cortex and improve functional recovery. Brain 2013;136(Pt 12):3561–77.
24. Takahashi K, Tanabe K, Ohnuki M, Narita M, Ichisaka T, Tomoda K, Yamanaka S. Induction of pluripotent stem cells from adult human fibroblasts by defined factors. Cell. 2007;131(5):861–72.
25. Takahashi K, Yamanaka S. Induction of pluripotent stem cells from mouse embryonic and adult fibroblast cultures by defined factors. Cell. 2006;126(4):663–76.
26. Briggs R, King TJ. Transplantation of living nuclei from blastula cells into enucleated frogs' eggs. Proc Natl Acad Sci U S A. 1952;38(5):455–63.
27. Gurdon JB. The developmental capacity of nuclei taken from intestinal epithelium cells of feeding tadpoles. J Embryol Exp Morphol. 1962;10:622–40.
28. Wilmut I, Schnieke AE, McWhir J, Kind AJ, Campbell KH. Viable offspring derived from fetal and adult mammalian cells. Nature. 1997;385(6619):810–3.
29. Cowan CA, Atienza J, Melton DA, Eggan K Nuclear reprogramming of somatic cells after fusion with human embryonic stem cells. Science. 2005;309(5739):1369–73.
30. Martin GR. Isolation of a pluripotent cell line from early mouse embryos cultured in medium conditioned by teratocarcinoma stem cells. Proc Natl Acad Sci U S A. 1981;78(12):7634–8.
31. Thomson JA, Itskovitz-Eldor J, Shapiro SS, Waknitz MA, Swiergiel JJ, Marshall VS, Jones JM. Embryonic stem cell lines derived from human blastocysts. Science. 1998;282(5391):1145–7.

32. Thomson JA, Kalishman J, Golos TG, Durning M, Harris CP, Becker RA, Hearn JP. Isolation of a primate embryonic stem cell line. Proc Natl Acad Sci U S A. 1995;92(17):7844–8.
33. Okita K, Ichisaka T, Yamanaka S. Generation of germline-competent induced pluripotent stem cells. Nature 2007;448(7151):313–7.
34. Zhao XY, Li W, Lv Z, Liu L, Tong M, Hai T, Hao J, Guo CL, Ma QW, Wang L, Zeng F, Zhou Q. iPS cells produce viable mice through tetraploid complementation. Nature. 2009;461(7260):86–90.
35. Boland MJ, Hazen JL, Nazor KL, Rodriguez AR, Gifford W, Martin G, Kupriyanov S, Baldwin KK. Adult mice generated from induced pluripotent stem cells. Nature. 2009;461(7260):91–4.
36. Zhao T, Zhang ZN, Rong Z, Xu Y. Immunogenicity of induced pluripotent stem cells. Nature. 2011;474 (7350):212–5.
37. Guha P, Morgan JW, Mostoslavsky G, Rodrigues NP, Boyd AS. Lack of immune response to differentiated cells derived from syngeneic induced pluripotent stem cells. Cell Stem Cell. 2013;12(4):407–12.
38. Araki R, Uda M, Hoki Y, Sunayama M, Nakamura M, Ando S, Sugiura M, Ideno H, Shimada A, Nifuji A, Abe M. Negligible immunogenicity of terminally differentiated cells derived from induced pluripotent or embryonic stem cells. Nature. 2013;494(7435):100–4.
39. Warren L, Manos PD, Ahfeldt T, Loh YH, Li H, Lau F, Ebina W, Mandal PK, Smith ZD, Meissner A, Daley GQ, Brack AS, Collins JJ, Cowan C, Schlaeger TM, Rossi DJ. Highly efficient reprogramming to pluripotency and directed differentiation of human cells with synthetic modified mRNA. Cell Stem Cell. 2010;7(5):618–30.
40. Kim D, Kim CH, Moon JI, Chung YG, Chang MY, Han BS, Ko S, Yang E, Cha KY, Lanza R, Kim KS. Generation of human induced pluripotent stem cells by direct delivery of reprogramming proteins. Cell Stem Cell. 2009;4(6):472–6.
41. Jia F, Wilson KD, Sun N, Gupta DM, Huang M, Li Z, Panetta NJ, Chen ZY, Robbins RC, Kay MA, Longaker MT, Wu JC. A nonviral minicircle vector for deriving human iPS cells. Nat Methods. 2010;7(3):197–9.
42. Nakagawa M, Koyanagi M, Tanabe K, Takahashi K, Ichisaka T, Aoi T, Okita K, Mochiduki Y, Takizawa N, Yamanaka S. Generation of induced pluripotent stem cells without Myc from mouse and human fibroblasts. Nat Biotechnol. 2008;26(1):101–6.
43. Zhang R, Zhang Z, Wang L, Wang Y, Gousev A, Zhang L, Ho KL, Morshead C, Chopp M. Activated neural stem cells contribute to stroke-induced neurogenesis and neuroblast migration toward the infarct boundary in adult rats. J Cereb Blood Flow Metab. 2004;24(4):441–8.
44. Vandeputte C, Reumers V, Aelvoet SA, Thiry I, De Swaef S, Van Den Haute C, Pascual-Brazo J, Farr TD, Vande Velde G, Hoehn M, Himmelreich U, Van Laere K, Debyser Z, Gijsbers R, Baekelandt V. Bioluminescence imaging of stroke-induced endogenous neural stem cell response. Neurobiol Dis. 2014;69:144-55.
45. Jin K, Wang X, Xie L, Mao XO, Zhu W, Wang Y, Shen J, Mao Y, Banwait S, Greenberg DA. Evidence for stroke-induced neurogenesis in the human brain. Proc Natl Acad Sci U S A. 2006;103(35):13198–202.
46. Marti-Fabregas J, Romaguera-Ros M, Gomez-Pinedo U, Martinez-Ramirez S, Jimenez-Xarrie E, Marin R, Marti-Vilalta JL, Garcia-Verdugo JM. Proliferation in the human ipsilateral subventricular zone after ischemic stroke. Neurology. 2010;74(5):357–65.
47. Kawai H, Yamashita T, Ohta Y, Deguchi K, Nagotani S, Zhang X, Ikeda Y, Matsuura T, Abe K. Tridermal tumorigenesis of induced pluripotent stem cells transplanted in ischemic brain. J Cereb Blood Flow Metab. 2010;30(8):1487–93.
48. Chen SJ, Chang CM, Tsai SK, Chang YL, Chou SJ, Huang SS, Tai LK, Chen YC, Ku HH, Li HY, Chiou SH Functional improvement of focal cerebral ischemia injury by subdural transplantation of induced pluripotent stem cells with fibrin glue. Stem Cells Dev. 2010;19(11):1757–67.
49. Zhang SC, Wernig M, Duncan ID, Brustle O, Thomson JA. In vitro differentiation of transplantable neural precursors from human embryonic stem cells. Nat Biotechnol. 2001;19(12):1129–33.

50. Vierbuchen T, Ostermeier A, Pang ZP, Kokubu Y, Sudhof TC, Wernig M. Direct conversion of fibroblasts to functional neurons by defined factors. Nature. 2010;463(7284):1035–41.
51. Pang ZP, Yang N, Vierbuchen T, Ostermeier A, Fuentes DR, Yang TQ, Citri A, Sebastiano V, Marro S, Sudhof TC, Wernig M. Induction of human neuronal cells by defined transcription factors. Nature. 2011;476(7359):220–3.
52. Lujan E, Chanda S, Ahlenius H, Sudhof TC, Wernig M. Direct conversion of mouse fibroblasts to self-renewing, tripotent neural precursor cells. Proc Natl Acad Sci U S A. 2012;109(7):2527–32.
53. Ring KL, Tong LM, Balestra ME, Javier R, Andrews-Zwilling Y, Li G, Walker D, Zhang WR, Kreitzer AC, Huang Y. Direct reprogramming of mouse and human fibroblasts into multipotent neural stem cells with a single factor. Cell Stem Cell. 2012;11(1):100–9.
54. Nori S, Okada Y, Yasuda A, Tsuji O, Takahashi Y, Kobayashi Y, Fujiyoshi K, Koike M, Uchiyama Y, Ikeda E, Toyama Y, Yamanaka S, Nakamura M, Okano H. Grafted human-induced pluripotent stem-cell-derived neurospheres promote motor functional recovery after spinal cord injury in mice. Proc Natl Acad Sci U S A. 2011;108(40):16825–30.
55. Kokaia Z, Martino G, Schwartz M, Lindvall O. Cross-talk between neural stem cells and immune cells: the key to better brain repair? Nat Neurosci. 2012;15(8):1078–87.
56. Jin K, Sun Y, Xie L, Mao XO, Childs J, Peel A, Logvinova A, Banwait S, Greenberg DA. Comparison of ischemia-directed migration of neural precursor cells after intrastriatal, intraventricular, or intravenous transplantation in the rat. Neurobiol Dis. 2005;18(2):366–74.
57. Li L, Jiang Q, Ding G, Zhang L, Zhang ZG, Li Q, Panda S, Lu M, Ewing JR, Chopp M. Effects of administration route on migration and distribution of neural progenitor cells transplanted into rats with focal cerebral ischemia, an MRI study. J Cereb Blood Flow Metab. 2010;30(3):653–62.
58. Bliss T, Guzman R, Daadi M, Steinberg GK. Cell transplantation therapy for stroke. Stroke. 2007;38 Suppl 2:817–26.
59. Savitz SI, Dinsmore J, Wu J, Henderson GV, Stieg P, Caplan LR. Neurotransplantation of fetal porcine cells in patients with basal ganglia infarcts: a preliminary safety and feasibility study. Cerebrovasc Dis (Basel, Switzerland). 2005;20(2):101–7.
60. Kondziolka D, Wechsler L, Goldstein S, Meltzer C, Thulborn KR, Gebel J, Jannetta P, DeCesare S, Elder EM, McGrogan M, Reitman MA, Bynum L. Transplantation of cultured human neuronal cells for patients with stroke. Neurology. 2000;55(4):565–9.
61. Smith EJ, Stroemer RP, Gorenkova N, Nakajima M, Crum WR, Tang E, Stevanato L, Sinden JD, Modo M. Implantation site and lesion topology determine efficacy of a human neural stem cell line in a rat model of chronic stroke. Stem Cells. 2012;30(4):785–96.
62. Modo M, Stroemer RP, Tang E, Patel S, Hodges H. Effects of implantation site of stem cell grafts on behavioral recovery from stroke damage. Stroke. 2002;33(9):2270–8.
63. Teixeira AI, Duckworth JK, Hermanson O. Getting the right stuff: controlling neural stem cell state and fate in vivo and in vitro with biomaterials. Cell Res. 2007;17(1):56–61.
64. Pluchino S, Zanotti L, Rossi B, Brambilla E, Ottoboni L, Salani G, Martinello M, Cattalini A, Bergami A, Furlan R, Comi G, Constantin G, Martino G. Neurosphere-derived multipotent precursors promote neuroprotection by an immunomodulatory mechanism. Nature. 2005;436(7048):266–71.
65. Wiltrout C, Lang B, Yan Y, Dempsey RJ, Vemuganti R. Repairing brain after stroke: A review on post-ischemic neurogenesis. Neurochem Int. 2007;50(7–8):1028–41.
66. Hess DC, Borlongan CV. Cell-based therapy in ischemic stroke. Expert Rev Neurother. 2008;8(8):1193–201.
67. Kojima T, Hirota Y, Ema M, Takahashi S, Miyoshi I, Okano H, Sawamoto K. Subventricular zone-derived neural progenitor cells migrate along a blood vessel scaffold toward the post-stroke striatum. Stem Cells. 2010;28(3):545–54.
68. Lappalainen RS, Narkilahti S, Huhtala T, Liimatainen T, Suuronen T, Närvänen A, Suuronen R, Hovatta O, Jolkkonen J. The SPECT imaging shows the accumulation of neural progenitor cells into internal organs after systemic administration in middle cerebral artery occlusion rats. Neurosci Lett. 2008;440(3):246–50.

69. Fischer UM, Harting MT, Jimenez F, Monzon-Posadas WO, Xue H, Savitz SI, Laine GA, Cox CS Jr. Pulmonary passage is a major obstacle for intravenous stem cell delivery: the pulmonary first-pass effect. Stem Cells Dev. 2009;18(5):683–92.
70. Mochizuki N, Moriyama Y, Takagi N, Takeo S, Tanonaka K. Intravenous injection of neural progenitor cells improves cerebral ischemia-induced learning dysfunction. Biol Pharm Bull. 2011;34(2):260–5.
71. Mezey E. Turning blood into brain: cells bearing neuronal antigens generated in vivo from bone marrow. Science. 2000;290(5497):1779–82.
72. Ishizaka S, Horie N, Satoh K, Fukuda Y, Nishida N, Nagata I. Intra-arterial cell transplantation provides timing-dependent cell distribution and functional recovery after stroke. Stroke. 2013;44(3):720–6.
73. Rosenblum S, Wang N, Smith TN, Pendharkar AV, Chua JY, Birk H, Guzman R. Timing of intra-arterial neural stem cell transplantation after hypoxia-ischemia influences cell engraftment, survival, and differentiation. Stroke. 2012;43(6):1624–31.
74. Mercier F, Kitasako JT, Hatton GI. Anatomy of the brain neurogenic zones revisited: fractones and the fibroblast/macrophage network. J Comp Neurol. 2002;451(2):170–88.
75. Zhang L, Li Y, Romanko M, Kramer BC, Gosiewska A, Chopp M, Hong K. Different routes of administration of human umbilical tissue-derived cells improve functional recovery in the rat after focal cerebral ischemia. Brain Res. 2012;1489:104–12.
76. Hicks AU, Lappalainen RS, Narkilahti S, Suuronen R, Corbett D, Sivenius J, Hovatta O, Jolkkonen J. Transplantation of human embryonic stem-cell-derived neural precursor cells and enriched environment after cortical stroke in rats: cell survival and functional recovery. Eur J Neurosci. 2009;29(3):562–74.
77. Uemura M, Refaat MM, Shinoyama M, Hayashi H, Hashimoto N, Takahashi J. Matrigel supports survival and neuronal differentiation of grafted embryonic stem-cell-derived neural precursor cells. J Neurosci Res. 2010;88(3):542–51.
78. Hoffman AS. Hydrogels for biomedical applications. Adv Drug Deliv Rev. 2002;54:3–12.
79. Zhong J, Chan A, Morad L, Kornblum HI, Fan G, Carmichael ST. Hydrogel matrix to support stem cell survival after brain transplantation in stroke. Neurorehabil Neural Repair. 2010;24(7):636–44.
80. Teng YD, Lavik EB, Qu X, Park KI, Ourednik J, Zurakowski D, Langer R, Snyder EY. Functional recovery following traumatic spinal cord injury mediated by a unique polymer scaffold seeded with neural stem cells. Proc Natl Acad Sci U S A. 2002;99(5):3024–9.
81. Park KI, Teng YD, Snyder EY. The injured brain interacts reciprocally with neural stem cells supported by scaffolds to reconstitute lost tissue. Nat Biotechnol. 2002;20(11):1111–7.
82. Jin K, Mao X, Xie L, Galvan V, Lai B, Wang Y, Gorostiza O, Wang X, Greenberg DA. Transplantation of human neural precursor cells in Matrigel scaffolding improves outcome from focal cerebral ischemia after delayed postischemic treatment in rats. J Cereb Blood Flow Metab. 2010;30(3):534–44.
83. Skop NB, Calderon F, Levison SW, Gandhi CD, Cho CH. Heparin crosslinked chitosan microspheres for the delivery of neural stem cells and growth factors for central nervous system repair. Acta Biomater. 2013;9(6):6834–43.
84. Drury JL, Mooney DJ. Hydrogels for tissue engineering: scaffold design variables and applications. Biomaterials. 2003;24(24):4337–51.
85. Place ES, Evans ND, Stevens MM. Complexity in biomaterials for tissue engineering. Nat Mater. 2009;8(6):457–70.
86. Delcroix GJ, Schiller PC, Benoit JP, Montero-Menei CN. Adult cell therapy for brain neuronal damages and the role of tissue engineering. Biomaterials. 2010;31(8):2105–20.
87. Jiang FX, Yurke B, Firestein BL, Langrana NA. Neurite outgrowth on a DNA crosslinked hydrogel with tunable stiffnesses. Ann Biomed Eng. 2008;36(9):1565–79.
88. Balgude AP, Yu X, Szymanski A, Bellamkonda RV. Agarose gel stiffness determines rate of DRG neurite extension in 3D cultures. Biomaterials. 2001;22:1077–84.
89. Willitis RK, Skornia SL. Effect of collagen gel stiffness on neurite extension. J Biomater Sci Polym Ed. 2004;15(12):1521–31.

90. Bible E, Chau DY, Alexander MR, Price J, Shakesheff KM, Modo M. The support of neural stem cells transplanted into stroke-induced brain cavities by PLGA particles. Biomaterials. 2009;30(16):2985–94.
91. Bible E, Dell'Acqua F, Solanky B, Balducci A, Crapo PM, Badylak SF, Ahrens ET, Modo M. Non-invasive imaging of transplanted human neural stem cells and ECM scaffold remodeling in the stroke-damaged rat brain by (19)F- and diffusion-MRI. Biomaterials. 2012;33(10):2858–71.
92. Fournier E, Passirani C, Montero-Menei CN, Benoit JP. Biocompatibility of implantable synthetic polymeric drug carriers: focus on brain biocompatibility. Biomaterials. 2003;24(19):3311–31.
93. Wong DY, Leveque JC, Brumblay H, Krebsbach PH, Hollister SJ, Lamarca F. Macroarchitectures in spinal cord scaffold implants influence regeneration. J Neurotrauma. 2008;25(8):1027–37.
94. Sokolova IB, Fedotova OR, Tsikunov SG, Polyntsev DG. Mesenchymal stem cells restore orientation and exploratory behavior of rats after brain injury. Bull Exp Biol Med. 2011;151(1):130–2.
95. Honma T, Honmou O, Iihoshi S, Harada K, Houkin K, Hamada H, Kocsis JD. Intravenous infusion of immortalized human mesenchymal stem cells protects against injury in a cerebral ischemia model in adult rat. Exp Neurol. 2006;199(1):56–66.
96. Chen J, Li Y, Wang L, Zhang Z, Lu D, Lu M, Chopp M. Therapeutic Benefit of Intravenous Administration of Bone Marrow Stromal Cells After Cerebral Ischemia in Rats. Stroke. 2001;32(4):1005–11.
97. Darsalia V, Allison SJ, Cusulin C, Monni E, Kuzdas D, Kallur T, Lindvall O, Kokaia Z. Cell number and timing of transplantation determine survival of human neural stem cell grafts in stroke-damaged rat brain. J Cereb Blood Flow Metab. 2011;31(1):235–42.
98. Reagan-Shaw S, Nihal M, Ahmad N. Dose translation from animal to human studies revisited. FASEB J. 2008;22(3):659–61.
99. Sharma V, McNeill JH. To scale or not to scale: the principles of dose extrapolation. Br J Pharmacol. 2009;157(6):907–21.
100. Bang OY, Lee JS, Lee PH, Lee G. Autologous mesenchymal stem cell transplantation in stroke patients. Ann Neurol. 2005;57(6):874–82.
101. Bliss TM, Andres RH, Steinberg GK. Optimizing the success of cell transplantation therapy for stroke. Neurobiol Dis. 2010;37(2):275–83.
102. Kondziolka D, Steinberg GK, Wechsler L, Meltzer CC, Elder E, Gebel J, DeCesare S, Jovin T, Zafonte R, Lebowitz J, Flickinger JC, Tong D, Marks MP, Jamieson C, Luu D, Bell-Stephens T, Teraoka J. Neurotransplantation for patients with subcortical motor stroke: a Phase 2 randomized trial. J Neurosurg. 2005;103(1):38–45.
103. Fisher M, Feuerstein G, Howells DW, Hurn PD, Kent TA, Savitz SI, Lo EH, Group S. Update of the stroke therapy academic industry roundtable preclinical recommendations. Stroke. 2009;40(6):2244–50.
104. Savitz SI, Chopp M, Deans R, Carmichael T, Phinney D, Wechsler L, Participants S. Stem cell therapy as an emerging paradigm for stroke (STEPS) II. Stroke. 2011;42(3):825–9.
105. Boltze J, Nitzsche B, Geiger KD, Schoon HA. Histopathological investigation of different MCAO modalities and impact of autologous bone marrow mononuclear cell administration in an ovine stroke model. Transl Stroke Res. 2011;2(3):279–93.
106. Platt SR, Holmes S, Howerth E, Duberstein J, Dove C, Kinder H, Wyatt E, Linville A, Lau V, Stice S, Hill W, Hess DC, West F. Development and characterization of a Yucatan miniature biomedical pig permanent middle cerebral artery occlusion stroke model. Exp Transl Stroke Med. 2014;6:5.
107. Tanaka Y, Imai H, Konno K, Miyagishima T, Kubota C, Puentes S, Aoki T, Hata H, Takata K, Yoshimoto Y, Saito N. Experimental model of lacunar infarction in the gyrencephalic brain of the miniature pig: neurological assessment and histological, immunohistochemical, and physiological evaluation of dynamic corticospinal tract deformation. Stroke. 2008;39(1):205–12.

108. Duberstein KJ, Platt SR, Holmes SP, Dove CR, Howerth EW, Kent M, Stice SL, Hill WD, Hess DC, West FD. Gait analysis in a pre- and post-ischemic stroke biomedical pig model. Physiol Behav. 2014;125:8–16.
109. Platt SR, Holmes SP, Howerth EW, Duberstein KJ, Dove CR, Kinder HA, Wyatt EL, Linville AV, Lau VW, Stice SL, Hill WD, Hess DC, West FD. Development and characterization of a Yucatan miniature biomedical pig permanent middle cerebral artery occlusion stroke model. Exp Transl Stroke Med. 2014;6(1):5.
110. Baltan S, Besancon EF, Mbow B, Ye Z, Hamner MA, Ransom BR. White matter vulnerability to ischemic injury increases with age because of enhanced excitotoxicity. J Neurosci. 2008;28(6):1479–89.
111. Mason GF, Pan JW, Chu WJ, Newcomer BR, Zhang Y, Orr R, Hetherington HP. Measurement of the tricarboxylic acid cycle rate in human grey and white matter in vivo by 1H-[13C] magnetic resonance spectroscopy at 4.1T. J Cereb Blood Flow Metab. 1999;19(11):1179–88.
112. Nakamura M, Imai H, Konno K, Kubota C, Seki K, Puentes S, Faried A, Yokoo H, Hata H, Yoshimoto Y, Saito N. Experimental investigation of encephalomyosynangiosis using gyrencephalic brain of the miniature pig: histopathological evaluation of dynamic reconstruction of vessels for functional anastomosis. Laboratory investigation. J Neurosurg Pediatr. 2009;3(6):488–95.
113. Kuluz JW, Prado R, He D, Zhao W, Dietrich WD, Watson B. New pediatric model of ischemic stroke in infant piglets by photothrombosis: acute changes in cerebral blood flow, microvasculature, and early histopathology. Stroke. 2007;38(6):1932–7.
114. Zhang K, Sejnowski TJ. A universal scaling law between gray matter and white matter of cerebral cortex. Proc Natl Acad Sci U S A. 2000;97(10):5621–6.
115. Watanabe H, Andersen F, Simonsen CZ, Evans SM, Gjedde A, Cumming P, DaNe XSG. MR-based statistical atlas of the Gottingen minipig brain. Neuroimage. 2001;14(5):1089–96.
116. Lind NM, Moustgaard A, Jelsing J, Vajta G, Cumming P, Hansen AK. The use of pigs in neuroscience: modeling brain disorders. Neurosci Biobehav Rev 2007;31(5):728–51.
117. Lowry W, Richter L, Yachechko R, Pyle A, Tchieu J, Sridharan R, Clark A, Plath K. Generation of human induced pluripotent stem cells from dermal fibroblasts. Proc Natl Acad Sci U S A. 2008;105(8):2883.
118. Huangfu D, Osafune K, Maehr R, Guo W, Eijkelenboom A, Chen S, Muhlestein W, Melton DA. Induction of pluripotent stem cells from primary human fibroblasts with only Oct4 and Sox2. Nat Biotechnol. 2008;26(11):1269–75.
119. Yu J, Vodyanik MA, Smuga-Otto K, Antosiewicz-Bourget J, Frane JL, Tian S, Nie J, Jonsdottir GA, Ruotti V, Stewart R. Induced pluripotent stem cell lines derived from human somatic cells. Science. 2007;318(5858):1917–20.
120. Stadtfeld M, Brennand K, Hochedlinger K. Reprogramming of pancreatic β cells into induced pluripotent stem cells. Curr Biol. 2008;18(12):890–894.
121. Sommer CA, Stadtfeld M, Murphy GJ, Hochedlinger K, Kotton DN, Mostoslavsky G. Induced pluripotent stem cell generation using a single lentiviral stem cell cassette. Stem Cells. 2009;27(3):543–9.
122. Anokye-Danso F, Trivedi CM, Juhr D, Gupta M, Cui Z, Tian Y, Zhang Y, Yang W, Gruber PJ, Epstein JA. Highly efficient miRNA-mediated reprogramming of mouse and human somatic cells to pluripotency. Cell Stem Cell. 2011;8(4):376–88.
123. Woltjen K, Michael IP, Mohseni P, Desai R, Mileikovsky M, Hämäläinen R, Cowling R, Wang W, Liu P, Gertsenstein M. piggyBac transposition reprograms fibroblasts to induced pluripotent stem cells. Nature. 2009;458(7239):766–70.
124. Somers A, Jean JC, Sommer CA, Omari A, Ford CC, Mills JA, Ying L, Sommer AG, Jean JM, Smith BW. Generation of transgene-free lung disease-specific human induced pluripotent stem cells using a single excisable lentiviral stem cell cassette. Stem Cells. 2010;28(10):1728–40.

125. Zhou W, Freed CR. Adenoviral gene delivery can reprogram human fibroblasts to induced pluripotent stem cells. Stem Cells. 2009;27(11):2667–74.
126. Okita K, Nakagawa M, Hyenjong H, Ichisaka T, Yamanaka S. Generation of mouse induced pluripotent stem cells without viral vectors. Science. 2008;322(5903):949–53.
127. Si-Tayeb K, Noto F, Sepac A, Sedlic F, Bosnjak Z, Lough J, Duncan S. Generation of human induced pluripotent stem cells by simple transient transfection of plasmid DNA encoding reprogramming factors. BMC Dev Biol. 2010;10(1):81.
128. Kim D, Kim CH, Moon JI, Chung YG, Chang MY, Han BS, Ko S, Yang E, Cha KY, Lanza R. Generation of human induced pluripotent stem cells by direct delivery of reprogramming proteins. Cell Stem Cell. 2009;4(6):472.
129. Zhou H, Wu S, Joo JY, Zhu S, Han DW, Lin T, Trauger S, Bien G, Yao S, Zhu Y. Generation of induced pluripotent stem cells using recombinant proteins. Cell Stem Cell. 2010;4(5):381.
130. Warren L, Manos PD, Ahfeldt T, Loh YH, Li H, Lau F, Ebina W, Mandal PK, Smith ZD, Meissner A. Highly efficient reprogramming to pluripotency and directed differentiation of human cells with synthetic modified mRNA. Cell Stem Cell. 2010;7(5):618–30.
131. Ambasudhan R, Talantova M, Coleman R, Yuan X, Zhu S, Lipton SA, Ding S. Direct reprogramming of adult human fibroblasts to functional neurons under defined conditions. Cell Stem Cell. 2011;9(2):113–8.
132. Han DW, Tapia N, Hermann A, Hemmer K, Hoing S, Arauzo-Bravo MJ, Zaehres H, Wu G, Frank S, Moritz S, Greber B, Yang JH, Lee HT, Schwamborn JC, Storch A, Scholer HR. Direct reprogramming of fibroblasts into neural stem cells by defined factors. Cell Stem Cell. 2012;10(4):465–72.
133. Kim J, Efe JA, Zhu S, Talantova M, Yuan X, Wang S, Lipton SA, Zhang K, Ding S. Direct reprogramming of mouse fibroblasts to neural progenitors. Proc Natl Acad Sci U S A. 2011;108(19):7838–43.
134. Marro S, Pang ZP, Yang N, Tsai MC, Qu K, Chang HY, Sudhof TC, Wernig M. Direct lineage conversion of terminally differentiated hepatocytes to functional neurons. Cell Stem Cell. 2011;9(4):374–82.
135. Sheng C, Zheng Q, Wu J, Xu Z, Wang L, Li W, Zhang H, Zhao XY, Liu L, Wang Z, Guo C, Wu HJ, Liu Z, He S, Wang XJ, Chen Z, Zhou Q. Direct reprogramming of Sertoli cells into multipotent neural stem cells by defined factors. Cell Res. 2012;22(1):208–18.
136. Thier M, Worsdorfer P, Lakes YB, Gorris R, Herms S, Opitz T, Seiferling D, Quandel T, Hoffmann P, Nothen MM, Brustle O, Edenhofer F. Direct conversion of fibroblasts into stably expandable neural stem cells. Cell Stem Cell. 2012;10(4):473–9.

Chapter 11
Preconditioning and Cell-Based Therapeutics

David C. Hess and Md Nasrul Hoda

One of the major barriers to cell therapy is the poor and incomplete engraftment of stem and progenitor cells after transplantation. Transplanted cells undergo apoptosis in the hostile environment of ischemic tissue of activated inflammatory cells such as neutrophils and macrophages. Mesenchymal stem cells (MSC), one of the most promising cell therapies, show limited survival and engraftment in the heart; after injection into the left ventricle of nude mice, less than 1 % survive 4 days after transplantation [1].

Early work in cell transplantation in the brain came in the Parkinson's disease field [2]. Studies showed that only about 5–20 % of transplanted fetal dopaminergic mesencephalic cells survived in the striatum, most undergoing apoptotic cell death in the first week after transplantation [3, 4]. Graft survival was even poorer in aged animals although the time course of cell death was similar to younger animals. Despite limited long-term engraftment, there has been long-term benefit in some patients up to 18 years after transplantation. However, some grafted fetal dopaminergic neurons show evidence of protein aggregates of alpha-synuclein that represent Lewy bodies indicating that the degenerative milieu of the brain has triggered the degenerative process in the grafted cells [2].

In terms of cell transplantation, stroke and other acute brain injuries differ from Parkinson's disease in that the injury is acute with a known time of onset and the underlying brain is not undergoing an active and ongoing neurodegenerative process. After the acute brain injury with a defined time of onset, there is an acute inflammatory response in the brain accompanied by a systemic inflammatory and

D. C. Hess (✉) · M. N. Hoda
Department of Neurology, Medical College of Georgia, Georgia Regent's University, 15th Street 1120, Augusta, GA 30912, USA
e-mail: dhess@gru.edu

M. N. Hoda
College of Allied Health Sciences, Georgia Regent's University, Augusta, GA 30912, USA

D. C. Hess (ed.), *Cell Therapy for Brain Injury,* DOI 10.1007/978-3-319-15063-5_11

immune response involving the spleen and other lymphoid tissue. Later, there is chronic inflammation with microglial activation. In the early hours and days after injury, there is oxidative stress; but within days there is also ongoing “remodeling” with early attempts at repair.

Transplantation of neural stem cells (NSC) into the area of the infarct in rodent models of focal cerebral ischemia shows a wide variability of graft survival. Hicks et al. found that less than 1 % of cells grafted and survived at 2 months after transplantation [5]. Darsalia et al. transplanted NSC into the striatum after stroke and found higher graft survival at 48 h (58 %) than at 6 weeks (27 %) [6]. Moreover, there was greater graft survival in the infarcted striatum than intact striatum (31 %) showing that the acute stroke environment may be more hospitable than normal brain [6]. Similar findings of improved graft survival in injured versus normal brains have been shown in transient global ischemia models. Transplantation of NSC into the hippocampal fissure or the periventricular areas near the corpus callosum was more efficient in “injured” than normal brain as integrated surviving cells engrafted in 60 % of lesioned animals (37/61) compared to only 25 % of nonlesioned controls (7/28). However, in the ischemic animals that showed engraftment, only about 19 % of the transplanted cells engrafted [7]. Other studies have found poorer survival of transplanted NSC in stroke than uninjured brain [8, 9]. Some of these differences relate to the experimental model tested, where and how the cells were transplanted, and how the cells were prepared.

Clearly, greater graft survival and function is important for the intracerebral delivery route where the goal is both “cell replacement” and direct trophic effects of the transplanted cells on the brain. However, the efficacy of some cell-based therapeutics may not require long-term engraftment and engraftment may not be desirable with the attendant risks of tumor formation and an ongoing inflammatory reaction with graft rejection. For example, with the intravenous route of cell transplantation with Multistem in the 24–48-h time window, the major mechanism of action is immunomodulation and the target organs are the spleen and lymphoid tissue with secondary effects on the brain [10]. In this case, the cells “do their job and leave.” However, even if long-term engraftment is not desired, greater “potency” of the cells in the short term will result in a greater therapeutic effect.

The potency of a stem cell is generally defined as its ability to differentiate into various cell types. For example, “totipotent cells” have the greatest differentiation potential followed by “pluripotent” cells, and then “multipotent” stem cells. However, the definition of “potency” more appropriate for cell therapy is the ability of the cells to repair or promote recovery and this may involve greater survival and integration in the tissue, the elaboration of trophic factors, the immunomodulatory effect on host immune and inflammatory responses, and the ability to promote angiogenesis and tissue repair. The “potency” of cell therapy can be increased by genetically modifying the cells. For example, MSC transfected with the heme oxygenase 1 (HO-1) or Akt genes are more effective in preclinical models than non-transfected cells [11, 12]. However, an easier and safer approach is to pretreat or “precondition” the cells with hypoxia or other agents such as IL-6 or minocycline that trigger protective pathways in the cells and increase their potency.

Preconditioning

Preconditioning is the application of a sublethal stimulus to a cell, tissue, organ such as the heart or brain, or whole animal in order to protect against a later lethal stimulus. This concept has been known since antiquity. King Mithridates IV of Pontus (132–63 B.C.), an implacable foe of the Romans, took small doses of poisons to protect himself against assassination attempts by poisoning, a common danger and fear in that era [13, 14]. "Mithridatism" refers to the use of small amounts of toxins or poisons to protect against a lethal dose. Paracelsus, a medieval physician, taught that the "dose makes the poison," a basic principle of toxicology [15, 16].

Preconditioning can be "hypoxic" or 'ischemic" or include other agents that are normally pro-inflammatory and harmful such as lipopolysaccharide (LPS). The definition of preconditioning is often widened to include agents that are not normally toxic or lethal such as pharmacological agents or growth factors such as erythropoietin (EPO). The fundamental principle is that preconditioning triggers endogenous protective pathways and elicits a "protective phenotype."

In 1986, Murry et al. first reported ischemic preconditioning in the heart when they demonstrated that brief repetitive 5-min periods of occlusion of the left anterior descending coronary artery in dogs protected against later longer durations of occlusion (40 min) and lethal ischemia [17]. This was extended to the brain in a transient global ischemia model when Kitigawa et al. reported that brief 2-min duration occlusions of both carotid arteries in the gerbil protected against CA1 neuronal cell in the hippocampus after later 5-min occlusions [18]. The durations and intervals of the ischemic stimulus were critical to the protection. This phenomenon was found in other areas of the brain, outside the hippocampus, and was termed "ischemic tolerance" [18, 19].

Hypoxia

Stem and progenitor cells reside in a "stem cell niche," an anatomical compartment enriched with blood vessels and glycoproteins [20]. Ordinarily, cells are cultured in conditions of "normoxia" of 21 % oxygen. However, stem cells reside in a niche where oxygen concentrations are far lower [21]. Using direct measurements of oxygen tensions in tissue or with mathematical modeling, oxygen tensions in MSC compartments are estimated to be 2–8 % oxygen, hematopoietic stem cell compartments (HSC) 1–6 % oxygen, and NSC <1–6 % oxygen. NSC reside in a very hypoxic environment in the subventricular zone (SVZ). Measurements of oxygen tension are as low as 0.55 % oxygen in various areas of the brain suggesting that low oxygen tensions exist in the SVZ although direct measurements from the SVZ have never been performed [21]. These conditions of hypoxia *in vivo,* termed "physiological normoxia," are much lower than the 21 % oxygen conditions of tissue culture normoxia.

Hypoxic Preconditioning of Cells

Since stem and progenitor cells normally reside in a "hypoxic" stem cell niche and they are often being transplanted in a hypoxic–ischemic tissue environment, culturing the cells in conditions of hypoxia might better prepare them for survival and engraftment after transplantation and improve their "trophic effects" on tissue. In 2003, Akita et al. reported that hypoxic preconditioning (HP) of peripheral blood mononuclear cells for 7 days increased their differentiation into endothelial progenitor cells (EPCs), increased secretion of vascular endothelial growth factor (VEGF), and increased their migratory potential and their efficacy *in vivo* at therapeutic vasculogenesis in a hind-limb ischemia model compared to cells treated under normoxia [22]. Other early studies showed that HP of peripheral bone marrow mononuclear cells in 2% oxygen for 24 h increased their resistance to oxidative stress *in vitro* and survival *in vivo* after intramuscular implantation and they were more effective at increasing microvascular density and blood blow in a hind-limb ischemia model at 28 days compared with cells treated under normoxia [23].

One of the first demonstrations of preconditioning of stem/progenitor cells in cardiac injury models was "pharmacological" preconditioning of skeletal myoblasts with diazoxide [24]. Diazoxide opens mitochondrial potassium channels, an underlying mechanism of preconditioning. Skeletal myoblasts represented a promising therapy for cardiac injury and heart failure but the major barrier was poor graft survival. Niagara et al. showed that preconditioning with diazoxide reduced apoptosis *in vitro,* increased the release of paracrine factors, increased graft survival, and improved cardiac function 4 weeks later as measured by echocardiography in a rodent myocardial infarction model [24].

MSC reside in close proximity to blood vessels in a "perivascular" niche in nearly all tissues and appear to be a subset of "pericytes" [25, 26]. Due to their ability to function as "trophic factories" or "drug stores" with immunomodulatory effects, MSC are a promising cell therapy [27]. One of the first studies of HP of MSC showed that culturing in 1–3% oxygen for 16–24 h, activated Akt, induced expression of cMet, a receptor for hepatocyte growth factor, increased migration, and increased their efficacy at restoring blood flow in a hind-limb ischemic model compared to cells treated in normoxia [28]. Other studies have confirmed that hypoxic conditioning of MSC increases their survival and function in a wide variety of *in vitro* and *in vivo* models [29–41]. HP "reverses" the aging of stem cells. HP of senescent adipose stromal stem cells reversed the decline in their angiogenic potential in an *in vivo* mouse hind-limb ischemia model [42].

Hypoxia Preconditioning in Brain Injury Models

HP is effective in NSC and in brain injury models. Embryonic stem cells differentiated into neural progenitor cells with retinoic acid (ES-NPCs) were more resistant to apoptosis and caspase 3 activation after exposure to hypoxia [43]. These same effects were seen with bone-marrow-derived MSC. HP increased secretion of EPO and upregulated expression of bcl-2, hypoxia-inducible factor (HIF-1α), erythropoietin receptor (EPOR), neurofilament (NF), and synaptophysin in ES-NPCs. The hypoxic effect was attenuated by blocking the EPOR R receptor and could be mimicked by EPO. These striking effects of HP *in vitro* were also seen in an *in vivo* stroke model. HP- ES-NPCs were transplanted by stereotactic implantation 2 days after MCA temporary suture occlusion in the infarct area and showed greater survival, enhanced neuronal differentiation, and facilitated long-term (35 days) functional recovery of the rats compared to nonconditioned cells.

HP of bone-marrow-derived MSC was also shown to be more effective than normoxic cells after intravenous delivery in a MCA rodent stroke model. In comparison to cells treated with normoxia, MSC exposed to 0.5 % oxygen upregulated HIF-1alpha and trophic/growth factors in BMSCs, including brain-derived neurotrophic factor (BDNF), glial cell-derived neurotrophic factor (GDNF), VEGF and its receptor FIK-1, EPO and its receptor EPOR, and stromal derived factor-1 (SDF-1) and its CXC chemokine receptor 4 (CXCR4). HP cells administered intravenously 24 h after a temporary 90-min middle cerebral artery (MCA) occlusion reduced microglial activation in the brain and improved functional outcome as measured by Rotarod at 15 days [44]. HP of MSC were also effective when these cells were delivered by an intranasal route after stroke. HP-MSC had greater migration to ischemic cortex, reduced infarct size, and improved functional outcome compared to nonconditioned MSC [45].

HP of human umbilical cord-derived CD34 cells ($hUCB^{34}$) has also been shown to be effective in stroke. In a three-vessel occlusion model in the rat (MCA and bilateral carotids), intracerebral transplantation of HP $hUCB^{34}$ led to greater engraftment, enhanced angiogenesis, facilitated proliferation of endogenous neural stem cells, promoted neurite outgrowth, and improved outcome in comparison to untreated (no hypoxia) $hUCB^{34}$. This effect was mediated, at least in part, by hypoxia-mediated upregulation of exchange protein activated by cAMP-1 (Epac1) [46].

The HP regimens are varied and include single brief episodes (of less than an hour), longer single episodes of 12–24 h in duration, [28, 43, 46], more prolonged hypoxia (0.5 %) for 24–72 h [44], and repeated brief (30 min) episodes of hypoxia [47]. Brief repeated episodes are more effective than single brief episodes (30 min) [47, 48]. However, perhaps due to ease of use, most studies have used continuous hypoxia rather than brief repeated episodes [49].

HP increases the survival and "potency" of a wide variety of stem and progenitor cells including PBMC, MSC, neural progenitor cells, and $hUCB^{34}$. A "potency assay" measures the biological activity of a cell therapy [50]. These assays may be *in vitro* such as an Elisa that measure levels of secreted trophic factors such as VEGF,

Table 11.1 Types of preconditioning of cells prior to transplantation

Hypoxia–ischemia	Cytokines	Trophic/growth factors	Drugs/pharmacological agents
Hypoxia (0.5 to 3% O2)	IL-6	EPO, SDF-1	Minocycline/doxycycline
Cobalt		Autologous stroke serum	Diazoxide

EPO erythropoietin, *SDF-1* stromal derived factor-1

or *in vitro* assays of migration or angiogenesis. *In vivo* assays such as angiogenesis, tissue protection, or facilitation of functional outcomes may be even more predictive of a response in humans although they tend to be "low throughput" (Tables 11.1 and 11.2).

Pathways of Protection

The mechanism by which hypoxia increases cell survival and "potency" involves HIF-1. HIF-1, discovered and characterized by Semenza, is a transcription factor involved in hypoxic adaptation and functions as a master regulator of oxygen homeostasis expressed in all metazoan species analyzed [51–53]. HIF-1 controls oxygen delivery by regulating angiogenesis and oxygen utilization, regulates glucose metabolism, and is involved in redox homeostasis. HIF-1 consists of HIF-1α and HIF-1β subunits, which each contain basic helix–loop–helix-PAS (bHLH-PAS) domains. After heterodimerization, they bind DNA leading to transcription of downstream genes such as VEGF, EPO, sodium–calcium, exchanger −1 (NCX-1). HIF binds to hypoxia-responsive elements and modulates up to 200 genes involved in angiogenesis, mitochondrial biogenesis, cell proliferation, and apoptosis [49, 54]. HIF cytoprotective pathways involve VEGF, EPO, and an HIF-1-sphingosine kinase-sphingosine 1 phosphate-CCl 2 (MCP-1) signaling pathway that mediates HP-induced ischemic tolerance in the brain [55, 56].

HIF-1α protein levels determine HIF-1 transcriptional activity as HIF-1β heterodimerizes with other bHLH-PAS proteins and are present in excess (Semenza et al. 1996). Under conditions of high oxygen tension, HIF-1 is bound by to Von Hippel–Lindau (VHL) protein which targets HIF1a for degradation by the proteasome. The binding of VHL to HIF-1 is dependent upon hydroxylation of a specific

Table 11.2 Hypoxic preconditioning increases "potency" of cells

In vitro	*In vivo*
Greater survival, reduced apoptosis	Increased cell survival and engraftment
Increased migration in migratory assays	Increased migration in the brain
Increased angiogenesis	Increased angiogenesis
	Reduction of tissue injury
	Facilitates functional recovery

proline residue in HIF-1α by the prolyl hydroxylase PHD2 which uses O_2 as a substrate. Therefore, its activity is inhibited under hypoxic conditions. One attractive target for pharmaceutical companies is to develop prolyl hydroxylase inhibitors (PHIs) to induce HIF activity for treatment of disorders in which HIF mediates protective responses. For example, a small molecule inhibitor of HIF prolyl hydroxylases was neuroprotective if administered prior to cerebral ischemia in an MCA occlusion model; however, it was not protective if given at the time of ischemia, limiting its clinical usefulness [57].

SDF-1/CXCR Axis

SDF-1/CXCL12 belongs to the CSC chemokine family and is involved in angiogenesis and stem and progenitor cell migration [58]. SDF-1 had two known receptors, CXCR4, the best known, and CXCR7. The CXCR4 receptor is expressed on lymphocytes, monocytes, HSC, and ES cells. SDF-1/CXCR4 is involved in progenitor and stem cell trafficking and homing to sites of injury. Hypoxia increases the expression of CXCR 4 and CXCR7 and the migration, adhesion, and survival of MSCs. CXCR4 mediates the increased migration while both CXCR4 and CXCR7 mediate adhesion and CXCR7 resistance to oxidative stress [35] This effect is mediated via HIF-1 and Akt.

Preconditioning of MSC with SDF-1 decreases their apoptosis, improves their survival and engraftment, and reduces infarction in an MI model [59]. Preconditioning EPCs with SDF-1 increases their angiogenic potential *in vitro* and *in vivo* in a hind-limb ischemia model [60]. To date, there are no studies of preconditioning cells with SDF-1 in a stroke or brain injury model.

Other "Preconditoning" Agents

IL-6

IL-6 is a proinflammatory cytokine and, as part of the acute phase response, is associated with recurrent stroke, although it is doubtful that IL-6 per se is causal in recurrent stroke [61, 62]. IL-6 promotes a survival pathway through activation of signal transducer and activator of transcription 3 (STAT3) and is neuroprotective in some stroke models [63]. NSC preconditioned with IL-6 and transplanted into mouse brain 6 h or 7 days after stroke increased angiogenesis and improved functional recovery compared to nonconditioned cells [8]. Transplantation at 6 h also reduced infarct size. These effects were abolished when the preconditioned cells were treated with a small interfering RNA to STAT 3.

Minocycline

Minocycline is an effective neuroprotectant and has been tested in early-phase clinical trials in stroke in humans [64–66]. Sakata et al. preconditioned NSC with minocycline and reported protection against *in vitro* oxygen glucose deprivation (OGD) [9] Minocycline preconditioning upregulated the transcription factor nuclear factor erythroid 2-related factor 2 (Nrf2) and the antioxidant genes, NQ01 and HO-1, and increased the secretion of the paracrine factors, BDNF, GDNF, nerve growth factor, and VEGF. Minocycline-preconditioned NSC also survived better in ischemic brain when transplanted at 6 h post stroke than nonconditioned cells. Moreover, the minocycline preconditioned NSC reduced infarct size and improved functional outcome at 28 days as measured by the rotarod and beam balance test. Treatment of the preconditioned cells with NRf2 small inhibitory RNA before transplantation blocked this neuroprotective effect. NSC preconditioning with doxycycline, another tetracycline similar to minocycline, also improved survival of NSC in a hypoxia reoxygenation model and also induced upregulation of Nrf2, similar to minocycline [67]. Minocycline is also effective at increasing graft survival when administered to the host. Intracerebral transplantation of porcine fetal neurons cells in the striatum resulted in no graft survival in any animal at 62 days; however, if the animals were treated daily with minocycline, cells engrafted in 40% of the animals [68].

Autologous Stroke Sera

Another approach is to precondition cells with autologous sera prior to transplantation. This approach leverages the reparative effects of trophic factors in the blood after stroke. In the STARTING clinical trial of autologous MSC in stroke patients, MSC will be *ex vivo* expanded in culture with autologous stroke patient sera obtained as soon as possible after the stroke and then administered IV [69].

Pre- and Postconditioning the Host Tissue

While there has been extensive work on preconditioning the donor cells prior to transplantation, there has been less work on "conditioning" the host or the recipient tissue. With this approach, the aim is to reduce the acute inflammatory response of the host that impairs engraftment and promotes the death of the graft. Since we are transplanting cells into tissue *after* an ischemic episode or potentially after a traumatic brain injury, the precise "temporal" term for this type of conditioning would be "postconditioning."

Remote Ischemic Conditioning

While direct pre- and postconditioning would require access and repetitive occlusion of the middle cerebral artery or other major cerebral artery in patients with stroke, conditioning can be applied at a "distance" or a "remote site" and still provide protection to a distant organ like the brain [70]. Remote ischemic conditioning (RIC) allows a more accessible site like the limb to be used. This can be accomplished with a simple tourniquet or blood pressure cuff repeatedly inflated and deflated on the arm or leg. The intervention is feasible, safe, and well tolerated. There is a large body of evidence that ischemic conditioning is effective in animal models for stroke at reducing infarct size and improving functional outcome when applied before (pre), during (per), and after ischemia–reperfusion (post) [70, 71]. This effect may be mediated by improved cerebral blood flow and cytoprotection of the organ. RIC also seems to dampen and modulate the immune response. There is already clinical trial data suggesting the efficacy of RIC in humans in a chronic conditioning setting to reduce recurrent stroke in high-risk patients with intracranial stenosis and in the prehospital setting in acute ischemic stroke [72, 73].

To date, there has been little published work on this approach. In a rodent MI model, remote ischemic postconditioning achieved by four cycles of 5-min occlusions of the abdominal aorta 1 week post MI, increased the retention of intravenously administered MSC 1 day later in the heart and improved cardiac function at 1 month as measured by echocardiography compared to controls [74]. The remote postconditioning increased SDF-1 in the ischemic heart and transiently in the blood; the beneficial effect of postconditioning was blocked by antibodies to CXCR4, suggesting that the SDF-1-CXCR4 axis was at least partially mediating the effect.

We propose that RIC be tested in preclinical models of stem cell transplantation. Testing with appropriate "sham" conditioning controls should be performed with intravenous, intra-arterial, and intracerebral delivery routes. Another related innovative approach would be to combine HP of transplanted stem/progenitor cells with RIC so that both the donor cells and the host would be conditioned. If preclinical models suggest efficacy, these approaches could be easily translated to patients (Fig. 11.1).

Summary

HP of stem and progenitor cells prior to transplantation increases their "potency" in terms of both *in vitro* and *in vivo* survival and function. One of the rationales for this approach is the finding that stem and progenitor cells normally reside in a "hypoxic" stem cell niche. Moreover, other "preconditioning agents" such as IL-6, minocycline, and SDF-1 represent alternative approaches. In the future, more attention will be given to "conditioning" the host to prepare the tissue to receive a transplant. Remote limb ischemic conditioning is a safe and feasible approach that might best be combined with hypoxic conditioning of transplanted cells to optimize cell-based therapeutics.

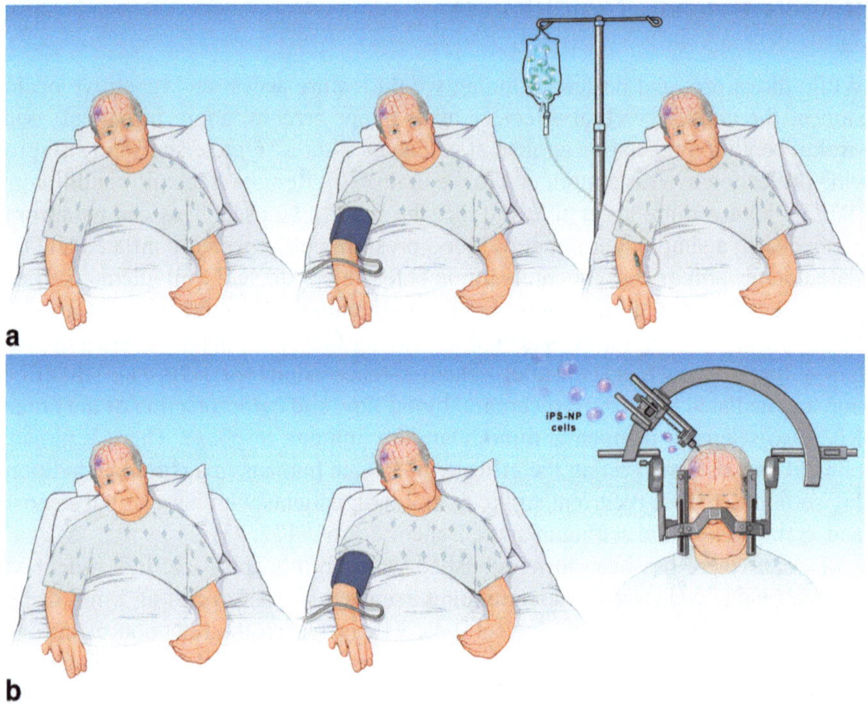

Fig. 11.1 **a** Remote limb conditioning (RIC) in patient with stroke. In **a,** the patient undergoes remote limb conditioning with a blood pressure cuff repeatedly inflated and deflated on the arm, 24–48 h after stroke (*middle pane*) The following day, bone-marrow-derived stem cells are administered intravenously (*right pane*). **b** In **b,** the RIC is applied at around day 5–7 post stroke, then the patient is taken to the operating room (OR, *right pane*) for stereotactic transplantation of induced pluripotent stem cells differentiated into neural progenitor cells (iPS-NP). The RIC could be continued in the OR and thereafter

References

1. Toma C, Pittenger MF, Cahill KS, Byrne BJ, Kessler PD. Human mesenchymal stem cells differentiate to a cardiomyocyte phenotype in the adult murine heart. Circulation. 2002;105(1):93–8.
2. Petit GH, Olsson TT, Brundin P. The future of cell therapies and brain repair: parkinson's disease leads the way. Neuropathol Appl Neurobio. 2014;40(1):60–70.
3. Sortwell CE, Camargo MD, Pitzer MR, Gyawali S, Collier TJ. Diminished survival of mesencephalic dopamine neurons grafted into aged hosts occurs during the immediate postgrafting interval. Exp Neurol. 2001;169(1):23–9.
4. Sortwell CE, Pitzer MR, Collier TJ. Time course of apoptotic cell death within mesencephalic cell suspension grafts: implications for improving grafted dopamine neuron survival. Exp Neurol. 2000;165(2):268–77.
5. Hicks AU, Lappalainen RS, Narkilahti S, Suuronen R, Corbett D, Sivenius J, et al. Transplantation of human embryonic stem-cell-derived neural precursor cells and enriched environment after cortical stroke in rats: cell survival and functional recovery. Eur J Neurosci. 2009;29(3):562–74.

6. Darsalia V, Allison SJ, Cusulin C, Monni E, Kuzdas D, Kallur T, et al. Cell number and timing of transplantation determine survival of human neural stem cell grafts in stroke-damaged rat brain. J Cereb Blood Flow Metab. 2011;31(1):235–42.
7. Rota Nodari L, Ferrari D, Giani F, Bossi M, Rodriguez-Menendez V, Tredici G, et al. Long-term survival of human neural stem cells in the ischemic rat brain upon transient immunosuppression. PLoS One. 2010;5(11):e14035.
8. Sakata H, Narasimhan P, Niizuma K, Maier CM, Wakai T, Chan PH. Interleukin 6-preconditioned neural stem cells reduce ischaemic injury in stroke mice. Brain. 2012;135(Pt 11):3298–310.
9. Sakata H, Niizuma K, Yoshioka H, Kim GS, Jung JE, Katsu M, et al. Minocycline-preconditioned neural stem cells enhance neuroprotection after ischemic stroke in rats. J Neurosci. 2012;32(10):3462–73.
10. Hess DC, Sila CA, Furlan AJ, Wechsler LR, Switzer JA, Mays RW. A double-blind placebo-controlled clinical evaluation of MultiStem for the treatment of ischemic stroke. Int J Stroke : Off J Int Stroke Soc. 2014;9(3):381–6.
11. Jiang Y, Chen L, Tang Y, Ma G, Shen C, Qi C, et al. HO-1 gene overexpression enhances the beneficial effects of superparamagnetic iron oxide labeled bone marrow stromal cells transplantation in swine hearts underwent ischemia/reperfusion: an MRI study. Basic Res Cardiol. 2010;105(3):431–42.
12. Lim SY, Kim YS, Ahn Y, Jeong MH, Hong MH, Joo SY, et al. The effects of mesenchymal stem cells transduced with Akt in a porcine myocardial infarction model. Cardiovasc Res. 2006;70(3):530–42.
13. Ring J, Gutermuth J. 100 years of hyposensitization: history of allergen-specific immunotherapy (ASIT). Allergy. 2011;66(6):713–24.
14. Valle G, Carmignani M, Stanislao M, Facciorusso A, Volpe AR. Mithridates VI Eupator of Pontus and mithridatism. Allergy. 2012;67(1):138–9; author reply 9–40.
15. Rozman KK, Doull J. Paracelsus, Haber and Arndt. Toxicology. 2001;160(1–3):191–6.
16. Waddell WJ. History of dose response. J Toxic Sci. 2010;35(1):1–8.
17. Murry CE, Jennings RB, Reimer KA. Preconditioning with ischemia: a delay of lethal cell injury in ischemic myocardium. Circulation. 1986;74(5):1124–36.
18. Kitagawa K, Matsumoto M, Kuwabara K, Tagaya M, Ohtsuki T, Hata R, et al. 'Ischemic tolerance' phenomenon detected in various brain regions. Brain Res. 1991;561(2):203–11.
19. Kitagawa K, Matsumoto M, Tagaya M, Hata R, Ueda H, Niinobe M, et al. 'Ischemic tolerance' phenomenon found in the brain. Brain Res. 1990;528(1):21–4.
20. Scadden DT. The stem cell niche as an entity of action. Nature. 2006;441(7097):1075–9.
21. Mohyeldin A, Garzon-Muvdi T, Quinones-Hinojosa A. Oxygen in stem cell biology: a critical component of the stem cell niche. Cell Stem Cell. 2010;7(2):150–61.
22. Akita T, Murohara T, Ikeda H, Sasaki K, Shimada T, Egami K, et al. Hypoxic preconditioning augments efficacy of human endothelial progenitor cells for therapeutic neovascularization. Lab Investig; J Tech Methods Pathol. 2003;83(1):65–73.
23. Kubo M, Li TS, Suzuki R, Shirasawa B, Morikage N, Ohshima M, et al. Hypoxic preconditioning increases survival and angiogenic potency of peripheral blood mononuclear cells via oxidative stress resistance. Am J Physiol Heart Circ Physiol. 2008;294(2):H590–5.
24. Niagara MI, Haider H, Jiang S, Ashraf M. Pharmacologically preconditioned skeletal myoblasts are resistant to oxidative stress and promote angiomyogenesis via release of paracrine factors in the infarcted heart. Circ Res. 2007;100(4):545–55.
25. Crisan M, Yap S, Casteilla L, Chen CW, Corselli M, Park TS, et al. A perivascular origin for mesenchymal stem cells in multiple human organs. Cell Stem Cell. 2008;3(3):301–13.
26. Lv FJ, Tuan RS, Cheung KM, Leung VY. Concise review: the surface markers and identity of human mesenchymal stem cells. Stem Cells. 2014;32(6):1408–19.
27. Caplan AI, Correa D. The MSC: an injury drugstore. Cell Stem Cell. 2011;9(1):11–5.
28. Rosova I, Dao M, Capoccia B, Link D, Nolta JA. Hypoxic preconditioning results in increased motility and improved therapeutic potential of human mesenchymal stem cells. Stem Cells. 2008;26(8):2173–82.

29. Fotia C, Massa A, Boriani F, Baldini N, Granchi D. Hypoxia enhances proliferation and stemness of human adipose-derived mesenchymal stem cells. Cytotechnology. 2014 May 6.
30. Jaussaud J, Biais M, Calderon J, Chevaleyre J, Duchez P, Ivanovic Z, et al. Hypoxia-preconditioned mesenchymal stromal cells improve cardiac function in a swine model of chronic myocardial ischaemia. Eur J Cardiothorac Surg. 2013;43(5):1050–7.
31. Chacko SM, Ahmed S, Selvendiran K, Kuppusamy ML, Khan M, Kuppusamy P. Hypoxic preconditioning induces the expression of prosurvival and proangiogenic markers in mesenchymal stem cells. Am J Physiol Cell Physiol. 2010;299(6):C1562–70.
32. Chang CP, Chio CC, Cheong CU, Chao CM, Cheng BC, Lin MT. Hypoxic preconditioning enhances the therapeutic potential of the secretome from cultured human mesenchymal stem cells in experimental traumatic brain injury. Clin Sci (Lond). 2013;124(3):165–76.
33. Hu X, Wei L, Taylor TM, Wei J, Zhou X, Wang JA, et al. Hypoxic preconditioning enhances bone marrow mesenchymal stem cell migration via Kv2.1 channel and FAK activation. Am J Physiol Cell Physiol. 2011;301(2):C362–72.
34. Leroux L, Descamps B, Tojais NF, Seguy B, Oses P, Moreau C, et al. Hypoxia preconditioned mesenchymal stem cells improve vascular and skeletal muscle fiber regeneration after ischemia through a Wnt4-dependent pathway. Mol Ther. 2010;18(8):1545–52.
35. Liu H, Xue W, Ge G, Luo X, Li Y, Xiang H, et al. Hypoxic preconditioning advances CXCR4 and CXCR7 expression by activating HIF-1alpha in MSCs. Biochem Biophys Res Commun. 2010;401(4):509–15.
36. Liu L, Gao J, Yuan Y, Chang Q, Liao Y, Lu F. Hypoxia preconditioned human adipose derived mesenchymal stem cells enhance angiogenic potential via secretion of increased VEGF and bFGF. Cell Biol Int. 2013;37(6):551–60.
37. Oh JS, Ha Y, An SS, Khan M, Pennant WA, Kim HJ, et al. Hypoxia-preconditioned adipose tissue-derived mesenchymal stem cell increase the survival and gene expression of engineered neural stem cells in a spinal cord injury model. Neurosci Lett. 2010;472(3):215–9.
38. Tsai CC, Yew TL, Yang DC, Huang WH, Hung SC. Benefits of hypoxic culture on bone marrow multipotent stromal cells. Am J Blood Res. 2012;2(3):148–59.
39. Wang JA, Chen TL, Jiang J, Shi H, Gui C, Luo RH, et al. Hypoxic preconditioning attenuates hypoxia/reoxygenation-induced apoptosis in mesenchymal stem cells. Acta Pharmacol Sin. 2008;29(1):74–82.
40. Watanabe S, Arimura Y, Nagaishi K, Isshiki H, Onodera K, Nasuno M, et al. Conditioned mesenchymal stem cells produce pleiotropic gut trophic factors. J Gastroenterol. 2014;49(2):270–82.
41. Yu J, Yin S, Zhang W, Gao F, Liu Y, Chen Z, et al. Hypoxia preconditioned bone marrow mesenchymal stem cells promote liver regeneration in a rat massive hepatectomy model. Stem Cell Res Ther. 2013;4(4):83.
42. De Barros S, Dehez S, Arnaud E, Barreau C, Cazavet A, Perez G, et al. Aging-related decrease of human ASC angiogenic potential is reversed by hypoxia preconditioning through ROS production. Mol Ther. 2013;21(2):399–408.
43. Theus MH, Wei L, Cui L, Francis K, Hu X, Keogh C, et al. *In vitro* hypoxic preconditioning of embryonic stem cells as a strategy of promoting cell survival and functional benefits after transplantation into the ischemic rat brain. Exp Neurol. 2008;210(2):656–70.
44. Wei L, Fraser JL, Lu ZY, Hu X, Yu SP. Transplantation of hypoxia preconditioned bone marrow mesenchymal stem cells enhances angiogenesis and neurogenesis after cerebral ischemia in rats. Neurobiol Dis. 2012;46(3):635–45.
45. Wei N, Yu SP, Gu X, Taylor TM, Song D, Liu XF, et al. Delayed intranasal delivery of hypoxic-preconditioned bone marrow mesenchymal stem cells enhanced cell homing and therapeutic benefits after ischemic stroke in mice. Cell Transpl. 2013;22(6):977–91.
46. Lin CH, Lee HT, Lee SD, Lee W, Cho CW, Lin SZ, et al. Role of HIF-1alpha-activated Epac1 on HSC-mediated neuroplasticity in stroke model. Neurobiol Dis. 2013;58:76–91.
47. Kim HW, Haider HK, Jiang S, Ashraf M. Ischemic preconditioning augments survival of stem cells via miR-210 expression by targeting caspase-8-associated protein 2. J Biol Chem. 2009;284(48):33161–8.

48. Haider H, Ashraf M. Preconditioning and stem cell survival. J Cardiovasc Transl Res. 2010;3(2):89–102.
49. Muscari C, Giordano E, Bonafe F, Govoni M, Pasini A, Guarnieri C. Priming adult stem cells by hypoxic pretreatments for applications in regenerative medicine. J Biomed Sci. 2013;20:63.
50. Stroncek DF, Jin P, Wang E, Jett B. Potency analysis of cellular therapies: the emerging role of molecular assays. J Transl Med. 2007;5:24.
51. Semenza GL. Hypoxia-inducible factors in physiology and medicine. Cell. 2012;148(3):399–408.
52. Wang GL, Semenza GL. Characterization of hypoxia-inducible factor 1 and regulation of DNA binding activity by hypoxia. J Biol Chem. 1993;268(29):21513–8.
53. Wang GL, Semenza GL. General involvement of hypoxia-inducible factor 1 in transcriptional response to hypoxia. Proc Natl Acad Sci U S A. 1993;90(9):4304–8.
54. Tsai YP, Wu KJ. Hypoxia-regulated target genes implicated in tumor metastasis. J Biomed Sci. 2012;19:102.
55. Stowe AM, Wacker BK, Cravens PD, Perfater JL, Li MK, Hu R, et al. CCL2 upregulation triggers hypoxic preconditioning-induced protection from stroke. J Neuroinflammation. 2012;9:33.
56. Wacker BK, Perfater JL, Gidday JM. Hypoxic preconditioning induces stroke tolerance in mice via a cascading HIF, sphingosine kinase, and CCL2 signaling pathway. J Neurochem. 2012;123(6):954–62.
57. Chen RL, Ogunshola OO, Yeoh KK, Jani A, Papadakis M, Nagel S, et al. HIF prolyl hydroxylase inhibition prior to transient focal cerebral ischaemia is neuroprotective in mice. J Neurochem. 2014 Jun 26. doi: 10.1111/jnc.12804.
58. Cencioni C, Capogrossi MC, Napolitano M. The SDF-1/CXCR4 axis in stem cell preconditioning. Cardiovasc Res. 2012;94(3):400–7.
59. Pasha Z, Wang Y, Sheikh R, Zhang D, Zhao T, Ashraf M. Preconditioning enhances cell survival and differentiation of stem cells during transplantation in infarcted myocardium. Cardiovasc Res. 2008;77(1):134–42.
60. Zemani F, Silvestre JS, Fauvel-Lafeve F, Bruel A, Vilar J, Bieche I, et al. Ex vivo priming of endothelial progenitor cells with SDF-1 before transplantation could increase their proangiogenic potential. Arterioscler Thromb Vasc Biol. 2008;28(4):644–50.
61. Welsh P, Lowe GD, Chalmers J, Campbell DJ, Rumley A, Neal BC, et al. Associations of proinflammatory cytokines with the risk of recurrent stroke. Stroke; J Cereb Circ. 2008;39(8):2226–30.
62. Whiteley W, Jackson C, Lewis S, Lowe G, Rumley A, Sandercock P, et al. Association of circulating inflammatory markers with recurrent vascular events after stroke: a prospective cohort study. Stroke; J Cereb Circ. 2011;42(1):10–6.
63. Jung JE, Kim GS, Chan PH. Neuroprotection by interleukin-6 is mediated by signal transducer and activator of transcription 3 and antioxidative signaling in ischemic stroke. Stroke; J Cereb Circ. 2011;42(12):3574–9.
64. Fagan SC, Waller JL, Nichols FT, Edwards DJ, Pettigrew LC, Clark WM, et al. Minocycline to improve neurologic outcome in stroke (MINOS): a dose-finding study. Stroke; J Cereb Circ. 2010;41(10):2283–7.
65. Hess DC, Fagan SC. Repurposing an old drug to improve the use and safety of tissue plasminogen activator for acute ischemic stroke: minocycline. Pharmacotherapy. 2010; 30(7 Pt 2):55S–61S.
66. Liao TV, Forehand CC, Hess DC, Fagan SC. Minocycline repurposing in critical illness: focus on stroke. Curr Top Med Chem. 2013;13(18):2283–90.
67. Malik YS, Sheikh MA, Zhu X. Doxycycline can stimulate cytoprotection in neural stem cells with oxygen-glucose deprivation-reoxygenation injury: a potential approach to enhance effectiveness of cell transplantation therapy. Biochem Biophys Res Commun. 2013;432(2):355–8.

68. Michel-Monigadon D, Nerriere-Daguin V, Leveque X, Plat M, Venturi E, Brachet P, et al. Minocycline promotes long-term survival of neuronal transplant in the brain by inhibiting late microglial activation and T-cell recruitment. Transplantation. 2010;89(7):816–23.
69. Kim SJ, Moon GJ, Chang WH, Kim YH, Bang OY, collaborators S-. Intravenous transplantation of mesenchymal stem cells preconditioned with early phase stroke serum: current evidence and study protocol for a randomized trial. Trials. 2013;14:317.
70. Hess DC, Hoda MN, Bhatia K. Remote Limb Preconditioning and Postconditioning: Will It Translate Into a Promising Treatment for Acute Stroke? Stroke. Stroke. 2013;44(4):1191–7.
71. Hoda MN, Siddiqui S, Herberg S, Periyasamy-Thandavan S, Bhatia K, Hafez SS, et al. Remote ischemic perconditioning is effective alone and in combination with intravenous tissue-type plasminogen activator in murine model of embolic stroke. Stroke. 2012;43(10):2794–9.
72. Meng R, Asmaro K, Meng L, Liu Y, Ma C, Xi C, et al. Upper limb ischemic preconditioning prevents recurrent stroke in intracranial arterial stenosis. Neurology. 2012;79(18):1853–61.
73. Hougaard KD, Hjort N, Zeidler D, Sorensen L, Norgaard A, Hansen TM, et al. Remote ischemic perconditioning as an adjunct therapy to thrombolysis in patients with acute ischemic stroke: a randomized trial. Stroke. 2014;45(1):159–67.
74. Jiang Q, Song P, Wang E, Li J, Hu S, Zhang H. Remote ischemic postconditioning enhances cell retention in the myocardium after intravenous administration of bone marrow mesenchymal stromal cells. J Mol Cell Cardiol. 2013;56:1–7.

Chapter 12
Tracking of Administered Progenitor Cells in Brain Injury and Stroke by Magnetic Resonance Imaging

Bhagelu R. Achyut and Ali S. Arbab

Introduction

Brain has limited regenerative potential in disease conditions such as neurodegeneration, stroke, and several other neural injuries. Vasculature density and blood supply to damaged brain areas are significantly reduced under pathological conditions. Therefore, possible attempts need to make the use of stem-cell-based therapies to regenerate vasculature and hence neural cells with functional similarities to brain cells. Cell therapies are intended to induce enhancement of neovascularization or prevention of vascular endothelial cell (EC) death by using endothelial progenitor cells (EPCs). Similarly, neural progenitor cells (NPC) can also be used to protect injured neurons or enhanced neurogenesis [1–3]. These progenitor cells are being collected from different sources such as peripheral blood, bone marrow (BM), umbilical cord blood (UCB), umbilical tissues, and embryonic tissues, such as embryonic stem cells and fetal subventricular zone (SVZ) NPCs [2–5]. Although most of the time administration of cells is through intravenous (IV) route, preclinical studies are also performed by injecting cells directly into either ipsi- or contra-lateral hemispheres [2, 3]. With the possibility of cell therapy in stroke or brain injury cases, investigators want to know the spatial or temporal migration of administered cells to the site of interests.

Different imaging modalities can be used to track administered cells. Various in vitro techniques are being used to tag these cells so that they can be detected using in vivo imaging. Different reporter genes have been introduced into cells and are detected by optical imager/fluorescent or confocal microcopy [6–10]. Exogenous optical or fluorescent tags such as quantum dots or other near infrared nanoparticles are being introduced into the cytoplasm of cells for optical imaging [11–15]. Nuclear medicine approaches have incorporated the sodium iodide symporter

A. S. Arbab (✉) · B. R. Achyut
Tumor Angiogenesis Laboratory, Department of Biochemistry and Molecular Biology, Cancer Center, Georgia Regents University, 1410 Laney Walker Blvd, Augusta, GA, 30912 USA
e-mail: aarbab@gru.edu

D. C. Hess (ed.), *Cell Therapy for Brain Injury*, DOI 10.1007/978-3-319-15063-5_12

(NIS) [16–18] or herpes simplex virus thymidine kinase (HSVtk) genes into cells and have used chelated radionuclides or positron emitters to track cells [19–22]. Indium-111-oxine and Technicium-99m chelates [23, 24] have been used to label cells to track by single photon emission computed tomography (SPECT) [25–28]. Positron emitting radioisotope 2-[18F]-fluoro-2-deoxy-D-glucose (18F-FDG) and copper 64 pyruvaldehyde-bis (N4-methylthiosemicarbazone) have been used in the in vitro labeling and subsequence tracking of labeled cells [29, 30] by positron emission tomography (PET).

Labeling cells with superparamagnetic iron oxide nanoparticles (superparamagnetic ironoxides, SPIO) or a paramagnetic contrast agent (gadolinium [Gd] or manganese [Mn]) allows for the possibility of detecting single cells or clusters of labeled cells within target tissues by magnetic resonance imaging (MRI) following either direct implantation or IV injection [31–37]. Various approaches have been developed to label cells with SPIO nanoparticles or soluble paramagnetic magnetic resonance (MR) contrast agents [31–33, 38–52]. This chapter focuses on how EPCs or NPCs can be labeled with SPIOs for cellular MRI (CMRI) and how these labeled cells are being tracked by CMRI. Cell tracking using other imaging modalities with Food and Drug Administration (FDA)-approved agents is also discussed.

Characterization of MRI Contrast Agents or Magnetic Nanoparticles Used in Cell Labeling for CMRI

MR contrast agents used to label cells can either exhibit properties of being paramagnetic or superparamagnetic. These agents alter the nuclear magnetic resonance (NMR) relaxation times of the water protons in solution or tissue known as T1, T2, and T2*. The spin–lattice or longitudinal relaxation time or T1 represents the exponential recovery of the proton spin to align with the external magnetic field. The spin–spin or transverse relaxation time or T2 is the exponential loss of coherence among the spins oriented at an angle to the static magnetic field due to interactions of the spins. The T2* (T2 star) is the loss of phase coherence of the spins in the external magnetic field and is a combination of magnetic field in-homogeneities and T2.

Paramagnetic Agents

Paramagnetism refers to the ability of a metal such as Mn, Gd, or iron to interact with water protons through dipole–dipole interaction with direct inner sphere effects resulting in a shortening of NMR relaxation times and is usually associated with enhancement (increase in signal intensity) on T1-weighted images. Gd chelates (i.e., GdDTPA, GdDOTA, or GdDO3A) and Mn chloride are paramagnetic contrast agents used in experimental and clinical studies. These agents tend to shorten T1 relaxation time greater than T2 and T2* of tissues.

Gd chelate-based contrast agents for cell labeling have been used to label cells ex vivo with limited results. Reports indicate modest T1 enhancement or in some cases no T1 enhancement when Gd-chelated agents are used for cell labeling [53–59]. Thus, the need remains to identify an agent that will exert a strong T1 effect, allowing the detection of cells in disease models where the cells need to be conspicuous versus surrounding tissues, where labeled cell numbers are low, or when concentration of Gd is low.

Giesel et al. [56] were able to label mesenchymal stem cells (MSCs) using a bifunctional gadoflurine M-Cy3.5 for both MRI and optical imaging. Gadoflurine M-Cy3.5 is designed with a hydrophilic tail that allows the agent to insert in the cell wall and then gets internalized into cytosol. Intracerebral implantation of 10^6 gadoflurine M-Cy3.5-labeled MSC allowed for clear visualization of cells in the rat brain on T1-weighted imaging at clinical relevant 1.5 T that could be confirmed by fluorescent microscopy. Anderson et al. used Gd fullerenol which has higher relativities than conventional Gd chelates, to label MSCs [49]. Gd fullerenol-labeled MSCs could be detected on 7T MRI following direct injection of 10^6 cells into the rat thigh. Gd fullerenol labeling decreased the stem cell proliferation initially suggesting that the agent may be altering mitochondrial function. Brekke et al. used a combination of Gd chelate with fluorescent tag to label cells and noted a significant decrease in proliferation and increase in reactive oxygen species with 24 h of incubation [60]. The transient negative effect of a Gd-based agent on cell proliferation used for cellular and molecular imaging will need further evaluation to ensure there is no long-term toxicity or ability of the cells to repair damage.

Mn chloride was the first paramagnetic contrast agent used in MRI and has been shown that it can be taken up by cells in vivo through calcium channels in the cell membrane. [52, 61, 62]. Aoki et al. have reported that lymphocytes could be labeled following incubation with Mn chloride [52]. MRI of cells in gelatin demonstrated increased signal intensity on T1-weighted images; however, it is not clear if there would be sufficient contrast enhancement to detect Mn-labeled cells in vivo by MRI. Recently, cells have been labeled with paramagnetic Mn oxide nanoparticles and the enhancement could be detected using standard T1-weighted imaging although further work is needed to fully understand the uptake and safety of this agent in stem cells [63]. Odaka et al. labeled mononuclear cells using Mn chloride and image them up to 21 days following intramuscular administration of labeled cells [64]. However, tracking of Mn-labeled cell following IV administration is yet to be feasible. This major drawback to use Mn as a MR contrast agent is its narrow therapeutic window and potential toxicity.

Superparamagnetic Agents

Superparamagnetic iron oxide nanoparticles are a family of MRI contrast agents that are presently being used to efficiently label cells for cellular imaging. There are various methods used to prepare SPIO nanoparticles, resulting in a wide range of physiochemical differences including core size (e.g., ultrasmall (U)SPIO), shape, mono- or oligocrystalline composition, and outer coating that may alter the ability to use these agents to label cells. The basic chemistry behind the formation of superparamagnetic iron oxide nanoparticles is a mixture of ferrous and ferric iron salts at alkaline pH with a coating (dextran or other types of coatings) that is actively stirred or sonicated resulting in magnetite containing various ratios of Fe_3O_4 to Fe_2O_3 in the crystals [65, 66].

The size of the (U)SPIO nanoparticles depends on the surface coating used and will determine whether the particle is monocrystalline (ferumoxtran) or consists of multiple or oligocrystalline such as ferumoxides [66]. Surface coatings on (U)SPIO nanoparticles may be various sized and surface charged molecules, including dextran and modified cross-linked dextran, dendrimers, starches, citrate, or viral particles [44, 66–85]. For several clinically approved SPIO nanoparticles (e.g., ferumoxides, ferumoxtran, and ferucarbotran), the coating is dextran, that is attached through electrostatic interaction to the iron core by hydrogen bonds between some of the dextran hydroxyl groups and the surface oxide groups of the iron core [86]. For SPIO nanoparticles (e.g., ferumoxides or ferucarbotran), the dextran coating links multiple iron oxide crystals together and they have a hydrodynamic diameter of about between 60 and 200 nm [86]. Recently, a semisynthetic carbohydrate non-dextran-coated ultrasmall SPION (USPIO), ferumoxytol, has been approved for the treatment of iron deficiency anemia in chronic kidney disease [87, 88]. Drs. Frank and Arbab's group has introduced ferumoxytol as cell-labeling agents and showed the efficiency in tracking administered cell by MRI [89, 90].

The coating molecules contribute to the surface charge or zeta potential of the (U)SPIO in water. The zeta potential or the average potential difference in millivolts existing between the surface of the (U)SPIO nanoparticles immersed in a conduction liquid (water) and the bulk of the liquid. Dextran-coated ferumoxide has a zeta potential of −32 mV while ferumoxtran that is coated with a shorter chain dextran has a measured zeta potential of −2.0 to 0 mV [91] and the near neutral surface potential of ferumoxtran possibly contributes to the long blood half-life compared to the larger SPIO nanoparticles. Zeta potentials of ferumoxytol is reported to be at -24.4 ± 9.32 mV in water [89].

In general, (U)SPIO nanoparticles will alter the T2/T2* of the surrounding tissue compared to the T1 relaxation times in part due to field gradients surrounding the nanoparticles resulting in a rapid dephasing of the protons in the environment. (U)SPIO nanoparticles effect on MRI signal intensities depend on various factors including particle size, hydrodynamic radius, concentration of particles within the voxel, image acquisition parameters, and whether the MR contrast agent is in solution or compartmentalized within a cell [35, 37, 92]. Long echo time T2-weighted spin echo pulse sequences or T2*-weighted gradient echo MR pulse sequences are usually used to detect the presence of (U)SPIO nanoparticles within tissues and these agents usually appear as hypointensities with or without associated susceptibility artifacts on the images. For MR cellular imaging, (U)SPIO nanoparticles are usually compartmentalized within endosomes or macropinosomes within the cytoplasma of cells and causing a decrease in the signal intensity of the target tissue on T2- and T2*-weighted images because of rapid dephasing of the water proton spins set up by the magnetic field gradients that develop around the magnetically labeled cells.

Methods of Labeling Cells with Magnetic Nanoparticles for CMRI

It is mentioned in the earlier section that the surface of most of the commercially available (U)SPIO nanoparticles is negatively charged. On the other hand, electric charge of cell surface is also negative. To make the nanoparticles efficiently taken up by the cells, surface charge of the nanoparticles should be modified. Investigators have modified the surface change of the nanoparticles by coating it with cationic materials or modified the surface of the coating by attaching membrane penetrable peptides. Commercially available transfection agents can also be used to modify the surface charge of the nanoparticles to facilitate the uptake by cells.

Modified Nanoparticles for Cell Labeling

Superparamagnetic iron oxide nanoparticles are being used to efficiently label cells for cellular MRI. There are various methods used to prepare SPIO and ultrasmall (U)SPIO nanoparticles, resulting in a wide range of physio-chemical properties including core size, shape, mono- or oligo-crystalline composition, and coatings that allow the SPIO nanoparticles to exist in a colloidal suspension in aqueous solutions. The types of coatings include dextran and modified cross-linked dextran, dendrimers, starches, citrate, or viral particles and are usually attached through electrostatic interactions with the surface of the iron oxide crystal core contributing to the hydrodynamic size and zeta potential of the SPIO nanoparticles. The zeta potential or the average potential difference in millivolts exists between the surface of the (U)SPIO nanoparticles immersed in a conduction liquid (water) and the bulk of the liquid. The SPIO nanoparticles have been characterized as either being anionic or cationic (positive or negative zeta potential) and which will determine the contrast agent's ability to interact with cell/plasma membrane.

Dendrimers are branched synthetic polymers with layered architecture that can be of various sizes or generation and have multiple applications including [93–96]: to compact DNA and transfect oligonucleotides into cells by binding to the plasma membrane and stimulating endocytosis. Adding generation 4.5 polyamindoamine (PAMAM) dendrimer as a coating of iron oxide nanoparticles resulted in the synthesis of magnetodendrimers (MD-100) [69]. Incubating mammalian cells with MD-100 for 1–2 days resulted in a wide variety of cells demonstrating on Prussian blue (PB) staining a remarkably high degree of intracellular labeling, with the cytoplasm containing large numbers of iron-containing vesicles or endosomes.

Modification to the dextran coating of USPIO nanoparticles cross-linking the dextran strands cross-linked iron oxide (CLIO) and then covalently attaching HIV-1 Tat proteins to the surface has allowed for efficient and effective labeling of non-phagocytic cells presumably through macropinocytosis [42]. Using MR imaging, homing of CLIO-Tat-labeled lymphocytes could be visualized in the liver and spleen in normal mice [97]. CLIO-Tat-labeled T cells have been used as adoptive transfer autoimmune diabetes mouse model and labeled cells have been shown to selectively home to specific antigens in B16 melanoma in mouse model by in vivo MRI [41, 98, 99]. CLIO-Tat or modification that include attaching to the dextran coat optical imaging agents (i.e., fluorescein isothiocyanate (FITC), CY5.5) is very efficient at labeling cells ex vivo [100]; CLIO-Tat use is relatively limited since it is a custom synthesized agent and not commercially available.

Conjugating antigen-specific internalizing monoclonal antibodies (MoAb) to the dextran coat of USPIO nanoparticles has facilitated the magnetic labeling of cells by clathrin-mediated endocytosis. [101–106]. The monoclonal antibody (OX-26) to the rat transferrin receptor was covalently attached to USPIO nanoparticles (MION-46L) and used to label rat progenitor oligodendrocytes (CG-4). MION-46L-OX-26-labeled rat CG-4 cells were directly implanted into spinal cords of myelin deficient rats and ex vivo MR images, obtained on days 10–14 after implantation, demonstrated excellent correlation between the hypointense regions and blooming artifacts caused by the presence of labeled cells and the degree of myelination in the spinal cord detected on immunohistochemistry. The results demonstrated that magnetically labeled cells would not interfere with the cell differentiation, migration along area of pathology, or the formation of myelin wraps around axons.

Viruses and viral shells are being explored as carriers for MRI contrast agents. Using hemagglutination virus of Japan (HVJ) envelope that encapsulated SPIO nanoparticles was found to label microglial cells in culture [44, 107, 108]. The HVJ SPIO-labeled cells were intra-cardiac injected and clusters of cells could be seen within 1 day following transplantation in the brains of mice. The HVJ SPIO particles were reportedly more efficient at labeling cells than combining dextran-coated SPIO with transfection agent, lipofectamine [44]; however, since HVJ envelopes are not commercially available, the use of this agent for labeling cells is limited.

Micron-sized iron oxide, commercially available particles or beads (MPIO), is also being used to label cells for cellular MRI studies in experimental models. These agents are from 0.3 to >5 µm in size containing greater than 60 % magnetite in a polymer coating that can include a fluorescent marker that allows for dual detection of labeled cells by MRI and fluorescent microscopy. MPIOs can be purchased with either terminal amines or carboxyl groups on the surface, thus allowing for chemical modification to attach peptides, ligands, or MoAb to specific targets and receptors on cells. MPIOs have been used to track macrophage infiltration in transplantation rejection, to monitor single-cell migration in tissues and to locate implanted stem cells in an area of myocardial infarction [46, 78, 109–111].

Transfection-Agent-Mediated Cell Labeling

Although several approaches for labeling cells with SPIO nanoparticles have been explored, most of the agents used were proprietary compounds, involved unique or complex synthesis, or biochemical modification of the dextran coat of the SPIO nanoparticles to stimulate endocytosis by cells. In 2002, a relatively straightforward approach was developed combining (U)SPIO nanoparticles (e.g., ferumoxtran and ferumoxides) with commonly available polycationic transfection agents to effectively label cells. Different commercially available transfection agents have been tried with varying results [31, 33, 34, 36, 38, 112, 113]. However, most of the commercially available transfection agents are toxic to the cells at relatively low doses and moreover these transfection agents are not FDA approved to be used in the clinics. By mixing two FDA-approved agents, ferumoxides (Feridex IV, Berlex, NJ) and protamine sulfate together form a complex that efficiently and effectively label stem cells [33, 34, 39, 114–116]. Protamine sulfate is an FDA-approved drug containing >60 % arginine for the treatment of heparin anticoagulation overdose. Cells are labeled with the ferumoxides–protamine sulfate (FePro) complex via macropinocytosis and can be imaged at clinically relevant MRI fields using standard imaging techniques. The concentration of iron in cells is dependent on nuclear to cytoplasmic ratio, the iron concentration in the nano- or μm-sized particles, iron content in media, incubation times, and method of endocytosis of the particles [33, 34, 46]. Unlabeled stem cells usually contain less that 0.1 pg of iron per cell, whereas labeled cells grown in suspension (i.e., hematopoietic stem cells (HSCs), T cells) contain 1–5 pg iron per cells and cells that adhere to culture dish (i.e., MSCs, human cervical cancer cells and macrophages) can take up from 5 to >20 pg iron per cell [33, 34, 39]. Labeling cells with ferumoxides do not alter the viability and functional capability of cells or the differential capacity of stem cells [33, 117]. Ferumoxides-protamine sulfate labeled embryonic, mesenchymal, hematopoietic, and neural stem cells (NSCs) showed similar rates of differentiation to different lineages, compared to control unlabeled cells [33, 112, 117, 118]. Janic et al. [119] have further improved the labeling procedures using ferumoxides-protamine sulfate and used extensively to label EPCs and track them in neovascularization in vascular diseases such as stroke.

Unfortunately, most of the FDA-approved SPIOs are not commercially available anymore. The manufacturers have stopped producing SPIOs due to nonviable commercial values. The good news is that there is another FDA-approved USPIO (ferumoxytol), which is being used in chronic renal failure patients. Drs. Frank and Arbab's group have utilized this USPIO and made complexes using protamine sulfate and heparin, all of them are FDA approved. The complexes are used to label different types of cells and tracked them by MRI [89, 90]. Recent articles indicated the safe handling of endocytosed ironoxides by the cells [120, 121]. Soon after phagocytosis, iron particles remain within endosomes and by days 3–5 these endosomes fuse with lysosomes and the ironoxides particles start disintegrating [121]. Pawelczyk et al. have shown that ironoxides labeling of HeLa and MSCs resulted in a transient decrease in TfR-1 mRNA and protein levels. In contrast, ironoxides labeling of primary macrophages resulted in an increase in TfR-1 mRNA but not in TfR-1 protein levels. Ferritin mRNA and protein levels increased transiently in labeled HeLa and macrophages but were sustained in MSCs [120]. Previously, we have shown that retention of iron in the labeled cells depends on the rate of division and metabolic activity of cells [34]. In rapidly growing cells, the intracellular iron completely disappeared by 5–8 divisions. On the other hand, the intracellular iron was observed after 6 weeks in cells, where cell division was almost inhibited.

Labeling and Tracking of Endothelial Progenitor Cells

The formation of blood vessels occurs by two mechanisms: vasculogenesis and angiogenesis. Vasculogenesis is the process where blood vessels are formed de novo by in situ differentiation of the primitive progenitors—i.e., angioblasts into mature ECs, which was thought to only take place during embryonic development [122]. In contrast, angiogenesis occurs during both the embryonic development and the postnatal life, and is defined as a process that gives rise to new blood vessels by proliferation and migration of preexisting, differentiated ECs [123]. Angiogenic factors, such as vascular endothelial growth factor (VEGF), stimulate angiogenesis by promoting activation, proliferation, sprouting, and migration of ECs and therefore allow for a rapid formation of new blood vessels [124]. ECs that contribute to neovasculatures can originate from sprouting and co-option of neighboring preexisting vessels [125]. However, there are emerging evidences indicating that BM-derived EPCs also contribute to the vasculogenesis [126]. A subpopulation of $CD34^+$ human HSCs identified by the cell-surface molecule $CD133^+$ ($AC133^+$) has been shown to be more specific for endothelial differentiation and angiogenesis [127, 128]. Additionally, published results also showed migration and incorporation of IV-administered $CD34^+/AC133^+$ cells in the neovasculatures [129]. EPCs collected from BM, peripheral, or cord blood have been used in different animal models to determine whether these cells have the capacity to become part of the neovasculatures in tissues.

Landscapes of $CD34^+/AC133^+$ Endothelial Progenitor Cells

Previous studies demonstrated the existence of circulating ECs in peripheral blood in various vascular diseases [130]. However, for a while, it was unclear whether these cells or their precursors play a role in postnatal vascular growth. The breakthrough came from the work by Asahara et al. (1997) who demonstrated the presence of $CD34^+$/VEGF receptor-2 (VEGFR2) + EPCs in human peripheral blood [126]. These cells gave rise to mature ECs in culture and were capable of incorporating into the sites of active neovascularization in animal models. This landmark work opened the possibility that in adults, endothelial stem or precursor cells may contribute to the formation of new blood vessels by vasculogenesis. Since then, researchers have been gaining significant insights into the postnatal neovascularization and the EPCs' origin, phenotype and function. However, the main factor hindering the EPCs research is the controversy on the identity of EPCs. Earlier studies defined EPCs as the cells co-expressing HSC marker CD34 and endothelial marker VEGFR2. Since subsequent work showed that some mature ECs also co-express CD34 and VEGFR2 and that CD34 was not an exclusive marker for hematopoietic cells, a novel CD133 glycoprotein was accepted as a more appropriate marker for immature progenitor cells [127, 131]. Glycosylated form of CD133 protein is expressed on HSCs but not on mature ECs, and it is recognized by AC133 monoclonal antibody. Peichev et al. (2000) suggested that a subset of circulating $CD34^+$ cells that are positive for both VEGFR-2 and AC133 represent a functional EPC population that plays a role in postnatal angiogenesis or vasculogenesis [131]. EPCs also share many cell-surface markers with ECs and with stem/progenitor cells of different tissues. However, currently it is customary to define EPCs as cells that are positive for AC133, CD34, and VEGFR2 markers, with the following distinction: $AC133^+$/$CD34^+$/$VEGFR2^+$ cells represent an immature, highly proliferative EPC population localized mainly in the BM, while AC133–/$CD34^+$/$VEGFR2^+$ cells are considered circulatory, more mature cells that are limited in their proliferative capacity [132]. In addition, these more mature cells also express some of the endothelial specific antigens such as platelet EC adhesion molecule 1 (PECAM-1 or CD31), E-selectin (CD62E) and VE-cadherin (CD144), chemokine receptor CXCR-4 (CD184) and have the ability to migrate in response to the CXCR-4 ligand, stromal cell-derived factor (SDF)-1α and VEGF. It is now generally accepted that new vessels can also be formed *via* recruitment of circulating EPCs.

EPCs have been identified mainly in the mononuclear cell fraction of peripheral blood, leukapheresis products, and in UCB [126], which represent immense therapeutic potential roles such as regenerative agent and imaging probe. Characterization and phenotypical expression of EPCs is still controversial and there is still no consensus on the EPC's definition. Some investigators pointed out that EPCs are CD45 marker negative and should express CD31, kinase insert domain receptor (KDR) (VEGFR2), VE cadherin and von Willebrand factor (vWF) [133]. These investigators usually collect cells from peripheral or cord blood mononuclear cells as an adherent cell population and propagate them in differentiating media containing high amount of fetal bovine serum (FBS). Dr. Arbab's group reported that cord blood (CB)-derived $AC133^+$ cells can be amplified by long-term in vitro expansion while preserving their angiogenic potential, which is critically important for developing EPC-based therapies [128, 134]. There are reports showing peripheral blood

$CD45^+$ angiogenic cells that show all the hallmarks of EPC as well as CD45 surface marker [135].

Indeed, studies in recent years demonstrated that IV-administered progenitors isolated from BM, peripheral blood, or UCB can home to ischemic sites, emphasizing the significance of the paracrine effect of lesion-secreted factors. To show the endothelial potential of cord blood $CD34^+/AC133^+$ cells, investigators have performed extensive in vitro and in vivo studies. Recent publications showed the potential of long- and short-term cultured UCB-derived $CD34^+/AC133^+$ EPCs to make tube-like structures in both in vitro and in vivo matrigel angiogenesis studies [128]. Despite of the significant amount of data available, controversy still remains on the identity and function of the putative EPC and its functional significance and contribution to vasculature growth and repair of damaged or degenerated brain areas. Gradually, the consensus on the putative EPC phenotype is arising, nevertheless further identification and characterization of novel, more specific EPC markers are warranted.

Endothelial Progenitor Cells in Vascular Integrity and Repair

The endogenous EPCs maintain vascular integrity and homeostasis by mediating response to vascular injury by inducing ECs regeneration, and hence promoting tissue neovascularization [136, 137]. Preclinical studies have shown that EPCs participate in neovascularization processes in ischemic organs, and hence their regulation could have therapeutic applications in vascular diseases [138, 139]. Recently, association between circulating EPCs and outcome in different subtypes of acute ischemic stroke was evaluated. The results showed that number of circulating EPCs is significantly lower in patients with large-vessel disease than in those with small-vessel disease. EPCs are indicator of vascular integrity in stroke, which is evident by low $CD133^+$ $CD34^+$ subset of EPCs in stroke patients. Low EPCs were associated with high intercellular adhesion molecule 1 (ICAM-1; marker of neuroinflammation) in those stroke patients [140]. Similarly, less number of EPCs on admission is an independent risk factor for poor 6-month outcome in patients with acute ischemic stroke [141]. After acute ischemic stroke, circulating EPC counts peaked at day 7 [142]. In patients with large artery atherosclerosis and small-vessel disease subtypes, higher counts were related to better outcome at 3 months [142]. In adult ischemic brain injury or stroke, transplanted EPCs reside to the ischemic injury core, and promote cerebral neovascularization and neuron progenitor cell migration and survival, and improve long-term neurobehavioral outcomes [143]. Integrity of the vascular endothelial monolayer is extremely important, since it represents a barrier between the blood and sub-endothelial matrix proteins. Thus, restricts the infiltration of inflammatory cells and controls vascular smooth muscle proliferation [139]. The cerebrovascular system is considered as dynamic entity that remodels according to patho-physiological conditions. As discussed before, under hypoxia/ischemia conditions, new blood vessel formation in adults has traditionally been understood to be the results from angiogenesis [123, 144] and vasculogenesis [122]. EPCs have shown great potential as neuroregenerative therapies as they regenerate into ECs, astrocytes, pericytes and smooth muscle cells, NSCs, oligodendrocytes, and neurons. EPCs migrate to injured tissue and participate in neovascularization, regeneration of the injured endothelium, which provide cells to differentiate into mature vascular ECs, and secrete pro-angiogenic growth factors [139]. The process of EPC homing to the site of injury is highly regulated that starts with detachment from the BM niche, migration into blood vessels, and reaching to homing site through the circulation. EPCs interact with the damaged endothelial monolayer in a similar way that leukocytes interact with activated ECs. Interestingly, adhesion molecules such as P-selectin, E-selectin, and β2-integrins have been identified as key regulators of EPC homing [139]. Investigators found that systemic administration of UCB-derived EPCs in adult mice resulted in significant protection against hypoxic/ischemic brain injury, with reduced infarct volume, decreased neutrophil infiltration, and increased focal blood flow [145]. Interestingly, this study also reported that circulating EPC levels were inversely correlated with cerebral infarction, but positively correlated with regional blood flow in hypoperfused areas of the brain after ischemia [145], suggesting that EPCs may be a predictor of the functional cerebral vasculature. In other study, authors reported that higher EPC levels were indicative of smaller volumes of acute lesion, final lesion, and lesion growth, and may serve as markers of acute phase stroke severity [146]. Cross talk between the brain and systemic responses in blood is increasingly suspected of playing critical roles in stroke. Authors showed that reactive astrocytes can release a damage-associated molecular-pattern molecule called high-mobility group box 1 (HMGB1) that promotes EPC-mediated neurovascular remodeling during stroke recovery. In a mouse model of focal cerebral ischemia, reactive astrocytes in the peri-infarct cortex upregulate HMGB1 at 14 days post stroke, along with an accumulation of endogenous EPCs[147].

Recently, studies have shown that $AC133^{+}$ stem/progenitor cells derived from hUCB improve structural and functional recovery in stroke models [140, 148, 149]. The study examined the effect of hUCB $AC133^{+}$ EPCs on stroke development and resolution in a middle cerebral artery occlusion (MCAo) rat model. Accumulation of transplanted cells in stroke-affected hemispheres and revealed that stroke volume decreased at a significantly higher rate exerted a therapeutic effect on the extent of tissue damage, regeneration, and time course of stroke resolution [148]. Administration of hUCB-derived cells significantly reduced ventricular volume and improved cerebral blood flow, which is histologically evidenced by enhanced expression of vWF and synaptophysin [149]. In mouse model, EPCs protect the brain against ischemic injury, promote neurovascular repair through SDF-1-mediated signaling pathways, and improve long-term neurobehavioral outcomes [150]. These studies indicate that EPCs possess tremendous regenerative potential. Application of UCB-derived EPCs could be used extensively for brain injury and stoke, if exploited correctly. However, these findings have subsequently been proven in animal models, clinical trials have not been encouraging. These discrepancies have limited translation of EPCs from bench to bedside [151].

AC133⁺ EPCs-Facilitated Mechanisms in Neovascularization

Higher expression of genes regulating angiogenesis was also observed in AC133$^+$ cells. Previous studies including ours have shown the mechanism associated with the EPCs migration to damage area. The recruitment of EPCs from BM is initiated by increased circulatory levels of factors such as VEGF, fibroblast growth factor (FGF), SDF-1α, granulocyte monocytes colony-stimulating factor (GM-CSF), osteopontin, etc., probably secreted by degenerated or damaged area [150, 152–155]. EPCs are known to express receptors for these aforementioned secreted factors, e.g., SDF-1 acts as a chemoattractant for EPCs migration due to abundant expression of CXCR4 receptors on cell membrane [155]. Moreover, EPCs were also shown to be attracted toward RANTES, which is an inflammatory cytokines [128, 156]. It was found that the CYP4A/F-20-HETE system is expressed in EPCs derived from human UCB and can act as both an autocrine and a paracrine regulatory factor [157]. HMGB1 up-regulation in post-ischemic brain could promote exogenous human peripheral blood-derived EPC-mediated stroke recovery by modulating paracrine function of EPCs in mice [158]. These factors activate BM microenvironment to switch from a dormant to a pro-angiogenic state and the process involves the activation of matrix metalloproteinase-9 (MMP-9) that releases BM stromal cells' membrane bound c-Kit (CD117) ligand. Generated soluble form of c-Kit ligand stimulates c-Kit positive EPCs to move from BM niche to the BM vascular zone and translocate to the circulation [159]. Tissue hypoxia present in ischemic vascular diseases is considered to be central to this paracrine mechanism and this ischemic effect was shown to be mediated by marked increase in VEGF and SDF-1α circulating levels [160]. VEGF and SDF-1α expressions are transcriptionally upregulated by tissue hypoxia-induced expression and/or activation of hypoxia inducible factor-1α (HIF-1α) [160]. In addition to the increase in their circulatory levels, VEGF and SDF-1α expressions are increased locally, within the hypoxic tissue itself that in turn stimulate recruitment of progenitor cells to the hypoxic site as well. Our previous results also indicated that the homing of EPCs into the neovessels was related to HIF-1α-induced SDF-1α expression [152]. Recent studies also demonstrated the correlation between VEGF and SDF-1α expression at the transcriptional and functional/effector level. VEGF was shown to upregulate SDF-1α and CXCR-4 molecules. In addition, it was demonstrated the possible synergistic effect between two cytokines by showing that without a concurrent VEGF signal; SDF-1 was insufficient in recruiting EPCs to the disease sites [161]. In addition, factors that regulate physiological angiogenesis can also play a role in EPC recruitment and mobilization [128].

Studies have also been done to investigate the expression of secreted molecules by the homing EPCs and cell–cell interaction products at the site of stroke. Authors studied the cell–cell interaction in neovascular recovery after stroke. HMGB1 and β-2 integrin signaling was found to play an important role in interactions between brain endothelium and EPCs, which is governed by reactive astrocytes in stroke [162]. Recently, elevated EPC and EC cells were found in hemorrhagic and ischemic stroke patients. VEGF, SDF-1, hepatocyte growth factor (HGF), and endothelin-1 (ET-1) were increased at plasma levels in the hemorrhagic patients. ET-1mRNA expression was increased in peripheral blood cells in the ischemic stroke patients. Significant correlations were observed between EPCs or ECs and Big ET-1 protein or mRNA levels in hemorrhagic stroke but not in the ischemic stroke patients. These results suggest that ET-1 may play a role in pathophysiology of stroke and subsequent EPC mobilization [163]. In other study involving early ischemic stroke patients, high ICAM-1 was associated with low $CD133^+$ $CD34^+$ subset of EPC. This suggests that biomarker of neuroinflammation may predict tissue injury and stroke severity in early ischemia [140]. Recently, VEGF, FGF-b, and platelet-derived growth factor (PDGF)-bb factors secreted by EPCs were found to enhance neuro-repair responses after cerebral ischemia in mice [164]. This indicates that administration of EPC-secreted factors could become a safe and effective cell-free option to be considered in future therapeutic strategies for stroke. Outgrowth ECs (OECs) from stroke patients present higher levels of pro-angiogenic factors such as C-C Motif Chemokine 2 (CCL2), inhibitor of DNA binding 3 (ID3), insulin-like growth factor 1 (IGF-1), MMP9, transforming growth factor receoptor 1 (TGF-BR1), TNFAIP2, tumor necrosis factor (TNF), and transforming growth factor beta 1 (TGFB1) at early stages, decreasing in mature OECs when they become more similar to mature microvascular ECs [165]. Collectively, these aforementioned studies have depicted the secretory response of EPCs with their microenvironment at the homing site. However, noninvasive imaging methods may provide for the localization of EPCs and HSCs within the injured areas, thus furthering the understanding of extracellular and stromal components required for incorporation of these cells into the neovasculatures.

Tracking of $CD34^+$/$AC133^+$ Endothelial Progenitor Cells

So far, the abundant of data has shown the involvement of EPCs in neovascularization process. Therefore, it is now required to monitor the neovascularization process and understand the involvement of EPCs using in vivo models of diseases. A major challenge in the development of cell-based therapies for stroke is to deliver optimal number of cells (therapeutic dose) to the site of lesions. In addition, CB-derived EPCs have potential use as a therapeutic and imaging probe [134]. Authors report that new technique with short incubation time using 100 μg/ml of Ferumoxides and 3 μg/ml of protamine sulfate is effective in labeling cells for cellular MRI [119]. Study by Varma et al. 2013 evaluated the dynamic biodistribution of systemically injected labeled hUCB-derived EPCs in animal model, which was monitored by In-111-oxine-based SPECT imaging [166]. Magnetically labeled cord blood EPCs can be in vitro expanded and cryopreserved for future use as MRI probes for monitoring the migration and incorporation to the sites of neovascularization [167]. Since, molecular and cellular imaging is essential in the determination of bioavailability and efficacy of various drugs and targeting agents, use of UCB cells is more efficient in such studies.

There is an argument that host macrophage will take up the dead iron-labeled cells after homing and incorporation into target tissues, and these macrophages along with dead cells will produce misleading low signal intensity on MRI or can show iron-positive cells on PB staining. With this report and with subsequent publications, Arbab et al. have shown that host macrophages (mouse) did not show any iron on PB staining [129, 152]. Moreover, iron-labeled EPCs also worked as histological marker, which was easily detected by PB. Surprisingly, host macrophage did not show any iron positivity even with local implantation of magnetically labeled EPCs.

In the following section, we have discussed the stroke studies that exploited EPCs as a MRI probe. MRI was used to monitor stroke development and resolution, as well as the migration and localization of administered magnetically labeled EPCs, 1, 7, and 14 days after the stroke onset [148]. T2-weighted images detected ischemic lesions that appeared as defined regions of signal hyperintensity in all animals at days 7 and 14 after the MCAo procedure (Fig. 12.1a) [148]. Animals receiving hUCB AC133^{+} EPCs, susceptibility-weighted imaging (SWI) of the same sections revealed signal hypointensity areas that resulted from accumulation of

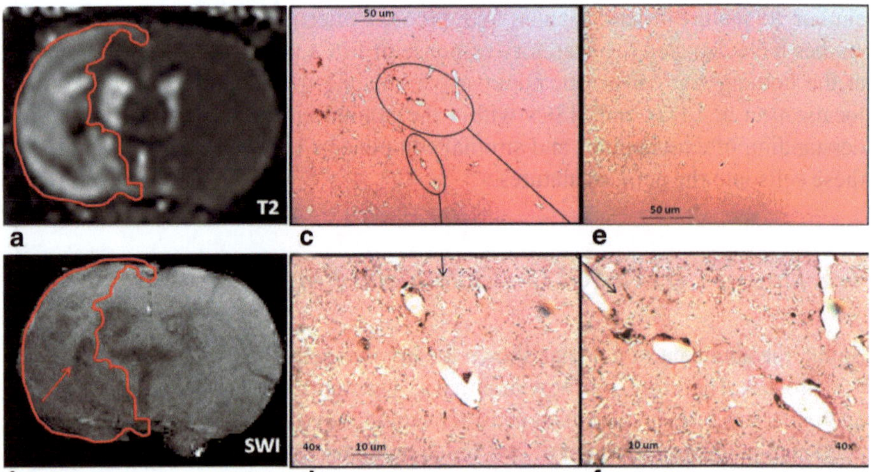

Fig. 12.1 Tracking of ferumoxide and protamine sulfate (FePro)-labeled human umbilical cord blood (hUCB)-derived AC133^{+} endothelial progenitor cells (EPCs) to brain ischemic lesions. FePro-labeled hUCB AC133^{+} EPCs were administered IV to the rats that had undergone middle cerebral artery occlusion 24 h earlier. Fourteen days after cell administration, magnetic resonance imaging (MRI) identified stroke lesions and demonstrated the presence of administered cells within the lesions. **a** T2 MRI maps depicting the stroke area as a hyperintense (*white*) region, *bordered in red*. **b** Susceptibility-weighted imaging (SWI) showing the accumulation of FePro-labeled cells in the stroke-affected hemisphere, within the same slice that is *bordered in red* in **a**. **c**, **d**, **f** 3,3-Diaminobenzidine enhanced Prussian blue staining confirmed the accumulation of FePro-labeled cells mainly in the ischemic boundary, within and around large thin blood vessels. **e** No cells were detected in the brains of control animals. Magnification: 10 (**c**, **e**) and 40 (**d**, **f**). Scale bars 50 μm (**c**, **e**) and 10 μm (**d**, **f**).

FePro-labeled transplanted cells (Fig. 12.1b) [148]. Staining by PB staining of corresponding tissue sections confirmed the presence of administered FePro-labeled cells that accumulated mainly within the ischemic boundary (IB) of the stroke lesion (Fig. 12.1c, d, f) [148]. Some cells were detected within and around the walls of large thin blood vessels that were indicative of neoangiogenesis (Fig. 12.1d, f). As expected, PB staining did not detect any cells in the contralateral hemispheres of control animals (Fig. 12.1e) [148]. Later, stroke tissue sections were stained with FITC-labeled tomato lectin (endothelial lining of blood vessels), which demonstrated the presence of large, thin blood vessels within the ipsilateral brain hemisphere. Further, we noticed that accumulation of transplanted and Dil-labeled cells were either colocalized with lectin or found in the vicinity of lectin-positive areas [148].

Further, MRI was also used in evaluating changes in stroke lesions over time. In the subacute phase (24 h after MCAo), bright zones observed on MRI were identified as ischemic lesions in both experimental and control groups. T2 maps constructed from T2-weighted images were used for measuring changes in stroke volume at different time points (Fig. 12.2a). Over the course of 15 days, stroke-affected areas decreased in size, and quantitative analysis demonstrated that the rate of shrinkage was higher in the animals receiving cells compared with the control animals. In the control group, stroke volume decreased by 43 % at day 7, whereas in the animals receiving cells this decrease was at 34 % of original volume calculated at day 1 after MCAo. Statistical analysis showed that by day 7, stroke-affected areas decreased in volume at a significantly higher rate in animals receiving cells compared with the control animals ($p < 0.05$). However, at day 14 after MCAo, a significant difference between control and experimental animals was not observed (Fig. 12.2b). An analysis of tissue sections obtained from animals that received

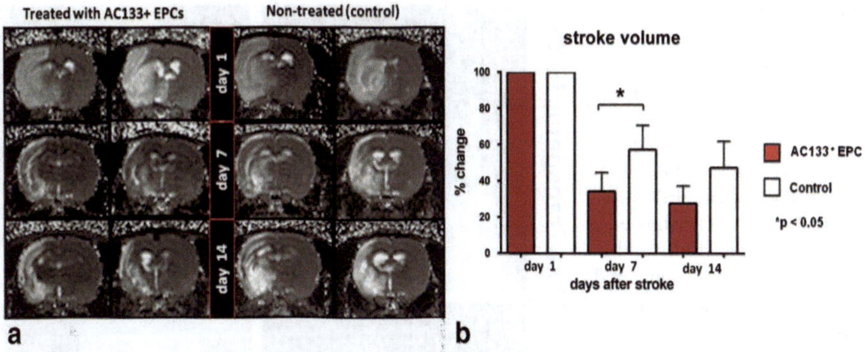

Fig. 12.2 Changes in stroke volume over time: magnetic resonance imaging analysis. Images of T2 maps constructed from T2-weighted images depict stroke lesions as hyperintense areas at days 1, 7, and 14 after middle cerebral artery occlusion in cell-treated and control animals. **a** Magnetic resonance T2-weighted images from two representative animals from each group (cell-treated and control). **b** Quantitative analysis of T2 maps revealed that over the course of 15 days, stroke-affected areas shrank at a significantly higher rate in animals that received ferumoxide and protamine sulfate-labeled human umbilical cord blood AC133 endothelial progenitor cells (*EPCs*) as compared with the control animals. Graph shows mean SD, $p < 0.05$.

hUCB AC133⁺ EPCs revealed strong positivity for vWF within the ischemic core and the IB [148]. Strong nestin activity was also observed within the ipsilateral SVZ. Interestingly, ipsilateral brain parenchyma within the SVZ and adjacent to ventricular walls exhibited stretches of nestin-positive cells that extended toward IB and IC, giving an impression of cells migrating toward the ischemic lesion [148].

Very recently our group also reported the use of EPCs as gene carrier and MRI imaging probes to target neovascularization [168]. The purposes of this study were to determine whether (1) IV-administered genetically transformed CB-derived EPC can carry hNIS and express transgene products and (2) accumulation of these administered EPC can be tracked by in vivo MRI and the expression of hNIS can be determined by in vivo Tc-99m SPECT. This study first time reported the use of CB-derived EPCs to carry a gene (hNIS), and the migration and the expression of gene products were determined by in vivo MRI as well as SPECT studies, respectively. These EPCs were used both as gene carrier and imaging probes. EPCs can be used to deliver therapeutic genes to the sites of lesions. The use of EPCs to carry therapeutic gene to the sites of neovascularization in different lesions are underway in our laboratory. Ultimate goal is to exploit EPCs as therapeutic agent to enhance neovascularization brain injury and stroke therapy. We have also used transgenic EPCs in stroke model and follow their accumulation by SPECT (Fig. 12.3). In conclusion, EPCs can effectively separate from peripheral blood, BM, and CB. With the established culture technique developed by our group, one can propagate the separated EPCs to many folds, which then can be manipulated ex vivo to carry contrast agents or reporter gene. Following systemic administration, EPCs can be tracked by different imaging modalities to the sites of active angiogenesis/vasculogenesis in brain injury and stroke.

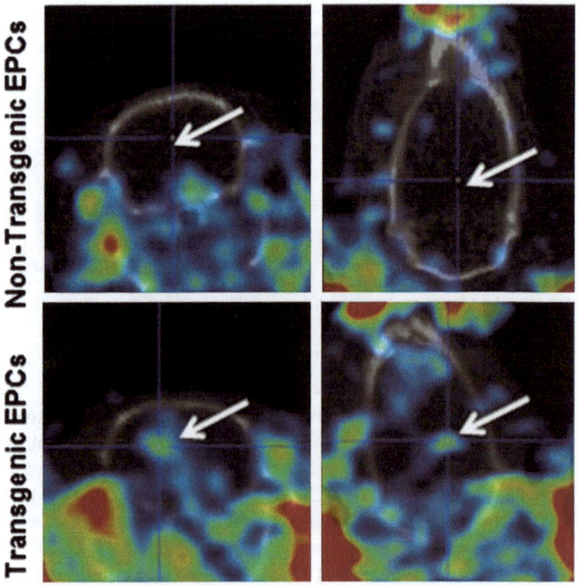

Fig. 12.3 SPECT images for tracking of IV-injected EPCs and transgene expression. Rats with stroke (MCAo) were injected with transgenic EPCs carrying the *hNIS* gene or control EPCs. Animals that received transgenic EPCs carrying the *hNIS* gene showed higher Tc-99m activity in the stroke area (*arrow*) compared with stroke rats receiving nontransgenic EPCs

Tracking of Neural Progenitor/Stem Cells

NSCs or NPCs are most evident adult-derived stem cell type for brain repair due to their tendency to develop into the required neural cells. NPCs are localized adjacent to ECs in the SVZ, subgranular zone of the dentate gyrus, and also the subependymal zone of the spinal cord in the adult rodent brain [169]. These regions are activated following an injury or stroke; however, this activation alone is not sufficient to elicit full functional repair [170]. Interestingly, NPCs derived from human fetal brain have also been utilized to treat experimental stroke in rodents [171]. First evidence about the role of NPCs as a regenerative agent for stroke was observed when homologous embryo brain cortex tissue was transplanted into hypoxic brain regions of rat [172]. These grafted cells containing NSCs established stable morphological connections with neighboring neurons and improved electrophysiological performance [173]. In addition to improved structural and functional behaviors, NPCs treatment reorganizes white matter in stroke that can be detected by MRI in rat. White matter reorganization was coextensive with increases of fractional anisotropy after stroke in the ischemic recovery regions compared to that in the IC region [174]. Similar to other cell-based therapies, NPCs administration route is also critical in stroke research and treatment modalities. High mortality with intra-arterial (IA) delivery (IA: 41 %; IC: 17 %; IV: 8 %) of NPCs poses a serious concern for using this route of administration. However, IA administration showed significantly increased migration, more diffuse distribution pattern, and a larger number of transplanted NPCs in the target brain compared to that of intracisternal (IC) or IV administration[175].

Several studies have been reported that deciphered the molecular mechanisms associated with the neuroregeneration by NPCs. For example, Liu et al. showed that miR17-92 cluster plays an important role in mediating NPCs function through sonic hedgehog (Shh) signaling pathway, which is involved in up-regulating miR17-92 cluster expression [176]. Same group reported that MiR-124a regulates proliferation of NPCs through Notch signaling pathway, which was discovered after profiling the SVZ of the brain with stroke [177]. In this study, transfection of NPCs with miR-124a significantly reduced progenitor cell proliferation and stimulated neuronal differentiation to neuroblasts, measured by an increase in the number of doublecortin positive cells [177]. The role of Notch pathway was deciphered by blocking it by siRNA against Notch or a gamma secretase inhibitor that significantly reduced Notch, NICD and Hes1 expression in stroke-induced cell proliferation [178]. Interestingly, SDF-1α secreted by human CD133-derived multipotent stromal cells promotes survival of NPCs through CXCR7 [179].

NPCs labeled with different ironoxides nanoparticles were administered in stroke and traumatic brain injury models and the migration and incorporation of NPCs in the lesions were determined by cellular MRI [180–182]. Following IC administration Dr. Chopp's group has demonstrated the spatial and temporal migration of NPC in stroke sites by MRI [181]. The administered cells migrated at a rate of 65 μm/h and administered labeled cells appeared at the site of lesion within 48 h following IC administration.

Conclusion

Brain is highly malleable due to ischemic insult, injury, and stroke. Endogenous angiogenesis neurogenesis and synaptogenesis are able to provide only partial functional recovery. Cell-based therapy offers potential therapeutic modalities to repair damaged brain compared to other classical therapies. Stroke studies involving experimental models have shown the evidence of stem cell migration to the lesion, survival, and improved differentiation. Both EPCs and NPCs have been associated with significantly improved behavioral outcomes due to induced angiogenesis and neurogenesis, respectively, in brain injury. The multiple mechanisms of action of EPCs and NPCs include the secretion of trophic factors, immunomodulation, and anti-inflammatory effects in the recovery process. At the same time, recent developments in cellular and molecular imaging modalities such as MRI have offered a noninvasive method to monitor the changes in damaged/recovered brain, before/after these cell-based therapies, respectively. All of this provides clear ideas of clinical trials that could be planned mostly using EPCs and NPCs administered through specific routes in stroke.

References

1. Yi BR, Kim SU, Choi KC. Development and application of neural stem cells for treating various human neurological diseases in animal models. Lab Anim Res. 2013;29(3):131–7.
2. Canazza A, et al. Experimental models of brain ischemia: a review of techniques, magnetic resonance imaging, and investigational cell-based therapies. Front Neurol. 2014;5:19.
3. Kalladka D, Muir KW. Brain repair: cell therapy in stroke. Stem Cells Cloning. 2014;7:31–44.
4. Sanberg PR, et al. Advantages and challenges of alternative sources of adult-derived stem cells for brain repair in stroke. Prog Brain Res. 2012;201:99–117.
5. Liu X, et al. Cell based therapies for ischemic stroke: from basic science to bedside. Prog Neurobiol. 2014;115:92–115.
6. Brazelton TR, Blau HM. Optimizing techniques for tracking transplanted stem cells in vivo. Stem Cells. 2005;23(9):1251–65.
7. Blits B, et al. Lentiviral vector-mediated transduction of neural progenitor cells before implantation into injured spinal cord and brain to detect their migration, deliver neurotrophic factors and repair tissue. Restor Neurol Neurosci. 2005;23(5–6):313–24.
8. Tanaka M, et al. In vivo visualization of cardiac allograft rejection and trafficking passenger leukocytes using bioluminescence imaging. Circulation. 2005;112 Suppl 9:I105–10.
9. Paulmurugan R, Gambhir SS. Novel fusion protein approach for efficient high-throughput screening of small molecule-mediating protein-protein interactions in cells and living animals. Cancer Res. 2005;65(16):7413–20.
10. Paulmurugan R, Gambhir SS. Firefly luciferase enzyme fragment complementation for imaging in cells and living animals. Anal Chem. 2005;77(5):1295–302.
11. Bruchez MP. Turning all the lights on: quantum dots in cellular assays. Curr Opin Chem Biol. 2005;9(5):533–7.
12. Akerman ME, et al. Nanocrystal targeting in vivo. Proc Natl Acad Sci U S A. 2002;99(20):12617–21.
13. Michalet X, et al. Quantum dots for live cells, in vivo imaging, and diagnostics. Science. 2005;307(5709):538–44.

14. Potapova IA, et al. Mesenchymal stem cells support migration, extracellular matrix invasion, proliferation, and survival of endothelial cells in vitro. Stem Cells. 2007;25(7):1761–8.
15. Rosen AB, et al. Finding fluorescent needles in the cardiac haystack: tracking human mesenchymal stem cells labeled with quantum dots for quantitative in vivo three-dimensional fluorescence analysis. Stem Cells. 2007;25(8):2128–38.
16. Dingli D, et al. Combined I-124 positron emission tomography/computed tomography imaging of NIS gene expression in animal models of stably transfected and intravenously transfected tumor. Mol Imaging Biol. 2006;8(1):16–23.
17. Kim DE, et al. Near-infrared fluorescent imaging of cerebral thrombi and blood-brain barrier disruption in a mouse model of cerebral venous sinus thrombosis. J Cereb Blood Flow Metab. 2005;25(2):226–33.
18. Chung JK. Sodium iodidesymporter: its role in nuclear medicine. J Nucl Med. 2002; 43(9):1188–200.
19. Yaghoubi SS, et al. Imaging progress of herpes simplex virus type 1 thymidine kinase suicide gene therapy in living subjects with positron emission tomography. Cancer Gene Ther. 2005;12(3):329–39.
20. Buursma AR, et al. 18F-FEAU as a radiotracer for herpes simplex virus thymidine kinase gene expression: in-vitro comparison with other PET tracers. Nucl Med Commun. 2006;27(1):25–30.
21. Hustinx R, et al. Imaging in vivo herpes simplex virus thymidine kinase gene transfer to tumour-bearing rodents using positron emission tomography and. Eur J Nucl Med. 2001;28(1):5–12.
22. Cao F, et al. In vivo visualization of embryonic stem cell survival, proliferation, and migration after cardiac delivery. Circulation. 2006;113(7):1005–14.
23. Zhang H, et al. Synthesis and evaluation of bombesin derivatives on the basis of pan-bombesin peptides labeled with indium-111, lutetium-177, and yttrium-90 for targeting bombesin receptor-expressing tumors. Cancer Res. 2004;64(18):6707–15.
24. Love C, et al. Diagnosing infection in the failed joint replacement: a comparison of coincidence detection 18F-FDG and 111In-labeled leukocyte/99mTc-sulfur colloid marrow imaging. J Nucl Med. 2004;45(11):1864–71.
25. Botti C, et al. Comparison of three different methods for radiolabelling human activated T lymphocytes. Eur J Nucl Med. 1997;24(5):497–504.
26. Read EJ, et al. In vivo traffic of indium-111-oxine labeled human lymphocytes collected by automated apheresis. J Nucl Med. 1990;31(6):999–1006.
27. Griffith KD, et al. In vivo distribution of adoptively transferred indium-111-labeled tumor infiltrating lymphocytes and peripheral blood lymphocytes in patients with metastatic melanoma. J Natl Cancer Inst. 1989;81(22):1709–17.
28. Fisher B, et al. Tumor localization of adoptively transferred indium-111 labeled tumor infiltrating lymphocytes in patients with metastatic melanoma. J Clin Oncol. 1989;7(2):250–61.
29. Adonai N, et al. Ex vivo cell labeling with 64Cu-pyruvaldehyde-bis(N4-methylthiosemicarbazone) for imaging cell trafficking in mice with positron-emission tomography. Proc Natl Acad Sci U S A. 2002;99(5):3030–5.
30. Hofmann M, et al. Monitoring of bone marrow cell homing into the infarcted human myocardium. Circulation. 2005;111(17):2198–202.
31. Frank JA, et al. Magnetic intracellular labeling of mammalian cells by combining (FDA-approved) superparamagnetic iron oxide MR contrast agents and commonly used transfection agents. Acad Radiol. 2002;9 Suppl 2:S484–7.
32. Frank JA, et al. Clinically applicable labeling of mammalian and stem cells by combining superparamagnetic iron oxides and transfection agents. Radiology. 2003;228(2):480–7.
33. Arbab AS, et al. Efficient magnetic cell labeling with protamine sulfate complexed to ferumoxides for cellular MRI. Blood. 2004;104(4):1217–23.
34. Arbab AS, et al. Characterization of biophysical and metabolic properties of cells labeled with superparamagnetic iron oxide nanoparticles and transfection agent for cellular MR imaging. Radiology. 2003;229(3):838–46.

35. Modo M, Hoehn M, Bulte JW. Cellular MR imaging. Mol Imaging. 2005;4(3):143–64.
36. Bulte JW, et al. Preparation of magnetically labeled cells for cell tracking by magnetic resonance imaging. Methods Enzymol. 2004;386:275–99.
37. Bulte JW, Kraitchman DL. Iron oxide MR contrast agents for molecular and cellular imaging. NMR Biomed. 2004;17(7):484–99.
38. Arbab AS, et al. Intracytoplasmic tagging of cells with ferumoxides and transfection agent for cellular magnetic resonance imaging after cell transplantation: methods and techniques. Transplantation. 2003;76(7):1123–30.
39. Arbab AS, et al. In vivo trafficking and targeted delivery of magnetically labeled stem cells. Hum Gene Ther. 2004;15(4):351–60.
40. Kircher MF, et al. A multimodal nanoparticle for preoperative magnetic resonance imaging and intraoperative optical brain tumor delineation. Cancer Res. 2003;63(23):8122–5.
41. Moore A, et al. Tracking the recruitment of diabetogenic CD8 + T-cells to the pancreas in real time. Diabetes. 2004;53(6):1459–66.
42. Josephson L, et al. High-efficiency intracellular magnetic labeling with novel superparamagnetic-Tat peptide conjugates. Bioconjug Chem. 1999;10(2):186–91.
43. Walczak P, et al. Instant MR labeling of stem cells using magnetoelectroporation. Magn Reson Med. 2005;54(4):769–74.
44. Toyoda K, et al. Effective magnetic labeling of transplanted cells with HVJ-E for magnetic resonance imaging. Neuroreport. 2004;15(4):589–93.
45. van den Bos EJ, et al. Improved efficacy of stem cell labeling for magnetic resonance imaging studies by the use of cationic liposomes. Cell Transplant. 2003;12(7):743–56.
46. Shapiro EM, Skrtic S, Koretsky AP. Sizing it up: cellular MRI using micron-sized iron oxide particles. Magn Reson Med. 2005;53(2):329–38.
47. Zheng H, et al. Novel potential neuroprotective agents with both iron chelating and amino acid-based derivatives targeting central nervous system neurons. Biochem Pharmacol. 2005;70(11):1642–52.
48. Vuu K, et al. Gadolinium-rhodamine nanoparticles for cell labeling and tracking via magnetic resonance and optical imaging. Bioconjug Chem. 2005;16(4):995–9.
49. Anderson SA, Lee KK, Frank JA. Gadolinium-fullerenol as a paramagnetic contrast agent for cellular imaging. Invest Radiol. 2006;41(3):332–8.
50. Rudelius M, et al. Highly efficient paramagnetic labelling of embryonic and neuronal stem cells. Eur J Nucl Med Mol Imaging. 2003;30(7):1038–44.
51. Aime S, et al. Targeting cells with MR imaging probes based on paramagnetic Gd(III) chelates. Curr Pharm Biotechnol. 2004;5(6):509–18.
52. Aoki I, et al. Cell labeling for magnetic resonance imaging with the T1 agent manganese chloride. NMR Biomed. 2006;19(1):50–9.
53. Crich SG, et al. Improved route for the visualization of stem cells labeled with a Gd-/Eu-chelate as dual (MRI and fluorescence) agent. Magn Reson Med. 2004;51(5):938–44.
54. Crich SG, et al. Visualization through magnetic resonance imaging of DNA internalized following "in vivo" electroporation. Mol Imaging. 2005;4(1):7–17.
55. Himmelreich U, et al. A responsive MRI contrast agent to monitor functional cell status. Neuroimage. 2006;32(3):1142–9.
56. Giesel FL, et al. Gadofluorine m uptake in stem cells as a new magnetic resonance imaging tracking method: an in vitro and in vivo study. Invest Radiol. 2006;41(12):868–73.
57. Modo M, et al. Tracking transplanted stem cell migration using bifunctional, contrast agent-enhanced, magnetic resonance imaging. Neuroimage. 2002;17(2):803–11.
58. Daldrup-Link HE, et al. Cell tracking with gadophrin-2: a bifunctional contrast agent for MR imaging, optical imaging, and fluorescence microscopy. Eur J Nucl Med Mol Imaging. 2004;31(9):1312–21.
59. Su W, et al. Synthesis and cellular uptake of a MR contrast agent coupled to an antisense peptide nucleic acid-cell- penetrating peptide conjugate. Contrast Media Mol Imaging. 2007;2(1):42–9.

60. Brekke C, et al. The in vitro effects of a bimodal contrast agent on cellular functions and relaxometry. NMR Biomed. 2007;20(2):77–89.
61. Wolf GL, et al. Contrast agents for magnetic resonance imaging. Magn Reson Annu. 1985:231–66.
62. Mendonca-Dias MH, Gaggelli E, Lauterbur PC. Paramagnetic contrast agents in nuclear magnetic resonance medical imaging. Semin Nucl Med. 1983;13(4):364–76.
63. Na HB, et al. Development of a T1 contrast agent for magnetic resonance imaging using MnO nanoparticles. Angew Chem Int Ed Engl. 2007;46(28):5397–401.
64. Odaka K, et al. In vivo tracking of transplanted mononuclear cells using manganese-enhanced magnetic resonance imaging (MEMRI). PLoS ONE. 2011;6(10):e25487.
65. Wang YX, Hussain SM, Krestin GP. Superparamagnetic iron oxide contrast agents: physicochemical characteristics and applications in MR imaging. Eur Radiol. 2001;11(11):2319–31.
66. Jung CW. Surface properties of superparamagnetic iron oxide MR contrast agents: ferumoxides, ferumoxtran, ferumoxsil. Magn Reson Imaging. 1995;13(5):675–91.
67. Hogemann D, et al. Improvement of MRI probes to allow efficient detection of gene expression. Bioconjug Chem. 2000;11(6):941–6.
68. Yeh TC, et al. Intracellular labeling of T-cells with superparamagnetic contrast agents. Magn Reson Med. 1993;30(5):617–25.
69. Bulte JW, et al. Magnetodendrimers allow endosomal magnetic labeling and in vivo tracking of stem cells. Nat Biotechnol. 2001;19(12):1141–7.
70. Ittrich H, et al. Labeling of mesenchymal stem cells with different superparamagnetic particles of iron oxide and detectability with MRI at 3T. Rofo. 2005;177(8):1151–63.
71. Mikhaylova M, Kim doK, Bobrysheva N, Osmolowsky M, Semenov V, Tsakalakos T, Muhammed M. Superparamagnetism of magnetite nanoparticles: dependence on surface modification. Langmuir. 2004;20(6):2472–7.
72. Hawrylak N, et al. Nuclear magnetic resonance (NMR) imaging of iron oxide-labeled neural transplants. Exp Neurol. 1993;121(2):181–92.
73. Yeh TC, et al. In vivo dynamic MRI tracking of rat T-cells labeled with superparamagnetic iron-oxide particles. Magn Reson Med. 1995;33(2):200–8.
74. Shen TT, et al. Magnetically labeled secretin retains receptor affinity to pancreas acinar cells. Bioconjug Chem. 1996;7(3):311–6.
75. Fleige G, et al. In vitro characterization of two different ultrasmall iron oxide particles for magnetic resonance cell tracking. Invest Radiol. 2002;37(9):482–8.
76. Kaufman CL, et al. Superparamagnetic iron oxide particles transactivator protein-fluorescein isothiocyanate particle labeling for in vivo magnetic resonance imaging detection of cell migration: uptake and durability. Transplantation. 2003;76(7):1043–6.
77. Koch AM, et al. Uptake and metabolism of a dual fluorochrome Tat-nanoparticle in HeLa cells. Bioconjug Chem. 2003;14(6):1115–21.
78. Ho C, Hitchens TK. A non-invasive approach to detecting organ rejection by MRI: monitoring the accumulation of immune cells at the transplanted organ. Curr Pharm Biotechnol. 2004;5(6):551–66.
79. Song H, Choi JS, Huh YM, Kim S, Jun YW, Suh JS, Cheon J. Surface modulation of magnetic nanocrystals in the development of highly efficient magnetic resonance probes for intracellular labeling. J Am Chem Soc. 2005;127(28):9992–3.
80. Schulze E, Ferrucci, Jr JT, Poss K, Lapointe L, Bogdanova A, Weissleder R. Cellular uptake and trafficking of a prototypical magnetic iron oxide label in vitro. Invest Radiol. 1995;30(10):604–10.
81. Bulte JW, et al. Monitoring stem cell therapy in vivo using magnetodendrimers as a new class of cellular MR contrast agents. Academic Radiology. 2002;9 Suppl 2:S332–5.
82. Smirnov P, et al. In vivo cellular imaging of magnetically labeled hybridomas in the spleen with a 1.5-T clinical MRI system. Magn Reson Med. 2004;52(1):73–9.
83. Riviere C, et al. Iron oxide nanoparticle-labeled rat smooth muscle cells: cardiac MR imaging for cell graft monitoring and quantitation. Radiology. 2005;235(3):959–67.

84. Brillet PY, et al. Evaluation of tumoral enhancement by superparamagnetic iron oxide particles: comparative studies with ferumoxtran and anionic iron oxide nanoparticles. Eur Radiol. 2005;15(7):1369–77.
85. Wilhelm C, et al. Intracellular uptake of anionic superparamagnetic nanoparticles as a function of their surface coating. Biomaterials. 2003;24(6):1001–11.
86. Jung CW, Jacobs P. Physical and chemical properties of superparamagnetic iron oxide MR contrast agents: ferumoxides, ferumoxtran, ferumoxsil. Magn Reson Imaging. 1995;13(5):661–74.
87. Balakrishnan VS, et al. Physicochemical properties of ferumoxytol, a new intravenous iron preparation. Eur J Clin Invest. 2009;39(6):489–96.
88. Schwenk MH. Ferumoxytol. a new intravenous iron preparation for the treatment of iron deficiency anemia in patients with chronic kidney disease. Pharmacotherapy. 2010;30(1):70–9.
89. Thu MS, et al. Self-assembling nanocomplexes by combining ferumoxytol, heparin and protamine for cell tracking by magnetic resonance imaging. Nat Med. 2012;18(3):463–7.
90. Gutova M, et al. Magnetic resonance imaging tracking of ferumoxytol-labeled human neural stem cells: studies leading to clinical use. Stem Cells Transl Med. 2013;2(10):766–75.
91. Kalish H, et al. Combination of transfection agents and magnetic resonance contrast agents for cellular imaging: relationship between relaxivities, electrostatic forces, and chemical composition. Magn Reson Med. 2003;50(2):275–82.
92. Taupitz M, Schmitz S, Hamm B. Superparamagnetic iron oxide particles: current state and future development. Rofo. 2003;175(6):752–65.
93. Sato N, et al. Pharmacokinetics and enhancement patterns of macromolecular MR contrast agents with various sizes of polyamidoamine dendrimer cores. Magn Reson Med. 2001;46(6):1169–73.
94. Yan GP, et al. Synthesis and evaluation of gadolinium complexes based on PAMAM as MRI contrast agents. J Pharm Pharmacol. 2005;57(3):351–7.
95. Bryant LH Jr, et al. Synthesis and relaxometry of high-generation (G = 5, 7, 9, and 10) PAMAM dendrimer-DOTA-gadolinium chelates. J Magn Reson Imaging. 1999;9(2):348–52.
96. Kobayashi H, et al. Macromolecular MRI contrast agents with small dendrimers: pharmacokinetic differences between sizes and cores. Bioconjug Chem. 2003;14(2):388–94.
97. Dodd CH, et al. Normal T-cell response and in vivo magnetic resonance imaging of T cells loaded with HIV transactivator-peptide-derived superparamagnetic nanoparticles. J Immunol Methods. 2001;256(1–2):89–105.
98. Moore A, et al. MRI of insulitis in autoimmune diabetes. Magn Reson Med. 2002;47(4):751–8.
99. Kircher MF, et al. In vivo high resolution three-dimensional imaging of antigen-specific cytotoxic T-lymphocyte trafficking to tumors. Cancer Res. 2003;63(20):6838–46.
100. Josephson L, et al. Near-infrared fluorescent nanoparticles as combined MR/optical imaging probes. Bioconjug Chem. 2002;13(3):554–60.
101. Moore A, et al. Human transferrin receptor gene as a marker gene for MR imaging. Radiology. 2001;221:244–250.
102. Moore A, et al. Measuring transferrin receptor gene expression by NMR imaging. Biochim Biophys Acta. 1998;1402(3):239–49.
103. Bulte JW, et al. Neurotransplantation of magnetically labeled oligodendrocyte progenitors: magnetic resonance tracking of cell migration and myelination. Proc Natl Acad Sci U S A. 1999;96(26):15256–61.
104. Bulte JW, et al. Specific MR imaging of human lymphocytes by monoclonal antibody-guided dextran-magnetite particles. Magn Reson Med. 1992;25(1):148–57.
105. Ahrens ET, et al. Receptor-mediated endocytosis of iron-oxide particles provides efficient labeling of dendritic cells for in vivo MR imaging. Magn Reson Med. 2003;49(6):1006–13.
106. Berry CC, et al. The influence of transferrin stabilised magnetic nanoparticles on human dermal fibroblasts in culture. Int J Pharm. 2004;269(1):211–25.

107. Miyoshi S, Flexman JA, Cross DJ, Maravilla KR, Kim Y, Anzai Y, Oshima J, Minoshima S. Transfection of neuroprogenitor cells with iron nanoparticles for magnetic resonance imaging tracking: cell viability, differentiation, and intracellular localization. Mol Imaging Biol. 2005;7(4):1–10.
108. Song Y, et al. Magnetic resonance imaging using hemagglutinating virus of Japan-envelope vector successfully detects localization of intra-cardially administered microglia in normal mouse brain. Neurosci Lett. 2006;395(1):42–5.
109. Hinds KA, et al. Highly efficient endosomal labeling of progenitor and stem cells with large magnetic particles allows magnetic resonance imaging of single cells. Blood. 2003;102(3):867–72.
110. Hill JM, et al. Serial cardiac magnetic resonance imaging of injected mesenchymal stem cells. Circulation. 2003;108(8):1009–14.
111. Wu YL, et al. In situ labeling of immune cells with iron oxide particles: an approach to detect organ rejection by cellular MRI. Proc Natl Acad Sci U S A. 2006;103(6):1852–7.
112. Suzuki Y, et al. In vitro comparison of the biological effects of three transfection methods for magnetically labeling mouse embryonic stem cells with ferumoxides. Magn Reson Med. 2007;57(6):1173–9.
113. Neri M, et al. Efficient in vitro labeling of human neural precursor cells with superparamagnetic iron oxide particles: relevance for in vivo cell tracking. Stem Cells. 2007;26(2):505–16.
114. Montet-Abou K, et al. Transfection agent induced nanoparticle cell loading. Mol Imaging. 2005;4(3):165–71.
115. Reynolds F, Weissleder R, Josephson L. Protamine as an efficient membrane-translocating peptide. Bioconjug Chem. 2005;16(5):1240–5.
116. Wu YJ, et al. In vivo leukocyte labeling with intravenous ferumoxides/protamine sulfate complex and in vitro characterization for cellular magnetic resonance imaging. Am J Physiol Cell Physiol. 2007;293(5):C1698–708.
117. Arbab AS, et al. Labeling of cells with ferumoxides-protamine sulfate complexes does not inhibit function or differentiation capacity of hematopoietic or mesenchymal stem cells. NMR Biomed. 2005;18(8):553–9.
118. Guzman R, et al. Long-term monitoring of transplanted human neural stem cells in developmental and pathological contexts with MRI. Proc Natl Acad Sci U S A. 2007;104(24):10211–6.
119. Janic B, et al. Optimization and validation of FePro cell labeling method. PLoS ONE. 2009;4(6):e5873.
120. Pawelczyk E, et al. Expression of transferrin receptor and ferritin following ferumoxides-protamine sulfate labeling of cells: implications for cellular magnetic resonance imaging. NMR Biomed. 2006;19(5):581–92.
121. Arbab AS, et al. A model of lysosomal metabolism of dextran coated superparamagnetic iron oxide (SPIO) nanoparticles: implications for cellular magnetic resonance imaging. NMR Biomed. 2005;18(6):383–9.
122. Risau W, Flamme I. Vasculogenesis. Annu Rev Cell Dev Biol. 1995;11(1):73–91.
123. Folkman J, Shing Y. Angiogenesis. J Biol Chem. 1992;267(16):10931–4.
124. Ellis LM, et al. Overview of angiogenesis: biologic implications for antiangiogenic therapy. Semin Oncol. 2001; 28(5 Suppl 16):94–104.
125. Zhang ZG, et al. Bone marrow-derived endothelial progenitor cells participate in cerebral neovascularization after focal cerebral ischemia in the adult mouse. Circ Res. 2002;90(3):284–8.
126. Asahara T, et al. Isolation of putative progenitor endothelial cells for angiogenesis. Science. 1997;275(5302):964–7.
127. Gehling UM, et al. In vitro differentiation of endothelial cells from AC133-positive progenitor cells. Blood. 2000;95(10):3106–12.
128. Janic B, et al. Human cord blood-derived AC133 + progenitor cells preserve endothelial progenitor characteristics after long term in vitro expansion. PLoS ONE. 2010;5(2):e9173.

129. Arbab AS, et al. Magnetic resonance imaging and confocal microscopy studies of magnetically labeled endothelial progenitor cells trafficking to sites of tumor angiogenesis. Stem Cells. 2006;24(3):671–8.
130. Hladovec J. Circulating endothelial cells as a sign of vessel wall lesions. Physiol Bohemoslov. 1978;27(2):140–4.
131. Peichev M, et al. Expression of VEGFR-2 and AC133 by circulating human CD34(+) cells identifies a population of functional endothelial precursors. Blood. 2000;95(3):952–8.
132. Khakoo AY, Finkel T. Endothelial progenitor cells. Annu Rev Med. 2005;56:79–101.
133. Ingram DA, et al. Identification of a novel hierarchy of endothelial progenitor cells using human peripheral and umbilical cord blood. Blood. 2004;104(9):2752–60.
134. Janic B, Arbab AS. Cord blood endothelial progenitor cells as therapeutic and imaging probes. Imaging Med. 2012;4(4):477–90.
135. Duda DG, et al. A protocol for phenotypic detection and enumeration of circulating endothelial cells and circulating progenitor cells in human blood. Nat Protoc. 2007;2(4):805–10.
136. Castillo-Melendez M, et al. Stem cell therapy to protect and repair the developing brain: a review of mechanisms of action of cord blood and amnion epithelial derived cells. Front Neurosci. 2013;7:194.
137. Patel SD, et al. Hematopoietic progenitor cells and restenosis after carotid endarterectomy. Stroke. 2012;43(6):1663–5.
138. Hristov M, Weber C. Endothelial progenitor cells in vascular repair and remodeling. Pharmacol Res. 2008;58(2):148–51.
139. Zampetaki A, Kirton JP, Xu Q. Vascular repair by endothelial progenitor cells. Cardiovasc Res. 2008;78(3):413–21.
140. Bogoslovsky T, et al. Circulating CD133 + CD34 + progenitor cells inversely correlate with soluble ICAM-1 in early ischemic stroke patients. J Transl Med. 2011;9:145.
141. Tsai NW, et al. The association between circulating endothelial progenitor cells and outcome in different subtypes of acute ischemic stroke. Clin Chim Acta. 2014;427:6–10.
142. Marti-Fabregas J, et al. Endothelial progenitor cells in acute ischemic stroke. Brain Behav. 2013;3(6):649–55.
143. Fan Y, et al. A modified culture medium increases blastocyst formation and the efficiency of human embryonic stem cell derivation from poor-quality embryos. J Reprod Dev. 2010;56(5):533–9.
144. Red-Horse K, et al. Endothelium-microenvironment interactions in the developing embryo and in the adult. Dev Cell. 2007;12(2):181–94.
145. Ohta T, et al. Administration of ex vivo-expanded bone marrow-derived endothelial progenitor cells attenuates focal cerebral ischemia-reperfusion injury in rats. Neurosurgery. 2006;59(3):679–86; (discussion 679-86).
146. Bogoslovsky T, et al. Endothelial progenitor cells correlate with lesion volume and growth in acute stroke. Neurology. 2010;75(23):2059–62.
147. Hayakawa K, et al. Astrocytic high-mobility group box 1 promotes endothelial progenitor cell-mediated neurovascular remodeling during stroke recovery. Proc Natl Acad Sci U S A. 2012;109(19):7505–10.
148. Iskander A, et al. Intravenous administration of human umbilical cord blood-derived ac133 + endothelial progenitor cells in rat stroke model reduces infarct volume: magnetic resonance imaging and histological findings. Stem Cells Transl Med. 2013;2(9):703–14.
149. Jiang Q, et al. MRI detects brain reorganization after human umbilical tissue-derived cells (hUTC) treatment of stroke in rat. PLoS ONE. 2012;7(8):e42845.
150. Fan Y, et al. Endothelial progenitor cell transplantation improves long-term stroke outcome in mice. Ann Neurol. 2010;67(4):488–97.
151. Lu C, et al. EPCs in vascular repair: how can we clear the hurdles between bench and bedside? Front Biosci (Landmark Ed). 2014;19:34–48.
152. Arbab AS, et al. Detection of migration of locally implanted AC133 + stem cells by cellular magnetic resonance imaging with histological findings. FASEB J. 2008;22(9):3234–46.

153. Finney MR, et al. Umbilical cord blood-selected CD133 + cells exhibit vasculogenic functionality in vitro and in vivo. CytoTherapy. 2010;12(1):67–78.
154. Kioi M, et al. Inhibition of vasculogenesis, but not angiogenesis, prevents the recurrence of glioblastoma after irradiation in mice. J Clin Invest. 2010;120(3):694–705.
155. Shichinohe H, et al. Role of SDF-1/CXCR4 system in survival and migration of bone marrow stromal cells after transplantation into mice cerebral infarct. Brain Res. 2007;1183:138–47.
156. Silverman MD, et al. Endothelial progenitor cell (EPC) recruitment in rheumatoid arthritis. FASEB J. 2007;21(5):A186–A6.
157. Guo AM, et al. The cytochrome P450 4A/F-20-hydroxyeicosatetraenoic acid system: a regulator of endothelial precursor cells derived from human umbilical cord blood. J Pharmacol Exp Ther. 2011;338(2):421–9.
158. Chen C, et al. Effect of HMGB1 on the paracrine action of EPC promotes post-ischemic neovascularization in mice. Stem Cells. 2014;32(10):2679–89.
159. Heissig B, et al. Recruitment of stem and progenitor cells from the bone marrow niche requires MMP-9 mediated release of kit-ligand. Cell. 2002;109(5):625–37.
160. Ceradini DJ, et al. Progenitor cell trafficking is regulated by hypoxic gradients through HIF-1 induction of SDF-1. Nat Med. 2004;10(8):858–64. Epub 2004 Jul 4.
161. Kollet O, et al. Rapid and efficient homing of human CD34(+)CD38(-/low)CXCR4(+) stem and progenitor cells to the bone marrow and spleen of NOD/SCID and NOD/SCID/B2m(null) mice. Blood. 2001;97(10):3283–91.
162. Hayakawa K, et al. Reactive astrocytes promote adhesive interactions between brain endothelium and endothelial progenitor cells via HMGB1 and beta-2 integrin signaling. Stem Cell Res. 2014;12(2):531–8.
163. Paczkowska E, et al. Increased circulating endothelial progenitor cells in patients with haemorrhagic and ischaemic stroke: the role of endothelin-1. J Neurol Sci. 2013;325(1–2):90–9.
164. Rosell A, et al. Factors secreted by endothelial progenitor cells enhance neurorepair responses after cerebral ischemia in mice. PLoS ONE. 2013;8(9):e73244.
165. Navarro-Sobrino M, et al. The angiogenic gene profile of circulating endothelial progenitor cells from ischemic stroke patients. Vasc Cell. 2013;5(1):3.
166. Varma NR, et al. Differential biodistribution of intravenously administered endothelial progenitor and cytotoxic T-cells in rat bearing orthotopic human glioma. BMC Med Imaging. 2013;13(1):17.
167. Janic B, et al. MRI tracking of FePro labeled fresh and cryopreserved long term in vitro expanded human cord blood AC133 + endothelial progenitor cells in rat glioma. PLOS ONE. 2012;7(5):e37577.
168. Varma NR, et al. Endothelial progenitor cells (EPCs) as gene carrier system for rat model of human glioma. PLoS ONE. 2012;7(1):e30310.
169. Gotts JE, Chesselet MF. Vascular changes in the subventricular zone after distal cortical lesions. Exp Neurol. 2005;194(1):139–50.
170. Nakayama D, et al. Injury-induced neural stem/progenitor cells in post-stroke human cerebral cortex. Eur J Neurosci. 2010;31(1):90–8.
171. Borlongan CV, et al. Bone marrow grafts restore cerebral blood flow and blood brain barrier in stroke rats. Brain Res. 2004;1010(1–2):108–16.
172. Polezhaev LV, Alexandrova MA. Transplantation of embryonic brain tissue into the brain of adult rats after hypoxic hypoxia. J Hirnforsch. 1984;25(1):99–106.
173. Polezhaev LV, et al. Morphological, biochemical and physiological changes in brain nervous tissue of adult intact and hypoxia-subjected rats after transplantation of embryonic nervous tissue. J Hirnforsch. 1985;26(3):281–9.
174. Jiang Q, et al. MRI detects white matter reorganization after neural progenitor cell treatment of stroke. Neuroimage. 2006;32(3):1080–9.

175. Li L, et al. Transplantation of marrow stromal cells restores cerebral blood flow and reduces cerebral atrophy in rats with traumatic brain injury: in vivo MRI study. J Neurotrauma. 2011;28(4):535–45.
176. Liu XS, et al. MicroRNA-17-92 cluster mediates the proliferation and survival of neural progenitor cells after stroke. J Biol Chem. 2013;288(18):12478–88.
177. Liu XS, et al. MicroRNA profiling in subventricular zone after stroke: MiR-124a regulates proliferation of neural progenitor cells through Notch signaling pathway. PLOS ONE. 2011;6(8):e23461.
178. Wang L, et al. The Notch pathway mediates expansion of a progenitor pool and neuronal differentiation in adult neural progenitor cells after stroke. Neuroscience. 2009;158(4):1356–63.
179. Bakondi B, et al. SDF-1alpha secreted by human CD133-derived multipotent stromal cells promotes neural progenitor cell survival through CXCR7. Stem Cells Dev. 2011;20(6):1021–9.
180. Zhu J, Zhou L, XingWu F. Tracking neural stem cells in patients with brain trauma. N Engl J Med. 2006;355(22):2376–8.
181. Zhang ZG, et al. Magnetic resonance imaging and neurosphere therapy of stroke in rat. Ann Neurol. 2003;53(2):259–63.
182. Zhang Z, et al. In vivo magnetic resonance imaging tracks adult neural progenitor cell targeting of brain tumor. Neuroimage. 2004;23(1):281–7.

Chapter 13
Biomaterials Application in Stem Cell Therapies for Stroke

Pouria Moshayedi and S. Thomas Carmichael

Stroke: Prevalence, Burden, and Pathophysiology

Epidemiology and Burden of Stroke

Stroke is among the leading causes of long-term disability in the world [1]. Currently, about 6.8 million Americans live with stroke. An expedited and more equipped emergency response has decreased stroke-related mortality by 13 % over a decade. Because of declining death rates, estimates are that stroke prevalence will increase to 10.2 million in the USA by 2030 [1]. Therefore, stroke is a significant and growing health-care problem.

The only available treatment for stroke is recombinant tissue plasminogen activator or tPA. This has to be administered within 4.5 h after the incidence of ischemic stroke in order to be effective. Even in specialized stroke centers, it is difficult to treat patients in this window, and the treatment rate for tPA is less than 10 % [1]. In addition hemorrhagic stroke, which accounts for 13–15 % of stroke cases [2], will not benefit from tPA. Even in those patients who receive tPA, there are always degrees of brain tissue damage. Statistics indicate only one third of stroke survivors are independent of health-care services when they are discharged from hospitals, and the rest are dependent on skilled nursing facilities, home health-care facilities, or hired help due to their high level of disabilities [1]. This indicates that the current medical care available leaves many patients in a chronically disabled state after stroke.

S. T. Carmichael (✉) · P. Moshayedi
Department of Neurology, David Geffen School of Medicine, University of California, Los Angeles, CA 90095, USA
e-mail: scarmichael@mednet.ucla.edu

P. Moshayedi
e-mail: pouria@cantab.net

D. C. Hess (ed.), *Cell Therapy for Brain Injury,* DOI 10.1007/978-3-319-15063-5_13

The brain mounts a series of limited regenerative processes in stroke, but these are incomplete. Roughly, 50 % of stroke patients suffer from hemiparesis and cognitive dysfunction 6 months after stroke, one third are not able to walk without assistance, and one fifth still have aphasia [3]. Added to this list of disabilities is blindness that is seen in 21 % of patients 3 months after stroke [4]. The cost of both acute and chronic stroke translates to an annual US$36.5 billion, directly through health-care delivery and indirectly through loss of productivity [1]. Therefore, discovering a treatment to improve stroke outcome will have an unparalleled impact on the lives of patients as well as the resultant economic burden.

Pathophysiology of Stroke and Self-Repair Mechanisms

Stroke is most commonly caused by a sudden and severe reduction in blood perfusion of the brain. Following the failure of energy-dependent processes, a subsequent cascade of events leads to cell death: membrane ionic and water imbalance, membrane depolarization, excitatory neurotransmitter release, influx of calcium, generation of oxygen free radicals, and ultimately disintegration of cellular membrane [5]. Following cell death, the inflammatory reaction is initiated by activation of local microglia and continues through an influx of blood-derived neutrophils and macrophages that are mobilized by proinflammatory cytokines and migration through a broken blood–brain barrier [6, 7]. By secreting reactive oxygen species, proapoptotic molecules and digestive enzymes, inflammatory cells will contribute to secondary injury that aggravates tissue damage and extends the boundaries of infarction [8]. In some cases, the upregulation of matrix metalloproteinase (MMP)-9 causes loosening of the blood–brain barrier, which in turn leads to secondary hemorrhage into an ischemic infarct, and it will consequently further damage the healing tissue [8].

The majority of stroke patients who survive the initial damage experience some spontaneous recovery in their impaired functions. This recovery response is caused by the potential of adult brain to undergo plasticity [9]. These compensatory changes are particularly taking place in the regions of the brain adjacent to and connected with infarcted core, an area known as peri-infarct tissue [10]. The recovery processes recapitulate many events observed in the developing nervous system, such as proliferation and migration of neural stem/progenitor cells (NSPCs) towards the brain tissue surrounding stroke [10, 11]. NSPCs orchestrate events to attenuate intrinsic inflammatory responses after injury and protect the brain tissue against further damage [12, 13]. They also secrete cytokines, chemokines, and growth factors to provide trophic support for the surviving neurons and glia [13, 14]. Through these functions, NSPCs may contribute to poststroke regenerative events by supporting local sprouting, new connection formations, modification of the existing synapses, and formation of new vessels. They can also differentiate into neurons and integrate into neural circuitry [15, 16], but this role appears to not significantly contribute to tissue repair after stroke.

The extent of long-term disabilities after stroke indicates the endogenous mechanisms of repair cause only a limited functional recovery. The most straightforward therapeutic strategy is thus to augment these endogenous mechanisms of repair [10], since those processes are already in place and contributing to some degrees of recovery. This could be done by enhancing proliferation and migration of endogenous NSPCs so that a larger number of cells would be available in the peri-infarct tissue [17–21]. But this approach has its own limitations, and it depends on the penetration of stimulating molecules, responsiveness of adult cells to external stimuli, and permissiveness of their migration path. Transplantation of NPSCs is an appealing alternative since it can directly deliver large numbers of well-characterized cells into the focus of stroke. This is particularly feasible since stroke is mainly a focal disease that can be targeted with cell transplantation. Cell therapy has therefore been well studied as a therapeutic option to promote neural repair after stroke [13, 16, 22].

Stem Cell Therapy for Stroke

Stem/progenitor cells have a promising record of inducing behavioral improvements and tissue repair in preclinical studies [16, 23, 24]. A true "stem cell" is one that can give rise to any tissue in the body, and a daughter stem cell with this same pluripotency. By this definition, "stem cell therapy" in stroke is not performed with stem cells, but instead more differentiated cells that can produce a more limited complement of downstream cells. There are many types of progenitor cells in use in the field of stroke preclinical research, originally derived from adult tissue sources, embryonic stem cells, or pluripotent stem cells. In a rough categorization, these can be grouped into NSPCs and non-NSPCs. They are derived from a variety of sources, such as human embryonic/fetal tissue [25], cell lines [26, 27], or somatic cells reprogrammed to induced pluripotent stem cells [28]. NSPCs are partially committed to a neural lineage and can therefore differentiate to neurons, astrocytes, and oligodendrocytes. They have the potential to restore building blocks of the brain tissue after they are lost in stroke [29, 30]. However, a rapid behavioral response to transplanted cells and a low rate of neuronal differentiation [31, 32] suggest that NSPCs lead to recovery by alternative mechanisms such as protecting survived tissue against further damage [33], modulating inflammatory response [34], and promoting endogenous neurogenesis and angiogenesis [12].

A second category of stem/progenitor cell that induces functional improvement after stroke are non-NSPCs [12, 16, 24]. These include umbilical cord blood stem cells, mesenchymal stem cells, multipotent adult progenitor cells, and mesenchymal cells derived from embryonic, fetal, or amniotic sources. These cell types promote recovery by secreting growth factors and augmenting endogenous mechanisms of recovery (see above).

With all the preclinical evidence supporting improved functional recovery, stem/progenitor cell therapies in stroke have been favorable candidates for clinical translation. Over the past 14 years, stem/progenitor cells have been applied in clinical stroke studies (see Table 13.1). The results, however, show a limited success and variability between implanted subjects, and therefore they are not as encouraging as the outcome of preclinical experiments. Besides inevitable interspecies differences of a biological response to transplanted cells, the failure may be partly rooted in differences between small animal models of stroke and human cases. Experimental stroke lesions are homogeneous and small, while patients in real clinical settings suffer from lesions in a variety of locations and dimensions that generally exceed experimental lesions in size. In fact, a larger stroke has been associated with a failed recovery [35] as it leads to inadequate delivery of cells and poor cell survival.

A review of experimental stroke studies reveals cell survival is still an ongoing challenge. Injection of transplanted stem/progenitor cells into the center of the stroke infarct is a reasonable choice since stem/progenitor cells potentiate endogenous repair processes that are taking place in the peri-infarct tissue. In addition, the stroke core is a potential space that can receive considerable volumes of injected cell suspension. However, the infarct core presents as an unfavorable recipient with lack of trophic support and continuous inflammation [36, 37]. As an alternative, cells can be injected into the intact tissue surrounding the infarction. However, an injection could be damaging to the very same tissue that undergoes repair. Therefore, we are yet to find a balance between safety and efficiency in cell therapy for stroke.

Researchers have explored two alternative approaches for a safe and efficacious delivery of stem/progenitor cells. Endovascular delivery of stem/progenitor cells, either intra-arterial or intravenous injection of stem/progenitor cells, has been subject of many experimental and some clinical studies. Experiments have shown the injected cells are attracted to the stroke site and promote functional recovery [35, 38]. While endovascular injection of cells presents a feasible method for multiple sessions of cell therapy, it seeds the entire body with transplanted cells. Although short-term safety for such a body-wide cell delivery has been established for few cell products [39–45], long-term safety is yet to be determined. Moreover, intravenous injection of cells leads to entrapment of cells into internal organs [46] and, in some cases, no migration to the brain lesions takes place [47]. Although a directed delivery is possible with intra-arterial injections, cell engraftment is still not ideal [38]. This low yield of cell delivery to the site of stroke can explain the limited efficiency of intravascular method in promoting recovery that is observed in few experiments directly comparing intravascular cell injection with a direct intraparenchymal cell delivery [47, 48]. These limitations in intravascular delivery of cells for treating ischemic stroke highlight the importance of promoting stem/progenitor cell survival and integration in a direct cell injection to the stroke core.

In this chapter, we focus on application of biomaterials as a feasible, flexible, and effective approach to localize the cells into the infarct core, isolate cells from inflammatory attack and harmful factors in the microenvironment of ischemic brain lesions, and further promote their differentiation and integration by providing them with a proper niche.

Table 13.1 Clinical studies on stem cell therapy for stroke. (Modified from [24])

Study	Phase	Type of cells	Route of injection	Time after stroke	Sample size	Results
Kondziolka et al., *Neurology* 2000	I	Cultured neuronal cells differentiated from a human teratocarcinoma cell line	Intracerebral	7–55 months	12	Treatment is safe and feasible
Kondziolka et al.,*J Neurosurg* 2005	II	Cultured neuronal cells differentiated from a human teratocarcinoma cell line	Intracerebral	1–6 years	14+4 controls	Functional improvements with no change in primary outcome measure
Savitz et al., *Cerebrovasc Dis* 2005	I	Dissociated cells from primordial porcine striatum	Intracerebral	1.5–10 years	5	Study stopped after cortical vein occlusion and seizures
Rabinovich et al., *Bull Exp Biol Med* 2005	I	Frozen samples from dissociated human fetal nervous and hemopoietic tissues	Subarachnoid space	4–24 months	10	No serious complication with potential functional improvement
Suarez-Monteagudo et al., *Restor Neurol Neurosci* 2009	I	Autologous bone marrow mononuclear cells	Intracerebral	1–10 years	5	Safe
Barbosa da Fonseca et al., *Exp Neurol* 2010	I	Labeled autologous bone marrow mononuclear cells	Intra-arterial	59–82 days	6	Safe and feasible, cell homing in the brain and other internal organs
Bang et al., *Ann Neurol* 2005	I/II	Cultured autologous human mesenchymal stem cells	Intravenous	<7 days	5+25 controls	Safe and feasible, potential functional benefit
Lee et al., *Stem Cells* 2010	I/II	Cultured autologous human mesenchymal stem cells	Intravenous	<7 days	16+36 controls	Safe and feasible, potential functional benefit
Battistella et al., *Regen Med* 2011	I	Labeled autologous bone marrow mononuclear cells	Intra-arterial	<90 days	6	Safe and feasible

Table 13.1 (continued)

Study	Phase	Type of cells	Route of injection	Time after stroke	Sample size	Results
Savitz et al., *Ann Neurol* 2011	I	Autologous bone marrow mononuclear cells	Intravenous	24–72 h	10	Safe and feasible
Honmou et al., *Brain* 2011	I	Autologous bone marrow mesenchymal stem cells	Intravenous	36–133 days	12	Safe and feasible
Moniche et al., *Stroke* 2012	I/II	Autologous bone marrow mononuclear cells	Intra-arterial	5–9 days	10+10 controls	Safe and feasible
Parsad et al., *Indian J Med Res* 2012	I	Autologous bone marrow mononuclear cells	Intravenous	7–30 days	11	Safe and feasible
Bhasin et al., *Cerebrovasc Dis Extra* 2011	I	Cultured autologous bone marrow mesenchymal cells	Intravenous	7–12 months	6+6 controls	Safe and feasible
Bhasin et al., *Clin Neurol Neurosurg* 2013	I/II	Autologous bone marrow mononuclear and mesenchymal cells	Intravenous	3–24 months	20+20 controls	Safe and improvement of modified Barthel Index

Biopolymer Hydrogels: Chemistry and Physics

Chemical Composition

Biomaterials have been widely used as scaffolds for cells transplanted to the central nervous system (CNS). They have been synthetized from either natural or synthetic materials (Table 13.2). Natural sources include proteins such as Matrigel [18], collagen [49, 50], fibronectin [51], and fibrin [52], polysaccharides such as chitosan [53], agarose [54], alginate [55] and methylcellulose [56], hyaluronan (HA) [14, 57] and acellular tissue matrix [58]. Many of these molecules are physiologically found in the extracellular matrix (ECM), and therefore their use has the advantage of promoting cell signaling through their binding sites for mammalian cells. In addition, chitosan and agarose have available functional groups that facilitate chemical alterations [59]. From a translational perspective, natural molecules are clinically used as dermal fillers, lubricants, wound sealants, and surgical sponges [60] which facilitate their approval for use from safety and regulatory standpoint.

On the other hand, by assembling a list of design criteria for certain scaffold applications, one can produce synthetic polymers with defined compositions, polymerization and degradation rates, and mechanochemical properties. Synthetic molecules are produced in consistent ways that reduce data variability. Moreover, by being biologically inert, synthetic molecules decrease interaction with inflammatory cells in the brain and improve implant biocompatibility. Poly-lactic-co-glycolic acid or PLGA [61, 62], poly (acrylonitrile)/poly(vinyl chloride) or PAN/PVC [61, 62], oligo (ethylene glycol) fumarate or OPF [65], poly(N-isopropylacrylamide) or PNiPAAM, and polyethylene glycols (PEGs) [66] are common examples of synthetic molecules used as scaffolds in the CNS. It is noteworthy that polyesters such as PLGA have been used as drug delivery vehicles, orthopedic fixation devices, and

Table 13.2 Biomaterials commonly used in therapies for the brain and spinal cord lesions

Natural	Synthetic
Collagen	Poly-lactic-co-glycolic acid (PLGA)
Fibrin	Poly (acrylonitrile)/poly(vinyl chloride) or PAN/PVC
Fibronectin	Oligo (ethylene glycol) fumarate (OPF)
Matrigel	Poly(N-isopropylacrylamide) or PNiPAAM
Chitosan	Polyethylene glycols (PEGs)
Agarose	
Alginate	
Methylcellulose	
Hyaluronan	
Acellular tissue matrix	

absorbable sutures [67, 68], and they have established a history of clinical safety which accelerates their approval in clinical trials.

It is important to know that limited chemical and biophysical interactions of the aforementioned molecules in the context of biological complexity of NSPCs and poststroke brain tissue often necessitate a combination of natural molecules, synthetic polymers, or a conglomerate of both to compose the optimized scaffold product for the best cell survival and integration following transplantation in stroke.

Physical Structure

Microspheres

Biological scaffolds can be further categorized based on their micro- and nanostructure. The earliest attempt to use scaffolds in order to improve cell viability goes back to the 1990s when scientists noticed adrenal chromaffin cells demonstrate a poor cell survival and efficacy upon transplantation to the experimental lesions of Parkinson's disease (PD). To improve the outcome of cell therapy, they injected cells attached to collagen-coated dextran or glass beads as microcarriers [69, 70]. This form of "particulated" matrix promoted cell survival and corrected for dopamine deficits in the striatum. Another particulated matrix, gelatin beads, produced under the name of Spheramine®, was shown to promote survival and function of human retinal pigmented epithelium cells in an experimental model of PD [71]. Alginate beads were able to protect encapsulated cells against Huntington's disease where survival of transplanted rat choroid plexus cells was improved following quinolinic acid (QA) injection [72]. Most relevant to our review, a scaffold of PLGA microspheres promoted integration of NSPCs following injection into the stroke brain cavity [36]. These studies demonstrated the importance of a viable three-dimensional (3D) matrix in transplanting cells to the brain lesions.

Besides scaffolding for cell transplantation, microspheres have been widely used to bypass the blood–brain barrier and deliver drugs, proteins, peptides, viral constructs, DNA, small interfering RNA (siRNA), and other therapeutics into the CNS. Based on their chemical properties, microspheres have a degradation rate that determines releasing pace of the compound encompassed within microparticles. This property ensures a reliable sustained release of an encapsulated molecule. Having multiple microspheres with different degradation rates gives a powerful tool to sequentially release growth factors and optimize tissue repair. For instance, to maximize proliferation of NSPCs as well as neuroprotection following stroke, it may be important to deliver sequential factors, such as epidermal growth factor (EGF) followed by release of the cytokine erythropoietin (EPO). The Shoichet group has achieved this goal by coating EPO-containing nanoparticles with a layer of poly(sebacic acid) that retards the release of EPO for about 7 days [56] and therefore optimizes tissue repair following stroke.

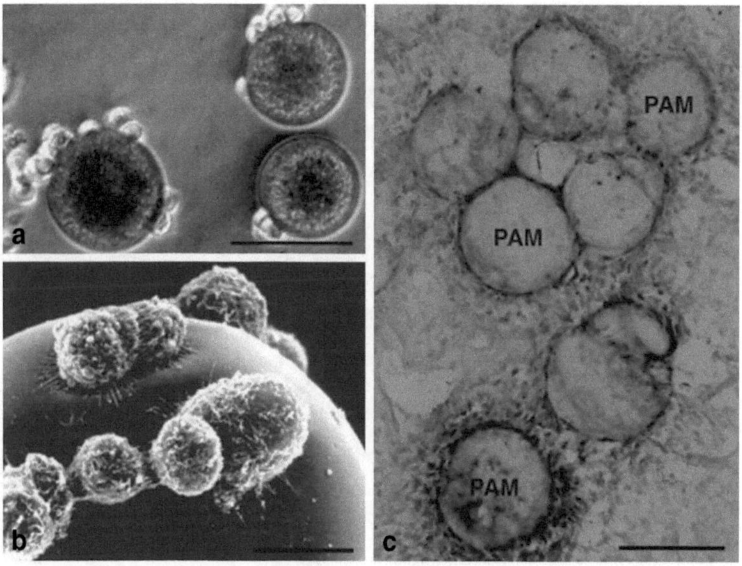

Fig. 13.1 Pheochromocytoma (PC)12 cells attach on pharmacologically active microcarriers (*PAM*) that release nerve growth factor (NGF). Optical (**a**) and surface electron (**b**) microscopy has documented adherence of PC12 cells on the *PAM* external surface. Two weeks after transplantation into the brain NGF-releasing *PAM* are photographed and stained for tyrosine hydroxylase (TH) positive cells (**c**). Scale bars **a, c:** 50 μm; **b:** 5 μm. (Adapted with permission from Tatard et al., Biomaterials 2005 and Delcroix et al., Biomaterials 2010)

The slow-release capability of microspheres can be combined with cell attachment on their outer surface (Fig. 13.). In fact, microspheres could be sintered at low temperature to form a stable 3D microenvironment for cell transplantation [15]. Survival of transplanted cells attached on a 3D matrix of microspheres could be further augmented when microspheres gradually release survival and differentiation factors and thereby exert synergistic effects of adhesion and growth factor signals on cells. This strategy has been employed in treating PD where glial-derived neurotrophic factor (GDNF)-loaded PLGA microspheres promoted the efficacy of embryonic ventral mesencephalon dopaminergic cells attached on their exterior surface [73].

Hydrogel Biopolymers

Hydrogels with their unique properties are suitable candidates for tissue engineering and promoting cell therapy. They are hydrophilic polymer networks made of long chains of monomers that are cross-linked and form a stable network. With their strong propensity for water, they absorb over ninefold of their dry weight. This makes hydrogels a highly porous and hydrated network (Fig. 13.2) that allows free diffusion of oxygen and nutrients and therefore support survival of encapsulated

Fig. 13.2 A hyaluronan hydrogel shown at lower (**a**) or higher (**b**) magnification using scanning electron microscope. Scale bars **a:** 50 µm; **b:** 20 µm. (Reproduced with permission from Hou et al., Neuroscience 2006)

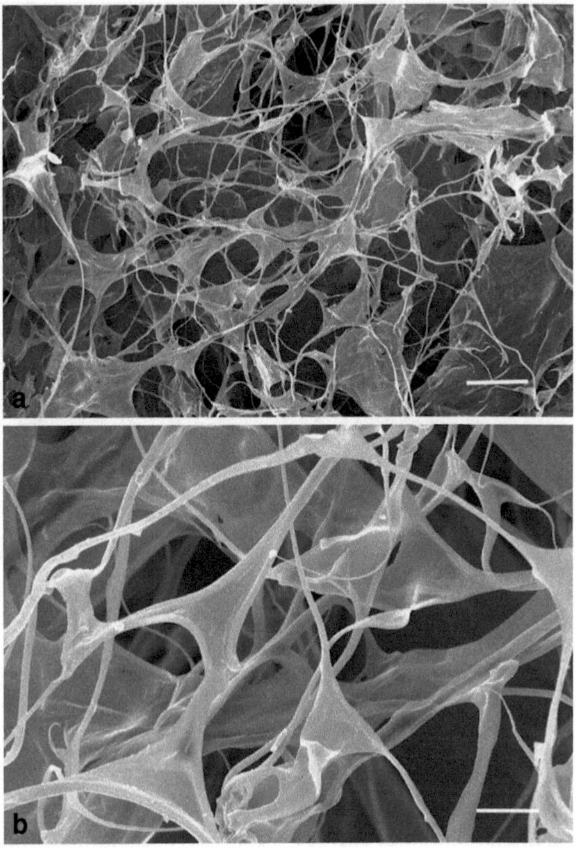

cells. It is noteworthy that the ECM of the brain consists of HA-based porous hydrogel network that lends support to neuronal, glial, and vascular elements of the brain and allows migration of different cell types, including NSPCs, in health and disease [74–76]. It is therefore compatible with the brain structure to use hydrogels as vehicles for cell delivery. The mechanics of their polymer network renders hydrogels as viscoelastic materials [77]. Their elastic property ensures they retain their shape after polymerization, and hence hydrogels will stay conformed to the boundaries of lesion following polymerization inside lesion cavity. The viscous component of their mechanics resists their flow out of lesion cavity, and, as a result, hydrogels will remain in place.

Polymerization Gelation, or polymerization, defines the process of changing from a liquid to solid state and producing a gel from a pre-gel solution. It includes formation of covalent cross-linking bonds between long polymer chains. This process generally involves reactive functional groups on the backbone chain such as acrylates, thiols, or polyesters. In contrast to chemical cross-linking, some polymers such as agarose, Matrigel, alginate, and collagen undergo physical cross-linking by formation of hydrogen bounds [61]. The degree of cross-linking will predict important parameters of hydrogels, such as mechanical properties, porosity, degradation rate, and functionality of hydrogels [61]. These parameters contribute to the function and efficacy of hydrogels in promoting survival and differentiation of encapsulated cells, as well as integration into the host brain tissue.

Following addition of cross-linkers, hydrogels reach their maximum strength after polymerization time has elapsed. The polymerization speed is controlled by the affinity of cross-linker to backbone functional groups, density of cross-linking points, and concentration of cross-linkers. Therefore, an optimum gelation time provides a unique opportunity to access the stroke core with a minimally invasive method—insertion of a fine needle—and injection of the hydrogel before gelation occurs. Gelation can also be initiated by environmental changes in temperature [78, 79] or pH [78–80], which provide further tools to provoke gelation following changes in temperature or pH of the lesion microenvironment. In situ gelation will allow hydrogels to conform to the boundaries of an irregularly bordered lesion and establish a proper contact with peri-infarct tissue [14], which is necessary for implant integration.

Stiffness and Stability Hydrogels can be synthetized in a variety of stiffnesses, but those aimed for application in the CNS need to match the brain in elasticity (see "Biopolymer Hydrogels: Promoting Integration"). Stiffer gels are generally associated with higher levels of cross-linking. Therefore, a degree of cross-linking is desired to produce gels with a compressive elastic modulus between 300–450 Pa that match the brain in its mechanical properties [81]. The extent of polymer cross-attachments also defines degree of polymer stability. Loosely attached polymers will lead to swelling. Swelling of implanted hydrogel, however, is not significant in vivo since the swelling gel is confined by counterforces of the surrounding brain tissue that resists increases in overall size [82].

Porosity The density of cross-linking points and length of cross-linking arms determine gel porosity or pore size, a crucial contributing factor to hydrogel function. The encapsulated cells need exchanging nutrients, oxygen, and waste with surrounding tissue by diffusion across borders of hydrogel; this exchange is mainly determined by hydrogel pore size. The ability of cells to migrate and axons to extend processes inside the gel also depends on size and connectivity of hydrogel pores. It is possible to guide formation of new connections by growing axons within a hydrogel with an interconnected network of pores [83]. In the case of slow-releasing hydrogels, the release rate is also predicted by mesh size, as well as size and polarity of molecules. In addition, ingrowth of blood vessels is integral to formation of any tissue larger than the diffusion limits of the extant vasculature. In stroke, the vasculature is distorted by the infarct, tissue adsorption, and vascular network remodeling. It would be desirable to induce neovascularization of a transplant so that it can achieve adequate blood supply.

Degradation Degradation is another important parameter in application of biopolymer matrices, and it leads to dissolution and disappearance of hydrogels. It is the result of breakage of cross-linking or labile bonds within polymers through hydrolytic or enzymatic cleavage [84–86]. Degradation has important functional implications for cell scaffolds as hydrogels can be synthetized to degrade with different speeds: A slow rate of degradation leads to gel digestion only after encapsulated NSPCs have developed their own matrix and are therefore ready to integrate into the host brain tissue. In the case of a quick gel digestion, early removal of a foreign object will attenuate inflammation and subsequent foreign body response. Therefore, optimizing degradation rate according to the pace of matrix deposition by encapsulated NSPCs will maximize tissue repair after stroke. In designing biopolymers, it is also important to know the degradation products to make sure they will be cleared away and are nontoxic leading to damage and inflammation. PLGA polymer, the most commonly used degradable synthetic polymers [66], is particularly safe from this aspect since the final degradation products are nontoxic molecules, CO_2 and H_2O [87]. This fact has contributed to the biocompatibility of PLGA constructs as carriers of NSPCs in the brain after stroke [36].

In the next two parts, we focus on interaction of hydrogel with encapsulated stem/progenitor cells to promote their survival and differentiation, and that of hydrogel with peri-infarct tissue to facilitate integration of transplanted biopolymer matrix to the surrounding normal brain tissue.

Biopolymer Hydrogels: Impact on Cell Survival and Differentiation

As discussed in "Stem Cell Therapy for Stroke," stem/progenitor cells are promising therapeutic options for stroke. However, poor survival of transplanted cells limits their efficacy. Transplanting stem/progenitor cells within a hydrogel matrix resembling their physiological niche improves cell survival through several mechanisms (Fig. 13.3) discussed below. Although the repair capabilities of stem/progenitor cells are not merely exercised through neural differentiation and integration to the host neural circuitry, using transplanted matrix to promote neural differentiation of stem/progenitor cells can further augment their repair potential. In addition, encapsulating multipotent stem/progenitor cells in a finely crafted biopolymer matrix may ensure cells will not take aberrant differentiation paths to produce tumors such as teratomas.

Promoting Transplanted Cell Survival

There is a multitude of factors contributing to low cell yield following transplantation to the stroke infarct core. We will show that biomaterial scaffolds improve cell survival by targeting mechanisms of cells death. An immediate example of technical difficulties that lead to poor transplant survival is seen in the injection of cells as a thin suspension of cells injected into the brain. The tissue retains some

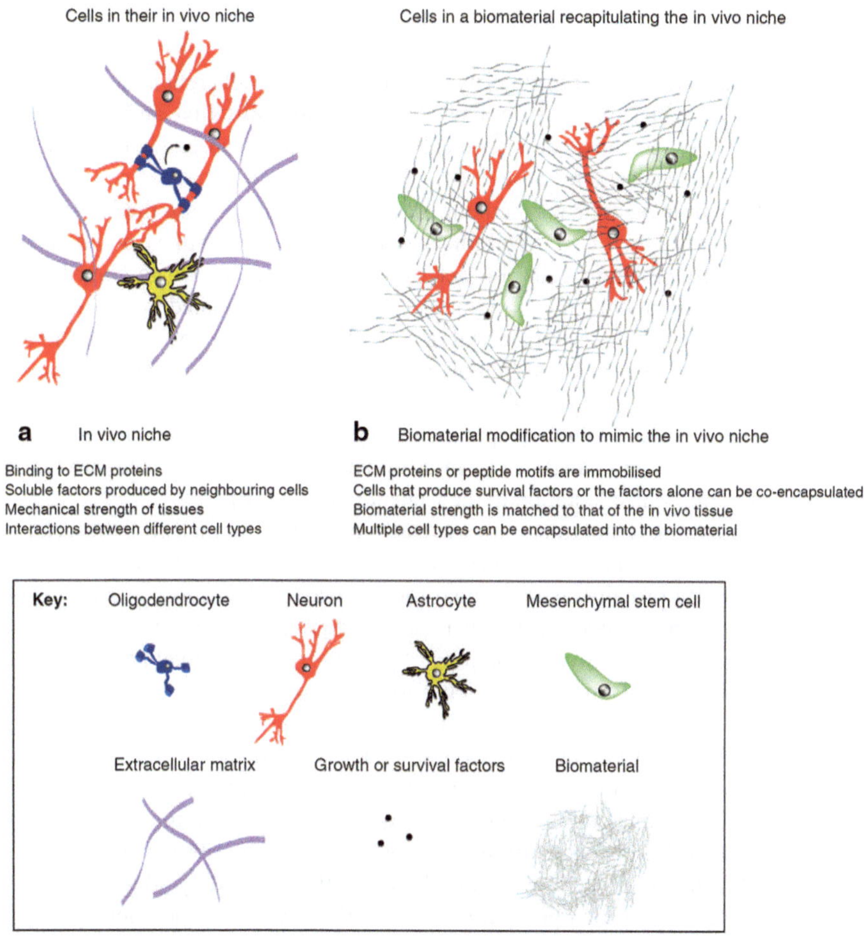

Fig. 13.3 Biomaterials provide a permissive microenvironment for transplanted cells by resembling the in vivo niche. **a** A combination of extracellular matrix (*ECM*) chemical and mechanical characteristics, soluble factors, and cell–cell interactions compose the in vivo niche that contribute to physiological functions of cells. **b** Biomaterial solutions for cell therapy in stroke mimic various mechanical, biochemical, and cellular components of in vivo cellular niche. (Reproduced with permission from [36])

but inevitably restricts part of the injected volume and part of cell suspension flows back to the pial surface and may diffuse into subarachnoid space. Mixing cells with biopolymer hydrogels that have viscosity equal or above the infarcted tissue improves retainability and targets entire cell population to the focus of stroke [14, 88]. This not only improves number of transplanted cells, and therefore promotes recovery, but also avoids distribution of stem/progenitor cells across the CNS, a safety concern in stem/progenitor cell clinical trials.

Adhesion

Besides hematologic cells, the majority of human cells are adherent to their underlying matrix. This adhesion plays a crucial role in keeping cells alive and functioning. In the early 1990s, scientists described cell apoptosis due to absence of cell anchorage, and they coined the term *anoikis* [89]. In cell therapies, detached cells are kept and injected within an aqueous medium. Upon transplantation into the infarct core, cells' attachment to the ECM is complicated due to ischemia damage and digestion by influx of inflammatory cells. In fact, anoikis has been shown to contribute to cell death in other models of CNS injuries [90]. Cell encapsulated in a particulated or hydrogel 3D matrix is provided with a cell-adhesive environment from the moment of injection. Many natural biopolymer gels, such as collagen and fibronectin [91], and HA [88], are naturally found in the stem/progenitor cell niche matrix and therefore engage directly with receptors on stem/progenitor cell. A different type of 3D matrix, fibronectin-coated PLGA microspheres, has also promoted survival of NSPCs following transplantation to the ischemic lesion [36].

Although synthetic polymers are generally nonadhesive for cells, they can be functionalized by covalently attached ECM proteins. In this approach, all the active sites of ECM proteins, which signal for survival, proliferation, migration, and differentiation, are included. Moreover, protein loss due to protein desorption is minimal [92]. However, covalent binding of proteins requires protein modifications that carries the risk of protein functional loss due to alterations in protein active sites, denaturation, or inaccessibility of active sites in a random orientation of molecules [62]. As a solution, the polypeptide sequence within active sites of proteins, which interacts with cell integrin receptors, has been identified and industrially synthetized with inclusion of an active moiety for attachment to the polymer backbone. Oligopeptide sequences from ECM protein, such as fibronectin (Arg-Gly-Asp or RGD) and laminin (Ile-Lys-Val-Ala-Val or IKVAV, and Tyr-Ile-Gly-Ser-Arg or YIGSR), have been incorporated in biomimetic solutions to produce hydrogels with the functional elements found in stem cell/progenitor niche and improve cell survival in transplant therapies.

Trophic Support and Protection from Inflammation

Stem or progenitor cells both in culture and in their in vivo niche are exposed to growth factors to promote their survival. Following transplantation into the infarct core, cells are placed in a microenvironment lacking supportive glial cells that produce survival factors. This picture becomes more complicated by presence of inflammatory cells secreting proapoptotic factors such as interleukin (IL)-1 and tumor necrosis factor (TNF)-α [93]. Biopolymer matrices promote cell survival by: (1) attenuating inflammatory response and (2) providing trophic support. Synthetic polymers generally do not interact with inflammatory cells, and they can therefore stealth encapsulated cells from inflammatory cells. Immunogenicity of transplanted stem/progenitor cells, due to interspecies genetic differences or genetic manipulation of cells, plays a major role in elimination of transplanted cells by the host immune system. Therefore, concealing cells from inflammatory cells will impede the host-versus-graft attack [94]. Some biopolymers, such as high molecular weight HA, are known to silence inflammation and protect encapsulated cells from inflammatory insult [88].

Cellular scaffolds can be further rendered prosurvival by releasing factors that suppress inflammation or induce cell survival. Simply mixing soluble molecules in hydrogel exposes cells to a transient and high concentration of molecules that could potentially exert toxic effects. A fast diffusion and degradation of soluble molecules leads to quickly diminishing concentrations of molecules that limit any potential benefits. To circumvent this problem, growth factors could be immobilized to the scaffold backbone by covalent linkage. Alternatively, a variety of sustained-release strategies could be employed. Cyclosporine A, a potent anti-inflammatory drug, has been incorporated into PLGA microspheres or HA–methylcellulose composites as drug reservoirs and slowly released over the brain cortex [95]. PLGA particles have been further used to deliver dexamethasone, another potent immunosuppressor, to the CNS [96]. The inflammatory response could be further suppressed by incorporating neutralizing molecules for reactive oxygen species (superoxide dismutase mimetic metalloporphyrin macromer or MnTPPyP-Acryl; [97]) or TNF-α antagonizing peptides [98].

Epidermal growth factor (EGF) and basic fibroblast growth factor (bFGF or FGF-2) are two molecules providing prosurvival and pro-proliferative support for transplanted NSPCs [99, 100]. The natural tendency of bFGF to attach to heparin has been utilized in heparin-modified HA gels to gradually release bFGF [101]. Incorporating bFGF in a PEG hydrogel has promoted survival of NSPCs [102] and human mesenchymal stem cells [103]. A combination of platelet-derived growth factor (PDGF) and neurotrophin (NT)-3 gradually releasing from fibrin scaffolds has induced survival of encapsulated neural progenitor cells [104]. In this example, a cell scaffold has significantly improved cell survival through synergistic effects of cell adhesives and trophic support. In another example of this synergistic effect, mesencephalon neuronal cells attached to PLGA microspheres received trophic support by sustained release of GDNF from microspheres [73].

Controlling Differentiation of Encapsulated Cells

The secretory profile of stem/progenitor cells and their differentiation, both mechanisms contributing to neural repair after stroke, are affected by their surrounding microenvironment. This includes the ECM composition of proteins, adjacent cells, humoral growth factors (such as brain-derived neurotrophic factor or BDNF, ciliary neurotrophic factor or CNTF, NT-3, PDGF, GDNF, and nerve growth factor or NGF), and biophysical properties of their neighboring matrix. These elements are present in the unique niche of neural stem or progenitor cells to cue them for

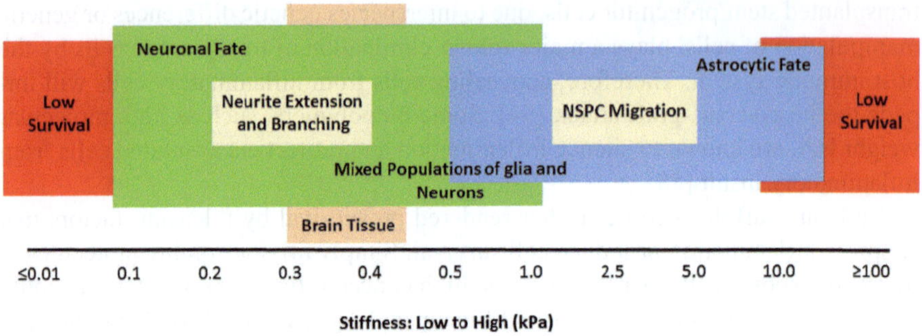

Fig. 13.4 Biopolymer stiffness determines survival, fate, and function of neural stem/progenitor cells (*NSPCs*). *NSPCs* do not survive well in a matrix with very low or very high stiffness. From those cells that survive at the lower stiffness, they differentiate more prominently towards a neuronal cell fate. In contrast, those growing at the higher stiffness tend to develop astrocytes. Neurite extension and *NSPC* migration also depend on biomaterial stiffness and maximize at lower and higher stiffnesses, respectively. (Original data published elsewhere [57, 81, 84, 102, 105–109]. Reproduced with permission from [61])

self-renewal and differentiation. Hence, any successful application of biologic scaffolds for cell delivery should consider mimicking a similar combination of factors.

A seminal study by Discher's group [105] revealed that focusing on biochemical factors as the sole regulators of stem cell differentiation is a simplistic view. They showed that the stiffness of the substrate matrix influences the differentiation of mesenchymal stem cells. More interestingly, the stiffness of tissues drives differentiation in resident populations of stem/progenitor cells by a very characteristic and specific manner, as when cultured on substrates mimicking bone, muscle, or brain in their stiffness, mesenchymal cells were inclined to make mature osteoblasts, myocytes, or neurons, respectively [105]. This indicates how tissue elasticity, besides biochemical factors, provides signals for renewal of specialized cells. Other groups have found the same concept in regulating NSPCs (Fig. 13.4). NSPCs' proliferation and self-renewal increase on hydrogels below 10 kPa (compressive modulus) and maximize around 3.5 kPa [110]. Neural differentiation of NSPCs, however, requires lower elasticities below 1 kPa that in fact resembles the mechanical properties of the brain [81]. Hydrogel scaffolds with their flexibility in degree of cross-linking offer a very powerful tool to optimize self-renewal and differentiation. For instance, NSPCs transplanted in a 3.5-kPa matrix will mount a proliferative response and following a slow degradation of hydrogel biopolymer and descent of stiffness to <1 kPa, their self-renewal slows and NSPCs start differentiating to neuron. Besides differentiation, other proregenerative functions of stem/progenitor cells such as their secretory profile are also controlled by matrix stiffness [111].

The incorporation of ECM proteins, such as collagen [112, 113], laminin [114], and fibronectin [112], has been used to mimic stem/progenitor cells niche and promote their differentiation. Stem/progenitor cells start to modify their microenvironment after transplantation and secrete their own ECM. Using cross-linkers amenable to cleavage by MMPs facilitates remodeling of hydrogel scaffolds by their encapsulated stem/progenitor cells [115] and promotes their differentiation and axon extension. Differentiation could also be promoted by other biophysical factors that affect cell–substrate interactions, such as micropatterned surfaces [116], addition of electrically charged monomer methacrylate [117], and forcing into distinctive cell morphologies [118].

Biopolymer matrices have been employed to present a range of growth factors to the population of encapsulated stem/progenitor cells and drive their proliferation and differentiation. This includes strategies to immobilize growth factors or incorporate them in a slow-releasing reservoir. Agarose-immobilized PDGF or chitosan-immobilized interferon-γ has led to differentiation of NSPCs to oligodendrocytes [119] or neurons [120], respectively. Alternatively, a sustained provision of NT-3 [121] or differentiation factor dibutyryl cyclic-adenosine monophosphate [122] from PLGA microspheres have induced neural differentiation in stem/progenitor cells. Differentiation has also been augmented by hydrogel delivery of CNTF, NT-3, PDGF, GDNF, or NGF [55, 104].

In the future, encapsulating cells within biopolymer scaffold with a selected set of factors to drive differentiation may be an alternative to transplanting pre-differentiated stem/progenitor cells. Instead, a scaffold with finely crafted elasticity, surface charge, micropattern structure, and ECM proteins composition together with timely release of appropriate proliferation and growth factors would be a viable approach to direct differentiation of encapsulated stem/progenitor cells for a prolonged time after transplantation.

Biopolymer Hydrogels: Promoting Integration

The integration of transplanted stem/progenitor cells into the host tissue is critical for tissue repair and necessary for any functional improvement following cell transplantation in stroke. Integration in the context of cell therapy is defined as "the ability of the transplanted cells to interact with the host tissue in a beneficial way" [66]. It depends on the ability of transplanted cells to reach and interact with the host tissue and likewise access of host cells to the implanted matrix. In situ gelation of a biopolymer matrix maximizes the contact between scaffold and the host tissue and consequently improves cell movement from and to the scaffold. Optimized porosity, chemistry, and elasticity, and inclusion of ECM proteins are important factors in scaffold design that promote cell migration and lead to assimilation of implanted matrix to the host tissue. In addition, mitigating poststroke inflammation and subsequent scar formation is crucial in scaffold integration, since astrocytic scar and associated chondroitin sulfate proteoglycans (CSPGs) may seal off the foreign object (in this instance the implant) from normal brain tissue and limit any interaction such as cell migration and axon extension. This fact highlights the importance of hydrogel biocompatibility to avoid a sustained inflammatory response and ensuing glial scar formation [123]. Angiogenesis is another landmark of integration that interconnects the host tissue to implanted matrix, supports growth of implant by providing oxygen and nutrients, and establishes trails for cell migration. In the following sections, we elaborate on biomaterial-based strategies to address different components of integration that improve cell therapies for stroke.

Inflammation, Glial Scar, and Foreign Body Reaction

Any nondegradable substrate in the brain leads to a constant activation of an inflammatory response that isolates the "foreign object" by forming an astrocytic scar [124]. This scar is a counter-integration phenomenon since it serves as a barrier for cell migration and axon growth through the scaffold-tissue interface [125]. Biopolymer hydrogels have several characteristics that improve their biocompatibility and integration in the brain.

Mechanical trauma during scaffold implantation causes an immediate inflammatory response that ultimately leads to the foreign body response against implants [126, 127]. This could be avoided in hydrogels by employing minimally invasive methods of gel injection. In addition, the high water content of hydrogels further improves their biocompatibility [61]. The size and shape of hydrogels are additional biophysical parameters that determine surface area available to inflammatory cells and therefore affects biocompatibility [127, 128]. The stiffness of the implanted matrix is another physical factor contributing to prolonged inflammation and implant isolation. Astrocytes and microglia, the main cells involved in inflammation in the brain, are primed by stiff substrates to produce proinflammatory molecules such as IL-1β and toll-like receptor (TLR)-4, respectively [123]. Implantation of hydrogels with elasticities higher than the brain's values incites inflammation in the early phase following implantation that leads to astrocytic scar in chronic phase [123]. This indicates scaffolds similar in stiffness to the brain will optimize differentiation in encapsulated stem/progenitor cells [110] as well as promoting integration to the surrounding brain tissue. Another solution to render hydrogels biocompatible is to make them degradable in the brain. Hydrogels provide support for encapsulated cells early after implantation and following growth, differentiation, and deposition of their own matrix, hydrogel is degraded by encapsulated or host tissue cells. Examples are HA [88] and collagen [66] gels that are degraded by cellular hyaluronidase and collagenase, respectively. It is important, however, to determine that hydrogel degradation products are nontoxic and cleared away.

In the pursuit of more biological solutions to improve biocompatibility, hydrogels have been loaded with a variety of anti-inflammatory factors (for details see "Trophic Support and Protection from Inflammation"). Chondroitinase ABC, a digestive enzyme for CSPG molecules, has also been used in implantation of neural stem cell (NSC)-loaded poly-ε-caprolactone scaffold to promote migration of NSCs into the injured CNS tissue [129]. In addition, high molecular weight HA gels through their anti-inflammatory effect can attenuate scar formation and improve integration of encapsulated cells [88]. This indicates biopolymer scaffolds, through their physical, chemical, and biological properties, attenuate inflammation and scar formation, and therefore promote integration of transplanted cells to the host brain tissue.

Angiogenesis

Oxygen and nutrients diffuse up to 150–250 μm from capillaries [130]. Therefore, any poststroke tissue reconstruction beyond these dimensions necessitates establishment of an environment with accessibility to blood perfusion. This requires formation of blood vessels reaching into the transplanted matrix. In fact, there is a correlation between transplanted cell survival and extent of new vessel formation [131]. Angiogenesis involves activation of endothelial cells by proangiogenic factors, such as vascular endothelial growth factor (VEGF), PDGF, FGF, hepatic growth factor (HGF), angiopoietin-1, and transforming growth factor-β or TGF-β. Endothelial cells then branch off from existing capillaries, proliferate, and migrate into the transplanted matrix [132, 133]. For successful angiogenesis, the transplanted matrix should support survival and proliferation of endothelial cells. In addition, the matrix has to allow remodeling by endothelial cells, forming cord-like structures which will mature into capillaries, and subsequently veins and arteries.

Hydrogels as highly hydrated and porous substrates are supportive for endothelial cell migration. A network of interconnecting and communicating pores facilitate angiogenesis and obviates the need for matrix digestion by endothelial cells. Angiogenesis could be further promoted by enriching the hydrogel environment and introducing elements of ECM matrix to the hydrogels, such as fibronectin [134] or fibronectin-derived synthetic polypeptide motifs [135]. Incorporation of slow-releasing proangiogenic factors, such as VEGF or bFGF, is another means to potentiate formation of new vessels inside a hydrogel matrix [136]. Migration of endothelial cells and remodeling of matrix involves secretion of MMPs, such as MMP-2 and MMP-9. Therefore, hydrogels with motifs recognized and cleaved by MMPs will have an additional proangiogenic effect. This could be achieved by including MMP-sensitive cross-linkers. A study combined both strategies by including VEGF tethered through a MMP-sensitive linkage and showed improved angiogenesis [118]. Some biopolymer hydrogels particularly support formation and stability of new vessels. HA gels are an example: While high molecular weight HA leads to endothelial cell quiescence, oligomeric HA, produced from degradation of HA gels, promotes proliferation and migration of endothelial cells by interacting with HA cells receptors, such as CD44, hyaluronan-mediated motility receptor (RHAMM), and TLR-4 [137]. In fact, HA hydrogel has been shown to promote angiogenesis upon implantation to the brain [138].

Inclusion of NSPCs adds to a proangiogenic benefit of hydrogels. NSPCs perceive lack of oxygen and nutrients by expressing the transcription factor hypoxia-induced factor-1 that leads to increased secretion of VEGF [139]. Secretory support of angiogenesis by NSPCs could be enhanced when they are encapsulated within a hydrogel with 40 kPa (compressive modulus) [111]; this is another example of how mechanical properties contribute to regenerative potential of cell scaffolds.

NSPC proangiogenic effects can be further enhanced by inclusion of endothelial cells in a co-transplantation approach. In fact, this combination of cells encapsulated in PLGA gels has improved angiogenesis at the center of traumatic CNS injury and promoted reestablishment of the blood–brain barrier [140]. The application of biopolymer matrices with optimized mechanical (i.e., porosity and elasticity), biochemical (i.e., degradability, inclusion of ECM elements and growth factors), and cellular composition will enhance formation of new vessels, which in turn contributes to survival and integration of transplanted cells.

Axonal Regeneration

Axon regeneration in stem/progenitor cell therapy after stroke marks a functional integration of transplanted cells with the host neural circuitry and could lead to subsequent behavioral recovery in the CNS. The regeneration of axons in the brain is very limited due to lack of supportive growth factors and ECM cues, and an inhibitory extracellular environment. The halted CNS axons demonstrate regenerative abilities once they are placed in the proregenerative environment of the peripheral nervous system that offers them a right combination of growth factors and ECM molecules [141, 142]. Therefore, engineering and implanting a permissive environment has the possibility of promoting axon growth, if have kept their intrinsic growth abilities, to regenerate.

The structure of biopolymer hydrogels, a porous and hydrated lattice, provides an advantage for axon regeneration through scaffolds; this is in contrast to the compact cellular arrangement of the astroglial scar, an impediment to axon growth. Movement of the growth cone, the growing end of axons, could be further facilitated by incorporating an interconnected network of pores. The elasticity of hydrogels is another important parameter in designing hydrogel polymer for axon growth since a softer hydrogel will promote axonal growth [143]. Implanting a hydrogel that resembles brain in its stiffness will reduce glial scar [88] that in turn facilitates axon growth through gel–tissue interface.

To further augment axon regeneration, it is important to know the molecules that stimulate or suppress the growth cone. HA found in the peri-neuronal net (an ECM structure around CNS neurons regulating synaptic activities [144, 145]) regulates axon pathfinding and fine-tunes the specific architecture of axon terminals in the brain [88]. Laminin is another example of a molecule whose presence is often co-localized with regenerating axons in the CNS [146]. Researchers have combined these two proregenerative elements by implanting an HA gel enriched with laminin-derived oligopeptides (IKVAV), and they observed axon regeneration through the gel placed into the injured rat brain [147]. High molecular weight HA gel can further facilitate axon growth by reducing glial scar in the adjacent tissue [148]. Implanting fibronectin mats into injured spinal cord promoted regeneration of axons along the direction of the mat fibers [149]. In this approach, axons benefited from combination of ECM molecules and topographical cues for growth and extension.

Delivery of growth factors provides surviving axons with trophic support to prevent their dieback. Growth factors also may potentially induce regeneration of different type of axonal pathways; examples are BDNF (rubrospinal, raphespinal, cerulospinal, and reticulospinal pathways), NT-3 (corticospinal and dorsal column sensory axons) [150], as well as NGF and GDNF (which support different types of axon including nociceptive pathways) [151–153]. These facts describe why incorporation of growth factors within biopolymers may support axon extension and regeneration into the scaffold [149].

Growth factors are often utilized as soluble and freely diffusing molecules. However, in neural development, there are gradients of growth factors, rather than homogeneous concentrations, which spatially guide axon growth [154]. Bioengineers have utilized microfluidic techniques or multiphoton lasers with labile protecting groups to create patterned gradients of growth factors and adhesion molecules within 3D biopolymer scaffolds (Fig. 13.5) to actively guide axon growth and cell migration [137, 155, 156]. The current technologies allow controlling molecular gradients across a micrometer scale [157] that is important for reconstructing tissues in a cellular resolution. Among the different types of hydrogels, agarose gels [158, 159] and PEG gels [160] have been most studied for a patterned gradient of molecules.

Axon growth is inhibited by a variety of anti-regenerative molecules that cause collapse of growth cone. CSPGs and myelin-associated inhibitory molecules (Nogo, oligodendrocyte myelin glycoprotein or OMgp, and myelin-associated glycoprotein or MAG) are two groups of molecules associated with the astrocytic scar that are upregulated after injury and prevent axon regeneration. Bioengineered approaches

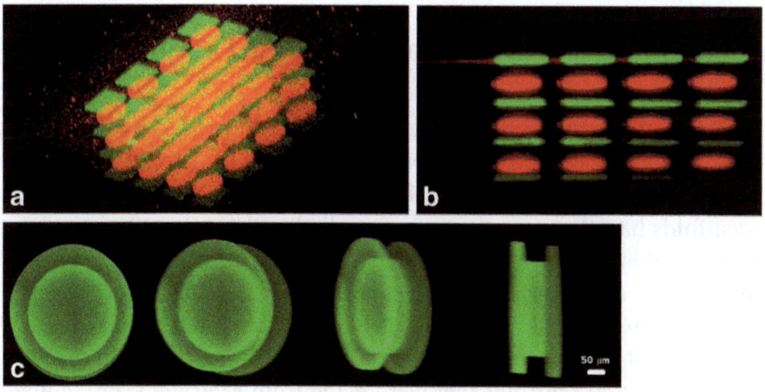

Fig.13.5 3D patterning of agarose (**a, b**) or PEG (**c**) hydrogels with chemically defined volumes in micrometer scale. Hydrogels modified with multiphoton-labile groups are activated through focusing multiphoton laser light to immobilize defined concentrations of peptides in a spatially defined arrangement. Oblique (**a**) and side (**b**) views of agarose gel show green fluorescence molecules patterned into 4×4×4 array of squares (ca. 60 µm per side) overpatterned by red fluorescence molecules in 4×4×4 array of circles (ca. 50 µm per side). The same method has been utilized to print green fluorescent molecules into PEG gel (**c**). Duration and intensity of laser light are two parameters determining the concentration of immobilized peptides. (Adapted with permission from Shoichet, Macromolecules 2010 and Wosnik and Shoichet, Chem Mater 2008)

to axon regeneration have envisaged several ways to oppose those inhibitory molecules. Incorporation of neutralizing antibodies for Nogo-66 receptors (surface molecules mediating anti-growth effects of myelin-associated inhibitory proteins) into HA gels has promoted axon regeneration in the injured spinal cord [161]. Moreover, hydrogel delivery system for the enzyme chondroitinase ABC promotes axon growth and recovery after spinal cord injury through removing inhibitory CSPG molecules [162]. In their report, Bellamkonda's group has improved the durability of the chondroitinase enzyme by thermostabilizing it using the sugar trehalose. Furthermore, delivering BDNF, besides its direct trophic support of axons, has an indirect effect on regeneration by attenuating astrocyte reactivity and reducing production of CSPG [163].

The delivery of anti-inflammatory drugs has a neuroprotective effect early after their administration through protecting axons against inflammatory insults and, in the long term, reduces astrocytic scar and therefore indirectly promotes axon regeneration in the CNS [164]. Methylprednisolone as an example of this group of drugs have been delivered locally through PLGA-based nanoparticles, and it reduced inflammation and glial scar after spinal cord injury [165]. The authors showed this bioengineered preparation of methylprednisolone was superior to systemic injection.

Combining biopolymer matrices with ECM molecules, growth factors, and anti-inhibitory molecules will target the impediments of axon growth in the infarcted brain and provide axons and transplanted cells with a permissive environment to grow and migrate. Inclusion of stem/progenitor cells may further promote axonal growth through cell-contact-mediated growth of axons.

Application of Bio-scaffolds for Stem Cell Therapy

Biopolymer scaffolds have been used as vehicles for the delivery of stem/progenitor cells in experimental models of ischemic stroke, and to date there are seven published reports available (Table 13.3). In those experiments, different combinations of bio-scaffolds have been used: one has applied a PLGA-based particulate matrix [36], and the other six have encapsulated their cells in hydrogel matrices [14, 49, 58, 166, 167]. In all the reports, NSPCs were used with one exception [167] that used bone marrow stromal cells. There were only two studies that used cells with human origin [58, 166]. Two studies used synthetic matrices [36, 167], and the others applied natural scaffolds that were made from Matrigel [166], collagen [49, 50], HA [14], or acellular tissue ECM [58]. All the experiments were performed in a middle cerebral artery occlusion model, except one that used cortical photothrombotic stroke [14].

In terms of functional benefit, only three papers reported improved motor outcome [49, 50, 166] and others did not test for that. The majority proved the benefit of bio-scaffold for improving cells survival. One study also tracked the transplanted cells by the means of magnetic resonance imaging [58].

Table 13.3 Application of biopolymers for neural stem/progenitor cell therapy of focal brain ischemic lesion. (Modified from Moshayedi and Carmichael, Biomatter 2013)

Article	Stroke model/species	Cell transplanted/density per animal	Hydrogel	Outcomes
Bible et al., *Biomaterials* 2009	MCAO/rat	NSC/3.15 × 10^5 cells in 30 ml	Fibronectin-coated PLGA particles	Descriptive analyses of cell survival, cell-scaffold-tissue integration, cell differentiation, angiogenesis and host inflammatory response
Jin et al., *J Cereb Blood Flow Metab* 2010	MCAO/rat	hES-NPC/6 × 10^6 cells in 50 ml	Matrigel	Reduction in lesion volume, improving cell survival, differentiation and behavioral indices
Osasai et al., *Neurosurg* 2010	MCAO/mouse	BMSC/5 × 10^5 in 125 ml	Mebiol from Ikeda Rika, Inc.	Promotes cell engraftment and neuronal differentiation, no functional benefit
Yu et al., *Anat Rec (Hoboken)* 2010	MCAO/rat	NSC/1.5 × 10^4in 5 ml	Collagen	Cell survival, NSC synapse formation and neurological severity score improved
Zhong et al., *Neurorehabil Neural Repair* 2010	PT/mouse	ES-NPC/10^5 cells in 7 ml	Hyaluronan/Heparin/Collagen	Improving cell survival and host inflammatory response, angiogenesis and astrocytic reactivity
Matsuse et al., *Tissue Eng Part A* 2011	MCAO/rat	MS-NSC/2 × 10^4 cells, 12 ml in stratum and 8 ml in cortex	Collagen with bFGF in gelatin microspheres	Infarct volume, cell survival and distribution, angiogenesis, number of host NSCs and motor behavior improved
Bible et al., *Biomaterials* 2012	MCAO/rat	hNSC /2.1–2.5 × 10^6 cells in 25–40 ml	Acellular ECM	Descriptive analyses of cell imaging by MRI, cell migration, differentiation and cell-host tissue interaction

BMSC bone marrow stromal cell, *ES* embryonic stem, *h* human, *MCAO* middle cerebral artery occlusion, *MS* mesenchymal stem, *NPC* neural precursor cell, *NSC* neural stem cell, *PLGA* poly (lactic-co-glycolic acid), *PT* photo-thrombotic, *ECM* extracellular matrix

Conclusion and Prospects

The brain tissue after stroke presents as a multifaceted inhibitory environment that opposes repair attempts due to levels of inflammation, lack of trophic support, destruction of viable matrix, and locally poor vasculature. Therefore, any bioengineered approach will likely fail unless it addresses these pathologies in the brain following stroke. Biopolymer scaffolds for stem/progenitor cells are a flexible tool with diverse capabilities for a combinatorial approach that simultaneously addresses multiple barriers for regeneration and benefits from synergistic effects of therapeutics towards an improved outcome. There are a handful of studies in transplantation of encapsulated cells that indicate a biopolymer matrix improves cell survival and, in few cases, promotes functional recovery.

With the current state of clinical trials on stem/progenitor cell therapy for stroke, with reports of mixed and often disappointing results, there might be a tendency to transition biomaterial-based cell therapies early into the clinical stage. There are, however, three important considerations to ensure that a sound and safe translation will take place:

1. The majority of experiments on biopolymer-encapsulated cells in stroke have not included functional recovery as their outcome measurement, or alternatively they might have not reported their negative results. This is important to perform extensive and blinded assessments on possible functional benefits of transplantation by comparing cells encapsulated in biopolymers versus cells in aqueous vehicle. Utilization of a comprehensive battery of tests is necessary to interrogate fine versus gross and proximal versus distal limb movements. It is important to communicate any negative finding to avoid spending resources on therapeutics with less promising outcomes.
2. The mere finding of functional benefit does not provide in-depth understanding of the mechanisms involved. Performing studies to elucidate mechanisms, through preferably gain of function and loss of function experimental paradigms, further ensures that the observed beneficial effect is genuine. In addition, understanding mechanisms will help to potentiate functional benefits and, in case of failed outcomes in clinical stage, will explain possible underlying reasons for failure and propose solutions.
3. In translation of a biological therapy, ensuring safety is of course the foremost important consideration. When biomaterials are involved, a consistently high level of quality in production and surveillance is of paramount importance. This becomes more difficult when a biopolymer scaffold has composed of several elements, such as polymer backbone, ECM molecules, growth factors, enzymes, drugs, cells, etc. It is crucial to ensure a high-quality standard for production and safety measures for each component, and ultimately for the combinational end product.

Acknowledgments PM and STC are partially supported by California Institute of Regenerative Medicine grant RT2–01881.

References

1. Go AS, Mozaffarian D, Roger VL, et al. Heart disease and stroke statistics-2014 update: a report from the American heart association. Circulation. 2014;129:e28–e92. doi: 10.1161/01.cir.0000441139.02102.80.
2. Go AS, Mozaffarian D, Roger VL, et al. Heart disease and stroke statistics-2013 update: a report from the American heart association. Circulation. 2013;127:e6–e245. doi: 10.1161/CIR.0b013e31828124ad
3. Kelly-Hayes M, Beiser A, Kase CS, et al. The influence of gender and age on disability following ischemic stroke: the Framingham study. J Stroke Cerebrovasc Dis. 2003;12:119–26. doi: 10.1016/S1052-3057(03)00042-9.
4. Ali M, Hazelton C, Lyden P, et al. (2012) Recovery from poststroke visual impairment: evidence from a clinical trials resource. Neurorehabil Neural Repair. doi: 10.1177/1545968312454683
5. Wechsler LR. Intravenous thrombolytic therapy for acute ischemic stroke. N Engl J Med. 2011;364:2138–46. doi: 10.1056/NEJMct1007370.
6. Dietrich WD, Prado R, Watson BD, Nakayama H. Middle cerebral artery thrombosis: acute blood-brain barrier consequences. J Neuropathol Exp Neurol. 1988;47:443–51.
7. Wang PY, Kao CH, Mui MY, Wang SJ. Leukocyte infiltration in acute hemispheric ischemic stroke. Stroke. 1993;24:236–40. doi: 10.1161/01.STR.24.2.236.
8. Lo EH, Dalkara T, Moskowitz MA. Mechanisms, challenges and opportunities in stroke. Nat Rev Neurosci. 2003;4:399–415. doi: 10.1038/nrn1106.
9. Murphy TH, Corbett D. Plasticity during stroke recovery: from synapse to behaviour. Nat Rev Neurosci. 2009;10:861–72. doi: 10.1038/nrn2735.
10. Carmichael ST. Cellular and molecular mechanisms of neural repair after stroke: making waves. Ann Neurol. 2006;59:735–42. doi: 10.1002/ana.20845.
11. Tsai PT, Ohab JJ, Kertesz N, et al. A critical role of erythropoietin receptor in neurogenesis and post-stroke recovery. J Neurosci. 2006;26:1269–74. doi: 10.1523/JNEUROSCI.4480-05.2006.
12. Bliss T, Andres R, Steinberg G. Optimizing the success of cell transplantation therapy for stroke. Neurobiol Dis. 2010;37:1–20. doi: 10.1016/j.nbd.2009.10.003.Optimizing.
13. Locatelli F, Bersano A, Ballabio E, et al. Stem cell therapy in stroke. Cell Mol Life Sci. 2009;66:757–72. doi: 10.1007/s00018-008-8346-1.
14. Zhong J, Chan A, Morad L, et al. Hydrogel matrix to support stem cell survival after brain transplantation in stroke. Neurorehabil Neural Repair. 2010;24:636–44. doi: 10.1177/1545968310361958.
15. Causa F, Netti PA, Ambrosio L. A multi-functional scaffold for tissue regeneration: the need to engineer a tissue analogue. Biomaterials. 2007;28:5093–9. doi: 10.1016/j.biomaterials.2007.07.030.
16. Luo Y. Cell-based therapy for stroke. J Neural Transm. 2011;118:61–74. doi: 10.1007/s00702-010-0478-4.
17. Espinera AR, Ogle ME, Gu X, Wei L. Citalopram enhances neurovascular regeneration and sensorimotor functional recovery after ischemic stroke in mice. Neuroscience. 2013;247:1–11. doi: 10.1016/j.neuroscience.2013.04.011.
18. Jin K, Sun Y, Xie L, et al. Post-ischemic administration of heparin-binding epidermal growth factor-like growth factor (HB-EGF) reduces infarct size and modifies neurogenesis after focal cerebral ischemia in the rat. J Cereb Blood Flow Metab. 2004;24:399–408. doi: 10.1097/00004647-200404000-00005.
19. Kim YR, Kim HN, Ahn SM, et al. Electroacupuncture promotes post-stroke functional recovery via enhancing endogenous neurogenesis in mouse focal cerebral ischemia. PLoS One. 2014;9:e90000. doi: 10.1371/journal.pone.0090000.
20. Ma M, Ma Y, Yi X, et al. Intranasal delivery of transforming growth factor-beta1 in mice after stroke reduces infarct volume and increases neurogenesis in the subventricular zone. BMC Neurosci. 2008;9:117. doi: 10.1186/1471-2202-9-117.

21. Popa-Wagner A, Stöcker K, Balseanu AT, et al. Effects of granulocyte-colony stimulating factor after stroke in aged rats. Stroke. 2010;41:1027–31. doi: 10.1161/STROKEAHA.109.575621.
22. Burns TC, Verfaillie CM, Low WC. Stem cells for ischemic brain injury: a critical review. J Comp Neurol. 2009;515:125–44. doi: 10.1002/cne.22038.
23. Bhasin A, Srivastava MVP, Kumaran SS, et al. Autologous mesenchymal stem cells in chronic stroke. Cerebrovasc Dis Extra. 2011;1:93–104. doi: 10.1159/000333381.
24. Lemmens R, Steinberg GK. Stem cell therapy for acute cerebral injury: what do we know and what will the future bring? Curr Opin Neurol. 2013;26:617–25. doi:10.1097/WCO.0000000000000023.
25. Seminatore C, Polentes J, Ellman D, et al. The postischemic environment differentially impacts teratoma or tumor formation after transplantation of human embryonic stem-cell-derived neural progenitors. Stroke. 2010;41:153–9. doi: 10.1161/STROKEAHA.109.563015.
26. Kondziolka D, Wechsler L, Goldstein S, et al. Transplantation of cultured human neuronal cells for patients with stroke. Neurology. 2000;55:565–9.
27. Kondziolka D, Steinberg GK, Wechsler L, et al. Neurotransplantation for patients with subcortical motor stroke: a phase 2 randomized trial. J Neurosurg. 2005;103:38–45. doi: 10.3171/jns.2005.103.1.0038.
28. Yu F, Li Y, Morshead C. Induced pluripotent stem cells for the treatment of stroke: the potential and the pitfalls. Curr Stem Cell Res Ther. 2013;8:407–14. doi: 10.2174/1574888X113089990052.
29. Buhnemann C, Scholz A, Bernreuther C, et al. Neuronal differentiation of transplanted embryonic stem-cell-derivedd precursors in stroke lesions of adult rats. Brain. 2006;129:3238–48. doi: 10.1093/brain/awl261.
30. Daadi MM, Li Z, Arac A, et al. Molecular and magnetic resonance imaging of human embryonic stem-cell-derived neural stem cell grafts in ischemic rat brain. Mol Ther. 2009;17:1282–91. doi: 10.1038/mt.2009.104.
31. Englund U, Bjorklund A, Wictorin K, et al. Grafted neural stem cells develop into functional pyramidal neurons and integrate into host cortical circuitry. Proc Natl Acad Sci U S A. 2002;99:17089–94. doi: 10.1073/pnas.252589099.
32. Song H, Stevens CF, Gage FH. Neural stem cells from adult hippocampus develop essential properties of functional CNS neurons. Nat Neurosci. 2002;5:438–45. doi: 10.1038/nn844.
33. Hicks A, Jolkkonen J. Challenges and possibilities of intravascular cell therapy in stroke. Acta Neurobiol Exp (Wars). 2009;69:1–11.
34. Lee ST, Chu K, Jung KH, et al. Anti-inflammatory mechanism of intravascular neural stem cell transplantation in haemorrhagic stroke. Brain. 2008;131:616–29. doi: 10.1093/brain/awm306.
35. Guzman R, Choi R, Gera A. Intravascular cell replacement therapy for stroke. Neurosurg Focus. 2008;24:1–10. doi: 10.3171/FOC/2008/24/3
36. Bible E, Chau DYS, Alexander MR, et al. The support of neural stem cells transplanted into stroke-induced brain cavities by PLGA particles. Biomaterials. 2009;30:2985–94. doi: 10.1016/j.biomaterials.2009.02.012.
37. Park KI, Teng YD, Snyder EY. The injured brain interacts reciprocally with neural stem cells supported by scaffolds to reconstitute lost tissue. Nat Biotechnol. 2002;20:1111–7. doi: 10.1038/nbt751.
38. Misra V, Lal A, El Khoury R, et al. Intra-arterial delivery of cell therapies for stroke. Stem Cells Dev. 2012;21:1007–15. doi: 10.1089/scd.2011.0612.
39. Bang OY, Lee JS, Lee PH, Lee G. Autologous mesenchymal stem cell transplantation in stroke patients. Ann Neurol. 2005;57:874–82. doi: 10.1002/ana.20501.
40. Barbosa da Fonseca LM, Gutfilen B Rosado de Castro PH, et al. Migration and homing of bone-marrow mononuclear cells in chronic ischemic stroke after intra-arterial injection. Exp Neurol. 2010;221:122–8. doi: 10.1016/j.expneurol.2009.10.010.
41. Battistella V, de Freitas GR, da Fonseca LMB, et al. Safety of autologous bone marrow mononuclear cell transplantation in patients with nonacute ischemic stroke. Regen Med. 2011;6:45–52. doi: 10.2217/rme.10.97.

42. Bhasin A, Padma Srivastava MV, Mohanty S, et al. Stem cell therapy: A clinical trial of stroke. Clin Neurol Neurosurg. 2013;115:1003–8. doi: 10.1016/j.clineuro.2012.10.015.
43. Moniche F, Gonzalez A, Gonzalez-Marcos J-R, et al. Intra-arterial bone marrow mononuclear cells in ischemic stroke: a pilot clinical trial. Stroke. 2012;43:2242–4. doi: 10.1161/STROKEAHA.112.659409.
44. Prasad K, Mohanty S, Bhatia R, et al. Autologous intravenous bone marrow mononuclear cell therapy for patients with subacute ischaemic stroke: a pilot study. Indian J Med Res. 2012;136:221–8.
45. Savitz SI, Misra V, Kasam M, et al. Intravenous autologous bone marrow mononuclear cells for ischemic stroke. Ann Neurol. 2011;70:59–69. doi: 10.1002/ana.22458.
46. Lappalainen RS, Narkilahti S, Huhtala T, et al. The SPECT imaging shows the accumulation of neural progenitor cells into internal organs after systemic administration in middle cerebral artery occlusion rats. Neurosci Lett. 2008;440:246–50. doi: 10.1016/j.neulet.2008.05.090.
47. Zawadzka M, Lukasiuk K, Machaj EK, et al. Lack of migration and neurological benefits after infusion of umbilical cord blood cells in ischemic brain injury. Acta Neurobiol Exp (Wars). 2009;69:46–51. doi: 6905 [pii].
48. Kawabori M, Kuroda S, Sugiyama T, et al. Intracerebral, but not intravenous, transplantation of bone marrow stromal cells enhances functional recovery in rat cerebral infarct: an optical imaging study. Neuropathology. 2012;32:217–26. doi: 10.1111/j.1440-1789.2011.01260.x.
49. Matsuse D, Kitada M, Ogura F, et al. Combined transplantation of bone marrow stromal cell-derived neural progenitor cells with a collagen sponge and basic fibroblast growth factor releasing microspheres enhances recovery after cerebral ischemia in rats. Tissue Eng Part A. 2011;17:1993–2004. doi: 10.1089/ten.tea.2010.0585.
50. Yu H, Cao B, Feng M, et al. Combinated transplantation of neural stem cells and collagen type I promote functional recovery after cerebral ischemia in rats. Anat Rec (Hoboken). 2010;293:911–7. doi: 10.1002/ar.20941.
51. Tate MC, Shear DA, Hoffman SW, et al. (2002) Fibronectin promotes survival and migration of primary neural stem cells transplanted into the traumatically injured mouse brain. Cell Transplant 11:283–95. doi: http://dx.doi.org/10.0000/096020198389933
52. Hyatt AJT, Wang D, van Oterendorp C, et al. (2014) Mesenchymal stromal cells integrate and form longitudinally-aligned layers when delivered to injured spinal cord via a novel fibrin scaffold. Neurosci Lett. doi: 10.1016/j.neulet.2014.03.023
53. Shi W, Nie D, Jin G, et al. BDNF blended chitosan scaffolds for human umbilical cord MSC transplants in traumatic brain injury therapy. Biomaterials. 2012;33:3119–26. doi: 10.1016/j.biomaterials.2012.01.009.
54. Miyoshi Y, Date I, Ohmoto T, Iwata H. Histological analysis of microencapsulated dopamine-secreting cells in agarose/poly (styrene sulfonic acid) mixed gel xenotransplanted into the brain. Exp Neurol. 1996;138:169–75. doi: 10.1006/exnr.1996.0055.
55. Wood MD, MacEwan MR, French AR, et al. Fibrin matrices with affinity-based delivery systems and neurotrophic factors promote functional nerve regeneration. Biotechnol Bioeng. 2010;106:970–9. doi: 10.1002/bit.22766.
56. Wang Y, Cooke MJ, Sachewsky N, et al. Bioengineered sequential growth factor delivery stimulates brain tissue regeneration after stroke. J Control Release. 2013;172:1–11. doi: 10.1016/j.jconrel.2013.07.032.
57. Seidlits SK, Khaing ZZ, Petersen RR, et al. The effects of hyaluronic acid hydrogels with tunable mechanical properties on neural progenitor cell differentiation. Biomaterials. 2010;31:3930–40. doi: 10.1016/j.biomaterials.2010.01.125.
58. Bible E, Dell'Acqua F, Solanky B, et al. Non-invasive imaging of transplanted human neural stem cells and ECM scaffold remodeling in the stroke-damaged rat brain by (19)F- and diffusion-MRI. Biomaterials. 2012;33:2858–71. doi: 10.1016/j.biomaterials.2011.12.033.
59. Pakulska MM, Ballios BG, Shoichet MS. Injectable hydrogels for central nervous system therapy. Biomed Mater. 2012;7:024101. doi: 10.1088/1748-6041/7/2/024101.
60. Johl SS, Burgett RA. Dermal filler agents: a practical review. Curr Opin Ophthalmol. 2006;17:471–9. doi: 10.1097/01.icu.0000243021.20499.4b.

61. Aurand ER, Lampe KJ, Bjugstad KB. Defining and designing polymers and hydrogels for neural tissue engineering. Neurosci Res. 2012;72:199–213. doi: 10.1016/j.neures.2011.12.005.
62. Kim H, Cooke MJ, Shoichet MS. Creating permissive microenvironments for stem cell transplantation into the central nervous system. Trends Biotechnol. 2012;30:55–63. doi: 10.1016/j.tibtech.2011.07.002.
63. Deng LX, Hu J, Liu N, et al. GDNF modifies reactive astrogliosis allowing robust axonal regeneration through Schwann cell-seeded guidance channels after spinal cord injury. Exp Neurol. 2011;229:238–50. doi: 10.1016/j.expneurol.2011.02.001.
64. Fouad K, Schnell L, Bunge MB, et al. Combining Schwann cell bridges and olfactory-ensheathing glia grafts with chondroitinase promotes locomotor recovery after complete transection of the spinal cord. J Neurosci. 2005;25:1169–78. doi: 10.1523/JNEUROSCI.3562-04.2005.
65. Rooney GE, Knight AM, Madigan NN, et al. Sustained delivery of dibutyryl cyclic adenosine monophosphate to the transected spinal cord via oligo [(polyethylene glycol) fumarate] hydrogels. Tissue Eng Part A. 2011;17:1287–302. doi: 10.1089/ten.tea.2010.0396.
66. Shoichet MS. Polymer scaffolds for biomaterials applications. Macromolecules. 2010;43:581–91. doi: 10.1021/ma901530r.
67. Lü J-M, Wang X, Marin-Muller C, et al. Current advances in research and clinical applications of PLGA-based nanotechnology. Expert Rev Mol Diagn. 2009;9:325–41. doi: 10.1586/erm.09.15.
68. Madhavan Nampoothiri K, Nair NR, John RP. An overview of the recent developments in polylactide (PLA) research. Bioresour Technol. 2010;101:8493–501. doi: 10.1016/j.biortech.2010.05.092.
69. Borlongan CV, Saporta S, Sanberg PR. Intrastriatal transplantation of rat adrenal chromaffin cells seeded on microcarrier beads promote long-term functional recovery in hemiparkinsonian rats. Exp Neurol. 1998;151:203–14. doi: 10.1006/exnr.1998.6790.
70. Cherksey BD, Sapirstein VS, Geraci AL. Adrenal chromaffin cells on microcarriers exhibit enhanced long-term functional effects when implanted into the mammalian brain. Neuroscience. 1996;75:657–64. doi: 10.1016/0306-4522(96)00262-X.
71. Stover NP, Watts RL. Spheramine for treatment of Parkinson's disease. Neurother. 2008;5:252–9. doi: 10.1016/j.nurt.2008.02.006.
72. Borlongan CV, Thanos CG, Skinner SJM, et al. Transplants of encapsulated rat choroid plexus cells exert neuroprotection in a rodent model of Huntington’s disease. Cell Transplant. 2008;16:987–92. doi: 10.3727/000000007783472426.
73. Tatard VM, Sindji L, Branton J(G), et al. Pharmacologically active microcarriers releasing glial cell line - derived neurotrophic factor: survival and differentiation of embryonic dopaminergic neurons after grafting in hemiparkinsonian rats. Biomaterials. 2007b;28:1978–88. doi: 10.1016/j.biomaterials.2006.12.021.
74. Bonneh-Barkay D, Wiley CA. Brain extracellular matrix in neurodegeneration. Brain Pathol. 2009;19:573–85. doi: 10.1111/j.1750-3639.2008.00195.x.
75. Quirico-Santos T, Fonseca CO, Lagrota-Candido J. Brain sweet brain: importance of sugars for the cerebral microenvironment and tumor development. Arq Neuropsiquiatr. 2010;68:799–803. doi: 10.1590/S0004-282X2010000500024.
76. Ruoslahti E. Brain extracellular matrix. Glycobiology. 1996;6:489–92. doi: 10.1093/glycob/6.5.489.
77. Moshayedi P, da F CL, Christ A, et al. Mechanosensitivity of astrocytes on optimized polyacrylamide gels analyzed by quantitative morphometry. J Phys Condens Matter. 2010;22:194114. doi: 10.1088/0953-8984/22/19/194114.
78. Nguyen MK, Lee DS. Injectable biodegradable hydrogels. Macromol Biosci. 2010;10:563–79. doi: 10.1002/mabi.200900402.
79. Ruel-Gariépy E, Leroux J-C. In situ-forming hydrogels-review of temperature-sensitive systems. Eur J Pharm Biopharm. 2004;58:409–26. doi: 10.1016/j.ejpb.2004.03.019.
80. Chiu YL, Chen SC, Su CJ, et al. pH-triggered injectable hydrogels prepared from aqueous N-palmitoyl chitosan: In vitro characteristics and in vivo biocompatibility. Biomaterials. 2009;30:4877–88. doi: 10.1016/j.biomaterials.2009.05.052.

81. Christ AF, Franze K, Gautier H, et al. Mechanical difference between white and gray matter in the rat cerebellum measured by scanning force microscopy. J Biomech. 2010;43:2986–92. doi: 10.1016/j.jbiomech.2010.07.002
82. Lampe KJ, Kern DS, Mahoney MJ, Bjugstad KB. The administration of BDNF and GDNF to the brain via PLGA microparticles patterned within a degradable PEG-based hydrogel: Protein distribution and the glial response. J Biomed Mater Res - Part A. 2011;96 A:595–607. doi: 10.1002/jbm.a.33011
83. Namba RM, Cole a a, Bjugstad KB, Mahoney MJ. Development of porous PEG hydrogels that enable efficient, uniform cell-seeding and permit early neural process extension. Acta Biomater. 2009;5:1884–97. doi: 10.1016/j.actbio.2009.01.036.
84. Lampe KJ, Mooney RG, Bjugstad KB, Mahoney MJ. Effect of macromer weight percent on neural cell growth in 2D and 3D nondegradable PEG hydrogel culture. J Biomed Mater Res—Part A. 2010b;94:1162–71. doi: 10.1002/jbm.a.32787.
85. Nicodemus GD, Bryant SJ. Cell encapsulation in biodegradable hydrogels for tissue engineering applications. Tissue Eng Part B Rev. 2008;14:149–65. doi: 10.1089/ten.teb.2007.0332.
86. Patterson J, Hubbell JA. Enhanced proteolytic degradation of molecularly engineered PEG hydrogels in response to MMP-1 and MMP-2. Biomaterials. 2010;31:7836–45. doi: 10.1016/j.biomaterials.2010.06.061.
87. Menei P, Montero-Menei C, Venier M-C, Benoit J-P. Drug delivery into the brain using poly (lactide-co-glycolide) microspheres. Expert Opin Drug Deliv. 2005;2:363–76. doi: 10.1517/17425247.2.2.363.
88. Moshayedi P, Carmichael ST. Hyaluronan, neural stem cells and tissue reconstruction after acute ischemic stroke. Biomatter. 2013;3:1–9.
89. Frisch SM, Francis H. Disruption of epithelial cell-matrix interactions induces apoptosis. J Cell Biol. 1994;124:619–26. doi: 10.1083/jcb.124.4.619.
90. Marchionini DM, Collier TJ, Camargo M, et al. Interference with anoikis-induced cell death of dopamine neurons: implications for augmenting embryonic graft survival in a rat model of Parkinson's disease. J Comp Neurol. 2003;464:172–9. doi: 10.1002/cne.10785.
91. Singh P, Schwarzbauer JE. Fibronectin and stem cell differentiation—lessons from chondrogenesis. J Cell Sci. 2012;125:3703–12. doi: 10.1242/jcs.095786.
92. Lu B, Smyth MR, O'Kennedy R. Oriented immobilization of antibodies and its applications in immunoassays and immunosensors. Analyst. 1996;121:29R–32R. doi: 10.1039/an996210029r.
93. Kelly S, Bliss TM, Shah AK, et al. Transplanted human fetal neural stem cells survive, migrate, and differentiate in ischemic rat cerebral cortex. Proc Natl Acad Sci U S A. 2004;101:11839–44. doi: 10.1073/pnas.0404474101.
94. Hoban DB, Newland B, Moloney TC, et al. The reduction in immunogenicity of neurotrophin overexpressing stem cells after intra-striatal transplantation by encapsulation inaninsitu gelling collagen hydrogel. Biomaterials. 2013;34:9420–9. doi: 10.1016/j.biomaterials.2013.08.073.
95. Caicco MJ, Cooke MJ, Wang Y, et al. A hydrogel composite system for sustained epicortical delivery of Cyclosporin A to the brain for treatment of stroke. J Control Release. 2013;166:197–202. doi: 10.1016/j.jconrel.2013.01.002.
96. Kim DH, Martin DC. Sustained release of dexamethasone from hydrophilic matrices using PLGA nanoparticles for neural drug delivery. Biomaterials. 2006;27:3031–7. doi: 10.1016/j.biomaterials.2005.12.021.
97. Cheung CY, McCartney SJ, Anseth KS. Synthesis of polymerizable superoxide dismutase mimetics to reduce reactive oxygen species damage in transplanted biomedical devices. Adv Funct Mater. 2008;18:3119–26. doi: 10.1002/adfm.200800566.
98. Lin CC, Metters AT, Anseth KS. Functional PEG-peptide hydrogels to modulate local inflammation inducedby the pro-inflammatory cytokine TNF[Alpha]. Biomaterials. 2009;30:4907–14. doi: 10.1016/j.biomaterials.2009.05.083.
99. Campos LS, Leone DP, Relvas JB, et al. Beta1 integrins activate a MAPK signalling pathway in neural stem cells that contributes to their maintenance. Development. 2004;131:3433–44. doi: 10.1242/dev.01199.

100. DeSilva DR, Jones EA, Favata MF, et al. Inhibition of mitogen-activated protein kinase kinase blocks T cell proliferation but does not induce or prevent anergy. J Immunol. 1998;160:4175–81.
101. Pike DB, Cai S, Pomraning KR, et al. Heparin-regulated release of growth factors in vitro and angiogenic response in vivo to implanted hyaluronan hydrogels containing VEGF and bFGF. Biomaterials. 2006;27:5242–51. doi: 10.1016/j.biomaterials.2006.05.018.
102. Mahoney MJ, Anseth KS. Contrasting effects of collagen and bFGF-2 on neural cell function in degradable synthetic PEG hydrogels. J Biomed Mater Res—Part A. 2007;81:269–78. doi: 10.1002/jbm.a.30970.
103. King WJ, Jongpaiboonkit L, Murphy WL. Influence of FGF2 and PEG hydrogel matrix properties on hMSC viability and spreading. J Biomed Mater Res—Part A. 2010b;93:1110–23. doi: 10.1002/jbm.a.32601.
104. Johnson PJ, Tatara A, Shiu A, Sakiyama-Elbert SE. Controlled release of neurotrophin-3 and platelet-derived growth factor from fibrin scaffolds containing neural progenitor cells enhances survival and differentiation into neurons in a subacute model of SCI. Cell Transplant. 2010;19:89–101. doi: 10.3727/096368909X477273.
105. Engler AJ, Sen S, Sweeney HL, Discher DE. Matrix elasticity directs stem cell lineage specification. Cell. 2006;126:677–89. doi: 10.1016/j.cell.2006.06.044.
106. Flanagan LA, Ju Y-E, Marg B, et al. Neurite branching on deformable substrates. Neuroreport. 2002;13:2411–5. doi: 10.1097/00001756-200212200-00007.
107. Hynes SR, Rauch MF, Bertram JP, Lavik EB. A library of tunable poly (ethylene glycol)/poly (L-lysine) hydrogels to investigate the material cues that influence neural stem cell differentiation. J Biomed Mater Res—Part A. 2009;89:499–509. doi: 10.1002/jbm.a.31987.
108. Lampe KJ, Bjugstad KB, Mahoney MJ. Impact of degradable macromer content in a poly (ethylene glycol) hydrogel on neural cell metabolic activity, redox state, proliferation, and differentiation. Tissue Eng Part A. 2010a;16:1857–66. doi: 10.1089/ten.tea.2009.0509.
109. Saha K, Keung AJ, Irwin EF, et al. Substrate modulus directs neural stem cell behavior. Biophys J. 2008;95:4426–38. doi: 10.1529/biophysj.108.132217.
110. Leipzig ND, Shoichet MS. The effect of substrate stiffness on adult neural stem cell behavior. Biomaterials. 2009;30:6867–78. doi: 10.1016/j.biomaterials.2009.09.002.
111. Abdeen A a, Weiss JB, Lee J, Kilian K. Matrix composition and mechanics directs proangiogenic signaling from mesenchymal stem cells. Tissue Eng Part A. 2014;2142:1–39. doi: 10.1089/ten.TEA.2013.0661.
112. Cooke MJ, Zahir T, Phillips SR, et al. Neural differentiation regulated by biomimetic surfaces presenting motifs of extracellular matrix proteins. J Biomed Mater Res—Part A. 2010;93:824–32. doi: 10.1002/jbm.a.32585.
113. Liu SQ, Tian Q, Hedrick JL, et al. Biomimetic hydrogels for chondrogenic differentiation of human mesenchymal stem cells to neocartilage. Biomaterials. 2010;31:7298–307. doi: 10.1016/j.biomaterials.2010.06.001.
114. Tate MC, García AJ, Keselowsky BG, et al. Specific beta1 integrins mediate adhesion, migration, and differentiation of neural progenitors derived from the embryonic striatum. Mol Cell Neurosci. 2004;27:22–31. doi: 10.1016/j.mcn.2004.05.001.
115. Park J, Lim E, Back S, et al. Nerve regeneration following spinal cord injury using matrix metalloproteinase-sensitive, hyaluronic acid-based biomimetic hydrogel scaffold containing brain-derived neurotrophic factor. J Biomed Mater Res—Part A. 2010;93:1091–9. doi: 10.1002/jbm.a.32519.
116. Wang W, Itaka K, Ohba S, et al. 3D spheroid culture system on micropatterned substrates for improved differentiation efficiency of multipotent mesenchymal stem cells. Biomaterials. 2009;30:2705–15. doi: 10.1016/j.biomaterials.2009.01.030.
117. Dadsetan M, Pumberger M, Casper ME, et al. The effects of fixed electrical charge on chondrocyte behavior. Acta Biomater. 2011;7:2080–90. doi: 10.1016/j.actbio.2011.01.012.
118. McBeath R, Pirone DM, Nelson CM, et al. Cell shape, cytoskeletal tension, and RhoA regulate stem cell lineage commitment. Dev Cell. 2004;6:483–95. doi: 10.1016/S1534-5807(04)00075-9.

119. Leipzig ND, Xu C, Zahir T, Shoichet MS. Functional immobilization of interferon-gamma induces neuronal differentiation of neural stem cells. J Biomed Mater Res—Part A. 2010;93:625–33. doi: 10.1002/jbm.a.32573.
120. Aizawa Y, Leipzig N, Zahir T, Shoichet M. The effect of immobilized platelet derived growth factor AA on neural stem/progenitor cell differentiation on cell-adhesive hydrogels. Biomaterials. 2008;29:4676–83. doi: 10.1016/j.biomaterials.2008.08.018.
121. Tatard VM, D'Ippolito G, Diabira S, et al. Neurotrophin-directed differentiation of human adult marrow stromal cells to dopaminergic-like neurons. Bone. 2007a;40:360–73. doi: 10.1016/j.bone.2006.09.013.
122. Kim H, Zahir T, Tator CH, Shoichet MS. Effects of dibutyryl cyclic-AMP on survival and neuronal differentiation of neural stem/progenitor cells transplanted into spinal cord injured rats. PLoS One. 2011;6(6):e21744. doi:10.1371/journal.pone.0021744
123. Moshayedi P, Ng G, Kwok JCF, et al. The relationship between glial cell mechanosensitivity and foreign body reactions in the central nervous system. Biomaterials. 2014;35:1–7. doi: 10.1016/j.biomaterials.2014.01.038
124. Polikov VS, Tresco PA, Reichert WM. Response of brain tissue to chronically implanted neural electrodes. J Neurosci Methods. 2005;148:1–18. doi: 10.1016/j.jneumeth.2005.08.015.
125. Fawcett JW, Asher RA. The glial scar and central nervous system repair. Brain Res Bull. 1999;49:377–91. doi: 10.1016/S0361-9230(99)00072-6.
126. Bjugstad KB, Lampe K, Kern DS, Mahoney M. Biocompatibility of poly (ethylene glycol)-based hydrogels in the brain: an analysis of the glial response across space and time. J Biomed Mater Res—Part A. 2010;95:79–91. doi: 10.1002/jbm.a.32809.
127. Fournier E, Passirani C, Montero-Menei CN, Benoit JP. Biocompatibility of implantable synthetic polymeric drug carriers: focus on brain biocompatibility. Biomaterials. 2003;24:3311–31. doi: 10.1016/S0142-9612(03)00161-3.
128. Anderson JM, Shive MS. Biodegradation and biocompatibility of PLA and PLGA microspheres. Adv Drug Deliv Rev. 2012;64:72–82. doi: 10.1016/j.addr.2012.09.004.
129. Hwang DH, Kim HM, Kang YM, et al. Combination of Multifaceted Strategies to Maximize the therapeutic benefits of neural stem cell transplantation for spinal cord repair. Cell Transplant. 2011;20:1361–79. doi: 10.3727/096368910X557155.
130. Pardue EL, Ibrahim S, Ramamurthi A. Role of hyaluronan in angiogenesis and its utility to angiogenic tissue engineering. Organogenesis. 2008;4:203–14. doi: 10.4161/org.4.4.6926.
131. Patel V, Joseph G, Patel A, et al. Suspension matrices for improved Schwann-cell survival after implantation into the injured rat spinal cord. J Neurotrauma. 2010;27:789–801. doi: 10.1089/neu.2008.0809.
132. Hayashi T, Deguchi K, Nagotani S, et al. Cerebral ischemia and angiogenesis. Curr Neurovasc Res. 2006;3:119–29. doi: 10.2174/156720206776875902.
133. Navaratna D, Guo S, Arai K, Lo EH. Mechanisms and targets for angiogenic therapy after stroke. Cell Adh Migr. 2009;3:216–23. doi: 10.4161/cam.3.2.8396.
134. Seidlits SK, Drinnan CT, Petersen RR, et al. Fibronectin-hyaluronic acid composite hydrogels for three-dimensional endothelial cell culture. Acta Biomater. 2011;7:2401–9. doi: 10.1016/j.actbio.2011.03.024.
135. Cui FZ, Tian WM, Hou SP, et al. Hyaluronic acid hydrogel immobilized with RGD peptides for brain tissue engineering. J Mater Sci Mater Med. 2006;17:1393–1401.
136. Preston M, Sherman LS (2011) Neural stem cell niches: roles for the hyaluronan-based extracellular matrix. Front Biosci (Schol Ed) 3:1165–79. doi: 10.2741/218
137. Moore K, MacSween M, Shoichet M. Immobilized concentration gradients of neurotrophic factors guide neurite outgrowth of primary neurons in macroporous scaffolds. Tissue Eng. 2006;12:267–78. doi: 10.1089/ten.2006.12.ft-42.
138. Hou S, Xu Q, Tian W, et al. The repair of brain lesion by implantation of hyaluronic acid hydrogels modified with laminin. J Neurosci Methods. 2005;148:60–70. doi: 10.1016/j.jneumeth.2005.04.016.
139. Harms KM, Li L, Cunningham LA. Murine neural stem/progenitor cells protect neurons against ischemia by HIF-1alpha-regulated VEGF signaling. PLoS One. 2010;5:e9767. doi: 10.1371/journal.pone.0009767.

140. McCreedy D a, Sakiyama-Elbert SE. Combination therapies in the CNS: engineering the environment. Neurosci Lett. 2012;519:115–21. doi: 10.1016/j.neulet.2012.02.025.
141. David S, Aguayo AJ. Axonal elongation into peripheral nervous system "bridges" after central nervous system injury in adult rats. Science. 1981;214:931–3. doi: 10.1126/science.6171034.
142. Richardson PM, McGuinness UM, Aguayo AJ. Axons from CNS neurons regenerate into PNS grafts. Nature. 1980;284:264–5. doi: 10.1038/284264a0.
143. Uibo R, Laidmäe I, Sawyer ES, et al. Soft materials to treat central nervous system injuries: evaluation of the suitability of non-mammalian fibrin gels. Biochim Biophys Acta. 2009;1793:924–30. doi: 10.1016/j.bbamcr.2009.01.007.
144. Bignami A, Asher R. Some observations on the localization of hyaluronic acid in adult, newborn and embryonal rat brain. Int J Dev Neurosci. 1992;10:45–57. doi: 10.1016/0736-5748(92)90006-L.
145. Bignami A, Asher R, Perides G. The extracellular matrix of rat spinal cord: a comparative study on the localization of hyaluronic acid, glial hyaluronate-binding protein, and chondroitin sulfate proteoglycan. Exp Neurol. 1992;117:90–3. doi: 10.1016/0014-4886(92)90115-7.
146. King VR, Alovskaya A, Wei DYT, et al. The use of injectable forms of fibrin and fibronectin to support axonal ingrowth after spinal cord injury. Biomaterials. 2010a;31:4447–56. doi: 10.1016/j.biomaterials.2010.02.018.
147. Wei YT, Tian WM, Yu X, et al. Hyaluronic acid hydrogels with IKVAV peptides for tissue repair and axonal regeneration in an injured rat brain. Biomed Mater. 2007;2:S142–S6. doi: 10.1088/1748-6041/2/3/S11.
148. Khaing ZZ, Milman BD, Vanscoy JE, et al. High molecular weight hyaluronic acid limits astrocyte activation and scar formation after spinal cord injury. J Neural Eng. 2011;8:046033. doi: 10.1088/1741-2560/8/4/046033.
149. King VR, Henseler M, Brown RA, Priestley JV. Mats made from fibronectin support oriented growth of axons in the damaged spinal cord of the adult rat. Exp Neurol. 2003;182:383–98. doi: 10.1016/S0014-4886(03)00033-5.
150. Lu P, Tuszynski MH. Growth factors and combinatorial therapies for CNS regeneration. Exp Neurol. 2008;209:313–20. doi: 10.1016/j.expneurol.2007.08.004.
151. Blesch A, Tuszynski MH. Cellular GDNF delivery promotes growth of motor and dorsal column sensory axons after partial and complete spinal cord transections and induces remyelination. J Comp Neurol. 2003;467:403–17. doi: 10.1002/cne.10934.
152. Ramer MS, Bradbury EJ, Michael GJ, et al. Glial cell line-derived neurotrophic factor increases calcitonin gene-related peptide immunoreactivity in sensory and motoneurons in vivo. Eur J Neurosci. 2003;18:2713–21. doi: 3012 [pii].
153. Tuszynski MH, Peterson DA, Ray J, et al. Fibroblasts genetically modified to produce nerve growth factor induce robust neuritic ingrowth after grafting to the spinal cord. Exp Neurol. 1994;126:1–14. doi: S0014488684710375 [pii].
154. Kennedy TE, Wang H, Marshall W, Tessier-Lavigne M. Axon guidance by diffusible chemoattractants: a gradient of netrin protein in the developing spinal cord. J Neurosci. 2006;26:8866–74. doi: 10.1523/JNEUROSCI.5191-05.2006.
155. Cao X, Shoichet MS. Defining the concentration gradient of nerve growth factor for guided neurite outgrowth. Neuroscience. 2001;103:831–40. doi: 10.1016/S0306-4522(01)00029-X.
156. Cao X, Shoichet MS. Investigating the synergistic effect of combined neurotrophic factor concentration gradients to guide axonal growth. Neuroscience. 2003;122:381–9. doi: 10.1016/j.neuroscience.2003.08.018.
157. Albrecht DR, Tsang VL, Sah RL, Bhatia SN. Photo- and electropatterning of hydrogel-encapsulated living cell arrays. Lab Chip. 2005;5:111–8. doi: 10.1039/b406953f.
158. Luo Y, Shoichet MS. Light-activated immobilization of biomolecules to agarose hydrogels for controlled cellular response. Biomacromolecules. 2004;5:2315–23. doi: 10.1021/bm0495811.
159. Wylie RG, Shoichet MS. Two-photon micropatterning of amines within an agarose hydrogel. J Mater Chem. 2008;18:2716. doi: 10.1039/b718431j.

160. Dertinger SKW, Jiang X, Li Z, et al. Gradients of substrate-bound laminin orient axonal specification of neurons. Proc Natl Acad Sci U S A. 2002;99:12542–7. doi: 10.1073/pnas.192457199.
161. Wei YT, He Y, Xu CL, et al. Hyaluronic acid hydrogel modified with nogo-66 receptor antibody and poly-L-lysine to promote axon regrowth after spinal cord injury. J Biomed Mater Res—Part B Appl Biomater. 2010;95:110–7. doi: 10.1002/jbm.b.31689.
162. Lee H, McKeon RJ, Bellamkonda RV. Sustained delivery of thermostabilized chABC enhances axonal sprouting and functional recovery after spinal cord injury. Proc Natl Acad Sci U S A. 2010a;107:3340–5. doi: 10.1073/pnas.0905437106.
163. Jain A, Kim YT, McKeon RJ, Bellamkonda RV. In situ gelling hydrogels for conformal repair of spinal cord defects, and local delivery of BDNF after spinal cord injury. Biomaterials. 2006;27:497–504. doi: 10.1016/j.biomaterials.2005.07.008.
164. Hurlbert RJ, Hamilton MG. Methylprednisolone for acute spinal cord injury: 5-year practice reversal. Can J Neurol Sci. 2008;35:41–5.
165. Kim Y, Caldwell JM, Bellamkonda RV. Nanoparticle-mediated local delivery of methylprednisolone after spinal cord injury. Biomaterials. 2009;30:2582–90. doi: 10.1016/j.biomaterials.2008.12.077.
166. Jin K, Mao X, Xie L, et al. Transplantation of human neural precursor cells in matrigel scaffolding improves outcome from focal cerebral ischemia after delayed postischemic treatment in rats. J Cereb Blood Flow Metab. 2010;30:534–44. doi: 10.1038/jcbfm.2009.219.
167. Osanai T, Kuroda S, Yasuda H, et al. Noninvasive transplantation of bone marrow stromal cells for ischemic stroke: preliminary study with a thermoreversible gelation polymer hydrogel. Neurosurgery. 2010;66:1140–7. doi: 10.1227/01.NEU.0000369610.76181.CF. (discussion 1147).
168. Lee JS, Hong JM, Moon GJ, et al. A long-term follow-up study of intravenous autologous mesenchymal stem cell transplantation in patients with ischemic stroke. Stem Cells. 2010b;28:1099–106. doi: 10.1002/stem.430.
169. Purcell EK, Seymour JP, Yandamuri S, Kipke DR. In vivo evaluation of a neural stem cell-seeded prosthesis. J Neural Eng. 2009;6:026005. doi: 10.1088/1741-2560/6/2/026005.
170. Zisch AH, Lutolf MP, Ehrbar M, et al. Cell-demanded release of VEGF from synthetic, bio-interactive cell ingrowth matrices for vascularized tissue growth. FASEB J. 2003;17:2260–2. doi: 10.1096/fj.02-1041fje.

Chapter 14
A Stem-Cell-Derived Cell-Free Therapy for Stroke: Moving Conditioned Medium into Clinical Trial

Brian H. Johnstone and Keith L. March

Introduction

Adult stem and progenitor cell-based therapies, including mesenchymal stem/stromal cells (MSC), have been shown to improve functional outcomes in many disease models and in early clinical trials of acute and chronic ischemic stroke [1–11]. It is becoming increasingly evident, however, that the mechanism by which multipotent MSC restore function of diseased and damaged tissues is minimally related to direct regeneration of tissues as originally postulated. Rather, over the last decade, accumulating evidence suggests that the primary mode of action of MSC is paracrine-mediated stimulation of the body's intrinsic repair processes and immune system suppression through secretion of trophic factors [12–16]. A mechanism of paracrine support to induce repair is consistent with observations that functional improvements are frequently reported in the absence of demonstrable integration of delivered cells at numbers sufficient to account for lost tissue replacement [17–21]. This is in contrast with a mass balance model of disease reversal, in which the magnitude of therapeutic effect directly correlates with the number of administered cells that reach the targeted tissues, stably engraft, possibly undergo division, and directly replace a sufficient quantity of lost or damaged tissue. The paracrine mechanism has the inherent advantage of signal amplification and, thus, requires lower cell doses to achieve therapeutic effects. Amplification occurs because each administered cell produces an abundance of trophic factors that elicit responses from the many cell types required for correcting the disease. Thus, even brief exposure of the

B. H. Johnstone (✉)
NeuroFx, Inc., 11650 Lantern Road, Fishers, IN 46038, USA
e-mail: bjohnstone@neurofx.co

K. L. March
Indiana Center for Vascular Biology and Medicine, Indianapolis, IN, USA

Vascular and Cardiac Center for Adult Stem Cell Therapy, Indiana University, Indianapolis, IN, USA

D. C. Hess (ed.), *Cell Therapy for Brain Injury*, DOI 10.1007/978-3-319-15063-5_14

target tissues to stem cells in the absence of stable engraftment produces therapeutic responses through paracrine induction of multiple beneficial effects, including (1) providing support to at-risk tissues in the border zone, thereby limiting expansion of the "dead zone," (2) reducing injury by suppressing or modulating inflammation, and (3) inducing repair through promoting recruitment and survival of endogenous stem and progenitor cells (Fig. 14.1).

The cumulative evidence supporting a paracrine mechanism of action of MSC does not preclude the involvement of direct cellular contact, which appears to be required in some circumstances, such as immunomodulation (reviewed in [22]). The local concentration of factors also may be enhanced by homing and retention of delivered cells in response to gradients of chemoattractants produced by injured tissues. Furthermore, it has been suggested that, following entrapment, therapeutic cells may respond to local stimuli by modulating the levels of secreted factors in beneficial ways [12, 20, 23, 24]. While certain signaling molecules (for example, tumor necrosis factor-alpha) have been shown to influence stem cell physiology in vitro, there is limited direct evidence for this phenomenon in situ, especially given the transient nature of this effect due to clearance of the bulk of administered cells [25].

It is possible that the potential of pluripotent and multipotent stem cells for replacing tissues will ultimately be realized; however, until that time, present cellular therapies with autologous or allogeneic MSC are providing medical benefits to previously intractable diseases. Understanding the actual potential and limitations of present cellular therapies will be key to gaining widespread adoption.

The Therapeutic Potential for Stroke of MSC-Derived Conditioned Medium

There is increasing interest in developing neurorestorative treatments for acute cerebrovascular injuries, of which acute ischemic stroke (AIS) is the most severe manifestation. Since the cascade of injury in AIS is mostly complete within 24–48 h, neuroprotection, to be most effective, must be started early following the injury, which is challenging in clinical practice [26]. On the other hand, in restorative treatments, the target of the therapy is to promote repair processes such as angiogenesis, neurogenesis, and synaptogenesis. The window for effective promotion of such restorative processes is not precisely known, but may extend to 1 week or 1 month or even longer [27, 28].

AIS involves destruction of multiple cell types including neurons, astrocytes, oligodendrocytes, endothelial cells, and pericytes. Therefore, regenerative strategies will address both neural elements and supportive structures such as blood vessels and glia. Evidence has recently emerged of endogenous repair mechanisms which are activated following cerebral ischemia. Nestin is upregulated in astrocytes after cerebral ischemia [10, 29]. In rodents, subventricular progenitor cells proliferate after middle cerebral artery occlusion (MCAO) and migrate to the striatum

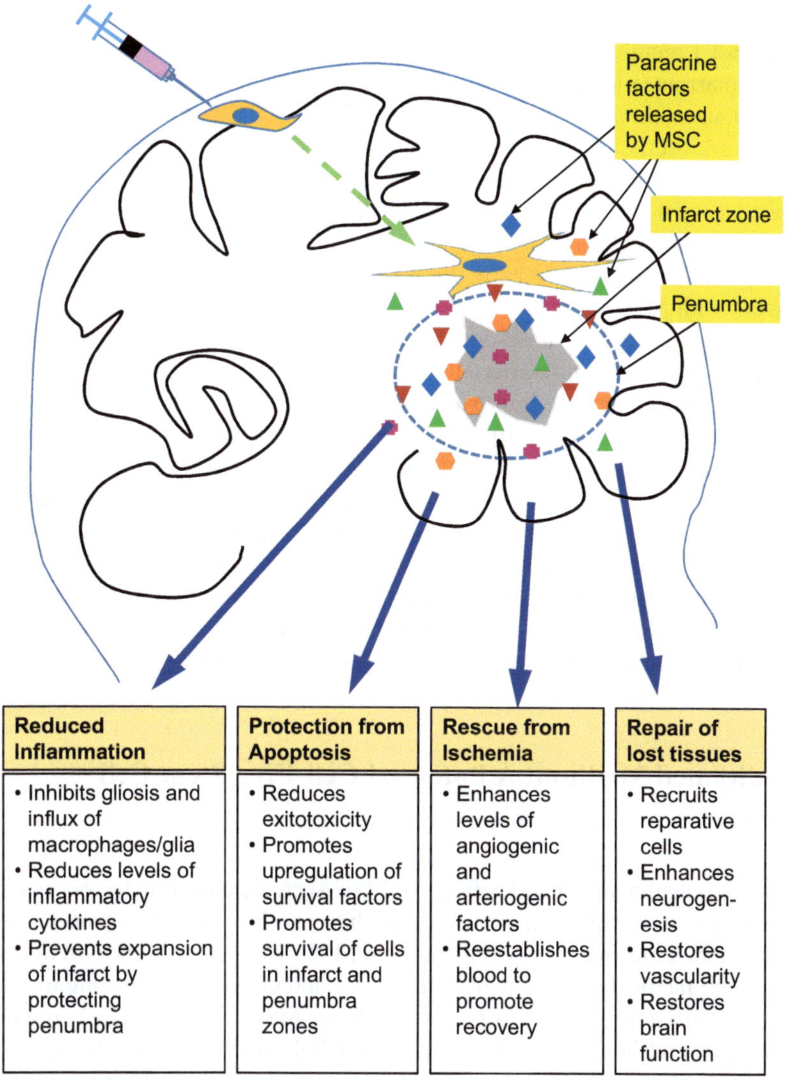

Fig. 14.1 The therapeutic potential of mesenchymal stem cells (MSC)-secreted factors for treating stroke. Following administration, MSC migrate to the site of stroke in response to chemoattractants, where they release paracrine factors that act to protect and promote repair of the infarcted region as well as surrounding penumbra. The aggregate action of individual paracrine factors that modulate discrete molecular pathways more completely addresses the complex pathophysiology of stroke compared to treatment with any one individual factor

where they contribute to the formation of striatal medium-sized spiny neurons and glial cells [10]. However, most of these neuroblasts undergo apoptosis and death. The subventricular progenitor cells continue to migrate to the striatum for at least 4 months after ischemia. This migration is directed by a gradient of stromal-derived factor (SDF)-1 upregulated in the ischemic tissue and CXCR4 expressed on the migrating neuroblasts [4]. SDF-1 is upregulated in astrocytes and endothelial cells for at least 1 month after cerebral ischemia, and also serves to direct the migration of bone marrow (BM)-derived cells involved in tissue repair [1]. Importantly, these endogenous repair mechanisms operating after cerebral ischemia are insufficient for recovery of lost tissues and deteriorated neurological functions. Stimulating this endogenous response with trophic factors is a practical approach directed to augmenting activity of these cellular targets, which are richest immediately (days) following stroke during processes of neural remodeling.

Several cell therapies in early clinical trials build on a large body of preclinical work with MSC isolated from a variety of mesodermal tissues. MSCs improve functional outcome in a dose-dependent fashion in rodent MCAO models when given intracerebrally, intra-arterially, or intravenously [2, 3, 30–35]. Intravenous transplantation effectively improves functional outcome even when given as late as 1 month after the insult [10]. In that instance, there was no reduction in infarct size; instead, there was a neurorestorative effect with increases in angiogenesis, neurogenesis and synaptogenesis in MSC-treated groups [1, 4]. The mechanism of action is not direct cell replacement, but rather the action of MSC as paracrine support factories, elaborating a transient burst of synergistic trophic and growth factors before being cleared [6]. Protective effects observed are anti-apoptosis of injured neural cells, reduced sensitivity to glutamate, suppressed levels of inflammatory molecules, and increased migration and survival of neuroprogenitors [8, 9, 33, 36].

Conditioned Medium: A Potential Cell-Free Stem Cell Therapy

Given the potential therapeutic effects of MSC-secreted paracrine factors as well as available experimental data, we and others have explored the therapeutic potential of conditioned medium (CM) from cultured MSC. This cell-free cocktail of trophic and growth factors has the potential to be safe and effective for treating many diseases currently under investigation as targets for MSC therapies. The combination of factors found in CM reproduces the beneficial effects of MSC. As opposed to delayed secretion in the context of cell therapy, the factors in CM are immediately available to modulate pathology, which is especially critical during the acute phases of diseases such as stroke. Repeated administration is also possible without the risk of rejection by the host. Finally, an approved drug based on this cocktail (especially if lyophilized) may be even more practical and cost-effective than cellular therapies for the reasons that maintenance of cell viability during storage and revival as well as the requirement to remove cryoprotectants (i.e., dimethyl sulfoxide, DMSO) before administration will be obviated.

The functional categorization of therapies based on CM lies between cell therapies and single-factor recombinant protein therapies. A variety of single-factor approaches have been proposed to repair and restore the brain after stroke. One initially promising approach tested hematopoietic growth factors, including erythropoietin or granulocyte colony-stimulating factor (G-CSF), to stimulate neurogenesis. However, both erythropoietin and G-CSF failed in clinical trials in acute stroke [36, 37]. These failures are consistent with the inability of individual agents, affecting discrete signaling pathways, to overcome the extensive damage to multiple cell types and pathways occurring in stroke, at least when administered in tolerable doses. For example, basic fibroblast growth factor (bFGF) was promising in initial studies but was not tolerable in clinical trials due to off-target effects of hypotension and renal toxicity [26].

A potential advantage of CM over single proteins is that the cocktail of factors function synergistically by modulating multiple pathways in concert at much lower (i.e., safer) doses of each protein. Toxicities of selected factors have been related to the relatively high levels required to achieve effects in the brain which leads to high systemic exposure. Most trophic and growth factors bind to different receptors to produce dissimilar and often opposing effects. For example, at relatively high concentrations, nerve growth factor (NGF) stimulates apoptosis through binding to its alternative low-affinity receptor $p75^{NTR}$, which is upregulated following ischemic cerebral injury [38]. Additionally, there is a sharp drop-off in effectiveness *above* optimal concentrations. The concentrations of individual therapeutic factors in an efficacious dose of adipose-derived stem cell (ASC)-CM range between 100- and 1000-fold lower than typically administered for single protein factors [39–41]. For example, a previous study in a rabbit model of peripheral vascular disease determined the optimal effect on revascularization of ischemic tissues was achieved with recombinant human vascular endothelial growth factor (VEGF) doses between 0.14 and 0.33 mg/kg body weight [41]. This is compared to an effective dose of CM used in a similar model that contained approximately 0.001 mg VEGF/kg body weight [39]. The lower levels of factors in CM provide for an increased safety factor at effective doses by minimizing off-target effects that may be noted at higher doses of single agents.

The exact form of factors comprising CM is not well understood; however, it is likely that the bulk is complexed with proteins or other macromolecules such as lipids and nucleic acids. One such complex could be microvesicles, such as exosomes, released from the cells in response to environmental cues. It was recently shown that exosomes isolated from MSC-CM potently induced functional recovery following experimental stroke in rats [42]. The exosomes acted to increase vascularity, neurogenesis, and neurite remodeling. A significant portion of the neuroplasticity observed could be attributed to a single microribonucleic acid (microRNA) species rather than protein factors [43]. The discovery that regulatory RNA species may be present in CM and that these activities could contribute significantly to the potency of MSC-CM through complementation of protein factors further substantiates the benefits of a cocktail of factors (as opposed to monotherapies) for treating stroke.

Neuroprotective and Neurorestorative Properties of MSC Secreted Factors

MSC-CM contains both neuroprotective and neurorestorative factors, and engages key molecular pathways involved in each of these processes (Table 14.1). These factors include NGF, insulin-like growth factor-1 (IGF-1), VEGF, glial cell-derived neurotrophic factor (GDNF), hepatocyte growth factor (HGF), brain-derived neurotrophic factor (BDNF), bFGF, and SDF-1 or CXCL12, which bind to cognate receptors expressed on many cell types to promote survival and cellular responses involved in protection and repair [1, 39, 40, 44–48]. Selective inactivation of either BDNF or IGF-1 before administration of ASC-CM in a neonatal model of stroke significantly reduced potency [40]. As expected, IGF-1 in this context acts to protect neurons against inducers of apoptosis via a mechanism involving IGF-1-induced activation of Akt [49]. Activation of Akt provides a pro-survival signal after stroke which attenuates glutamate excitotoxicity and upregulates p-cAMP(adenosine 5'-monophosphate) response element-binding protein (CREB)/BDNF leading to increased cell survival, neurogenesis, and synaptogenesis [50–53]. Further protection is afforded by attenuation of excitotoxicity through stimulating Akt and mitogen-activated protein kinase/extracellular signal-regulated kinase (MAPK/ERK) pathways by multiple factors in CM, including BDNF, HGF, and VEGF [39, 40, 49, 54]. Dendrite formation and synaptogenesis in cultured rata hippocampal neurons were enhanced by soluble factors produced by umbilical- and BM-derived MSC. This activity was partially mediated by BDNF in the CM [55].

Neuritogenesis in PC12 cells was also enhanced by treatment with ASC-CM, which was due in part to NGF activation of AMP-activated kinase α (AMPK-α) in vitro [56, 57]. AMPK is the central energy sensor switch, and a key anti-inflammatory regulator of lipid metabolizing enzymes [58]. AMPK activation promotes neuritogenesis and axogenesis during metabolic stress and its loss in supporting structures like oligodendrocytes and astrocytes exacerbates the disease [56, 59, 60]. In the vasculature, AMPK-α1 is upstream to Akt and their cross talk promotes VEGF-induced angiogenesis [61].

Demonstrating the Therapeutic Potential of MSC-CM in Experimental Models of Stroke

Perhaps the earliest demonstration of MSC-CM potency in an animal disease model was by Kinnaird and Epstein [62]. The CM from BM-MSC promoted reperfusion of ischemic tissue in a mouse model of peripheral vascular disease. Treated tissues exhibited reduced muscular atrophy, enhanced vascularity, and reduced necrosis. A number of hypoxia-inducible pro-angiogenic and pro-survival factors were present in the CM and neutralization of two (VEGF and bFGF) only partially reduced the ability of the mixture to promote endothelial and smooth muscle cell migration. A related study by Rehman et al. demonstrated ASC expression of hypoxia-inducible

Table 14.1 Neuroregenerative and neuroprotective factors secreted by MSC

Functional category	Factor	Mechanisms of action
Angiogenic/arteriogenic		
	VEGF	Endothelial cell mitogen and chemoattractant
	bFGF	Endothelial cell mitogen and chemoattractant promotes remodeling
	HGF	Mobilizes vascular and parenchyma forming cells
	GDNF	Cooperates with VEGF to enhance angiogenesis
Pro-survival		
	VEGF	Blocks endothelial cell apoptosis induced by factors such as TGF-b
	HGF	Blocks apoptosis through activation of Akt
	IGF-1	Blocks apoptosis through activation of Akt
	BDNF	Protects against glutamate excitotoxicity
Neurotrophic		
	NGF	Stimulates neurogenesis
	GDNF	Protects many different cell types from apoptosis
	bFGF	Promotes synaptogenesis
Anti-inflammatory/Immunomodulators		
	PGE2	Blocks stimulation of many different inflammatory cells
	IDO	Blocks B cell antibody production. Downregulates IFN-γ
	HGF	Blocks activation/proliferation of B and NK cells
	TSG-6	Suppresses NF-kB stimulation
	IL-1Ra	IL-1 antagonist
	sTNF-R	Traps TNF-α to prevent stimulation of cells
Recruitment/Mobilization		
	SDF1	Chemoattractant through CXCR4 receptor expressed on many cells
	HGF	Pro-migratory (also known as scatter factor)
	G-CSF	Promotes cell mobilization

IDO indoleamine 2,3-dioxygenase, *PGE2* prostaglandin E-2, *HGF* hepatocyte growth factor, *SDF-1* stromal-derived factor-1, *TSG-6* tumor necrosis alpha stimulating gene-6, *IL-1Ra* interleukin-1 receptor antagonist, *BDNF* brain-derived neurotrophic factor, *GDNF* glial-derived neurotrophic factor, *NGF* nerve growth factor, *VEGF* vascular endothelial growth factor, *IGF-1* insulin-like growth factor-1, *bFGF* basic fibroblast growth factor, *G-CSF* granulocyte-colony stimulating factor, *TGF* tumor growth factor, *STNF* soluble tumor necrosis factor, *IFN* interferon, *NK* natural killer

trophic factors, including bFGF, VEGF, and HGF, and observed rapid positive effects after administration in a murine hind limb ischemia model [45]. Subsequently, it was determined that knockdown of HGF expression by siRNA severely attenuated ASC-mediated reperfusion in the same model [63]. In an attempt to increase

BM-MSC survival and, thus increase the effective cell dose, the laboratory of Victor Dzau generated cells expressing the transgene for Akt [64]. The transgenic cells showed enhanced ability to promote repair of ischemic myocardium; it was determined that this effect was due to enhanced paracrine activity and that similar effects could be demonstrated by provision of the CM alone [13].

The CM from BM-MSC and ASC have been shown to protect and repair brain tissues in rodent models of focal and global ischemic stroke. A global ischemia neonatal rat hypoxia-ischemia encephalopathy (HIE) model was used to test ASC-CM in a neonatal model of hypoxia-ischemia (HI) injury. A single intravenous injection at up to 36 h post injury significantly reduced brain volume loss and promoted functional recovery at 1 week and 3 months [40]. Neutralization of either IGF-1 or BDNF before injection partially reduced this effect. The salutary effects of ASC-CM were related to protection from glutamate excitotoxicity and protection of neurons from ischemic insult. Similarly, a rat model of stroke induced by MCAO was treated with ASC-CM infused directly into the lateral ventricle [5]. Continuous infusion for 8 days beginning at 1 week after surgery reduced infarct volume and enhanced motor control at up to 15 days following MCAO surgery. This activity was attributed to enhanced microvascularity in the infarct zone as well as reduced cell death and inflammation with treatment. Short-term protection (e.g., 24 h) of a single intracerebroventricular injection of concentrated ASC-CM following temporary MCAO with reperfusion in a murine model was effective in reducing infarct volume and edema [7]. In a study that has implications for treatment of stroke victims with an autologous product, Tsai et al. evaluated CM from BM-MSC isolated from either normal healthy rats or animals that had experienced an induced stroke [30]. Both CM exhibited similar effectiveness in enhancing functional recovery when delivered intravenously. The effect was associated with increased neurogenesis and reduced inflammation within the infarct zone.

Toward Development and Testing of Clinical-Grade MSC-CM

The cumulative data from preclinical studies strongly support the therapeutic potential of MSC-CM for treating stroke as well as other diseases. As with any unproven investigational agent, regulatory approval by governing bodies will be required before commencing trials in humans. These agencies include the US Food and Drug Administration (FDA) and the European Medicines Agency (EMA) as well as local Institutional Review Boards (IRB), and Competent Authorities (CA) in the European Union. This section focuses on the regulatory pathway for new drug approval in the USA.

Possible Regulatory Pathways for MSC-CM

In the USA, an investigational new drug application (IND) is required for nearly all new drugs entering clinical trials. The IND comprises three sections: chemistry and manufacturing controls (CMC), clinical study design, and nonclinical studies. The nonclinical studies section mainly concerns safety and toxicity in animals using the clinically intended route of administration and a product very similar, if not identical, to that which will be used in the clinic. This section typically includes a description of efficacy studies in relevant disease models. The CMC section pertains to manufacturing processes and quality control systems for ensuring consistency and the absence of potentially deleterious agents in the final product. Each of the sections of the IND must provide reviewers with a sufficient amount of detail to determine the potential safety of any product before allowing evaluation in humans.

The regulatory route for licensure of an eventual drug based on MSC-CM will likely require a Biological License Application (BLA) as opposed to a New Drug Application (NDA), the latter which generally pertains to drugs of well-defined composition. Within the FDA there are two centers responsible for oversight and approval of new drugs. The Center for Biologics Evaluation and Research (CBER) and the Center for Drug Evaluation and Research (CDER). Jurisdictional oversight of biologics generally falls to CBER; with important exceptions for less complex entities, such as monoclonal antibodies and recombinant proteins. Therefore, the complexity of MSC-CM, whether whole or partially fractionated, likely will place it under the review of CBER.

Establishing Good Manufacturing Process for Producing MSC-CM

The FDA, as well as other entities charged with overseeing aspects of biologic and cell therapies, has issued guidance documents detailing the requirements that must be fulfilled to create a product with appropriate characteristics (i.e., the CMC component of an IND) [65–72]. This section presents an overview of the guidelines for reagents and materials of acceptable quality for good manufacturing practices (GMP) production of MSC-CM to be used in the clinic.

Sources of Donor Tissues and Cells

Of foremost importance to the manufacturing process is the nature and qualification of cell sources. The FDA has constructed careful guidelines for screening donors for human cell and tissue products (HCT/P) to minimize the chance of introducing adventitious agents into the final drug product [67]. Donors must be qualified for minimal risk of transmitting diseases through the use of both a comprehensive health history questionnaire and blood testing. Infectious agents of concern include human immunodeficiency viruses (HIV), hepatitis viruses, human T-lymphotrophic virus (HTLV), cytomegalovirus (CMV), *Treponema pallidum* (causative agent of syphilis), *Trypanosoma cruzi* (causative agent of Chagas disease), West Nile virus, and prion proteins. Following qualification of the donor, additional screening is performed periodically during the manufacturing process.

Ancillary Materials Used in Manufacturing MSC-CM

Ancillary materials (AM), also referred to as ancillary reagents, are substances used in manufacturing of the product but not anticipated to be present in the final product. The control and appropriate qualification standards for AM are intended to reduce the chance of introducing substances into the final product which may transmit disease or induce toxicity in recipients. Generally, AM used for manufacturing should be well characterized and have a certificate of analysis (CoA) containing detailed information regarding source materials and quality testing. An important caveat for MSC-CM is that current somatic cell therapy manufacturing guidance documents were contemplated for a cellular product and not for the medium conditioned by cells. Whereas it is possible to reduce the levels of AM in the final product through washing of the cells, some level of reagents used in the manufacture of MSC-CM will remain a part of the final product, regardless of the processes used. For example, cells grown in a nutrient-rich medium containing proteins, such as animal serum or recombinant growth factors, could be washed thoroughly before transferring to basal medium to remove the majority of these proteins. However, there is invariable carry-through of medium components associated with the cells. Partial purification through ultrafiltration or other means may further reduce the quantities of these agents, but the extent will depend on the nature of the substance and the pore size of the ultrafiltration medium. Thus, certain AM used for the manufacture of MSC-CM will become part of the final product.

The US Pharmacopoeia (USP) monograph on AM (USP<1043>) as well as the International Committee for Harmonization (ICH) guidance provide useful guidelines for selecting appropriate materials to be used in cell manufacturing, and these guidelines would appear to pertain to MSC-CM [65, 69]. Examples of components include enzymes used for cell liberation, media components, disposables employed for cell isolation and separation, and culture flasks. The first tier of AM qualification describes the optimal characteristics of reagent possessing the lowest potential risk of having a negative impact on safety and efficacy of the final product (Table 14.2). These reagents are highly qualified and generally are approved for human use. The next level (tier 2) are generally regarded as low-risk reagents and materials produced in full compliance with current good manufacturing practices (cGMPs), as demonstrated by appropriate documentation in a CoA. Further down on the list and possibly unacceptable for incorporation into cell product manufacturing due to unknown risks of introducing undesirable substances are tier 3 (research or diagnostic grade) AM or the highest-risk tier 4 AM that are not well characterized.

In vitro culture of mammalian cells at commercial scales of production at present requires the addition of blood components to the medium to supply necessary nutrients to promote growth. Animal serum components are considered a tier 4, high-risk AM. Classically, bovine serum has been used; however, the potential exists for transmission of the causative agents of bovine spongiform encephalitis and other diseases. Similarly, zoonotic disease transmission is a potential problem for blood products from ovine and equine species. Certified animal blood products that fully meet cGMP requirements by virtue of thorough screening and sourcing from closed herds are available from a number of different companies. An alternative to animal products is human blood-derived components, such as serum, albumin, and platelet lysate obtained from qualified donors. These products have the potential to be safer and may be viewed as more desirable than animal products [73, 74]. However, concerns exist regarding the availability of supplies at quantities necessary to prevent disruptions in manufacturing as well as consistency between batches of human-derived components.

Required Testing at Stages During Manufacture and of the Final Product

To ensure safety and maintenance of conditions conducive to the desired final product, FDA guidelines require extensive testing for sterility, purity, potency, identity, and viability throughout the manufacturing process. Implementation of these assays into the manufacturing process is critical for quality control and to determine consistency of product between lots. Purity of the product and intermediates pertains particularly to monitoring absence of pyrogens and endotoxins to maintain safety for recipients or preventing deleterious effects on activity or stability of the final product.

Identity in the case of MSC-CM pertains to both the phenotype of cells used to produce the mixture and protein composition. In regard to the latter, it may not be practical to determine the quantity and identity of every protein in the complex mixture. Rather, the presence and abundance of factors known to produce bioactivity will be key criteria. Evaluation of potency of the resulting final product will involve establishment of appropriate in vitro and in vivo *assays* capable of quantitatively assessing activity. Finally, microbial contamination (including fungi, bacteria, and virus) must be monitored to ensure that these contaminants will not be inadvertently administered to patients, especially with parenteral dosing.

Once a final product is produced for clinical evaluation, MSC-CM must be qualified for stability during storage to ensure that each of the qualities listed above are maintained in the product to be administered to humans. The length of period required for testing depends on the interval between manufacturing and use in clinical trials. Forced degradation by storage at temperatures elevated beyond the intended storage temperature is generally used to accelerate testing of stability. In addition, the absence of negative interactions with storage containers must be evaluated to ensure that adsorption or inactivation by the container or closure device does not occur.

Table 14.2 Risk stratification of ancillary materials

Tier	Risk	Suitability for manufacturing	Qualification level	Examples	Notes
1	Low	Optimal	High	Drugs or nutrients approved for use in humans	Should be provided in sterile packaging
				Licensed biologic	
				Approved medical device	
				Intended for use as implantable material	
2	Low	Very good providing relevant documentation is provided	High	The following reagents which have been produced in compliance with cGMP guidelines	Excludes most animal products
				Drug	
				Biologic	
				Medical device	
3	Moderate	Appropriate qualification must be performed before including in manufacturing	Good to moderate	Products intended for in vitro diagnostic use:	
				Antibodies	
				Growth factors and cytokines	
4	High	Extensive qualification must be performed to ensure absence, removal, or inactivation of toxic and infectious agents	Minimal to none	Relatively uncharacterized substances and most animal products:	
				Animal tissues and fluids	
				Research grade reagents	

Examples of MSC-CM Manufactured in Compliance with FDA Standards

Development of therapeutic MSC-CM in compliance with the conditions listed above will require careful attention to all steps and components used in manufacturing. For instance, the medium used for conditioning must be devoid of components not suitable for use in humans. Many common research-grade media comprise amino acids, vitamins, and inorganic salts from incompletely characterized sources and, thus, the presence of potentially toxic contaminants are not rigorously tested. Furthermore, media are often buffered with 4-(2-hydroxyethyl)-1-piperazineethanesulfonic acid (HEPES) and contain phenol red as a pH indicator. These compounds are not approved for use in humans. It is possible to formulate a medium possessing the essentially the same components through the use of drugs and supplements that are approved for human use. This strategy was adopted in a recent study that examined ASC-CM produced with a medium comprising approved substances or those produced in compliance with regulations [39]. The resulting ASC-CM accelerated reperfusion of ischemic tissues in the murine peripheral vascular disease model. Even this strategy for producing clinical-grade ASC-CM, controls and testing throughout the production process are still required to ensure that the final product possesses the required characteristics.

Development and Qualification of an FDA-Compliant ASC-CM

NeuroFx, Inc. (Indianapolis, IN, USA) is developing a drug product (NFx-101) comprising substantially xenogen-depleted (through washing and use of protein-free components in the final conditioning step) and partially fractionated CM derived from culturing human ASC. Through interactions with regulatory and CMC consultants as well as FDA, the company is building a commercial-scale manufacturing process that is fully compliant with cGMP and FDA standards.

A flow diagram describing the process for cGMP- and good tissue practices (GTP)-compliant manufacturing of NFx-101 is presented in Fig. 14.2. Adipose tissues are obtained under an IRB approved protocol from consenting, qualified adult donors. The tissue is processed to isolate stromal vascular fraction (SVF) from adipose tissues by enzymatic digestion using GMP-certified reagents. The SVF is then characterized with regard to cell phenotype and viability and cryopreserved as the master bank. Passaged, plastic adherent cells, which are predominantly ASC, are cryopreserved as working stocks. Cells from working stocks are expanded in a proprietary complete medium to high passage after which they are washed extensively before adding a GMP-compliant basal medium for conditioning. The CM is collected, filtered, and concentrated. Each lot of NFx-101 product is evaluated for sterility, endotoxin, residual human DNA, composition, and potency.

Standards will be set for cell identity to determine whether a change in cell type occurs during manufacturing and, thus, deviate significantly from cells used to assess safety and efficacy. Most commonly, flow cytometric analysis of cell surface phenotypic proteins are used to establish identity. The process developed by NeuroFx incorporates these analyses at critical points in manufacturing; however, this only ensures that the cell type remains consistent during production of NFx-101, but does not ensure consistency of the final product. The final partially purified CM product must also be characterized. Since efficacy of cellular therapies is related to paracrine effect, assays to measure selective factors or their function are commonly employed in potency assays to determine release criteria. The FDA has indicated its

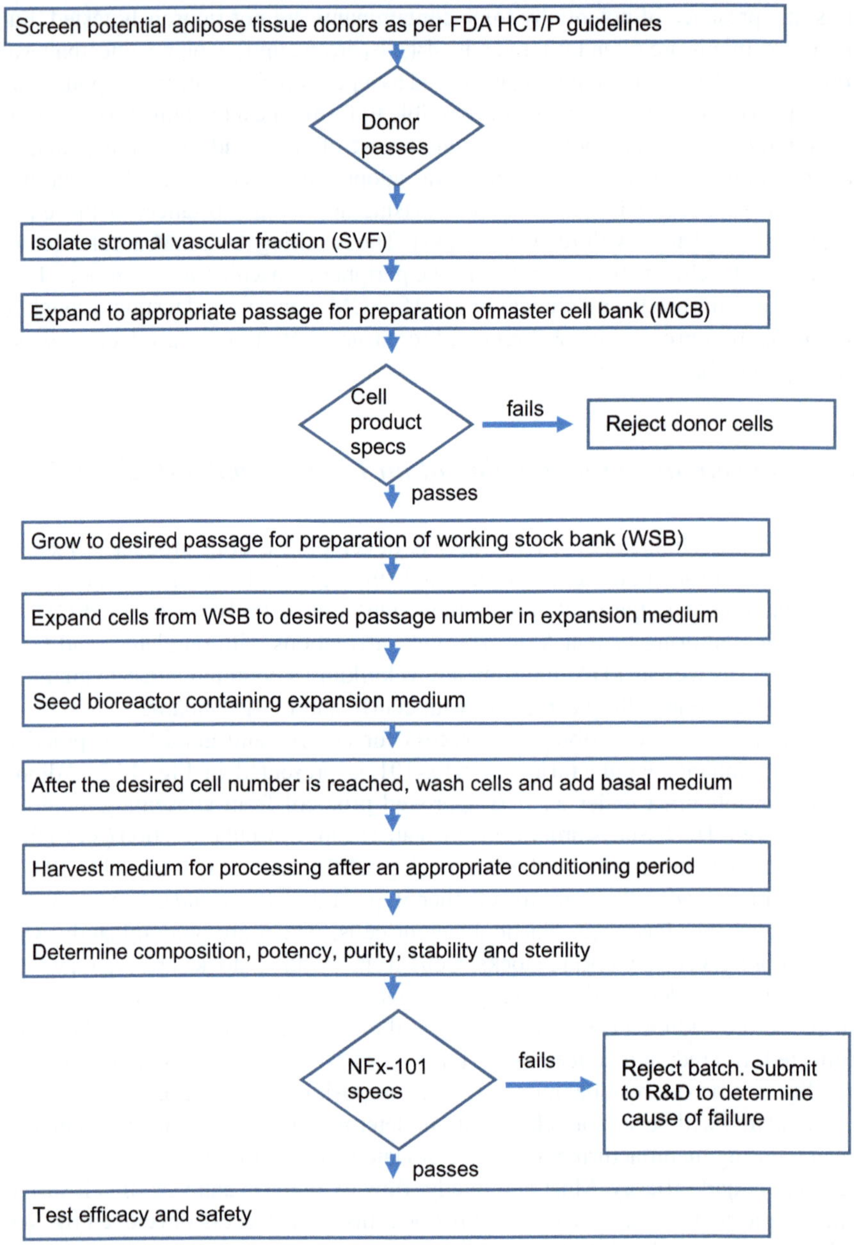

Fig. 14.2 Schematic of the screening and manufacturing process used to produce NFx-101. The flow diagram describes the process for preparing clinical-grade conditioned medium that is compliant with pertinent regulatory guidelines. *specs* specifications, *R&D* research and development, *HCT/P* human cell and tissue products, *FDA* Food and Drug Administration

willingness to accept this limited assessment for cell therapies now in testing [75]; even though it is not known how closely the factors produced ex vivo are representative in quantity or composition of those produced in the uncontrolled environment following administration to patients. This is in contrast with CM production in a controlled environment using carefully characterized reagents and materials, where control of lot-to-lot variability is feasible. In regards to establishment of release criteria, we have demonstrated in cerebral ischemic disease models that potency of the product is mainly dependent on only a few key factors. Thus, qualification based on these factors will establish sufficiently robust release criteria.

Concluding Remarks and Perspectives

The use of MSC-CM has the potential to be a clinically relevant therapy that overcomes potential issues with cell therapies of tumorigenesis, embolism, and rejection. Furthermore, it is amenable to an off-the-shelf preparation that is immediately accessible for administration, does not require special attention to maintaining cell viability during prolonged storage or preparation for administration, and is not associated with potentially toxic additives (e.g., DMSO) that must be removed at the time of use. Providing that the appropriate care is taken to design a robust and consistent production process, MSC-CM may be considered a safe and effective alternative to MSC. Based on our new understanding of the mechanism of action of cell therapies, it may be considered that CM has already undergone extensive clinical testing, albeit with less control over the actual composition or relative levels of factors secreted. We look forward to testing these assumptions using clinical-grade material, first in animals and then in humans. Ultimately, the goal is to gain approval for this new therapy in order to address the unmet medical need for approved therapies to treat the large population of stroke patients with limited therapeutic options.

References

1. Chen J, Li Y, Katakowski M, Chen X, Wang L, Lu D, et al. Intravenous bone marrow stromal cell therapy reduces apoptosis and promotes endogenous cell proliferation after stroke in female rat. J Neurosci Res. 2003;73(6):778–86.
2. Chen J, Li Y, Wang L, Lu M, Zhang X, Chopp M. Therapeutic benefit of intracerebral transplantation of bone marrow stromal cells after cerebral ischemia in rats. J Neurol Sci. 2001;189(1–2):49–57.
3. Chen J, Li Y, Wang L, Zhang Z, Lu D, Lu M, et al. Therapeutic benefit of intravenous administration of bone marrow stromal cells after cerebral ischemia in rats. Stroke. 2001;32(4):1005–11.
4. Chen J, Zhang ZG, Li Y, Wang L, Xu YX, Gautam SC, et al. Intravenous administration of human bone marrow stromal cells induces angiogenesis in the ischemic boundary zone after stroke in rats. Circ Res. 2003;92(6):692–9.

5. Cho YJ, Song HS, Bhang S, Lee S, Kang BG, Lee JC, et al. Therapeutic effects of human adipose stem-cell-conditioned medium on stroke. J Neurosci Res. 2012;90(9):1794–802.
6. Chopp M, Li Y. Treatment of neural injury with marrow stromal cells. Lancet Neurol. 2002;1(2):92–100.
7. Egashira Y, Sugitani S, Suzuki Y, Mishiro K, Tsuruma K, Shimazawa M, et al. The conditioned medium of murine and human adipose-derived stem cells exerts neuroprotective effects against experimental stroke model. Brain Res. 2012;1461:87–95.
8. Ikegame Y, Yamashita K, Hayashi S, Mizuno H, Tawada M, You F, et al. Comparison of mesenchymal stem cells from adipose tissue and bone marrow for ischemic stroke therapy. Cytotherapy. 2011;13(6):675–85.
9. Kang SK, Lee DH, Bae YC, Kim HK, Baik SY, Jung JS. Improvement of neurological deficits by intracerebral transplantation of human adipose tissue-derived stromal cells after cerebral ischemia in rats. Exp Neurol. 2003;183(2):355–66.
10. Shen LH, Li Y, Chen J, Zacharek A, Gao Q, Kapke A, et al. Therapeutic benefit of bone marrow stromal cells administered 1 month after stroke. J Cereb Blood Flow Metab. 2007;27(1):6–13.
11. Yang YC, Liu BS, Shen CC, Lin CH, Chiao MT, Cheng HC. Transplantation of adipose tissue-derived stem cells for treatment of focal cerebral ischemia. Curr Neurovasc Res. 2011;8(1):1–13.
12. Caplan AI, Dennis JE. Mesenchymal stem cells as trophic mediators. J Cell Biochem. 2006;98(5):1076–84.
13. Gnecchi M, Zhang Z, Ni A, Dzau VJ. Paracrine mechanisms in adult stem cell signaling and therapy. Circulation Res. 2008;103(11):1204–19.
14. Ratajczak MZ, Kucia M, Jadczyk T, Greco NJ, Wojakowski W, Tendera M, et al. Pivotal role of paracrine effects in stem cell therapies in regenerative medicine: can we translate stem-cell-secreted paracrine factors and microvesicles into better therapeutic strategies? Leukemia. 2012;26(6):1166–73.
15. Prockop DJ. Concise review: two negative feedback loops place mesenchymal stem/stromal cells at the center of early regulators of inflammation. Stem Cells. 2013;31(10):2042–6.
16. Salgado AJ, Reis RL, Sousa NJ, Gimble JM. Adipose tissue derived stem cells secretome: soluble factors and their roles in regenerative medicine. Curr Stem Cell Res Ther. 2010;5(2):103–10.
17. Chen X, Li Y, Wang L, Katakowski M, Zhang L, Chen J, et al. Ischemic rat brain extracts induce human marrow stromal cell growth factor production. Neuropathology. 2002;22(4):275–9.
18. Kang SK, Jun ES, Bae YC, Jung JS. Interactions between human adipose stromal cells and mouse neural stem cells *in vitro*. Brain Res Dev Brain Res. 2003;145(1):141–9.
19. Shen LH, Li Y, Chen J, Cui Y, Zhang C, Kapke A, et al. One-year follow-up after bone marrow stromal cell treatment in middle-aged female rats with stroke. Stroke. 2007;38(7):2150–6.
20. Phinney DG, Prockop DJ. Concise review: mesenchymal stem/multipotent stromal cells: the state of transdifferentiation and modes of tissue repair–current views. Stem Cells. 2007;25(11):2896–902.
21. Prockop DJ, Oh JY. Medical therapies with adult stem/progenitor cells (MSCs): a backward journey from dramatic results *in vivo* to the cellular and molecular explanations. J Cell Biochem. 2012;113(5):1460–9.
22. Murphy MB, Moncivais K, Caplan AI. Mesenchymal stem cells: environmentally responsive therapeutics for regenerative medicine. Exp Mol Med. 2013;45:e54.
23. Prockop DJ, Kota DJ, Bazhanov N, Reger RL. Evolving paradigms for repair of tissues by adult stem/progenitor cells (MSCs). J Cell Mol Med. 2010;14(9):2190–9.
24. Lee RH, Pulin AA, Seo MJ, Kota DJ, Ylostalo J, Larson BL, et al. Intravenous hMSCs improve myocardial infarction in mice because cells embolized in lung are activated to secrete the anti-inflammatory protein TSG-6. Cell Stem Cell. 2009;5(1):54–63.

25. Danchuk S, Ylostalo JH, Hossain F, Sorge R, Ramsey A, Bonvillain RW, et al. Human multipotent stromal cells attenuate lipopolysaccharide-induced acute lung injury in mice via secretion of tumor necrosis factor-alpha-induced protein 6. Stem Cell Res Ther. 2011;2(3):27.
26. Hossmann KA. The two pathophysiologies of focal brain ischemia: implications for translational stroke research. J Cereb Blood Flow Metab. 2012;32(7):1310–6.
27. Cramer SC. An overview of therapies to promote repair of the brain after stroke. Head Neck. 2011;33 Suppl 1:S5–7.
28. Cramer SC, Riley JD. Neuroplasticity and brain repair after stroke. Curr Opin Neurol. 2008;21(1):76–82.
29. Li Y, Chopp M. Temporal profile of nestin expression after focal cerebral ischemia in adult rat. Brain Res. 1999;838(1–2):1–10.
30. Tsai MJ, Tsai SK, Hu BR, Liou DY, Huang SL, Huang MC, et al. Recovery of neurological function of ischemic stroke by application of conditioned medium of bone marrow mesenchymal stem cells derived from normal and cerebral ischemia rats. J Biomed Sci. 2014;21:5.
31. Leong WK, Henshall TL, Arthur A, Kremer KL, Lewis MD, Helps SC, et al. Human adult dental pulp stem cells enhance poststroke functional recovery through non-neural replacement mechanisms. Stem Cells Transl Med. 2012;1(3):177–87.
32. Leu S, Lin YC, Yuen CM, Yen CH, Kao YH, Sun CK, et al. Adipose-derived mesenchymal stem cells markedly attenuate brain infarct size and improve neurological function in rats. J Transl Med. 2010;8:63.
33. Li D, Fang Y, Wang P, Shan W, Zuo Z, Xie L. Autologous transplantation of adipose-derived mesenchymal stem cells attenuates cerebral ischemia and reperfusion injury through suppressing apoptosis and inducible nitric oxide synthase. Int J Mol Med. 2012;29(5):848–54.
34. Li Y, Chen J, Wang L, Lu M, Chopp M. Treatment of stroke in rat with intracarotid administration of marrow stromal cells. Neurology. 2001;56(12):1666–72.
35. Ramos-Cabrer P, Justicia C, Wiedermann D, Hoehn M. Stem cell mediation of functional recovery after stroke in the rat. PloS One. 2010;5(9):e12779.
36. Voulgari-Kokota A, Fairless R, Karamita M, Kyrargyri V, Tseveleki V, Evangelidou M, et al. Mesenchymal stem cells protect CNS neurons against glutamate excitotoxicity by inhibiting glutamate receptor expression and function. Exp Neurol. 2012;236(1):161–70.
37. Ehrenreich H, Weissenborn K, Prange H, Schneider D, Weimar C, Wartenberg K, et al. Recombinant human erythropoietin in the treatment of acute ischemic stroke. Stroke. 2009;40(12):e647–56.
38. Lee R, Kermani P, Teng KK, Hempstead BL. Regulation of cell survival by secreted proneurotrophins. Science. 2001;294(5548):1945–8.
39. Bhang SH, Lee S, Shin JY, Lee TJ, Jang HK, Kim BS. Efficacious and clinically relevant conditioned medium of human adipose-derived stem cells for therapeutic angiogenesis. Mol Ther. 2014;22(4):862–72.
40. Wei X, Du Z, Zhao L, Feng D, Wei G, He Y, et al. IFATS collection: The conditioned media of adipose stromal cells protect against hypoxia-ischemia-induced brain damage in neonatal rats. Stem Cells. 2009;27(2):478–88.
41. Takeshita S, Zheng LP, Brogi E, Kearney M, Pu LQ, Bunting S, et al. Therapeutic angiogenesis. A single intraarterial bolus of vascular endothelial growth factor augments revascularization in a rabbit ischemic hind limb model. J Clin Invest. 1994;93(2):662–70.
42. Xin H, Li Y, Cui Y, Yang JJ, Zhang ZG, Chopp M. Systemic administration of exosomes released from mesenchymal stromal cells promote functional recovery and neurovascular plasticity after stroke in rats. J Cereb Blood Flow Metab. 2013;33(11):1711–5.
43. Xin H, Li Y, Liu Z, Wang X, Shang X, Cui Y, et al. MiR-133b promotes neural plasticity and functional recovery after treatment of stroke with multipotent mesenchymal stromal cells in rats via transfer of exosome-enriched extracellular particles. Stem Cells. 2013;31(12):2737–46.
44. Liu N, Zhang Y, Fan L, Yuan M, Du H, Cheng R, et al. Effects of transplantation with bone marrow-derived mesenchymal stem cells modified by Survivin on experimental stroke in rats. J Transl Med. 2011;9:105.

45. Rehman J, Traktuev D, Li J, Merfeld-Clauss S, Temm-Grove CJ, Bovenkerk JE, et al. Secretion of angiogenic and antiapoptotic factors by human adipose stromal cells. Circulation. 2004;109(10):1292–8.
46. Song M, Mohamad O, Gu X, Wei L, Yu SP. Restoration of intracortical and thalamocortical circuits after transplantation of bone marrow mesenchymal stem cells into the ischemic brain of mice. Cell Transplant. 2013;22(11):2001–15.
47. Wei L, Fraser JL, Lu ZY, Hu X, Yu SP. Transplantation of hypoxia preconditioned bone marrow mesenchymal stem cells enhances angiogenesis and neurogenesis after cerebral ischemia in rats. Neurobiol Dis. 2012;46(3):635–45.
48. Zhang J, Li Y, Chen J, Yang M, Katakowski M, Lu M, et al. Expression of insulin-like growth factor 1 and receptor in ischemic rats treated with human marrow stromal cells. Brain Res. 2004;1030(1):19–27.
49. Wei X, Zhao L, Zhong J, Gu H, Feng D, Johnstone BH, et al. Adipose stromal cells-secreted neuroprotective media against neuronal apoptosis. Neurosci Lett. 2009;462(1):76–9.
50. Hardingham GE, Fukunaga Y, Bading H. Extrasynaptic NMDARs oppose synaptic NMDARs by triggering CREB shut-off and cell death pathways. Nat Neurosci. 2002;5(5):405–14.
51. Kawano T, Morioka M, Yano S, Hamada J, Ushio Y, Miyamoto E, et al. Decreased akt activity is associated with activation of forkhead transcription factor after transient forebrain ischemia in gerbil hippocampus. J Cereb Blood Flow Metab. 2002;22(8):926–34.
52. Yano S, Morioka M, Fukunaga K, Kawano T, Hara T, Kai Y, et al. Activation of Akt/protein kinase B contributes to induction of ischemic tolerance in the CA1 subfield of gerbil hippocampus. J Cereb Blood Flow Metab. 2001;21(4):351–60.
53. Yoshii A, Constantine-Paton M. BDNF induces transport of PSD-95 to dendrites through PI3K-AKT signaling after NMDA receptor activation. Nat Neurosci. 2007;10(6):702–11.
54. Lu S, Lu C, Han Q, Li J, Du Z, Liao L, et al. Adipose-derived mesenchymal stem cells protect PC12 cells from glutamate excitotoxicity-induced apoptosis by upregulation of XIAP through PI3-K/Akt activation. Toxicology. 2011;279(1–3):189–95.
55. Alder J, Kramer BC, Hoskin C, Thakker-Varia S. Brain-derived neurotrophic factor produced by human umbilical tissue-derived cells is required for its effect on hippocampal dendritic differentiation. Dev Nerobiol. 2012;72(6):755–65.
56. Tan B, Luan Z, Wei X, He Y, Wei G, Johnstone BH, et al. AMP-activated kinase mediates adipose stem-cell-stimulated neuritogenesis of PC12 cells. Neuroscience. 2011;181:40–7.
57. Zhao L, Wei X, Ma Z, Feng D, Tu P, Johnstone BH, et al. Adipose stromal cells-conditional medium protected glutamate-induced CGNs neuronal death by BDNF. Neurosci Lett. 2009;452(3):238–40.
58. Nath N, Khan M, Paintlia MK, Singh I, Hoda MN, Giri S. Metformin attenuated the autoimmune disease of the central nervous system in animal models of multiple sclerosis. J Immunol. 2009;182(12):8005–14.
59. Giri S, Khan M, Nath N, Singh I, Singh AK. The role of AMPK in psychosine mediated effects on oligodendrocytes and astrocytes: implication for Krabbe disease. J Neurochem. 2008;105(5):1820–33.
60. Williams T, Courchet J, Viollet B, Brenman JE, Polleux F. AMP-activated protein kinase (AMPK) activity is not required for neuronal development but regulates axogenesis during metabolic stress. Proc Natl Acad Sci U S A. 2011;108(14):5849–54.
61. Stahmann N, Woods A, Spengler K, Heslegrave A, Bauer R, Krause S, et al. Activation of AMP-activated protein kinase by vascular endothelial growth factor mediates endothelial angiogenesis independently of nitric-oxide synthase. J Biol Chem. 2010;285(14):10638–52.
62. Kinnaird T, Stabile E, Burnett MS, Lee CW, Barr S, Fuchs S, et al. Marrow-derived stromal cells express genes encoding a broad spectrum of arteriogenic cytokines and promote *in vitro* and *in vivo* arteriogenesis through paracrine mechanisms. Circ Res. 2004;94(5):678–85.
63. Cai L, Johnstone BH, Cook TG, Liang Z, Traktuev D, Cornetta K, et al. Suppression of hepatocyte growth factor production impairs the ability of adipose-derived stem cells to promote ischemic tissue revascularization. Stem Cells. 2007;25(12):3234–43.

64. Noiseux N, Gnecchi M, Lopez-Ilasaca M, Zhang L, Solomon SD, Deb A, et al. Mesenchymal stem cells overexpressing Akt dramatically repair infarcted myocardium and improve cardiac function despite infrequent cellular fusion or differentiation. Mol Ther. 2006;14(6):840–50.
65. Atouf F, Provost NM, Rosenthal FM. Standards for ancillary materials used in cell-and tissue-based therapies. BioProcess Int. 2013;11:8.
66. United States Food and Drug Administration: eligibility determination for donors of human cells, tissues, and cellular and tissue-based products. Final rule. Fed Regist. 2004;69(101):29785–834.
67. Unites States Food and Drug Administration, Center for Biologics Evaluation and Research. Guidance for industry: eligibility determination for donors of human cells, tissues, and cellular and tissue-based products. 2008. http://www. fda. gov/BiologicsBloodVaccines/GuidanceComplianceRegulatoryInformation/Guidances/Xenotransplantation/ucm074131.htm. Accessed 21 Feb 2015.
68. Unites States Food and Drug Administration, Center for Biologics Evaluation and Research. Guidance for industry: source animal, product, preclinical, and clinical issues concerning the use of Xenotransplantation products in humans. 2003. http://www.fda.gov/BiologicsBloodVaccines/GuidanceComplianceRegulatoryInformation/Guidances/Xenotransplantation/ucm074354.htm. Accessed 21 Feb 2015.
69. International Conference of Harmonization of Technical Requirements for Registration of Pharmaceuticals for Human Use. Q5D: derivation and characterisation of cell substrates used for production of biotechnological/biological products. 1997. http://www.ich.org/products/guidelines/quality/quality-single/article/derivation-and-characterisation-of-cell-substrates-used-for-production-of-biotechnologicalbiologica.html. Accessed 21 Feb 2015.
70. Service UPH. PHS guideline on infectious disease issues in xenotransplantation. MMWR Recommendations and Reports. 2001;50:1–46.
71. Center of Biological Research and Evaluation: Department of Health and Human Services. Points to consider in the characterization of cell lines used to produce biologicals. 1993. www.fda.gov/downloads/BiologicsBloodVaccines/SafetyAvailability/UCM162863.pdf. Accessed 21 Feb 2015.
72. Unites States Food and Drug Administration, Center for Biologics Evaluation and Research. Guidance for FDA reviewers and sponsors: content and review of chemistry, manufacturing, and control (CMC) information for human somatic cell therapy investigational new drug applications (INDs). 2008. www. fda. gov/BiologicsBloodVaccines/GuidanceComplianceRegulatoryInformation/Guidances/Xenotransplantation/ucm074131.htm. Accessed 21 Feb 2015.
73. Jung S, Panchalingam KM, Rosenberg L, Behie LA. Ex vivo expansion of human mesenchymal stem cells in defined serum-free media. Stem Cells Int. 2012;2012:123030.
74. Jung S, Panchalingam KM, Wuerth RD, Rosenberg L, Behie LA. Large-scale production of human mesenchymal stem cells for clinical applications. Biotechnol Appl Biochem. 2012;59(2):106–20.
75. Fink DW, Bauer SR. stem-cell-based therapies: food and drug administration product and pre-clinical regulatory considerationsed. In: Stem cell anthology: from stem cell biology, tissue engineering, cloning, regenerative medicine and biology. London: Academic; 2009. p. 379.

Chapter 15
Pathophysiology of Traumatic Brain Injury: Rationale and Role for Cellular Therapies

George Paul Liao and Charles S. Cox, Jr.

Traumatic brain injury (TBI) is a significant societal burden with limited treatment options. More than 90,000 adults in the United States annually sustain severe TBI resulting in long-term disability [1]. The prevalence of disabled survivors has been estimated to be as high as 5.3 million, or 2 % of the population in the United States [2]. Internationally, in nations such as China and Brazil, the prevalence of disabled survivors of TBI has been reported as being 5 times higher [3].

TBI has two phases, with the initial primary injury occurring from the transmission of mechanical forces through the cranium to the brain tissue. Immediate tissue damage occurs through various mechanisms including coup-contrecoup lesions along impact vectors as well as shearing injuries from rotational forces that cause diffuse axonal injury often identified on neuroimaging as hemorrhages at the gray-white junction. The secondary injury of TBI is the pathophysiologic sequelae of the initial impact, which occurs in the minutes to days and months beyond the primary injury. This focal or diffuse secondary injury occurs at every level of the brain and can be manifest from alterations in gene expression to increases in vascular permeability and loss of cerebral vascular autoregulation. While the neuroinflammatory response and vascular permeability associated with the secondary injury are often global, the pattern of lesions is heterogeneous and cause location specific neuromuscular deficits. While primary injury causes immediate tissue damage and neuronal death, secondary injury can be self-propagating with continued chronic injury evidenced by postmortem hippocampal DNA fragmentation detectable up to 12 months after head injury [4]. Activated microglia have been detected using positron emission tomography (PET) ligand [11C](R)PK11195 in brains of human patients up to 17 years after injury.

Cellular therapy offers the ability to sense and respond to a number of physiological and inflammatory signals in the injured brain. This chapter will review the complex pathophysiology of severe traumatic brain injury, discuss management

C. S. Cox, Jr. (✉) · G. P. Liao
6431 Fannin Street, MSB 5.236, Houston, TX 77030, USA
e-mail: Charles.S.Cox@uth.tmc.edu

D. C. Hess (ed.), *Cell Therapy for Brain Injury*, DOI 10.1007/978-3-319-15063-5_15

and provide an overview of current therapeutics as well as the rationale and role of cellular therapy.

Pathophysiology of Traumatic Brian Injury

Primary TBI The primary injury refers to the kinetic energy transfer that disrupts brain tissue and results in organ dysfunction. The direct impact of parenchymal tissue against bone leads to neuronal and vascular injury. Intracranial bleeding is a component of the primary injury and can be intra or extra-axial. Contusions or hematomas are intra-axial lesions that are most commonly found in the orbitofrontal or anterior temporal brain parenchyma, but can vary in location, contributing to the heterogeneity of presentation, prognosis and outcome in TBI. Bilateral lesions may arise from coup-contrecoup brain motion following impact. Extra-axial bleeding occurs in epidural, subdural and subarachnoid locations. Epidural hematomas can expand rapidly, causing brain compression, but are often associated with good recovery when promptly diagnosed and evacuated. Subdural hematomas and subarachnoid hematomas are associated with worse prognosis and represent cortical injury with the disruption of draining bridging veins and vessels coursing on the pia matter respectively. While epidural hematomas account for 3 % of extra-axial bleeding, subdural hematomas are most prevalent, occurring in almost 50 % of TBIs [5]. Apoptosis, evidenced by Bcl-2 gene expression and DNA fragmentation were most frequently seen in areas of subdural hematomas in rat models of TBI [6–8]. Diffuse axonal injury (DAI) describes the disruption of the fine axonal processes from rotational forces during the primary injury and is associated with poor prognosis and level of consciousness. DAI may be delayed up to 12 h and can be modeled in preclinical studies with rotational injury in the sagittal plane [9].

Clinical Presentation There are few reliable findings on physical examination that can assess the global severity of insult to the central nervous system following TBI. The Glasgow Coma Scale (GCS) remains one of the most used clinical tools to evaluate the severity of injury initially and throughout the critical stages of recovery. GCS scores range from mild (15-13), moderate (12-9), to severe (3–8). Patients in the severe category require intensive care monitoring and therapy aimed at both supporting and treating the brain and its intimately involved pulmonary, hepatic and reticuloendotheial systems.

Systemic Implications of TBI Isolated traumatic brain injury is associated with distant secondary organ dysfunction. In a weight drop in rodent models, TBI was associated with axonal injury detectable in the spinal cord as low as the thoracic lumbar levels [10]. The hypothalamic-pituitary-adrenal axis is also affected by TBI and is manifest as symptoms of dysautonomia such as sympathetic storms, autonomic dysreflexia and paroxysmal autonomic dystonia, present in 8–33 % of TBI patients. TBI patients will often have a complicated autonomic profile with contributions from pressors to maintain cerebral perfusion pressure and the presence of systemic alcohol, which decreases sympatho-adrenal activation [11, 12]. Cognitive as well as systemic autonomic dysfunction may be exacerbated by increased activity of the COMT (Catechol-O-methyltransferase) enzyme which metabolizes catecholamies at synaptic junctions [13, 14]. In rodents, studies show improved working memory with alpha-1 adrenergic receptor blocking agents such as prazosin working through the cAMP-responsive element binding protein mediated pathway [15].

The lungs are the most common non-neurological organ system that manifests secondary injury (23 %) [16]. In rodent models, brain trauma leads to enhanced lung inflammation and acute lung injury due to infiltrating macrophages and neutrophils leading to the release of leukotriene B4 [17]. Outside the CNS, TBI impacts renal and hepatic function, which alters drug clearance [18]. Lastly, the spleen has been identified as a key non-neurologic organ following TBI. Studies using rats undergoing immediate splenectomy following TBI had downregulation of MAPK-NFκB signaling, reduced proinflammatory cytokines, less brain edema and improved cognition [19, 20].

Secondary TBI In the brain, the delayed secondary injury includes inflammation, ischemia and hypoxia, which causes free radical, excitatory neurotoxin and calcium release, which in turn leads to mitochondrial failure and apoptosis [5, 21–25]. The secondary injury processes culminate in dendritic and synaptic degeneration and impaired function in populations such as the mature granular neurons in the dentate gyrus of the hippocampus [26]. Following TBI, the brain is no longer immunologically privileged as inflammatory cells including polymorphonuclear leukocytes and mononuclear phagocytes from the systemic circulation cross the blood brain barrier (BBB) and activate microglia and astrocytes [27]. In rats, early CD3 T-lymphocyte and microglia activation can occur as early as 30 min post injury, reaching maximum levels at 45 min to 3 h after trauma [28]. Activated cells appear histologically as microglial stars with fibrillary process activation with increased expression of GFAP (glial fibrillary acidic protein) and participate in penumbral proliferation and perivascular cuffing [29, 30].

Neuroinflammation post TBI is a balance of cellular infiltration and secretion of inflammatory molecules and growth factors that together exacerbate cell injury, but also are necessary for regeneration (Fig. 15.1) [3, 31, 32]. Using in vitro studies, murine embryonic stem cells mixed with cerebral tissue extracts from injured mice amplified apoptosis but also increased morphologic axonal growth along with increased expression of the intermediate filament nestin and neuronal structural protein MAP2 [33]. In rats, the number of inflammatory M1 vs. anti-inflammatory M2 microglia/macrophage fluctuate following TBI with the rise of the M1 phenotype from the first 2 weeks in the cortex, striatum, corpus callosum correlating with white matter injury via SMI-32 neurofilament staining [34]. This phenotype can be identified both morphologically as well as by surface markers such as FcγRII/III (M1) and CD206 (M2) [35].

Extracellular matrix destabilization occurs along with cellular dysfunction. The concentration of several matrix metalloproteinase (MMP) proteins are elevated post TBI in the brain tissue, CSF and blood samples of patients with poor neurological outcomes [36, 37]. As glial cells in the area of injury die, they form scars that may inhibit the survival of the remaining neurons and function of the local neural network.

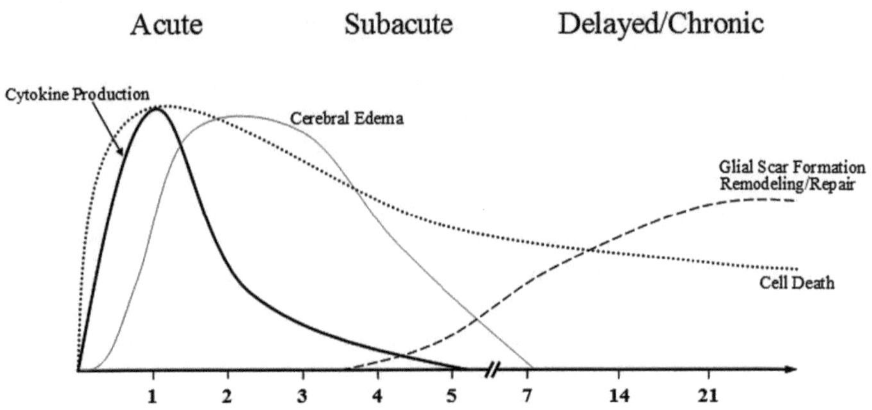

Fig. 15.1 Pathophysiological time course following TBI. (Reproduced from Walker et al. [32])

The secondary injury also affects the cerebral vasculature and includes hypoperfusion, vasospasm and hypoxia. Injured brains are susceptible to hypoxic-ischemic states due to dysfunctional cerebral autoregulation and commonly encountered post traumatic vasospasm, both of which may contribute to further neurologic deterioration [38]. However, tissue perfusion and oxygenation does not always correlate with the severity of injury suggested by computed tomography or ICP measurements [39]. At the tissue level, angiogenesis occurs post injury, and is mediated by both the resident mature endothelial cells via capillary outgrowth but also by bone marrow or peripheral endothelial progenitor cells (EPC) arriving within 24 h in either a primary neovascular or supportive role [40].

Intracranial Dynamics Brain edema, which classically peaks approximately 48–72 h after the initial trauma, is the primary target for neurointensive care. Edema results from both primary cellular death and tissue osmotic dysregulation. The severity of edema is currently monitored clinically by neurologic clinical exam, neuroimaging and intracranial pressure monitoring, the latter of which is used to guide therapy for adequate perfusion and oxygen delivery to the brain. Direct brain tissue oxygen tension monitoring and microdialysis may also be used along with intracranial pressure monitoring as multimodal therapy to optimize cerebral blood flow, oxygen and energy substrate delivery [41].

Three volumes comprise the brain and include the 1300 mL of brain parenchyma, 110 mL of blood and 65 mL of cerebral spinal fluid (CSF) [42]. The Monro-Kellie doctrine describes the intracranial dynamics following injury. Change in one of the three volumes must be compensated by the reduction in another component. Without volume changes, the ICP, which is normally at or below 10 mmHg, rises [43].

The brain parenchyma is mostly water, which is not compressible [44]. Intracranial blood is two-thirds venous and one-third arterial and is controlled via autoregulation and is influenced by arterial pressure, PO_2 and PCO_2. Cerebral perfusion pressure (CPP) defined as mean arterial pressure minus ICP is used as a surrogate for cerebral blood flow, and 50–70 mmHg is the clinical range recommended for optimal blood and oxygen delivery to the brain. CSF is produced at 10–20 cc/hr, and is taken up through arachnoid villi and plays an important role in the transportation of and clearance of metabolites.

Following traumatic brain injury, volume derangements occur to each of the three brain volumes. The brain parenchyma volume increases due to vasogenic and cellular edema [45]. In vitro studies suggest that mechanical stretch can activate cation channels in astrocytes that exacerbate cytotoxic edema [46]. Up to one-third of TBI patients have abnormal cerebral autoregulation. As the result of the loss of cerebral autoregulation, cerebral blood flow and capillary hydrostatic pressure increases, exacerbating edema and ICP. CSF circulation and clearance of metabolites is also impaired following TBI.

Clinical Neurointensive Care The current management of the injured brain following TBI still remains largely supportive with much of the focus on optimizing cerebral perfusion by managing cerebral edema and intracranial pressures [5, 47]. Clinical care guidelines have been developed and are associated with improved outcomes when used by critical care teams [48–50]. However, these practices lack robust randomized control trials. Currently, the management strategy escalates in intensity in a tiered fashion with first tier treatments typically including sedation, establishing an ICP threshold, cerebral perfusion monitoring, neuromuscular blockade, CSF drainage and hyperosmolar therapy [51]. Second tier treatments include hyperventilation, barbiturates for pharmacological coma with electroencephalogram monitoring for burst suppression, hypothermia and surgical decompression.

First tiered therapy begins with sedation, which is a combination of anesthetics and analgesics. Propofol is a commonly used anesthetic used for quick clearance required for frequent neurological tests but can cause myocardial depression as part of an infusion syndrome and may not reduce the cerebral ischemic burden [52]. Norepinephrine and phenylephrine are commonly used pressors to maintain adequate cerebral perfusion pressure because they have the least effect on cerebral vasomotor tone, but overaggressive hypertension may increase the risk of acute respiratory distress syndrome [5]. Fever increases the cerebral metabolic burden and increases the ICP and is treated aggressively with a low index of suspicion for infection and atelectasis. Hyperosmolar therapy includes Mannitol, which is administered at 0.5–1 g/kg and produces effect within 15–30 min. This can be administered every 6 h to a target serum osmolaritiy of 310–320 Osm/L. In addition to lowering the intracranial pressure, Mannitol also has been shown to improve cerebral blood flow (CBF) [53]. 23 % hypertonic saline can be used for hyperacute ICP elevations and for herniation syndromes and can reduce the ICP by up to 50 % within minutes and produce a durable response over hours. Sodium chloride and sodium acetate can be used as a mixture to minimize hyperchloremic metabolic acidosis.

Second tier options include barbiturates and surgical decompression. Barbiturate coma reduces the cerebral metabolic rate as well as the ICP but also has numerous systemic risks, including hypotension, hypocalcemia, hepatic renal dysfunction, sepsis, and ileus. Furthermore, the long term outcome for barbiturate coma is unknown [54, 55]. The use of mild induced hypothermia (body temperature between 32–35 °C) has produced mixed results with some studies suggesting no benefit while others suggesting modest benefit [56]. Hypothermia does exacerbate electrolyte disorders, arrhythmia and infections. The DECRA (Decompressive Craniectomy in patients with severe traumatic brain injury) trial examined outcomes for patients undergoing decompressive craniectomies for elevated intracranial pressures. Although decompression lowers intracranial pressure and decreases length of stay, the investigators cited worse long term outcomes. However, the trial did not include a significant population undergoing craniectomies, namely those with space occupying hematomas or those undergoing unilateral craniectomies [57]. The international multicenter RESCUEicp (Randomised Evaluation of Surgery with Craniectomy for Uncontrollable Elevation of Intra-Cranial Pressure) has been designed to compare surgical decompression versus medical management alone.

Experimental therapies aim to improve the monitoring and therapeutic response to the post injured brain. A controversial issue in neurocritical care is in regard to the practice of directed therapy to maintain ICP levels below 20 mmHg. The multicentered randomized Benchmark Evidence from South American Trials: Treatment of Intracranial Pressures (BEST TRIP) study reported no difference in functional/cognitive outcome, mortality, median ICU stay, and serious adverse events between maintaining ICP at or below 20 mmHg to imaging and clinical examination alone [58, 59]. Proponents of continued ICP and CPP monitoring suggest that the study used practices that varied from established guidelines and did not specifically look into ICP monitor use for the management of intracranial hypertension, thereby limiting external validity and generalizability. ICP is an indicator of injury severity but the operational process of measuring, interpreting and making treatment decisions is complex and outcome measures such as mortality fail to address the specific contribution of ICP directed care [60]. Recent evidence looking specifically at large databases and studies following the Brain Trauma Foundation (BTF) guidelines suggest that ICP monitoring contributed to improved outcomes [61–63].

The debate regarding ICP monitoring and outcome has led investigators to seek additional, multimodal approaches to assess the physiological status of the injured brain. Multimodal monitoring includes brain oxygen monitoring (currently considered a level III clinical practice guideline recommendation) and microdialysis (not yet endorsed as a guideline). Poor short term outcome is associated with hypoxia measured by $pBrO_2$ (partial pressure of oxygen in brain tissue) independent of elevated ICP, low CPP and injury severity [64]. The multicentered Phase II BOOST 2 (Brain Tissue Oxygen Monitoring in Traumatic Brian Injury) trial (ClinicalTrials.gov NCT00974259), estimated to complete in 2014 will evaluate whether $pBrO_2$ levels below the critical threshold of 20 mmHg can be reduced with monitoring, in addition to the evaluation of safety, feasibility and GOSE (Glasgow Outcome Scale-Extended) scores 6 months post injury.

Microdialysis has the ability to provide information regarding the metabolic status of penumbral brain tissue, and includes real-time glucose, lactate, glycerol and glutamate measurements although robust randomized clinical trials have not yet been pursued. Studies have suggested that metabolic derangements can be detected by microdialysis prior to increases in ICP [65]. Investigators have also demonstrated that metabolic crisis, defined by brain glucose <0.8 mmol/L and lactate/pyruvate ratio >25 can occur at an incidence of 74 % despite adequate resuscitation and controlled ICP [66]. In rodent models, lactate levels are elevated at the site of injury after TBI [67]. Furthermore, microdialysis has been used to detect inadequate glucose levels in the brain with the use of strict systemic glycemic control [68].

While the use of $pBrO_2$ monitoring and microdialysis has not been widely adopted in clinical use, these two devices provide investigators valuable tools beyond simple ICP measurements when evaluating emerging therapeutics. Combined microdialysis and positron emission tomography in patients following severe TBI demonstrated that metabolic crisis can even be present without cerebral ischemia as measured by oxygen extraction fraction and cerebral venous oxygen content [69].

Long Term Outcome The long term sequelae of TBI are difficult to measure in terms of outcomes. Outcome measures of mortality and function are often used in clinical trials, but are often nonspecific in regard to the therapeutic strategy under investigation. In recent years, imaging has become an important outcome measure for TBI. Several regions of the brain are sensitive to TBI and include the hippocampus [4, 70, 71]. Long term changes in hippocampal areas such as the dentate gyrus and CA1 can stem from newborn neuron death after TBI which affects memory and causes learning deficits [72, 73]. Corpus callosum volume loss has also been demonstrated in humans and has been topographically correlated with neuropsychological outcomes (Fig. 15.2) [74]. Many studies seek to determine if early clinical intervention can translate to long term improvements. The average

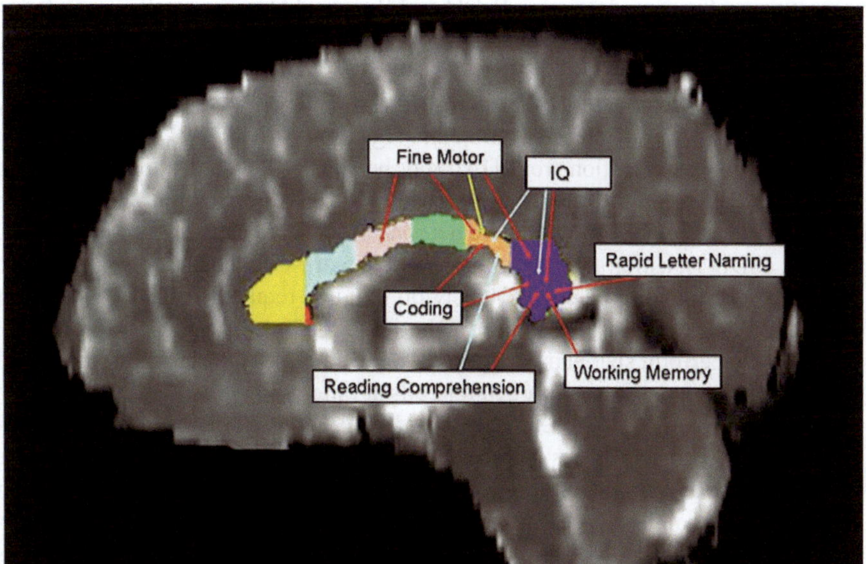

Fig. 15.2 Changes in the macro- and microstructural metrics across the callosum volume seen on diffusion tensor MRI can be topographically correlated to neuropsychological outcomes. The architecture and integrity of the high density fiber tracts of the corpus callosum are frequently altered by TBI. Fractional anisotropy (*red arrows*) was more frequently correlated with motor and cognitive outcomes. The mid-sagittal area (*yellow arrow*) of the isthmus correlated only with fine motor scores. (Reproduced from [74])

ICP during the acute neurointensive has been used as an early target for therapy in hopes that this indicator can correlate with long term outcome. Studies have reported that average ICP during the first 48 h do not correlate with 6 month functional nor neuropsychological outcomes [75]. However, these studies generally do not reflect continuous monitoring trends, number of spikes and waveforms, and are thus likely limited by design.

Endogenous Neurogenesis

As early as the 1990's, investigators observed regeneration in areas of the adult mammalian brain, notably by the subependymal cells located in the subventricular zone (SVZ) following hypophysectomy [76–79]. In 2000, investigators reported the isolation of neuronal progenitor cells from human samples from the dentate gyrus [80]. In 2005, neural stem cells were isolated from the cerebral cortex of rats post TBI [81].

Proliferating and differentiating cells have been identified in the ventricular subependymal and subgranular zones of the hippocampal dentate gyrus in rat brains following TBI [82–86]. Following injury, the cells in the dentate gyrus proliferate to 3–4 fold in number, peaking at 2 days and returning to baseline at 35 days [85–87]. In rat models, TBI has been shown to promote the maturation of immature neurons. In rat models using BrDU and fluorogold tracking, cells generated in the dentate gyrus have been shown to functionally integrate via retrograde axonal tracking [88].

In mice, brain remodeling after controlled cortical injury (CCI) occurs secondary to neuronal and astrocyte proliferation and signaling activity [89]. Studies suggest that Wallerian degeneration from DAI may be sign of reorganization and healing rather than cell death [90]. NPC cells from the SVZ have been tagged and tracked as they migrate and develop into mature neurons in mice brains [91]. Stromal derived factor-1 (SDF-1), a migratory factor involved in extra-CNS homing of regenerative cells has been found to be endogenously present and required for the migration of progenitor cells from the SVZ in post TBI mice [92, 83]. In transgenic mice models, the ability to perform spatial memory tasks was eliminated by ganciclovir-mediated ablation of nestin expressing progenitor cells at time of injury [94].

Neurogenesis has also been seen outside the SVZ and hippocampus. Perilesional local neurogenesis has also been demonstrated in rats, as evidenced by the transitioning of staining from the neuroblast microtubule associated doublecortin (DCX) to the more mature NeuN neural marker [95–98]. Recently neurogenesis has been demonstrated by positive neural stem cell marker staining from samples of perilesional cortices of adult humans undergoing surgical procedures for traumatic brain injury [99, 100]. Unlike rodent models, the migration of progenitor cells from the SVZ or hippocampus to injured sites has not yet been demonstrated in humans and is likely due to the separation of the SVZ from the ependyma by a hypocellular gap [3, 101, 102]. Thus the ultimate translational relevance of the rodent data is not certain.

Therapies

The delayed nature of secondary injury allows for the possibility of potential intervention, and has been a major focus in TBI research [103]. Since the 1990's, preclinical and clinical trials have struggled to bring novel molecular therapeutics to the bedside [104, 105]. The complexity of the post traumatic brain for targeted treatment can be demonstrated by preclinical studies of time sequence gene expression in post TBI mice, which included more than 80 genes and 24 expression sequence tags for transcription factors, signal transduction genes and inflammatory proteins [106]. Unfortunately, no specific treatment has been shown to halt or reverse neuronal death following TBI. Since the early 2000's, investigators have explored using stem cells as potential therapy in various forms and routes of delivery.

Targets

Excitotoxicity Glutamate has been identified as major neuroexcitatory amino acid that exacerbates cell injury, such as astrocyte swelling following traumatic brain injury [107–109]. Elevated levels can be identified in the CSF post TBI [110]. Research regarding glutamate antagonism began in the 1990's [111–113]. Recently, the adenosine A2A receptor, found in cells such as bone marrow derived cells has been associated with increased glutamate levels following TBI. Post TBI, glutamate levels were reduced by adenosine A2A receptor inactivation or in knockout mouse models, along with reduced proinflammatory cytokines such as IL-1 and TNFα [114]. Valproate is an antiepileptic drug that has multiple targets including the GABA, sodium channel, glycogen and histone pathways, and has been demonstrated in rats to protect the BBB, reduce neural damage and improve cognition [115]. Topiramate is another antiepileptic drug that has been clinically used to reduce glutamate release after TBI in humans and a Phase II clinical trial designed to determine whether this drug can prevent epilepsy post injury is currently ongoing [116]. Investigators have even explored the benefit of caffeine and alcohol with their potential neuroexcitatory modulating mechanisms [117, 118].

Oxidative Stress Neuroprotection with improvements in behavior has been demonstrated in rodent models for antioxidants such as deferoxamine, selenium, alpha-Phenyl-tert-N-butyl nitrone (PBN), and NXY-059 [119–122]. More recently, (-)-Epigallocatechin-3-gallate (EGCG) in green tea has been shown in post TBI rats to preserve neuronal stem cells [123]. In mice, CAPE (Caffeic phenol acid ester) improved the BBB via an antioxidant oxidant pathway [124]. Edaravone, an antioxidant and neuroprotectant scavenges NO, protects the BBB and reduced CA3 neuronal loss, apoptosis and astrocyte/glial activation in rats [125–131]. When given to a small number of TBI patients (n = 17), jugular bulb measurements demonstrated decreased reactive oxidative species [125]. Edaravone has been investigated in controlled trials for stroke but the optimal dose and therapeutic window has not yet been established [132]. Determining the optimal dosage and therapeutic window will undoubtedly be crucial in the design of a human TBI trial.

Blood Brain Barrier The BBB has been a target of interest in TBI due to its participation in the neuroinflammatory process and the development of cerebral edema. Post injury supplementation of the endogenously expressed cyclophilin A (a protein involved in endotheial cell activation and inflammation) reduced BBB permeability 24 hr post injury in rats [133]. TIMP metallopeptidase inhibitor 3 (TIMP-3) is a MMP inhibitor that stabilizes and improves BBB integrity in animal models [134, 135]. Progesterone has been demonstrated in preclinical studies to promote endothelial progenitor cell (EPC) mediated vascular remodeling, downregulate the inflammatory cascade and decrease cerebral edema [136]. Sulforaphane (isothiocyanate) in cruciferous vegetables attenuate aquaporin-4 (AQP4) loss and improves the BBB [137]. Cannabinoid type 2 receptor agonists have been shown in mice models of TBI to improve BBB permeability and reduce macrophage/microglial activation and neuronal degeneration [138]. Citicoline, a naturally endogenous compound found to be effective in preclinical trials of BBB protection was not found to significantly improve function nor cognitive outcomes in human Phase III trials [139, 140].

Signaling Pathways Many signaling pathways have been implicated in TBI. Erk pathway has been described as an important extracellular signal pathway in preclinical models of TBI [141]. Animal models have associated increased transcription factors such as CREB (cAMP) following TBI with changes in behavior [142]. Strategies involving histones can preserve Akt signaling, decreasing apoptosis and has been shown to increase nestin expression [143, 144]. Phosphodiesterase targeting strategies have also been explored in TBI research. PDE-4 treatment targeting the cAMP pathway improves histopathological outcomes and decreases inflammation [145, 146]. The transforming growth factor-beta (TGF-β) pathway has been implicated in mice TBI models to be linked with the Runt-related transcription factor-1 (Runx1) to promote activation and proliferation in the dentate gyrus [147]. Progesterone has been shown to regulate apoptotic protein expression in the dentage gyrus in rats but also has been to increase vasculogenesis [136, 148, 149]. Despite preclinical success, the Phase III ProTECT trial using intravenous progesterone in the acute post-TBI period was halted for futility in late 2013 (NCT00822900). The international SyNAPSe study is another Phase III progesterone clinical trial for acute TBI trial that finished enrollment in late 2013 with results now pending (NCT01143064).

Growth Factors Growth factors are an attractive target to be either endogenously augmented or delivered exogenously to aid in the regenerative process. Intraventricular infusion of fibroblast growth factor (FGF) intro rats post TBI has been shown to increase neurogenesis in the SVZ [150]. Nerve growth factor (NGF) has been used to promote astrocyte migration and is associated with reduced apoptosis following TBI in rats. Intraventricular delivery of epidermal growth factor (EGF) has been shown to be neuroprotective in rats [151]. Vascular endothelial growth factor (VEGF) was shown to increase de novo hippocampal neurogenesis more than proliferation of resident neuroblasts in rats [152]. In mice intraventricular delivery augmented neurogenesis and angiogenesis and reduced post TBI lesion volume [153]. Gold salts have been used via local perilesion injections to reduce inflammation and apoptosis via VEGF and FGF [154, 155]. The neutrophic/mitogenic protein S100B made by astrocytes has been demonstrated to increase hippocampal neurogenesis as well as improve cognition in post TBI rats but has classically been correlated with poor outcomes when found in human CSF [156, 157]. Recently, P75, a small molecule ligand has been shown to bind neurotrophin receptors on neuronal precursors and enhance their regenerative properties in rats [158]. A randomized double blinded pilot study by Liu and colleagues is currently underway investigating repeated intranasal delivery of NGF for acute TBI (NCT01212679).

Neuronal Architecture Cyclosporin A was shown in the 1990's to protect axons following TBI in rats by inhibiting calcium induced mitochondrial damage [159, 160]. Preinjury inhibition of calpain, a calcium influx mediated cysteine protease preserves axonal integrity following TBI in rats [161]. Inhibitory myelin molecules such as Nogo-A has been implicated in axonal sprouting post injury, but monoclonal antibodies against this protein did not appear to act through sprouting nor cell loss protection, but still improved cognition in animal models [162]. Myelin-associated glycoprotein is another inhibitor of axonal growth, and investigators found improved sensorimotor function following intraventricular administration in rats [163]. Cyclosporin is currently being investigated in a Phase II trial.

Post-TBI Neuroinflammation The neuroinflammatory state following traumatic brain injury is complex and likely an evolving process of dynamic cytokine expression. For example, TNF has been shown to be toxic to the neuronal stem cell proliferation phase, but not during differentiation. In fact, IFNγ enhances neuronal stem cell differentiation and neurite outgrowth [164]. This may explain why Anti-TNF or anti-IL-6 strategies have not been found to improve acute edema or motor or cognitive function in rats [165]. In preclinical studies, strategies to neutralize IL-1 have been shown to improve edema, tissue loss and cognition in mice [166, 167]. Studies have shown the local mileu in the rat brain in the first 48 h following TBI is highly proinflammatory, with elevated levels of IL-1b, IL-6, and TNFα, along with the presence of microglia and macrophages (Figs. 15.3 and 15.4) [168].

The proinflammatory environment can serve as a target of cell therapy but may limit drug efficacy or even exacerbate injury. Purely pharmacologic anti-inflammatory strategies may also interfere with complex reparative inflammatory pathways. Bortezomib a selective proteasome inhibitor used in multiple sclerosis was found to have neuroprotective properties in animal models and was associated with decreased NFkB expression [169]. TSG-6, a multifunctional immunomodulator reduces neutrophil extravasation and BBB leakage in animal models [170]. Ibuprofen has been

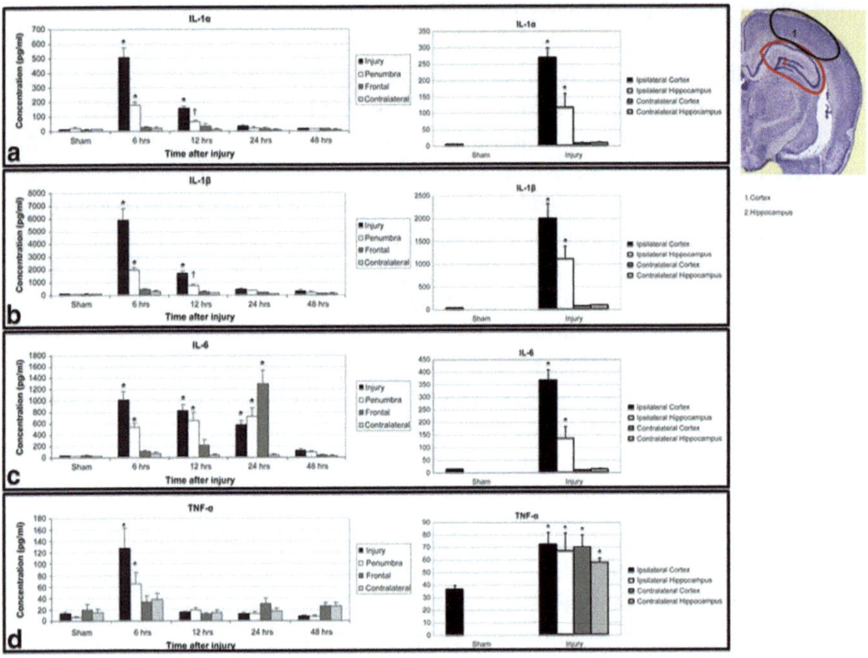

Fig. 15.3 Locoregional cytokine response following TBI in a rat model. Proinflammatory cytokines can be seen to be elevated in the brain as early as 6 h and lasting up to 24 h. IL-6 can be detected at high levels in the frontal lobes away from the impact site at 24 h. Similarly, TNF-α is elevated in a global fashion 6 h after injury compared to the local increase in the other cytokines. Note—brain section taken from Paxinos and Watson 2005. (Reproduced from Harting et al. [168])

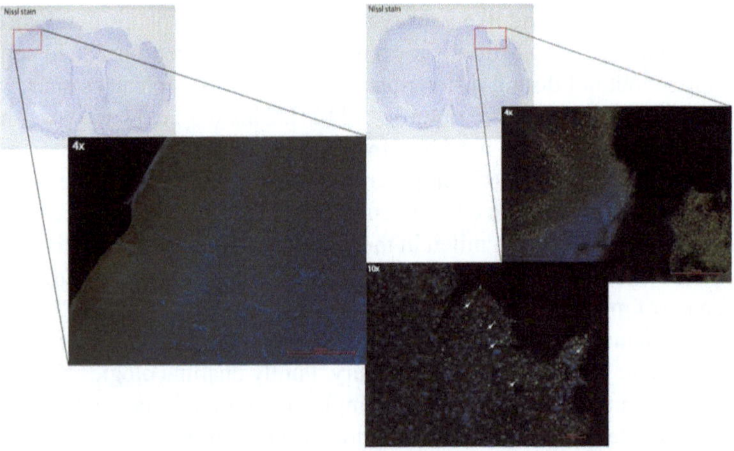

Fig. 15.4 TBI exposes the otherwise immune privileged cortex to infiltrating macrophages and microglia that participate in the neuroinflammatory response. Brain tissue sections were isolated after TBI, and incubated anti-CD68 antibodies (*green*) to identify macrophages/microglia. Numerous CD68 positive cells (not seen in the contralateral hemisphere) were identified in the area of injury. (Reproduced from Harting et al. [216])

demonstrated to improve outcome of transplanted stem cells in animal models [171, 172]. COX-2, a protein involved in the generation of inflammation mediating prostaglandins was found to be exclusively expressed in rat neurons but not astroglia and may have protective roles in only certain neurons [173, 174]. The administration of COX-2 inhibitors in a rat TBI model was found to improve cognition but worsen motor function. Thus the use of selective anti-inflammatory drugs may not be specific enough to truly target neuroprotection. Anti-inflammatory drugs such as rolipram may unfortunately increase bleeding in animal models [175]. Inhibition of single proinflammatory cascades of overlapping/redundant signal transduction pathways has not proven successful.

Neurovasculature Investigators have also explored strategies of optimizing the neurovascular niche for resident stem cell activation and function in response to hypoxia following TBI. The hypoxia induced factor (HIF-1α) pathway has downstream components that include VEGF, SDF-1, brain derived neurotrophic factor (BDNF), tyrosine kinase receptor TrkB and associated co-receptor Nrp-1 as well as chemokine receptor CXCR4 and nitric oxide (NO) [176]. NO donor DETA/NONOate delivered via intraperitoneal injection in rats, improved proliferation, survival, and differentiation of resident neuronal stem cells [177]. Statins have also been shown to induce angiogenesis, reduce neurologic deficits, increase neuronal survival and hippocampal synaptogenesis induced angiogenesis in rats [178, 179]. In human studies, statin therapy for 10 days following moderate to severe TBI was found to reduce TNFα levels at 72 h post injury as well as disability scores at time points up to 6 months [180]. Erythropoietin is being investigated in a Phase III trial as a subcutaneous injection in patients with severe TBI under the hypothesis that secondary injury can be improved through optimizing oxygen delivery.

Biomarkers

TBI researchers have used biomarkers ranging from genes to clinical bedside measurements to evaluate injury pathophysiology as well as treatment outcomes [181, 182]. Gene expression via RNA production associated with astrocytes, phagocytes, microglia and immune-reactive cells have been described as potential biomarkers in animal models [183–188].

Biomarkers can help diagnose specific mechanisms of injury, such as BBB compromise following blast injury [189]. Elevated serum N-acetyl-asparate has been associated with DAI as well as continued secondary injury in post TBI rats [190, 191]. Microdialysis has also been used to identify axonal injury using levels of total tau and amyloid beta proteins [192, 193].

Metabolic intermediates and breakdown products have been investigated as potential biomarkers. Glycerol has been used as a marker of cell degradation and lysis [194]. Elevated plasma bilirubin, a byproduct of heme oxygenase mediated breakdown of heme and acts at free radical scavenger in mice and humans [195, 196]. Some investigators have advocated a systems biology approach for biomarker discovery, using protein network interactions as a way to screen for potential markers [197]. Common data elements have been advocated for biomarkers to improve the preclinical and clinical investigation of promising therapeutics [198, 199]. Glial to neuron ratio GFAP/ubiquitin ratio correlates to the degree of TBI on imaging such as focal versus diffuse injury patterns [200]. Biomarkers may also help identify certain cellular injury patterns that may not be identified on neuroimaging when evaluating outcomes of potential therapeutics [201].

Biomarkers have also been used to prognosticate outcomes. In human studies, increased serum IL-6, ceruloplasmin and copper levels were associated with eventual elevations in ICP following TBI [115, 202, 203]. High constant levels of CSF alpha synuculein in patients with ventriculostomies at 8 days has been associated with poor outcome [204]. Spectrin breakdown products in the CSF of patients with TBI peak at 2–3 days and indicate cytoskeletal injury and may predict injury severity and mortality [205, 206]. Studies suggest that biomarkers including serum and CSF GFAP, and CSF SBDP145 can improve the prognostic ability of scoring systems such as IMPACT [207].

Cell Therapy

In the 1990's, Povlishock and colleagues suggested that the multiple deleterious mechanisms activated by TBI may require multiple therapies [208]. Thus single agents may be a limited and naïve strategy, even when combined. Investigators in the 2000's began to explore the potential for stem cell therapy in neurotrauma and for CNS regeneration [209–212]. Cell therapy is an appealing therapeutic option as cells are capable of sensing and responding to environmental signals to potentially target multiple mechanisms in a sustained manner in a system where regeneration has classically thought to be limited [213]. Strategies have included both endogenous and exogenous stem cells [214]. In addition to NSC, numerous stem cell populations such as embryonic (ESC), hematopoietic (HSC) and mesenchymal (MSC) have been explored as potential cell therapy options [215–217]. In the following three decades, these cells have been applied in protocols to replace, repair, or enhance function of the post traumatic brain [218–222].

Systemic Delivery Systemic delivery of stem cells has been explored using both intraarterial and intravenous routes. A number of stem cell types, including neuronal, mesenchymal and fetal associated cells have been utilized for systemic delivery. MSCs and NSCs have been delivered using an intraarterial strategy via the internal carotid artery for TBI in animal models but has not found popularity in human trials due to concern of ischemic embolic events (Fig. 15.5) [221, 223–225]. Intravenous delivery of cells encounter the pulmonary first pass effect described by Fischer et al (Fig. 15.6) [221, 226, 227]. The majority of mesenchymal cells are sequestered in the lung and eventually cleared by the spleen and rarely reach the brain. In fact, less than 1 % of green fluorescent protein labeled mesenchymal

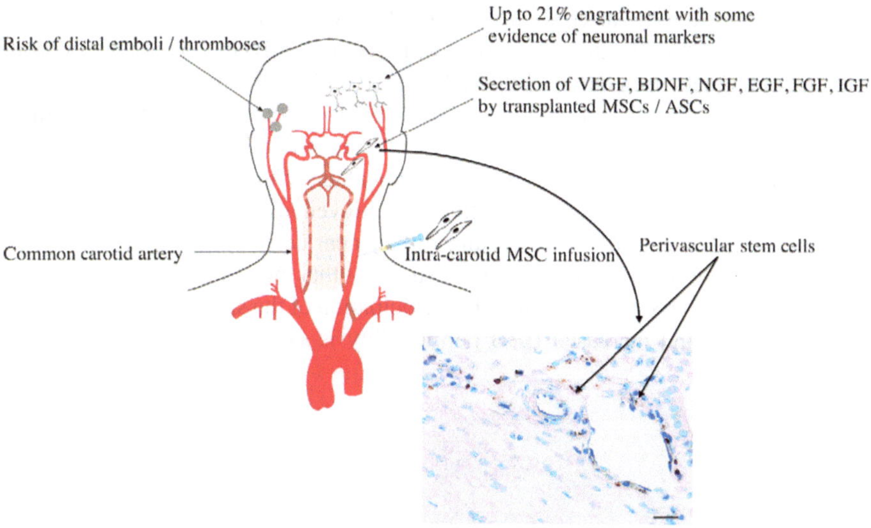

Fig. 15.5 Intraarterial introduction of stem cells. Intraarterial delivery of cells increases potential engraftment percentage due to the avoidance of the sequestration and clearance issues associated with intravenous delivery. However, the risk of distal emboli and thrombosis from intraarterial delivery can be devastating and thus this route has not been widely adopted. (Reproduced from Walker et al. [221])

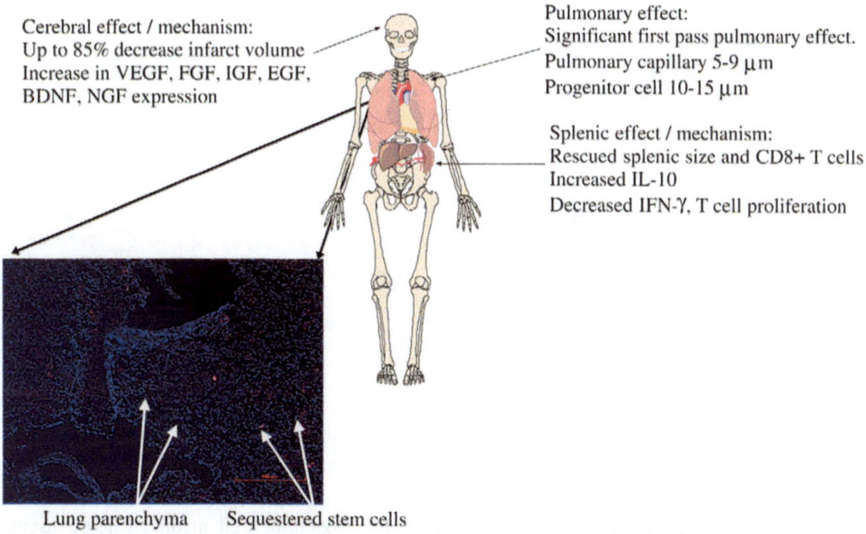

Fig. 15.6 The first pass pulmonary effect plays a significant role in intravenous strategies of cell therapy. Cells labeled with quantum dots (*stained red*) delivered intravenously are sequestered in the pulmonary capillaries. Less than 1 % of green fluorescent protein labeled mesenchymal stem cells reached the arterial circulation. (Reproduced from Walker et al. [227])

stem cells reached the arterial circulation. However, certain cell types have been reported to localize in the brain after intravenous delivery. Investigators using superparamagnetic iron oxide (SPIO) labeling of endothelial progenitor cells (EPC) delivered intravenously into rats have been able to detect EPCs in areas of injury on MRI and observed associated improvements in cerebral perfusion on computed tomography [228]. Nevertheless, both preclinical and clinical studies have demonstrated improvements in the brain following intravenous cell therapy suggesting a therapeutic effect that does not require direct local delivery [229, –231]. Rodent studies demonstrating BBB preservation suggests that the effect of intravenous cell therapy to the injured brain requires the participation of the spleen (Fig. 15.7) [232]. Cortical injury and increased BBB permeability is associated with a decrease in

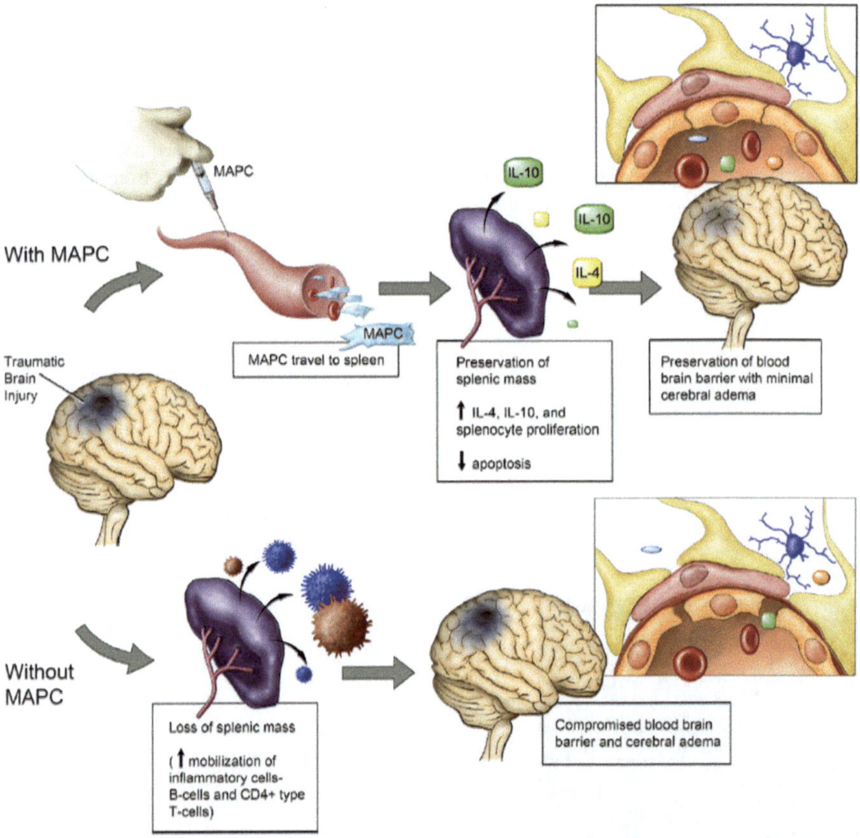

Fig. 15.7 Proposed mechanism of Multipotent Adult Progenitor Cell mediated neurovascular protection post-TBI and the splenic interaction. TBI results in decreased splenic mass as well as increased blood brain barrier permeability. The administration of Multipotent Adult Progenitor Cells increases CD4+ splenocyte proliferation and the production of anti-inflammatory cytokines resulting in the preservation of the cerebral microvasculature and the blood brain barrier. (Reproduced from Walker et al. [232])

splenic mass. Following the intravenous delivery of Multipotent Adult Progenitor Cells (MAPC), CD4+ splenocyte proliferation increased, along with the production of anti-inflammatory cytokines.

Fetal associated progenitor cells are also candidates for cell therapy and include umbilical cord blood and Wharton's Jelly as well as components of the placenta. Cultured human umbilical cord blood (HUCB) cells have been demonstrated to produce cytokines and chemokines that include IL-8, MCP-1 and IL-1a. Intravenous HUCB cells reduced neurologic deficits in rats after TBI [223, 234]. Prelabeled human fetal neural progenitors have been delivered both intravenously and locally into TBI rats with the therapeutic effects likely due to angiogenesis and reduced astrogliosis rather than cell replacement [235].

The systemic delivery of stem cells has been safely used in various disease models. In preclinical and clinical trials, MSC systemic toxicity has been well studied. Adenosine A2A receptors found on bone marrow-derived cells has been implicated in mouse TBI models to involve in glutamate and inflammatory cytokine release, which was shown to lead to acute lung injury [236]. Thrombosis in stem cell transplantation has been described to cause sinusoidal obstructive syndrome of the liver (SOS) in 50–60 % of patients, with the severe form causing up to 84.3 % in mortality [237]. However, infusional toxicities associated with autologous therapy have not been demonstrated in human trials, particularly in a Phase I pediatric TBI safety trial using intravenously delivered BMMNC's [230].

Replacement Strategies/Direct Delivery Stereotactic transplantation of various stem cell types, including neuronal, mesenchymal, embryonic as well as induced pluripotent (IPS) stem cells into rodent brains have been shown to rescue CA3 neurons, improve cognition and neuromotor function (Fig. 15.8) [238–245]. Studies in mice where NSC injections were delivered into different ipsilateral and contralateral locations suggested that local delivery or migration

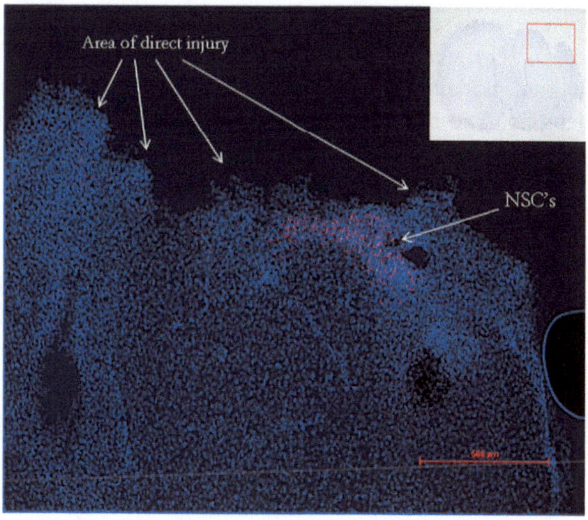

Fig. 15.8 Immunohistochemistry demonstrates the persistence of NSCs 48 h after local delivery to the penumbra of the cortical injury in a rat model of TBI. (Reproduced from [240])

of cells may not be necessary to produce functional improvements [238, 246]. In practice, the transplantation of cells directly into the injured brain has been challenging. The neuroinflammatory environment following TBI has been shown to be hostile, leading to protracted effectiveness and early loss of exogenous cells [247–250].

Neural stem cells have even been described as biologic minipumps used for their trophic, migratory and secretory ability [251, 252]. Transplanted human NSCs have shown to produce glial derived neurotrophic factor (GDNF) resulting in axonal growth following fluid percussion TBI in rats [253]. Gene therapy such as the use of NGF transfection has also been explored to further augment the production of growth factors by modified stem cells being introduced into animal TBI models [241, 254–257].

Paracrine and Systemic Effects MSCs have been proposed as an ideal candidate for TBI directed cell therapy due to the ability of these cells to have paracrine and systemic regenerative and anti-inflammatory effects. In vitro co-culture studies demonstrated improved NSC proliferation and expression of GFAP towards astrocyte differentiation. Human adipose tissue derived MSCs have also been demonstrated to support native NSCs in vitro [258]. Direct injection of MSCs into the injured cortex in mice promoted anti-inflammatory cytokine expression [259]. Intracranial delivery of MSCs has been applied to rodent TBI models with improved celluar as well as functional outcomes [260–263]. Contralateral cerebroventricular introduction of human umbilical cord MSCs in cyclosporine A immunosuppressed mice have been shown to increase lesional BDNF levels, decreased glial scar, improved non-phagocytic to phagocytic macrophage ratio and improved neurologic function [263]. Intrathecal MSCs have been shown in rat models to enhance the neuroprotection of NSCs via direct cell contact [264].

Embryonic stem cells (ESC) have also been used in direct transplantation. Mouse ESCs have been shown to differentiate into GABAnergic neurons as well as astrocytes [265]. However, the use of ESCs in preclinical trials has not been widely adopted due to reports that murine ESCs transplanted into rats produced tumors [266].

Direct transplantation appears to be an attractive delivery method in preclinical animal models. However, translating delivery to human applications can be challenging. The difference between the human brain post injury and animal models has been well documented. Animal heterogeneity also exists in preclinical trials [267]. Most preclinical animal models involve a unifocal injury. Human TBI injury patterns are often multifocal, and present a challenge for the local delivery of therapeutics. Green fluorescent protein labeled neural progenitor cells injected into the contralateral ventricle of mice 1 week post injury have been shown to migrate to the injured site and were detectable up to 3 months [268]. This migratory chain however, has not been demonstrated in human studies [101, 102]. The fact that TBI has multiple foci of injury if the cells don't migrate makes direct transplantation an unattractive approach. NSCs injected locally in other human trials demonstrate little migratory capacity.

Probable Pleiotropic Mechanisms of Action of Cell Therapy

Mesenchymal Stem Cells (MSC) have been studied as a delivery mechanism for therapeutic molecules in both systemic and paracrine strategies rather than just cell replacement or transdifferentiaton [269]. MSCs have been isolated from peripheral blood, bone marrow and even adipose tissue. As an example of stem cells responding to their inflammatory environment, human bone derived MSCs cultured with extracts of TBI were induced to produce growth factors [270]. Intravenous delivery of human MSCs in rats has been shown to increase cerebral levels of NGF, BDNF, NT-3 (neurotropin-3) early after TBI and reduced apoptotic activity [271, 272]. Bone marrow derived MSCs transfected with BDNF, then given intravenously to rats showed increased BDNF in the CSF and had improved immune tolerance [273]. MAPCs used in rodent models of TBI have been shown to interact with splenocytes and T regulatory cells in the promotion of M2 regenerative microglia and the down-regulation of pro-inflammatory macrophages and cytokines (Fig. 15.9) [274, 275]. The secondary injury associated with TBI leads to invariable brain volume loss over time. However, conventional MRI studies on children receiving intravenous autologous BMMNCs have demonstrated brain volume preservation up to 6 months following injury (Fig. 15.10) [230].

Transdifferentiation Non-neuronal progenitor cells had been proposed to act by means of transdifferentiating into neuronal cells after in vitro and in vivo studies demonstrated expression of neuronal markers. Recent studies have suggested that in vitro transdifferentiation may not actually be occurring in vivo and that the expression of neuronal markers are likely due to induced cellular stress [276]. Cells delivered intravenously rarely localize to the brain. Investigators introducing untransdifferentiated versus transdifferentiated umbilical mesenchymal cells directly into the area of injury in a rat TBI model found that the former was able to improve cognitive function and tissue morphology as well as increase neurotrophin expression [248]. Thus, the current pre-clinical strategy is to use progenitor cells, such as bone marrow derived MSCs introduced via direct or intravenous transplantation to protect native cells by creating a more favorable regenerative environment rather than inducing neuronal differentiation [277].

Other Strategies to Augment Engraftment and Function

Bioengineered constructs have been applied to help cell delivery, targeting and survival. NSCs can migrate along non-stereotypical routes for great anatomic (but short absolute distances) in immature animal models, a process that may be augmented with laminin or fibronectin based scaffolds or biodegradable nano-fibers [278, 279]. Investigators have described the concept of a biobridge that is initially comprised of transplanted stem cells that evolve to a tract with an abundance of

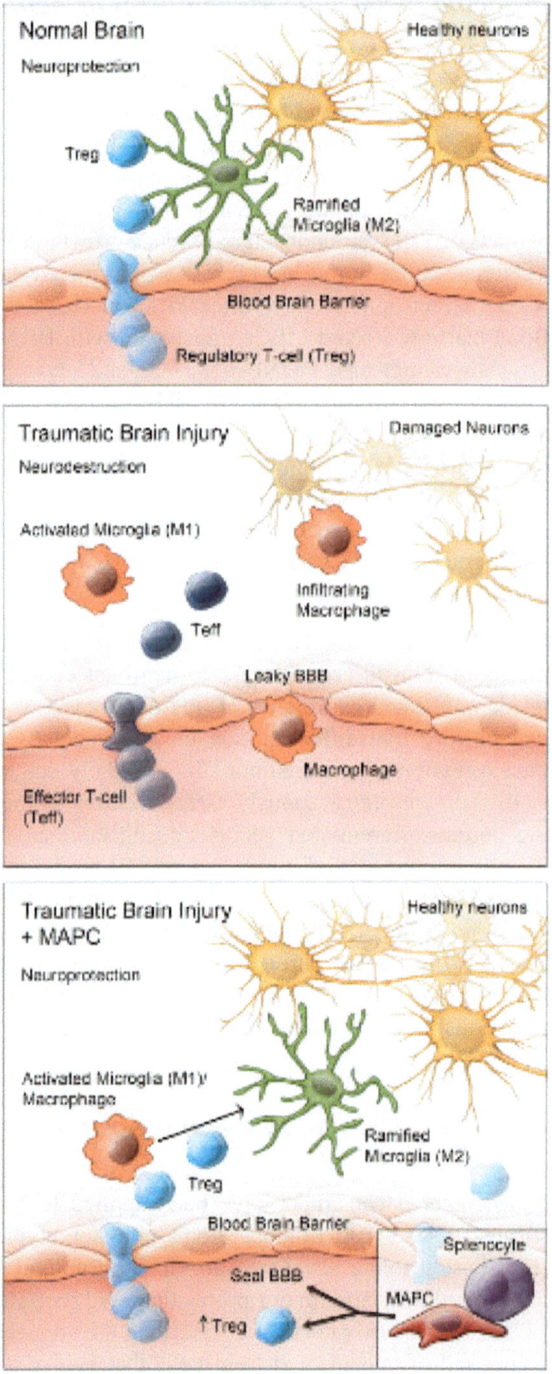

Fig. 15.9 The neuroinflammatory response following TBI involves the infiltration of neurodestructive activated M1 microglia and effector T cells across a compromised blood brain barrier. Multipotent adult progenitor cells (MAPCs), a type of MSC interacts with splenocytes, and increases anti-inflammatory cytokine production, promoting regulatory T cell activity and proliferation, which restores the blood brain barrier and shifts the neuroinflammatory state in the brain towards a M2 regenerative state. (Reproduced from [275])

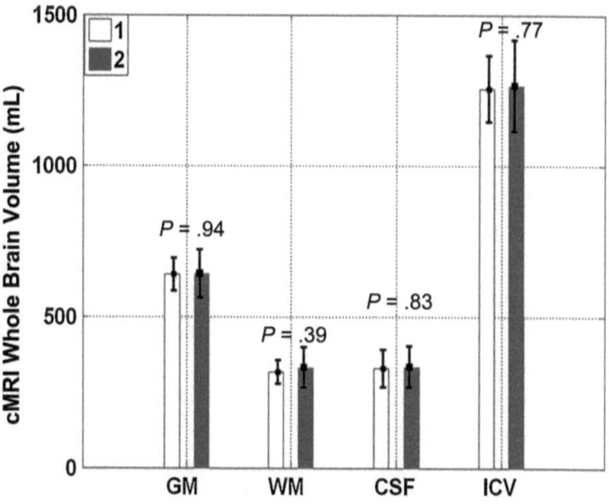

Fig. 15.10 Conventional MRI (cMRI) of various brain volumes at 1 month (*white bar*) and 6 months (*gray bar*) following severe TBI in 10 pediatric patients receiving intravenous autologous bone marrow derived mononuclear cells (BMMNC). Post TBI brains are expected to undergo volume loss, but those receiving BMMNCs demonstrated preservation of brain volume over 6 months following injury. Gray matter (*GM*), White matter (*WM*), Intracranial volume (*ICV*). (Reproduced from [230])

matrix metalloproteinases, which facilitates the migration of other cells [280]. NSC induced production of connexion 43 may alter the surrounding tissue architecture [281]. NSCs interact with brain microvascular endothelial cells (BMECs) and have been used to form a model of the BBB in vitro [282]. Scaffolds can be used in vitro to model the in vivo interaction between potential cell therapies with the extracellular matrix as well as co-cultured astrocytes [283]. Scaffolds have been used to enhance the therapeutic potential of progenitor cells [227, 284–287]. FDG labeled human MSCs were placed into collagen scaffolds that are then implanted into injured rat brains. This strategy was shown to improve cell survival and energy uptake [288].

The priming and genetic enhancement of cells being used as therapeutic agents has also been investigated. GDNF enhanced neuroprogenitor cells delivered into the brain exhibit improved neuronal differentiation in rats with enhanced cognitive recovery [289]. Primed human fetal neural stem cells with agents such as progesterone improved cognition in rats when delivered post TBI [290]. Hypoxia has been shown to prime MSCs to produce growth factors such as VEGF when the cells are delivered intravenously [291, 292]. Pre-clinical animal studies have also looked at dual therapy of using intravenous G-CSF in combination with intravenously delivered human umbilical cord or bone marrow derived MSCs to maximize treatment efficacy [229, 293].

Although the intravenous route does not appear to result in the delivery of cells into the post TBI brain, strategies have been developed to increase the intracranial localization. MRI guided ultrasonic disruption of the BBB has been used as a potential strategy for the targeted local delivery of stem cells [294]. An encapsulated approach has also been used to improve biodelivery [295].

Conclusion

TBI remains challenging in both the understanding of the pathophysiology and the development of therapeutics. As the injured brain progresses from neuroinflammatory degradation to regeneration, numerous molecular and cellular processes occur that can influence outcome. While many pharmaceutical agents have been developed and tested in preclinical trials, their contribution to the complex and evolving inflammasome following TBI has not materialized in any successful clinical trials. The advancement of preclinical studies for TBI using cell therapy has been slow due to the heterogeneity of injury models, outcome measures and assays. Thus, the International Society for Cellular Therapy (ISCT) recently has recently advocated validating and strengthening standardized assays to improve the reproducibility and consistency of such data [296].

Cellular therapy offers a pharmaceutical bioreactor that can sense and interact with the inflammasome. Studies have demonstrated that inflammatory M1 to regenerative M2 phenotypic shifts can be initiated by progenitor cells. Once delivered, the fate of various stem cells needs to be verified in preclinical models. Tracking of transplanted stem cells is challenging due to issues such as auto-fluorescence and phagocytosis [250]. MicroPET and SPIO MRI can be used to monitor transplanted cells [297–300] and may be translated to humans trials [301].

Translating cell therapy from preclinical to clinical trials requires many questions to be answered. Off the shelf heterologous cells have been proposed for TBI therapy. Exogenous cells introduced either locally or systemically have not shown to engraft or survive, and thus the optimal dosing time and regimen remains unanswered. In rats, investigators have suggested that functional and behavior improvements are maximized when bone marrow derived MSCs are delivered at 7 days after injury into the corpus callosum [240, 302, 303]. Human cytokine assays demonstrate that the systemic window of therapy against pro-inflammatory cytokines can be beyond 48–72 h.

Finally, any therapy for TBI can only be deemed successful if clinical trials can demonstrate improvements in outcome. Neuroimaging has demonstrated promising volume preservation with cell therapy, but the impact of these findings to cognitive and functional outcomes are unknown and require further understanding about the role of cell therapy in the regenerative process following TBI. In the short term, a multimodal approach that includes drugs such as neurostimulants, neurorehabilitation and immunomodulatory cell therapy likely offers the best strategy to maximize recovery potential.

References

1. Rose VL. NIH issues consensus statement on the rehabilitation of persons with traumatic brain injury. Am Fam Physician. 1999;59(4):1051–3.
2. Thurman DJ, Alverson C, Dunn KA, Guerrero J, Sniezek JE. Traumatic brain injury in the United States: a public health perspective. J Head Trauma Rehabil. 1999;14(6):602–15.

3. Richardson RM, Singh A, Sun D, Fillmore HL, Dietrich DW, 3rd, Bullock MR. Stem cell biology in traumatic brain injury: effects of injury and strategies for repair. J Neurosurg. 2010;112(5):1125–38. doi:10.3171/2009.4.JNS081087.
4. Williams S, Raghupathi R, MacKinnon MA, McIntosh TK, Saatman KE, Graham DI. In situ DNA fragmentation occurs in white matter up to 12 months after head injury in man. Acta Neuropathol. 2001;102(6):581–90.
5. Frattalone AR, Ling GS. Moderate and severe traumatic brain injury: pathophysiology and management. Neurosurg Clin N Am. 2013;24(3):309–19. doi:10.1016/j.nec.2013.03.006.
6. Alessandri B, Nishioka T, Heimann A, Bullock RM, Kempski O. Caspase-dependent cell death involved in brain damage after acute subdural hematoma in rats. Brain Res. 2006;1111(1):196–202. doi:10.1016/j.brainres.2006.06.105.
7. Clark RS, Chen J, Watkins SC, Kochanek PM, Chen M, Stetler RA, Loeffert JE, Graham SH. Apoptosis-suppressor gene bcl-2 expression after traumatic brain injury in rats. J Neurosci (the official journal of the Society for Neuroscience). 1997;17(23):9172–82.
8. Zhang X, Chen J, Graham SH, Du L, Kochanek PM, Draviam R, Guo F, Nathaniel PD, Szabo C, Watkins SC, Clark RS. Intranuclear localization of apoptosis-inducing factor (AIF) and large scale DNA fragmentation after traumatic brain injury in rats and in neuronal cultures exposed to peroxynitrite. J Neurochem. 2002;82(1):181–91.
9. Davidsson J, Risling M. A new model to produce sagittal plane rotational induced diffuse axonal injuries. Front Neurol. 2011;2:41. doi:10.3389/fneur.2011.00041.
10. Czeiter E, Pal J, Kovesdi E, Bukovics P, Luckl J, Doczi T, Buki A. Traumatic axonal injury in the spinal cord evoked by traumatic brain injury. J Neurotrauma. 2008;25(3):205–13. doi:10.1089/neu.2007.0331.
11. Gottesfeld Z, Moore AN, Dash PK. Acute ethanol intake attenuates inflammatory cytokines after brain injury in rats: a possible role for corticosterone. J Neurotrauma. 2002;19(3):317–26. doi:10.1089/089771502753594882.
12. Woolf PD, Cox C, Kelly M, McDonald JV, Hamill RW. Alcohol intoxication blunts sympatho-adrenal activation following brain injury. Alcohol Clin Exp Res. 1990;14(2):205–9.
13. Kobori N, Clifton GL, Dash PK. Enhanced catecholamine synthesis in the prefrontal cortex after traumatic brain injury: implications for prefrontal dysfunction. J Neurotrauma. 2006;23(7):1094–102. doi:10.1089/neu.2006.23.1094.
14. Redell JB, Dash PK. Traumatic brain injury stimulates hippocampal catechol-O-methyl transferase expression in microglia. Neurosci Lett. 2007;413(1):36–41. doi:10.1016/j.neulet.2006.11.060.
15. Kobori N, Hu B, Dash PK. Altered adrenergic receptor signaling following traumatic brain injury contributes to working memory dysfunction. Neuroscience. 2011;172:293–302. doi:10.1016/j.neuroscience.2010.10.048.
16. Zygun DA, Kortbeek JB, Fick GH, Laupland KB, Doig CJ. Non-neurologic organ dysfunction in severe traumatic brain injury. Crit Care Med. 2005;33(3):654–60.
17. Kalsotra A, Zhao J, Anakk S, Dash PK, Strobel HW. Brain trauma leads to enhanced lung inflammation and injury: evidence for role of P4504Fs in resolution. J Cereb Blood Flow Metab (official journal of the International Society of Cerebral Blood Flow and Metabolism). 2007;27(5):963–74. doi:10.1038/sj.jcbfm.9600396.
18. Kalsotra A, Turman CM, Dash PK, Strobel HW. Differential effects of traumatic brain injury on the cytochrome p450 system: a perspective into hepatic and renal drug metabolism. J Neurotrauma. 2003;20(12):1339–50. doi:10.1089/089771503322686139.
19. Chu W, Li M, Li F, Hu R, Chen Z, Lin J, Feng H. Immediate splenectomy down-regulates the MAPK-NF-kappaB signaling pathway in rat brain after severe traumatic brain injury. J Trauma Acute Care Surg. 2013;74(6):1446–53. doi:10.1097/TA.0b013e31829246ad.
20. Li M, Li F, Luo C, Shan Y, Zhang L, Qian Z, Zhu G, Lin J, Feng H. Immediate splenectomy decreases mortality and improves cognitive function of rats after severe traumatic brain injury. J Trauma. 2011;71(1):141–7. doi:10.1097/TA.0b013e3181f30fc9.
21. Enriquez P, Bullock R. Molecular and cellular mechanisms in the pathophysiology of severe head injury. Curr Pharm Des. 2004;10(18):2131–43.

22. Jain KK. Neuroprotection in traumatic brain injury. Drug Discov Today. 2008;13(23–24):1082–9. doi:10.1016/j.drudis.2008.09.006.
23. Povlishock JT, Katz DI. Update of neuropathology and neurological recovery after traumatic brain injury. The J Head Trauma Rehabil. 2005;20(1):76–94.
24. Sahuquillo J, Poca MA, Amoros S. Current aspects of pathophysiology and cell dysfunction after severe head injury. Curr Pharm Des. 2001;7(15):1475–503.
25. Schubert A, Emory L. Cellular mechanisms of brain injury and cell death. Curr Pharm Des. 2012;18(38):6325–30.
26. Gao X, Deng P, Xu ZC, Chen J. Moderate traumatic brain injury causes acute dendritic and synaptic degeneration in the hippocampal dentate gyrus. PloS one. 2011;6(9):e24566. doi:10.1371/journal.pone.0024566.
27. Giulian D, Chen J, Ingeman JE, George JK, Noponen M. The role of mononuclear phagocytes in wound healing after traumatic injury to adult mammalian brain. J Neurosci (the official journal of the Society for Neuroscience). 1989;9(12):4416–29.
28. Czigner A, Mihaly A, Farkas O, Buki A, Krisztin-Peva B, Dobo E, Barzo P. Kinetics of the cellular immune response following closed head injury. Acta Neurochir. 2007;149(3):281–9. doi:10.1007/s00701-006-1095-8.
29. Adams JH, Doyle D, Ford I, Gennarelli TA, Graham DI, McLellan DR. Diffuse axonal injury in head injury: definition, diagnosis and grading. Histopathology. 1989;15(1):49–59.
30. Silver J, Miller JH. Regeneration beyond the glial scar. Nat Rev Neurosci. 2004;5(2):146–56. doi:10.1038/nrn1326.
31. Morganti-Kossmann MC, Rancan M, Stahel PF, Kossmann T. Inflammatory response in acute traumatic brain injury: a double-edged sword. Curr Opin Crit Care. 2002;8(2):101–5.
32. Walker PA, Harting MT, Baumgartner JE, Fletcher S, Strobel N, Cox CS Jr. Modern approaches to pediatric brain injury therapy. J Trauma. 2009;67(2 Suppl):S120–7. doi:10.1097/TA.0b013e3181ad323a.
33. Bentz K, Molcanyi M, Schneider A, Riess P, Maegele M, Bosche B, Hampl JA, Hescheler J, Patz S, Schafer U. Extract derived from rat brains in the acute phase following traumatic brain injury impairs survival of undifferentiated stem cells and induces rapid differentiation of surviving cells. Cell Physiol Biochem (international journal of experimental cellular physiology, biochemistry, and pharmacology). 2010;26(6):821–30. doi:10.1159/000323991.
34. Wang G, Zhang J, Hu X, Zhang L, Mao L, Jiang X, Liou AK, Leak RK, Gao Y, Chen J. Microglia/macrophage polarization dynamics in white matter after traumatic brain injury. J Cereb Blood Flow Metab (official journal of the International Society of Cerebral Blood Flow and Metabolism). 2013b. doi:10.1038/jcbfm.2013.146.
35. Bedi SS, Smith P, Hetz RA, Xue H, Cox CS. Immunomagnetic enrichment and flow cytometric characterization of mouse microglia. J Neurosci Methods. 2013a. doi:10.1016/j.jneumeth.2013.07.017.
36. Roberts DJ, Jenne CN, Leger C, Kramer AH, Gallagher CN, Todd S, Parney IF, Doig CJ, Yong VW, Kubes P, Zygun D. Association between the cerebral inflammatory and matrix metalloproteinase responses after severe traumatic brain injury in humans. J Neurotrauma. 2013a. doi:10.1089/neu.2012.2842.
37. Roberts DJ, Jenne CN, Leger C, Kramer AH, Gallagher CN, Todd S, Parney IF, Doig CJ, Yong VW, Kubes P, Zygun DA. Prospective evaluation of the temporal matrix metalloproteinase response after severe traumatic brain injury in humans. J Neurotrauma. 2013b. doi:10.1089/neu.2012.2841.
38. Oertel M, Boscardin WJ, Obrist WD, Glenn TC, McArthur DL, Gravori T, Lee JH, Martin NA. Posttraumatic vasospasm: the epidemiology, severity, and time course of an underestimated phenomenon: a prospective study performed in 299 patients. J Neurosurg. 2005;103(5):812–24. doi:10.3171/jns.2005.103.5.0812.
39. Katsnelson M, Mackenzie L, Frangos S, Oddo M, Levine JM, Pukenas B, Faerber J, Dong C, Kofke WA, le Roux PD. Are initial radiographic and clinical scales associated with subsequent intracranial pressure and brain oxygen levels after severe traumatic brain injury? Neurosurgery. 2012;70(5):1095–105; discussion 1105. doi:10.1227/NEU.0b013e318240c1ed.

40. Xiong Y, Mahmood A, Chopp M. Angiogenesis, neurogenesis and brain recovery of function following injury. Curr Opin Investig Drugs. 2010;11(3):298–308.
41. Bouzat P, Sala N, Payen JF, Oddo M. Beyond intracranial pressure: optimization of cerebral blood flow, oxygen, and substrate delivery after traumatic brain injury. Ann Intensive Care. 2013;3(1):23. doi:10.1186/2110-5820-3-23.
42. Manz HJ. Pathophysiology and pathology of elevated intracranial pressure. Pathobiol Annu. 1979;9:359–81.
43. Wijdicks EFM, Wijdicks EFM. The practice of emergency and critical care neurology. New York: Oxford University Press; 2010.
44. Doczi T. Volume regulation of the brain tissue–a survey. Acta Neurochir. 1993;121(1–2):1–8.
45. Barzo P, Marmarou A, Fatouros P, Hayasaki K, Corwin F. Contribution of vasogenic and cellular edema to traumatic brain swelling measured by diffusion-weighted imaging. J Neurosurg. 1997;87(6):900–7. doi:10.3171/jns.1997.87.6.0900.
46. Di X, Goforth PB, Bullock R, Ellis E, Satin L. Mechanical injury alters volume activated ion channels in cortical astrocytes. Acta Neurochir Suppl. 2000;76:379–83.
47. Clausen T, Bullock R. Medical treatment and neuroprotection in traumatic brain injury. Curr Pharm Des. 2001;7(15):1517–32.
48. Adelson PD, Bratton SL, Carney NA, Chesnut RM, du Coudray HE, Goldstein B, Kochanek PM, Miller HC, Partington MD, Selden NR, Warden CR, Wright DW, American Association for Surgery of T, Child Neurology S, International Society for Pediatric N, International Trauma A, Critical Care S, Society of Critical Care M, World Federation of Pediatric I, Critical Care S. Guidelines for the acute medical management of severe traumatic brain injury in infants, children, and adolescents. Chapter 1: Introduction. Pediatr Crit Care Med (a journal of the society of critical care medicine and the world federation of pediatric intensive and critical care societies). 2003;4(3 Suppl):S2–4. doi:10.1097/01.CCM.0000066600.71233.01.
49. Suarez JI, Zaidat OO, Suri MF, Feen ES, Lynch G, Hickman J, Georgiadis A, Selman WR. Length of stay and mortality in neurocritically ill patients: impact of a specialized neurocritical care team. Crit Care Med. 2004;32(11):2311–7.
50. Varelas PN, Conti MM, Spanaki MV, Potts E, Bradford D, Sunstrom C, Fedder W, Hacein Bey L, Jaradeh S, Gennarelli TA. The impact of a neurointensivist-led team on a semiclosed neurosciences intensive care unit. Crit Care Med. 2004;32(11):2191–8.
51. Sakellaridis N, Pavlou E, Karatzas S, Chroni D, Vlachos K, Chatzopoulos K, Dimopoulou E, Kelesis C, Karaouli V. Comparison of mannitol and hypertonic saline in the treatment of severe brain injuries. J Neurosurg. 2011;114(2):545–8. doi:10.3171/2010.5.JNS091685.
52. Johnston AJ, Steiner LA, Chatfield DA, Coleman MR, Coles JP, Al-Rawi PG, Menon DK, Gupta AK. Effects of propofol on cerebral oxygenation and metabolism after head injury. Br J Anaesth. 2003;91(6):781–6.
53. Scalfani MT, Dhar R, Zazulia AR, Videen TO, Diringer MN. Effect of osmotic agents on regional cerebral blood flow in traumatic brain injury. J Crit Care. 2012;27(5):526 e7–12. doi:10.1016/j.jcrc.2011.10.008.
54. Bochicchio GV, Bochicchio K, Nehman S, Casey C, Andrews P, Scalea TM. Tolerance and efficacy of enteral nutrition in traumatic brain-injured patients induced into barbiturate coma. JPEN J Parenter Enteral Nutr. 2006;30(6):503–6.
55. Eisenberg HM, Frankowski RF, Contant CF, Marshall LF, Walker MD. High-dose barbiturate control of elevated intracranial pressure in patients with severe head injury. J Neurosurg. 1988;69(1):15–23. doi:10.3171/jns.1988.69.1.0015.
56. Sydenham E, Roberts I, Alderson P. Hypothermia for traumatic head injury. Cochrane Database Syst Rev. 2009;(2):CD001048. doi:10.1002/14651858.CD001048.pub4.
57. Cooper DJ, Rosenfeld JV, Murray L, Arabi YM, Davies AR, D'Urso P, Kossmann T, Ponsford J, Seppelt I, Reilly P, Wolfe R, Investigators DT, Australian, New Zealand Intensive Care Society Clinical Trials G. Decompressive craniectomy in diffuse traumatic brain injury. N Engl J Med. 2011;364(16):1493–502. doi:10.1056/NEJMoa1102077.
58. Chesnut RM, Temkin N, Carney N, Dikmen S, Rondina C, Videtta W, Petroni G, Lujan S, Pridgeon J, Barber J, Machamer J, Chaddock K, Celix JM, Cherner M, Hendrix T, Global

Neurotrauma Research G. A trial of intracranial-pressure monitoring in traumatic brain injury. N Engl J Med. 2012;367(26):2471–81. doi:10.1056/NEJMoa1207363.
59. Melhem S, Shutter L, Kaynar A. A trial of intracranial pressure monitoring in traumatic brain injury. Crit Care. 2014;18(1):302. doi:10.1186/cc13713.
60. Le Roux P. Intracranial pressure after the BEST TRIP trial: a call for more monitoring. Curr Opin Crit Care. 2014;20(2):141–7. doi:10.1097/MCC.0000000000000078.
61. Alali AS, Fowler RA, Mainprize TG, Scales DC, Kiss A, de Mestral C, Ray JG, Nathens AB. Intracranial pressure monitoring in severe traumatic brain injury: results from the American College of Surgeons Trauma Quality Improvement Program. J Neurotrauma. 2013;30(20):1737–46. doi:10.1089/neu.2012.2802.
62. Gerber LM, Chiu YL, Carney N, Hartl R, Ghajar J. Marked reduction in mortality in patients with severe traumatic brain injury. J Neurosurg. 2013;119(6):1583–90. doi:10.3171/2013.8.JNS13276.
63. Talving P, Karamanos E, Teixeira PG, Skiada D, Lam L, Belzberg H, Inaba K, Demetriades D. Intracranial pressure monitoring in severe head injury: compliance with Brain Trauma Foundation guidelines and effect on outcomes: a prospective study. J Neurosurg. 2013;119(5):1248–54. doi:10.3171/2013.7.JNS122255.
64. Oddo M, Levine JM, Mackenzie L, Frangos S, Feihl F, Kasner SE, Katsnelson M, Pukenas B, Macmurtrie E, Maloney-Wilensky E, Kofke WA, LeRoux PD. Brain hypoxia is associated with short-term outcome after severe traumatic brain injury independently of intracranial hypertension and low cerebral perfusion pressure. Neurosurgery. 2011;69(5):1037–45; discussion 1045. doi:10.1227/NEU.0b013e3182287ca7.
65. Belli A, Sen J, Petzold A, Russo S, Kitchen N, Smith M. Metabolic failure precedes intracranial pressure rises in traumatic brain injury: a microdialysis study. Acta Neurochir. 2008;150(5):461–9; discussion 470. doi:10.1007/s00701-008-1580-3.
66. Stein NR, McArthur DL, Etchepare M, Vespa PM. Early cerebral metabolic crisis after TBI influences outcome despite adequate hemodynamic resuscitation. Neurocrit Care. 2012;17(1):49–57. doi:10.1007/s12028-012-9708-y.
67. Chen T, Qian YZ, Rice A, Zhu JP, Di X, Bullock R. Brain lactate uptake increases at the site of impact after traumatic brain injury. Brain Res. 2000;861(2):281–7.
68. Oddo M, Schmidt JM, Carrera E, Badjatia N, Connolly ES, Presciutti M, Ostapkovich ND, Levine JM, Le Roux P, Mayer SA. Impact of tight glycemic control on cerebral glucose metabolism after severe brain injury: a microdialysis study. Crit Care Med. 2008;36(12):3233–8. doi:10.1097/CCM.0b013e31818f4026.
69. Vespa P, Bergsneider M, Hattori N, Wu HM, Huang SC, Martin NA, Glenn TC, McArthur DL, Hovda DA. Metabolic crisis without brain ischemia is common after traumatic brain injury: a combined microdialysis and positron emission tomography study. J Cereb Blood Flow Metab (official journal of the International Society of Cerebral Blood Flow and Metabolism). 2005;25(6):763–74. doi:10.1038/sj.jcbfm.9600073.
70. Colicos MA, Dash PK. Apoptotic morphology of dentate gyrus granule cells following experimental cortical impact injury in rats: possible role in spatial memory deficits. Brain Res. 1996;739(1–2):120–31.
71. Colicos MA, Dixon CE, Dash PK. Delayed, selective neuronal death following experimental cortical impact injury in rats: possible role in memory deficits. Brain Res. 1996;739 (1–2):111–9.
72. Atkins CM. Decoding hippocampal signaling deficits after traumatic brain injury. Transl Stroke Res. 2011;2(4):546–55. doi:10.1007/s12975-011-0123-z.
73. Gao X, Deng-Bryant Y, Cho W, Carrico KM, Hall ED, Chen J. Selective death of newborn neurons in hippocampal dentate gyrus following moderate experimental traumatic brain injury. J Neurosci Res. 2008a;86(10):2258–70. doi:10.1002/jnr.21677.
74. Ewing-Cobbs L, Prasad MR, Swank P, Kramer L, Cox CS Jr, Fletcher JM, Barnes M, Zhang X, Hasan KM. Arrested development and disrupted callosal microstructure following pediatric traumatic brain injury: relation to neurobehavioral outcomes. NeuroImage. 2008;42(4):1305–15. doi:10.1016/j.neuroimage.2008.06.031.

75. Badri S, Chen J, Barber J, Temkin NR, Dikmen SS, Chesnut RM, Deem S, Yanez ND, Treggiari MM. Mortality and long-term functional outcome associated with intracranial pressure after traumatic brain injury. Intensive Care Med. 2012;38(11):1800–9. doi:10.1007/s00134-012-2655-4.
76. Chirumamilla S, Sun D, Bullock MR, Colello RJ. Traumatic brain injury induced cell proliferation in the adult mammalian central nervous system. J Neurotrauma. 2002;19(6):693–703. doi:10.1089/08977150260139084.
77. Reynolds BA, Weiss S. Generation of neurons and astrocytes from isolated cells of the adult mammalian central nervous system. Science. 1992;255(5052):1707–10.
78. Romanko MJ, Rola R, Fike JR, Szele FG, Dizon ML, Felling RJ, Brazel CY, Levison SW. Roles of the mammalian subventricular zone in cell replacement after brain injury. Prog Neurobiol. 2004;74(2):77–99. doi:10.1016/j.pneurobio.2004.07.001.
79. Scott DE, Hansen SL. Post-traumatic regeneration, neurogenesis and neuronal migration in the adult mammalian brain. Va Med Q (VMQ). 1997;124(4):249–61.
80. Roy NS, Wang S, Jiang L, Kang J, Benraiss A, Harrison-Restelli C, Fraser RA, Couldwell WT, Kawaguchi A, Okano H, Nedergaard M, Goldman SA. In vitro neurogenesis by progenitor cells isolated from the adult human hippocampus. Nat Med. 2000;6(3):271–7. doi:10.1038/73119.
81. Itoh T, Satou T, Hashimoto S, Ito H. Isolation of neural stem cells from damaged rat cerebral cortex after traumatic brain injury. Neuroreport. 2005;16(15):1687–91.
82. Christie KJ, Turnley AM. Regulation of endogenous neural stem/progenitor cells for neural repair-factors that promote neurogenesis and gliogenesis in the normal and damaged brain. Front Cell Neurosci. 2012;6:70. doi:10.3389/fncel.2012.00070.
83. Dash PK, Mach SA, Moore AN. Enhanced neurogenesis in the rodent hippocampus following traumatic brain injury. J Neurosci Res. 2001;63(4):313–9.
84. Gao X, Enikolopov G, Chen J. Direct isolation of neural stem cells in the adult hippocampus after traumatic brain injury. J Neurotrauma. 2008b;25(8):985–95. doi:10.1089/neu.2008.0460.
85. Rice AC, Khaldi A, Harvey HB, Salman NJ, White F, Fillmore H, Bullock MR. Proliferation and neuronal differentiation of mitotically active cells following traumatic brain injury. Exp Neurol. 2003;183(2):406–17.
86. Sun D, Colello RJ, Daugherty WP, Kwon TH, McGinn MJ, Harvey HB, Bullock MR. Cell proliferation and neuronal differentiation in the dentate gyrus in juvenile and adult rats following traumatic brain injury. J Neurotrauma. 2005;22(1):95–105. doi:10.1089/neu.2005.22.95.
87. Yu TS, Zhang G, Liebl DJ, Kernie SG. Traumatic brain injury-induced hippocampal neurogenesis requires activation of early nestin-expressing progenitors. J Neurosci (the official journal of the Society for Neuroscience). 2008;28(48):12901–12. doi:10.1523/JNEUROSCI.4629-08.2008.
88. Sun D, McGinn MJ, Zhou Z, Harvey HB, Bullock MR, Colello RJ. Anatomical integration of newly generated dentate granule neurons following traumatic brain injury in adult rats and its association to cognitive recovery. Exp Neurol. 2007;204(1):264–72. doi:10.1016/j.expneurol.2006.11.005.
89. Kernie SG, Erwin TM, Parada LF. Brain remodeling due to neuronal and astrocytic proliferation after controlled cortical injury in mice. J Neurosci Res. 2001;66(3):317–26.
90. Singleton RH, Zhu J, Stone JR, Povlishock JT. Traumatically induced axotomy adjacent to the soma does not result in acute neuronal death. J Neurosci (the official journal of the Society for Neuroscience). 2002;22(3):791–802.
91. Salman H, Ghosh P, Kernie SG. Subventricular zone neural stem cells remodel the brain following traumatic injury in adult mice. J Neurotrauma. 2004;21(3):283–92. doi:10.1089/089771504322972077.
92. Itoh T, Satou T, Ishida H, Nishida S, Tsubaki M, Hashimoto S, Ito H. The relationship between SDF-1alpha/CXCR4 and neural stem cells appearing in damaged area after traumatic brain injury in rats. Neurol Res. 2009a;31(1):90–102. doi:10.1179/174313208X332995.

93. Moon Y, Kim JY, Kim WR, Kim HJ, Jang MJ, Nam Y, Kim K, Kim H, Sun W. Function of ezrin-radixin-moesin proteins in migration of subventricular zone-derived neuroblasts following traumatic brain injury. Stem Cells. 2013;31(8):1696–705. doi:10.1002/stem.1420.
94. Blaiss CA, Yu TS, Zhang G, Chen J, Dimchev G, Parada LF, Powell CM, Kernie SG. Temporally specified genetic ablation of neurogenesis impairs cognitive recovery after traumatic brain injury. J Neurosci (the official journal of the Society for Neuroscience). 2011;31(13):4906–16. doi:10.1523/JNEUROSCI.5265-10.2011.
95. Ahmed AI, Shtaya AB, Zaben MJ, Owens EV, Kiecker C, Gray WP. Endogenous GFAP-positive neural stem/progenitor cells in the postnatal mouse cortex are activated following traumatic brain injury. J Neurotrauma. 2012;29(5):828–42. doi:10.1089/neu.2011.1923.
96. Itoh T, Satou T, Nishida S, Hashimoto S, Ito H. Immature and mature neurons coexist among glial scars after rat traumatic brain injury. Neurol Res. 2007;29(7):734–42.
97. Itoh T, Imano M, Nishida S, Tsubaki M, Nakayama T, Mizuguchi N, Yamanaka S, Tabuchi M, Munakata H, Hashimoto S, Ito A, Satou T. Appearance of neural stem cells around the damaged area following traumatic brain injury in aged rats. J Neural Transm. 2013;120(3):361–74. doi:10.1007/s00702-012-0895-7.
98. Urrea C, Castellanos DA, Sagen J, Tsoulfas P, Bramlett HM, Dietrich WD. Widespread cellular proliferation and focal neurogenesis after traumatic brain injury in the rat. Restor Neurol Neurosci. 2007;25(1):65–76.
99. Sgubin D, Aztiria E, Perin A, Longatti P, Leanza G. Activation of endogenous neural stem cells in the adult human brain following subarachnoid hemorrhage. J Neurosci Res. 2007;85(8):1647–55. doi:10.1002/jnr.21303.
100. Zheng W, Zhuge Q, Zhong M, Chen G, Shao B, Wang H, Mao X, Xie L, Jin K. Neurogenesis in adult human brain after traumatic brain injury. J Neurotrauma. 2013. doi:10.1089/neu.2010.1579.
101. Quinones-Hinojosa A, Sanai N, Soriano-Navarro M, Gonzalez-Perez O, Mirzadeh Z, Gil-Perotin S, Romero-Rodriguez R, Berger MS, Garcia-Verdugo JM, Alvarez-Buylla A. Cellular composition and cytoarchitecture of the adult human subventricular zone: a niche of neural stem cells. J Comp Neurol. 2006;494(3):415–34. doi:10.1002/cne.20798.
102. Sanai N, Berger MS, Garcia-Verdugo JM, Alvarez-Buylla A. Comment on "Human neuroblasts migrate to the olfactory bulb via a lateral ventricular extension". Science. 2007;318(5849):393; author reply 393. doi:10.1126/science.1145011.
103. McHugh GS, Engel DC, Butcher I, Steyerberg EW, Lu J, Mushkudiani N, Hernandez AV, Marmarou A, Maas AI, Murray GD. Prognostic value of secondary insults in traumatic brain injury: results from the IMPACT study. J Neurotrauma. 2007;24(2):287–93. doi:10.1089/neu.2006.0031.
104. Doppenberg EM, Bullock R. Clinical neuro-protection trials in severe traumatic brain injury: lessons from previous studies. J Neurotrauma. 1997;14(2):71–80.
105. Reinert MM, Bullock R. Clinical trials in head injury. Neurolo Res. 1999;21(4):330–8.
106. Kobori N, Clifton GL, Dash P. Altered expression of novel genes in the cerebral cortex following experimental brain injury. Brain Res Mol Brain Res. 2002;104(2):148–58.
107. Bullock R, Zauner A, Woodward JJ, Myseros J, Choi SC, Ward JD, Marmarou A, Young HF. Factors affecting excitatory amino acid release following severe human head injury. J Neurosurg. 1998;89(4):507–18. doi:10.3171/jns.1998.89.4.0507.
108. Koura SS, Doppenberg EM, Marmarou A, Choi S, Young HF, Bullock R. Relationship between excitatory amino acid release and outcome after severe human head injury. Acta Neurochir Suppl. 1998;71:244–6.
109. Zauner A, Bullock R, Kuta AJ, Woodward J, Young HF. Glutamate release and cerebral blood flow after severe human head injury. Acta Neurochir Suppl. 1996;67:40–4.
110. Yamamoto T, Rossi S, Stiefel M, Doppenberg E, Zauner A, Bullock R, Marmarou A. CSF and ECF glutamate concentrations in head injured patients. Acta Neurochir Suppl. 1999;75:17–9.
111. Bullock R, Kuroda Y, Teasdale GM, McCulloch J. Prevention of post-traumatic excitotoxic brain damage with NMDA antagonist drugs: a new strategy for the nineties. Acta Neurochir Suppl. 1992;55:49–55.

112. Maxwell WL, Bullock R, Landholt H, Fujisawa H. Massive astrocytic swelling in response to extracellular glutamate–a possible mechanism for post-traumatic brain swelling? Acta Neurochir Suppl. 1994;60:465–7.
113. Myseros JS, Bullock R. The rationale for glutamate antagonists in the treatment of traumatic brain injury. Ann N Y Acad Sci. 1995;765:262–71; discussion 298.
114. Dai SS, Zhou YG, Li W, An JH, Li P, Yang N, Chen XY, Xiong RP, Liu P, Zhao Y, Shen HY, Zhu PF, Chen JF. Local glutamate level dictates adenosine A2A receptor regulation of neuroinflammation and traumatic brain injury. J Neurosci (the official journal of the Society for Neuroscience). 2010;30(16):5802–10. doi:10.1523/JNEUROSCI.0268-10.2010.
115. Dash PK, Zhao J, Hergenroeder G, Moore AN. Biomarkers for the diagnosis, prognosis, and evaluation of treatment efficacy for traumatic brain injury. Neurotherapeutics (the journal of the American Society for Experimental NeuroTherapeutics). 2010b;7(1):100–14. doi:10.1016/j.nurt.2009.10.019.
116. Alves OL, Doyle AJ, Clausen T, Gilman C, Bullock R. Evaluation of topiramate neuroprotective effect in severe TBI using microdialysis. Ann N Y Acad Sci. 2003;993:25–34; discussion 48–53.
117. Dash PK, Moore AN, Moody MR, Treadwell R, Felix JL, Clifton GL. Post-trauma administration of caffeine plus ethanol reduces contusion volume and improves working memory in rats. J Neurotrauma. 2004b;21(11):1573–83. doi:10.1089/neu.2004.21.1573.
118. Li W, Dai S, An J, Li P, Chen X, Xiong R, Liu P, Wang H, Zhao Y, Zhu M, Liu X, Zhu P, Chen JF, Zhou Y. Chronic but not acute treatment with caffeine attenuates traumatic brain injury in the mouse cortical impact model. Neuroscience. 2008;151(4):1198–207. doi:10.1016/j.neuroscience.2007.11.020.
119. Clausen F, Marklund N, Lewen A, Hillered L. The nitrone free radical scavenger NXY-059 is neuroprotective when administered after traumatic brain injury in the rat. J Neurotrauma. 2008;25(12):1449–57. doi:10.1089/neu.2008.0585.
120. Long DA, Ghosh K, Moore AN, Dixon CE, Dash PK. Deferoxamine improves spatial memory performance following experimental brain injury in rats. Brain Res. 1996;717(1–2):109–17.
121. Marklund N, Clausen F, Lewen A, Hovda DA, Olsson Y, Hillered L. alpha-Phenyl-tert-N-butyl nitrone (PBN) improves functional and morphological outcome after cortical contusion injury in the rat. Acta Neurochir. 2001;143(1):73–81.
122. Yeo JE, Kang SK. Selenium effectively inhibits ROS-mediated apoptotic neural precursor cell death in vitro and in vivo in traumatic brain injury. Biochim Biophys Acta. 2007;1772(11–12):1199–210. doi:10.1016/j.bbadis.2007.09.004.
123. Itoh T, Imano M, Nishida S, Tsubaki M, Mizuguchi N, Hashimoto S, Ito A, Satou T. (-)-Epigallocatechin-3-gallate increases the number of neural stem cells around the damaged area after rat traumatic brain injury. J Neural Transm. 2012;119(8):877–90. doi:10.1007/s00702-011-0764-9.
124. Zhao J, Pati S, Redell JB, Zhang M, Moore AN, Dash PK. Caffeic acid phenethyl ester protects blood-brain barrier integrity and reduces contusion volume in rodent models of traumatic brain injury. J Neurotrauma. 2012;29(6):1209–18. doi:10.1089/neu.2011.1858.
125. Dohi K, Satoh K, Mihara Y, Nakamura S, Miyake Y, Ohtaki H, Nakamachi T, Yoshikawa T, Shioda S, Aruga T. Alkoxyl radical-scavenging activity of edaravone in patients with traumatic brain injury. J Neurotrauma. 2006;23(11):1591–9. doi:10.1089/neu.2006.23.1591.
126. Dohi K, Satoh K, Nakamachi T, Yofu S, Hiratsuka K, Nakamura S, Ohtaki H, Yoshikawa T, Shioda S, Aruga T. Does edaravone (MCI- 186) act as an antioxidant and a neuroprotector in experimental traumatic brain injury? Antioxid Redox Signal. 2007;9(2):281–7. doi:10.1089/ars.2007.9.ft-12.
127. Itoh T, Satou T, Nishida S, Tsubaki M, Hashimoto S, Ito H. The novel free radical scavenger, edaravone, increases neural stem cell number around the area of damage following rat traumatic brain injury. Neurotox Res. 2009b;16(4):378–89. doi:10.1007/s12640-009-9081-6.

128. Miyamoto K, Ohtaki H, Dohi K, Tsumuraya T, Nakano H, Kiriyama K, Song D, Aruga T, Shioda S. Edaravone increases regional cerebral blood flow after traumatic brain injury in mice. Acta Neurochir Suppl. 2013a;118:103–9. doi:10.1007/978-3-7091-1434-6_18.
129. Miyamoto K, Ohtaki H, Dohi K, Tsumuraya T, Song D, Kiriyama K, Satoh K, Shimizu A, Aruga T, Shioda S. Therapeutic time window for edaravone treatment of traumatic brain injury in mice. BioMed Res Int. 2013b;2013:379206. doi:10.1155/2013/379206.
130. Satoh K, Ikeda Y, Shioda S, Tobe T, Yoshikawa T. Edarabone scavenges nitric oxide. Redox Rep (communications in free radical research). 2002;7(4):219–22. doi:10.1179/135100002125000587.
131. Wang GH, Jiang ZL, Li YC, Li X, Shi H, Gao YQ, Vosler PS, Chen J. Free-radical scavenger edaravone treatment confers neuroprotection against traumatic brain injury in rats. J Neurotrauma. 2011;28(10):2123–34. doi:10.1089/neu.2011.1939.
132. Lapchak PA. A critical assessment of edaravone acute ischemic stroke efficacy trials: is edaravone an effective neuroprotective therapy? Expert Opin Pharmacother. 2010;11(10):1753–63. doi:10.1517/14656566.2010.493558.
133. Redell JB, Zhao J, Dash PK. Acutely increased cyclophilin a expression after brain injury: a role in blood-brain barrier function and tissue preservation. J Neurosci Res. 2007;85(9):1980–8. doi:10.1002/jnr.21324.
134. Menge T, Zhao Y, Zhao J, Wataha K, Gerber M, Zhang J, Letourneau P, Redell J, Shen L, Wang J, Peng Z, Xue H, Kozar R, Cox CS Jr, Khakoo AY, Holcomb JB, Dash PK, Pati S. Mesenchymal stem cells regulate blood-brain barrier integrity through TIMP3 release after traumatic brain injury. Sci Transl Med. 2012;4(161):161ra150. doi:10.1126/scitranslmed.3004660.
135. Tejima E, Guo S, Murata Y, Arai K, Lok J, van Leyen K, Rosell A, Wang X, Lo EH. Neuroprotective effects of overexpressing tissue inhibitor of metalloproteinase TIMP-1. J Neurotrauma. 2009;26(11):1935–41. doi:10.1089/neu.2009-0959.
136. Li Z, Wang B, Kan Z, Zhang B, Yang Z, Chen J, Wang D, Wei H, Zhang JN, Jiang R. Progesterone increases circulating endothelial progenitor cells and induces neural regeneration after traumatic brain injury in aged rats. J Neurotrauma. 2012;29(2):343–53. doi:10.1089/neu.2011.1807.
137. Zhao J, Moore AN, Clifton GL, Dash PK. Sulforaphane enhances aquaporin-4 expression and decreases cerebral edema following traumatic brain injury. J Neurosci Res. 2005;82(4):499–506. doi:10.1002/jnr.20649.
138. Amenta PS, Jallo JI, Tuma RF, Elliott MB. A cannabinoid type 2 receptor agonist attenuates blood-brain barrier damage and neurodegeneration in a murine model of traumatic brain injury. J Neurosci Res. 2012;90(12):2293–305. doi:10.1002/jnr.23114.
139. Baskaya MK, Dogan A, Rao AM, Dempsey RJ. Neuroprotective effects of citicoline on brain edema and blood-brain barrier breakdown after traumatic brain injury. J Neurosurg. 2000;92(3):448–52. doi:10.3171/jns.2000.92.3.0448.
140. Zafonte RD, Bagiella E, Ansel BM, Novack TA, Friedewald WT, Hesdorffer DC, Timmons SD, Jallo J, Eisenberg H, Hart T, Ricker JH, Diaz-Arrastia R, Merchant RE, Temkin NR, Melton S, Dikmen SS. Effect of citicoline on functional and cognitive status among patients with traumatic brain injury: Citicoline Brain Injury Treatment Trial (COBRIT). JAMA (the journal of the American Medical Association). 2012;308(19):1993–2000. doi:10.1001/jama.2012.13256.
141. Dash PK, Mach SA, Moore AN. The role of extracellular signal-regulated kinase in cognitive and motor deficits following experimental traumatic brain injury. Neuroscience. 2002;114(3):755–67.
142. Dash PK, Moore AN, Dixon CE. Spatial memory deficits, increased phosphorylation of the transcription factor CREB, and induction of the AP-1 complex following experimental brain injury. J Neurosci (the official journal of the Society for Neuroscience). 1995;15(3 Pt 1):2030–9.
143. Lu J, Frerich JM, Turtzo LC, Li S, Chiang J, Yang C, Wang X, Zhang C, Wu C, Sun Z, Niu G, Zhuang Z, Brady RO, Chen X. Histone deacetylase inhibitors are neuroprotective and

preserve NGF-mediated cell survival following traumatic brain injury. Proc Natl Acad Sci U S A. 2013;110(26):10747–52. doi:10.1073/pnas.1308950110.

144. Wang G, Jiang X, Pu H, Zhang W, An C, Hu X, Liou AK, Leak RK, Gao Y, Chen J. Scriptaid, a novel histone deacetylase inhibitor, protects against traumatic brain injury via modulation of PTEN and AKT pathway: scriptaid protects against TBI via AKT. Neurotherapeutics (the journal of the American Society for Experimental NeuroTherapeutics). 2013a;10(1):124–42. doi:10.1007/s13311-012-0157-2.
145. Atkins CM, Oliva AA Jr, Alonso OF, Pearse DD, Bramlett HM, Dietrich WD. Modulation of the cAMP signaling pathway after traumatic brain injury. Exp Neurol. 2007;208(1):145–58. doi:10.1016/j.expneurol.2007.08.011.
146. Titus DJ, Sakurai A, Kang Y, Furones C, Jergova S, Santos R, Sick TJ, Atkins CM. Phosphodiesterase inhibition rescues chronic cognitive deficits induced by traumatic brain injury. J Neurosci (the official journal of the Society for Neuroscience). 2013;33(12):5216–26. doi:10.1523/JNEUROSCI.5133-12.2013.
147. Logan TT, Villapol S, Symes AJ. TGF-beta superfamily gene expression and induction of the Runx1 transcription factor in adult neurogenic regions after brain injury. PloS one. 2013;8(3):e59250. doi:10.1371/journal.pone.0059250.
148. Barha CK, Ishrat T, Epp JR, Galea LA, Stein DG. Progesterone treatment normalizes the levels of cell proliferation and cell death in the dentate gyrus of the hippocampus after traumatic brain injury. Exp Neurol. 2011;231(1):72–81. doi:10.1016/j.expneurol.2011.05.016.
149. Yao XL, Liu J, Lee E, Ling GS, McCabe JT. Progesterone differentially regulates pro- and anti-apoptotic gene expression in cerebral cortex following traumatic brain injury in rats. J Neurotrauma. 2005;22(6):656–68. doi:10.1089/neu.2005.22.656.
150. Sun D, Bullock MR, McGinn MJ, Zhou Z, Altememi N, Hagood S, Hamm R, Colello RJ. Basic fibroblast growth factor-enhanced neurogenesis contributes to cognitive recovery in rats following traumatic brain injury. Exp Neurol. 2009;216(1):56–65. doi:10.1016/j.expneurol.2008.11.011.
151. Sun D, Bullock MR, Altememi N, Zhou Z, Hagood S, Rolfe A, McGinn MJ, Hamm R, Colello RJ. The effect of epidermal growth factor in the injured brain after trauma in rats. J Neurotrauma. 2010;27(5):923–38. doi:10.1089/neu.2009.1209.
152. Lee C, Agoston DV. Vascular endothelial growth factor is involved in mediating increased de novo hippocampal neurogenesis in response to traumatic brain injury. J Neurotrauma. 2010;27(3):541–53. doi:10.1089/neu.2009.0905.
153. Thau-Zuchman O, Shohami E, Alexandrovich AG, Leker RR. Vascular endothelial growth factor increases neurogenesis after traumatic brain injury. J Cereb Blood Flow Metab (official journal of the International Society of Cerebral Blood Flow and Metabolism). 2010;30(5):1008–16. doi:10.1038/jcbfm.2009.271.
154. Larsen A, Kolind K, Pedersen DS, Doering P, Pedersen MO, Danscher G, Penkowa M, Stoltenberg M. Gold ions bio-released from metallic gold particles reduce inflammation and apoptosis and increase the regenerative responses in focal brain injury. Histochem Cell Biol. 2008;130(4):681–92. doi:10.1007/s00418-008-0448-1.
155. Pedersen MO, Larsen A, Pedersen DS, Stoltenberg M, Penkova M. Metallic gold treatment reduces proliferation of inflammatory cells, increases expression of VEGF and FGF, and stimulates cell proliferation in the subventricular zone following experimental traumatic brain injury. Histol Histopathol. 2009;24(5):573–86.
156. Kleindienst A, Ross Bullock M. A critical analysis of the role of the neurotrophic protein S100B in acute brain injury. J Neurotrauma. 2006;23(8):1185–200. doi:10.1089/neu.2006.23.1185.
157. Kleindienst A, McGinn MJ, Harvey HB, Colello RJ, Hamm RJ, Bullock MR. Enhanced hippocampal neurogenesis by intraventricular S100B infusion is associated with improved cognitive recovery after traumatic brain injury. J Neurotrauma. 2005;22(6):645–55. doi:10.1089/neu.2005.22.645.
158. Shi J, Longo FM, Massa SM. A small molecule P75 ligand protects neurogenesis after traumatic brain injury. Stem Cells. 2013. doi:10.1002/stem.1516.

159. Buki A, Okonkwo DO, Povlishock JT. Postinjury cyclosporin A administration limits axonal damage and disconnection in traumatic brain injury. J Neurotrauma. 1999;16(6):511–21.
160. Okonkwo DO, Buki A, Siman R, Povlishock JT. Cyclosporin A limits calcium-induced axonal damage following traumatic brain injury. Neuroreport. 1999;10(2):353–8.
161. Buki A, Farkas O, Doczi T, Povlishock JT. Preinjury administration of the calpain inhibitor MDL-28170 attenuates traumatically induced axonal injury. J Neurotrauma. 2003;20(3):261–8. doi:10.1089/089771503321532842.
162. Lenzlinger PM, Shimizu S, Marklund N, Thompson HJ, Schwab ME, Saatman KE, Hoover RC, Bareyre FM, Motta M, Luginbuhl A, Pape R, Clouse AK, Morganti-Kossmann C, McIntosh TK. Delayed inhibition of Nogo-A does not alter injury-induced axonal sprouting but enhances recovery of cognitive function following experimental traumatic brain injury in rats. Neuroscience. 2005;134(3):1047–56. doi:10.1016/j.neuroscience.2005.04.048.
163. Thompson HJ, Marklund N, LeBold DG, Morales DM, Keck CA, Vinson M, Royo NC, Grundy R, McIntosh TK. Tissue sparing and functional recovery following experimental traumatic brain injury is provided by treatment with an anti-myelin-associated glycoprotein antibody. Eur J Neurosci. 2006;24(11):3063–72. doi:10.1111/j.1460-9568.2006.05197.x.
164. Wong G, Goldshmit Y, Turnley AM. Interferon-gamma but not TNF alpha promotes neuronal differentiation and neurite outgrowth of murine adult neural stem cells. Exp Neurol. 2004;187(1):171–7. doi:10.1016/j.expneurol.2004.01.009.
165. Marklund N, Keck C, Hoover R, Soltesz K, Millard M, LeBold D, Spangler Z, Banning A, Benson J, McIntosh TK. Administration of monoclonal antibodies neutralizing the inflammatory mediators tumor necrosis factor alpha and interleukin -6 does not attenuate acute behavioral deficits following experimental traumatic brain injury in the rat. Restor Neurol Neurosci. 2005;23(1):31–42.
166. Clausen F, Hanell A, Bjork M, Hillered L, Mir AK, Gram H, Marklund N. Neutralization of interleukin-1beta modifies the inflammatory response and improves histological and cognitive outcome following traumatic brain injury in mice. Eur J Neurosci. 2009;30(3):385–96. doi:10.1111/j.1460-9568.2009.06820.x.
167. Clausen F, Hanell A, Israelsson C, Hedin J, Ebendal T, Mir AK, Gram H, Marklund N. Neutralization of interleukin-1beta reduces cerebral edema and tissue loss and improves late cognitive outcome following traumatic brain injury in mice. Eur J Neurosci. 2011;34(1):110–23. doi:10.1111/j.1460-9568.2011.07723.x.
168. Harting MT, Jimenez F, Adams SD, Mercer DW, Cox CS Jr. Acute, regional inflammatory response after traumatic brain injury: implications for cellular therapy. Surgery. 2008;144(5):803–13. doi:10.1016/j.surg.2008.05.017.
169. Qu C, Mahmood A, Ning R, Xiong Y, Zhang L, Chen J, Jiang H, Chopp M. The treatment of traumatic brain injury with velcade. J Neurotrauma. 2010;27(9):1625–34. doi:10.1089/neu.2010.1359.
170. Watanabe J, Shetty AK, Hattiangady B, Kim DK, Foraker JE, Nishida H, Prockop DJ. Administration of TSG-6 improves memory after traumatic brain injury in mice. Neurobiol Dis. 2013;59:86–99. doi:10.1016/j.nbd.2013.06.017.
171. Wallenquist U, Holmqvist K, Hanell A, Marklund N, Hillered L, Forsberg-Nilsson K. Ibuprofen attenuates the inflammatory response and allows formation of migratory neuroblasts from grafted stem cells after traumatic brain injury. Restor Neurol Neurosci. 2011. doi:10.3233/RNN-2011-606.
172. Wallenquist U, Holmqvist K, Hanell A, Marklund N, Hillered L, Forsberg-Nilsson K. Ibuprofen attenuates the inflammatory response and allows formation of migratory neuroblasts from grafted stem cells after traumatic brain injury. Restor Neurol Neurosci. 2012;30(1):9–19. doi:10.3233/RNN-2011-0606.
173. Dash PK, Mach SA, Moore AN. Regional expression and role of cyclooxygenase-2 following experimental traumatic brain injury. J Neurotrauma. 2000;17(1):69–81.
174. Kunz T, Marklund N, Hillered L, Oliw EH. Cyclooxygenase-2, prostaglandin synthases, and prostaglandin H2 metabolism in traumatic brain injury in the rat. J Neurotrauma. 2002;19(9):1051–64. doi:10.1089/089771502760341965.

175. Atkins CM, Kang Y, Furones C, Truettner JS, Alonso OF, Dietrich WD. Postinjury treatment with rolipram increases hemorrhage after traumatic brain injury. J Neurosci Res. 2012;90(9):1861–71. doi:10.1002/jnr.23069.
176. Madri JA. Modeling the neurovascular niche: implications for recovery from CNS injury. J Physiol Pharmacol (an official journal of the Polish Physiological Society). 2009;60(Suppl 4):95–104.
177. Lu D, Mahmood A, Zhang R, Copp M. Upregulation of neurogenesis and reduction in functional deficits following administration of DEtA/NONOate, a nitric oxide donor, after traumatic brain injury in rats. J Neurosurg. 2003;99(2):351–61. doi:10.3171/jns.2003.99.2.0351.
178. Lu D, Goussev A, Chen J, Pannu P, Li Y, Mahmood A, Chopp M. Atorvastatin reduces neurological deficit and increases synaptogenesis, angiogenesis, and neuronal survival in rats subjected to traumatic brain injury. J Neurotrauma. 2004;21(1):21–32. doi:10.1089/089771504772695913.
179. Lu D, Qu C, Goussev A, Jiang H, Lu C, Schallert T, Mahmood A, Chen J, Li Y, Chopp M. Statins increase neurogenesis in the dentate gyrus, reduce delayed neuronal death in the hippocampal CA3 region, and improve spatial learning in rat after traumatic brain injury. J Neurotrauma. 2007;24(7):1132–46. doi:10.1089/neu.2007.0288.
180. Sanchez-Aguilar M, Tapia-Perez JH, Sanchez-Rodriguez JJ, Vinas-Rios JM, Martinez-Perez P, de la Cruz-Mendoza E, Sanchez-Reyna M, Torres-Corzo JG, Gordillo-Moscoso A. Effect of rosuvastatin on cytokines after traumatic head injury. J Neurosurg. 2013;118(3):669–75. doi:10.3171/2012.12.JNS121084.
181. Hergenroeder GW, Redell JB, Moore AN, Dash PK. Biomarkers in the clinical diagnosis and management of traumatic brain injury. Mol Diagn Ther. 2008;12(6):345–58. doi:10.2165/1250444-200812060-00002.
182. Wang KK, Ottens AK, Liu MC, Lewis SB, Meegan C, Oli MW, Tortella FC, Hayes RL. Proteomic identification of biomarkers of traumatic brain injury. Expert Rev Proteomics. 2005;2(4):603–14. doi:10.1586/14789450.2.4.603.
183. Dash PK, Kobori N, Moore AN. A molecular description of brain trauma pathophysiology using microarray technology: an overview. Neurochem Res. 2004a;29(6):1275–86.
184. Israelsson C, Wang Y, Kylberg A, Pick CG, Hoffer BJ, Ebendal T. Closed head injury in a mouse model results in molecular changes indicating inflammatory responses. J Neurotrauma. 2009;26(8):1307–14. doi:10.1089/neu.2008-0676.
185. Matzilevich DA, Rall JM, Moore AN, Grill RJ, Dash PK. High-density microarray analysis of hippocampal gene expression following experimental brain injury. J Neurosci Res. 2002;67(5):646–63.
186. Rall JM, Matzilevich DA, Dash PK. Comparative analysis of mRNA levels in the frontal cortex and the hippocampus in the basal state and in response to experimental brain injury. Neuropathol Appl Neurobiol. 2003;29(2):118–31.
187. Redell JB, Liu Y, Dash PK. Traumatic brain injury alters expression of hippocampal microRNAs: potential regulators of multiple pathophysiological processes. J Neurosci Res. 2009;87(6):1435–48. doi:10.1002/jnr.21945.
188. Redell JB, Moore AN, Ward NH, 3rd, Hergenroeder GW, Dash PK. Human traumatic brain injury alters plasma microRNA levels. J Neurotrauma. 2010;27(12):2147–56. doi:10.1089/neu.2010.1481.
189. Arun P, Abu-Taleb R, Oguntayo S, Tanaka M, Wang Y, Valiyaveettil M, Long JB, Zhang Y, Nambiar MP. Distinct patterns of expression of traumatic brain injury biomarkers after blast exposure: role of compromised cell membrane integrity. Neurosci Lett. 2013. doi:10.1016/j.neulet.2013.07.047.
190. Alessandri B, al-Samsam R, Corwin F, Fatouros P, Young HF, Bullock RM. Acute and late changes in N-acetyl-aspartate following diffuse axonal injury in rats: an MRI spectroscopy and microdialysis study. Neurol Res. 2000;22(7):705–12.
191. Al-Samsam RH, Alessandri B, Bullock R. Extracellular N-acetyl-aspartate as a biochemical marker of the severity of neuronal damage following experimental acute traumatic brain injury. J Neurotrauma. 2000;17(1):31–9.

192. Marklund N, Blennow K, Zetterberg H, Ronne-Engstrom E, Enblad P, Hillered L. Monitoring of brain interstitial total tau and beta amyloid proteins by microdialysis in patients with traumatic brain injury. J Neurosurg. 2009;110(6):1227–37. doi:10.3171/2008.9.JNS08584.
193. Marklund N, Farrokhnia N, Hanell A, Vanmechelen E, Enblad P, Zetterberg H, Blennow K, Hillered L. Monitoring of beta-Amyloid dynamics after human traumatic brain injury. J Neurotrauma. 2013. doi:10.1089/neu.2013.2964.
194. Clausen T, Alves OL, Reinert M, Doppenberg E, Zauner A, Bullock R. Association between elevated brain tissue glycerol levels and poor outcome following severe traumatic brain injury. J Neurosurg. 2005;103(2):233–8. doi:10.3171/jns.2005.103.2.0233.
195. Chen J, Tu Y, Connolly EC, Ronnett GV. Heme oxygenase-2 protects against glutathione depletion-induced neuronal apoptosis mediated by bilirubin and cyclic GMP. Curr Neurovasc Res. 2005;2(2):121–31.
196. Dohi K, Satoh K, Ohtaki H, Shioda S, Miyake Y, Shindo M, Aruga T. Elevated plasma levels of bilirubin in patients with neurotrauma reflect its pathophysiological role in free radical scavenging. In Vivo. 2005;19(5):855–60.
197. Feala JD, Abdulhameed MD, Yu C, Dutta B, Yu X, Schmid K, Dave J, Tortella F, Reifman J. Systems biology approaches for discovering biomarkers for traumatic brain injury. J Neurotrauma. 2013;30(13):1101–16. doi:10.1089/neu.2012.2631.
198. Maas AI, Harrison-Felix CL, Menon D, Adelson PD, Balkin T, Bullock R, Engel DC, Gordon W, Orman JL, Lew HL, Robertson C, Temkin N, Valadka A, Verfaellie M, Wainwright M, Wright DW, Schwab K. Common data elements for traumatic brain injury: recommendations from the interagency working group on demographics and clinical assessment. Arch Phys Med Rehabil. 2010;91(11):1641–9. doi:10.1016/j.apmr.2010.07.232.
199. Manley GT, Diaz-Arrastia R, Brophy M, Engel D, Goodman C, Gwinn K, Veenstra TD, Ling G, Ottens AK, Tortella F, Hayes RL. Common data elements for traumatic brain injury: recommendations from the biospecimens and biomarkers working group. Arch Phys Med Rehabil. 2010;91(11):1667–72. doi:10.1016/j.apmr.2010.05.018.
200. Mondello S, Jeromin A, Buki A, Bullock R, Czeiter E, Kovacs N, Barzo P, Schmid K, Tortella F, Wang KK, Hayes RL. Glial neuronal ratio: a novel index for differentiating injury type in patients with severe traumatic brain injury. J Neurotrauma. 2012;29(6):1096–104. doi:10.1089/neu.2011.2092.
201. Mondello S, Papa L, Buki A, Bullock MR, Czeiter E, Tortella FC, Wang KK, Hayes RL. Neuronal and glial markers are differently associated with computed tomography findings and outcome in patients with severe traumatic brain injury: a case control study. Crit Care. 2011;15(3):R156. doi:10.1186/cc10286.
202. Dash PK, Redell JB, Hergenroeder G, Zhao J, Clifton GL, Moore A. Serum ceruloplasmin and copper are early biomarkers for traumatic brain injury-associated elevated intracranial pressure. J Neurosci Res. 2010;88(8):1719–26. doi:10.1002/jnr.22336.
203. Hergenroeder GW, Moore AN, McCoy JP Jr, Samsel L, Ward NH, 3rd, Clifton GL, Dash PK. Serum IL-6: a candidate biomarker for intracranial pressure elevation following isolated traumatic brain injury. J Neuroinflammation. 2010;7:19. doi:10.1186/1742-2094-7-19.
204. Mondello S, Buki A, Italiano D, Jeromin A. alpha-Synuclein in CSF of patients with severe traumatic brain injury. Neurology. 2013;80(18):1662–8. doi:10.1212/WNL.0b013e3182904d43.
205. Farkas O, Polgar B, Szekeres-Bartho J, Doczi T, Povlishock JT, Buki A. Spectrin breakdown products in the cerebrospinal fluid in severe head injury–preliminary observations. Acta Neurochir. 2005;147(8):855–61. doi:10.1007/s00701-005-0559-6.
206. Mondello S, Robicsek SA, Gabrielli A, Brophy GM, Papa L, Tepas J, Robertson C, Buki A, Scharf D, Jixiang M, Akinyi L, Muller U, Wang KK, Hayes RL. alphaII-spectrin breakdown products (SBDPs): diagnosis and outcome in severe traumatic brain injury patients. J Neurotrauma. 2010;27(7):1203–13. doi:10.1089/neu.2010.1278.
207. Czeiter E, Mondello S, Kovacs N, Sandor J, Gabrielli A, Schmid K, Tortella F, Wang KK, Hayes RL, Barzo P, Ezer E, Doczi T, Buki A. Brain injury biomarkers may improve the

predictive power of the IMPACT outcome calculator. J Neurotrauma. 2012;29(9):1770–8. doi:10.1089/neu.2011.2127.
208. Povlishock JT, Buki A, Koiziumi H, Stone J, Okonkwo DO. Initiating mechanisms involved in the pathobiology of traumatically induced axonal injury and interventions targeted at blunting their progression. Acta Neurochiru Suppl. 1999;73:15–20.
209. Brodhun M, Bauer R, Patt S. Potential stem cell therapy and application in neurotrauma. Exp Toxicol Pathol (official journal of the Gesellschaft fur Toxikologische Pathologie). 2004;56(1–2):103–12. doi:10.1016/j.etp.2004.04.004.
210. Longhi L, Zanier ER, Royo N, Stocchetti N, McIntosh TK. Stem cell transplantation as a therapeutic strategy for traumatic brain injury. Transpl Immunol. 2005;15(2):143–8. doi:10.1016/j.trim.2005.09.003.
211. Webber DJ, Minger SL. Therapeutic potential of stem cells in central nervous system regeneration. Curr Opin Investig Drugs. 2004;5(7):714–9.
212. Xiong Y, Mahmood A, Chopp M. Emerging treatments for traumatic brain injury. Expert Opin Emerg Drugs. 2009;14(1):67–84. doi:10.1517/14728210902769601.
213. Jain KK. Cell therapy for CNS trauma. Mol Biotechnol. 2009;42(3):367–76. doi:10.1007/s12033-009-9166-8.
214. Kulbatski I, Mothe AJ, Nomura H, Tator CH. Endogenous and exogenous CNS derived stem/progenitor cell approaches for neurotrauma. Curr Drug Targets. 2005;6(1):111–26.
215. Axell MZ, Zlateva S, Curtis M. A method for rapid derivation and propagation of neural progenitors from human embryonic stem cells. J Neurosci Methods. 2009;184(2):275–84. doi:10.1016/j.jneumeth.2009.08.015.
216. Harting MT, Baumgartner JE, Worth LL, Ewing-Cobbs L, Gee AP, Day MC, Cox CS Jr. Cell therapies for traumatic brain injury. Neurosurg Focus. 2008;24(3–4):E18. doi:10.3171/FOC/2008/24/3-4/E17.
217. Royo NC, Schouten JW, Fulp CT, Shimizu S, Marklund N, Graham DI, McIntosh TK. From cell death to neuronal regeneration: building a new brain after traumatic brain injury. J Neuropathol Exp Neurol. 2003;62(8):801–11.
218. Chopp M, Mahmood A, Lu D, Li Y. Editorial. Mesenchymal stem cell treatment of traumatic brain injury. J Neurosurg. 2009;110(6):1186–8. doi:10.3171/2008.10.JNS081254.
219. Gold EM, Su D, Lopez-Velazquez L, Haus DL, Perez H, Lacuesta GA, Anderson AJ, Cummings BJ. Functional assessment of long-term deficits in rodent models of traumatic brain injury. Regen Med. 2013;8(4):483–516. doi:10.2217/rme.13.41.
220. Li Y, Chopp M. Marrow stromal cell transplantation in stroke and traumatic brain injury. Neurosci Lett. 2009;456(3):120–3. doi:10.1016/j.neulet.2008.03.096.
221. Walker PA, Shah SK, Harting MT, Cox CS Jr. Progenitor cell therapies for traumatic brain injury: barriers and opportunities in translation. Disease Model Mech. 2009;2(1–2):23–38. doi:10.1242/dmm.001198.
222. Walker PA, Harting MT, Shah SK, Cox CS. Current trends in cell therapy for pediatric acquired brain injury. Minerva Pediatr. 2010;62(1):91–106.
223. Lu D, Li Y, Wang L, Chen J, Mahmood A, Chopp M. Intraarterial administration of marrow stromal cells in a rat model of traumatic brain injury. J Neurotrauma. 2001;18(8):813–19. doi:10.1089/089771501316919175.
224. Lundberg J, Le Blanc K, Soderman M, Andersson T, Holmin S. Endovascular transplantation of stem cells to the injured rat CNS. Neuroradiology. 2009;51(10):661–7. doi:10.1007/s00234-009-0551-6.
225. Lundberg J, Sodersten E, Sundstrom E, Le Blanc K, Andersson T, Hermanson O, Holmin S. Targeted intra-arterial transplantation of stem cells to the injured CNS is more effective than intravenous administration: engraftment is dependent on cell type and adhesion molecule expression. Cell Transplant. 2012;21(1):333–43. doi:10.3727/096368911X576036.
226. Fischer UM, Harting MT, Jimenez F, Monzon-Posadas WO, Xue H, Savitz SI, Laine GA, Cox CS Jr. Pulmonary passage is a major obstacle for intravenous stem cell delivery: the pulmonary first-pass effect. Stem Cells Dev. 2009;18(5):683–92. doi:10.1089/scd.2008.0253.

227. Walker PA, Aroom KR, Jimenez F, Shah SK, Harting MT, Gill BS, Cox CS Jr. Advances in progenitor cell therapy using scaffolding constructs for central nervous system injury. Stem Cell Rev. 2009;5(3):283–300. doi:10.1007/s12015-009-9081-1.
228. Chen X, Yin J, Wu X, Li R, Fang J, Chen R, Zhang B, Zhang W. Effects of magnetically labeled exogenous endothelial progenitor cells on cerebral blood perfusion and microvasculature alterations after traumatic brain injury in rat model. Acta Radiol. 2013;54(3):313–23. doi:10.1258/ar.2012.120605.
229. Bakhtiary M, Marzban M, Mehdizadeh M, Joghataei MT, Khoei S, Tondar M, Mahabadi VP, Laribi B, Ebrahimi A, Hashemian SJ, Modiry N, Mehrabi S. Combination of stem cell mobilized by granulocyte-colony stimulating factor and human umbilical cord matrix stem cell: therapy of traumatic brain injury in rats. Iran J Basic Med Sci. 2011;14(4):327–39.
230. Cox CS Jr, Baumgartner JE, Harting MT, Worth LL, Walker PA, Shah SK, Ewing-Cobbs L, Hasan KM, Day MC, Lee D, Jimenez F, Gee A. Autologous bone marrow mononuclear cell therapy for severe traumatic brain injury in children. Neurosurgery. 2011;68(3):588–600. doi:10.1227/NEU.0b013e318207734c.
231. Zhang R, Liu Y, Yan K, Chen L, Chen XR, Li P, Chen FF, Jiang XD. Anti-inflammatory and immunomodulatory mechanisms of mesenchymal stem cell transplantation in experimental traumatic brain injury. J Neuroinflammation. 2013;10(1):106. doi:10.1186/1742-2094-10-106.
232. Walker PA, Shah SK, Jimenez F, Gerber MH, Xue H, Cutrone R, Hamilton JA, Mays RW, Deans R, Pati S, Dash PK, Cox CS Jr. Intravenous multipotent adult progenitor cell therapy for traumatic brain injury: preserving the blood brain barrier via an interaction with splenocytes. Exp Neurol. 2010;225(2):341–52. doi:10.1016/j.expneurol.2010.07.005.
233. Lu D, Sanberg PR, Mahmood A, Li Y, Wang L, Sanchez-Ramos J, Chopp M. Intravenous administration of human umbilical cord blood reduces neurological deficit in the rat after traumatic brain injury. Cell Transplant. 2002;11(3):275–81.
234. Newman MB, Willing AE, Manresa JJ, Sanberg CD, Sanberg PR. Cytokines produced by cultured human umbilical cord blood (HUCB) cells: implications for brain repair. Exp Neurol. 2006;199(1):201–8. doi:10.1016/j.expneurol.2006.04.001.
235. Skardelly M, Gaber K, Burdack S, Scheidt F, Hilbig H, Boltze J, Forschler A, Schwarz S, Schwarz J, Meixensberger J, Schuhmann MU. Long-term benefit of human fetal neuronal progenitor cell transplantation in a clinically adapted model after traumatic brain injury. J Neurotrauma. 2011;28(3):401–14. doi:10.1089/neu.2010.1526.
236. Dai SS, Wang H, Yang N, An JH, Li W, Ning YL, Zhu PF, Chen JF, Zhou YG. Plasma glutamate-modulated interaction of A2AR and mGluR5 on BMDCs aggravates traumatic brain injury-induced acute lung injury. J Exp Med. 2013;210(4):839–51. doi:10.1084/jem.20122196.
237. Kansu E. Thrombosis in stem cell transplantation. Hematology. 2012;17(Suppl 1):S159–62. doi:10.1179/102453312X13336169156735.
238. Boockvar JA, Schouten J, Royo N, Millard M, Spangler Z, Castelbuono D, Snyder E, O'Rourke D, McIntosh T. Experimental traumatic brain injury modulates the survival, migration, and terminal phenotype of transplanted epidermal growth factor receptor-activated neural stem cells. Neurosurgery. 2005;56(1):163–71; discussion 171.
239. Conti L, Reitano E, Cattaneo E. Neural stem cell systems: diversities and properties after transplantation in animal models of diseases. Brain Pathol. 2006;16(2):143–54. doi:10.1111/j.1750-3639.2006.00009.x.
240. Harting MT, Sloan LE, Jimenez F, Baumgartner J, Cox CS Jr. Subacute neural stem cell therapy for traumatic brain injury. J Surg Res. 2009;153(2):188–94. doi:10.1016/j.jss.2008.03.037.
241. Philips MF, Mattiasson G, Wieloch T, Bjorklund A, Johansson BB, Tomasevic G, Martinez-Serrano A, Lenzlinger PM, Sinson G, Grady MS, McIntosh TK. Neuroprotective and behavioral efficacy of nerve growth factor-transfected hippocampal progenitor cell transplants after experimental traumatic brain injury. J Neurosurg. 2001;94(5):765–74. doi:10.3171/jns.2001.94.5.0765.

242. Riess P, Zhang C, Saatman KE, Laurer HL, Longhi LG, Raghupathi R, Lenzlinger PM, Lifshitz J, Boockvar J, Neugebauer E, Snyder EY, McIntosh TK. Transplanted neural stem cells survive, differentiate, and improve neurological motor function after experimental traumatic brain injury. Neurosurgery. 2002;51(4):1043–52; discussion 1052–1044.
243. Schouten JW, Fulp CT, Royo NC, Saatman KE, Watson DJ, Snyder EY, Trojanowski JQ, Prockop DJ, Maas AI, McIntosh TK. A review and rationale for the use of cellular transplantation as a therapeutic strategy for traumatic brain injury. J Neurotrauma. 2004;21(11):1501–38. doi:10.1089/neu.2004.21.1501.
244. Shear DA, Tate MC, Archer DR, Hoffman SW, Hulce VD, Laplaca MC, Stein DG. Neural progenitor cell transplants promote long-term functional recovery after traumatic brain injury. Brain Res. 2004;1026(1):11–22. doi:10.1016/j.brainres.2004.07.087.
245. Wennersten A, Meier X, Holmin S, Wahlberg L, Mathiesen T. Proliferation, migration, and differentiation of human neural stem/progenitor cells after transplantation into a rat model of traumatic brain injury. J Neurosurg. 2004;100(1):88–96. doi:10.3171/jns.2004.100.1.0088.
246. Shear DA, Tate CC, Tate MC, Archer DR, LaPlaca MC, Stein DG, Dunbar GL. Stem cell survival and functional outcome after traumatic brain injury is dependent on transplant timing and location. Restor Neurol Neurosci. 2011;29(4):215–25. doi:10.3233/RNN-2011-0593.
247. Bakshi A, Keck CA, Koshkin VS, LeBold DG, Siman R, Snyder EY, McIntosh TK. Caspase-mediated cell death predominates following engraftment of neural progenitor cells into traumatically injured rat brain. Brain Res. 2005;1065(1–2):8–19. doi:10.1016/j.brainres.2005.09.059.
248. Hong SQ, Zhang HT, You J, Zhang MY, Cai YQ, Jiang XD, Xu RX. Comparison of transdifferentiated and untransdifferentiated human umbilical mesenchymal stem cells in rats after traumatic brain injury. Neurochem Res. 2011;36(12):2391–400. doi:10.1007/s11064-011-0567-2.
249. Molcanyi M, Riess P, Bentz K, Maegele M, Hescheler J, Schafke B, Trapp T, Neugebauer E, Klug N, Schafer U. Trauma-associated inflammatory response impairs embryonic stem cell survival and integration after implantation into injured rat brain. J Neurotrauma. 2007;24(4):625–37. doi:10.1089/neu.2006.0180.
250. Molcanyi M, Bosche B, Kraitsy K, Patz S, Zivcak J, Riess P, Majdoub FE, Hescheler J, Goldbrunner R, Schafer U. Pitfalls and fallacies interfering with correct identification of embryonic stem cells implanted into the brain after experimental traumatic injury. J Neurosci Methods. 2013;215(1):60–70. doi:10.1016/j.jneumeth.2013.02.012.
251. Chen HI, Bakshi A, Royo NC, Magge SN, Watson DJ. Neural stem cells as biological minipumps: a faster route to cell therapy for the CNS? Curr Stem Cell Res Ther. 2007;2(1):13–22.
252. Hagan M, Wennersten A, Meijer X, Holmin S, Wahlberg L, Mathiesen T. Neuroprotection by human neural progenitor cells after experimental contusion in rats. Neurosci Lett. 2003;351(3):149–52.
253. Wang E, Gao J, Yang Q, Parsley MO, Dunn TJ, Zhang L, DeWitt DS, Denner L, Prough DS, Wu P. Molecular mechanisms underlying effects of neural stem cells against traumatic axonal injury. J Neurotrauma. 2012;29(2):295–312. doi:10.1089/neu.2011.2043.
254. Hwang DH, Jeong SR, Kim BG. Gene transfer mediated by stem cell grafts to treat CNS injury. Expert Opin Biol Ther. 2011;11(12):1599–610. doi:10.1517/14712598.2011.631908.
255. Ma H, Yu B, Kong L, Zhang Y, Shi Y. Transplantation of neural stem cells enhances expression of synaptic protein and promotes functional recovery in a rat model of traumatic brain injury. Mol Med Rep. 2011;4(5):849–56. doi:10.3892/mmr.2011.510.
256. Ugoya SO, Tu J. Bench to bedside of neural stem cell in traumatic brain injury. Stem Cells Int. 2012;2012:141624. doi:10.1155/2012/141624.
257. Watson DJ, Longhi L, Lee EB, Fulp CT, Fujimoto S, Royo NC, Passini MA, Trojanowski JQ, Lee VM, McIntosh TK, Wolfe JH. Genetically modified NT2N human neuronal cells mediate long-term gene expression as CNS grafts in vivo and improve functional cognitive outcome following experimental traumatic brain injury. J Neuropathol Exp Neurol. 2003;62(4):368–80.

258. Kang SK, Jun ES, Bae YC, Jung JS. Interactions between human adipose stromal cells and mouse neural stem cells in vitro. Brain Res Dev Brain Res. 2003;145(1):141–9.
259. Galindo LT, Filippo TR, Semedo P, Ariza CB, Moreira CM, Camara NO, Porcionatto MA. Mesenchymal stem cell therapy modulates the inflammatory response in experimental traumatic brain injury. Neurol Res Int. 2011;2011:564089. doi:10.1155/2011/564089.
260. Mahmood A, Lu D, Yi L, Chen JL, Chopp M. Intracranial bone marrow transplantation after traumatic brain injury improving functional outcome in adult rats. J Neurosurg. 2001;94(4):589–95. doi:10.3171/jns.2001.94.4.0589.
261. Mahmood A, Wu H, Qu C, Xiong Y, Chopp M. Effects of treating traumatic brain injury with collagen scaffolds and human bone marrow stromal cells on sprouting of corticospinal tract axons into the denervated side of the spinal cord. J Neurosurg. 2013;118(2):381–9. doi:10.3171/2012.11.JNS12753.
262. Nichols JE, Niles JA, Dewitt D, Prough D, Parsley M, Vega S, Cantu A, Lee E, Cortiella J. Neurogenic and neuro-protective potential of a novel subpopulation of peripheral blood-derived CD133+ABCG2+CXCR4+ mesenchymal stem cells: development of autologous cell-based therapeutics for traumatic brain injury. Stem Cell Res Ther. 2013;4(1):3. doi:10.1186/scrt151.
263. Zanier ER, Montinaro M, Vigano M, Villa P, Fumagalli S, Pischiutta F, Longhi L, Leoni ML, Rebulla P, Stocchetti N, Lazzari L, De Simoni MG. Human umbilical cord blood mesenchymal stem cells protect mice brain after trauma. Crit Care Med. 2011;39(11):2501–10. doi:10.1097/CCM.0b013e31822629ba.
264. Walker PA, Harting MT, Jimenez F, Shah SK, Pati S, Dash PK, Cox CS Jr. Direct intrathecal implantation of mesenchymal stromal cells leads to enhanced neuroprotection via an NFkappaB-mediated increase in interleukin-6 production. Stem Cells Dev. 2010;19(6):867–76. doi:10.1089/scd.2009.0188.
265. Becerra GD, Tatko LM, Pak ES, Murashov AK, Hoane MR. Transplantation of GABAergic neurons but not astrocytes induces recovery of sensorimotor function in the traumatically injured brain. Behav Brain Res. 2007;179(1):118–25. doi:10.1016/j.bbr.2007.01.024.
266. Riess P, Molcanyi M, Bentz K, Maegele M, Simanski C, Carlitscheck C, Schneider A, Hescheler J, Bouillon B, Schafer U, Neugebauer E. Embryonic stem cell transplantation after experimental traumatic brain injury dramatically improves neurological outcome, but may cause tumors. J Neurotrauma. 2007;24(1):216–25. doi:10.1089/neu.2006.0141.
267. Marklund N, Hillered L. Animal modelling of traumatic brain injury in preclinical drug development: where do we go from here? Br J Pharmacol. 2011;164(4):1207–29. doi:10.1111/j.1476-5381.2010.01163.x.
268. Wallenquist U, Brannvall K, Clausen F, Lewen A, Hillered L, Forsberg-Nilsson K. Grafted neural progenitors migrate and form neurons after experimental traumatic brain injury. Restor Neurol Neurosci. 2009;27(4):323–34. doi:10.3233/RNN-2009-0481.
269. Azari MF, Mathias L, Ozturk E, Cram DS, Boyd RL, Petratos S. Mesenchymal stem cells for treatment of CNS injury. Curr Neuropharmacol. 2010;8(4):316–23. doi:10.2174/157015910793358204.
270. Chen X, Katakowski M, Li Y, Lu D, Wang L, Zhang L, Chen J, Xu Y, Gautam S, Mahmood A, Chopp M. Human bone marrow stromal cell cultures conditioned by traumatic brain tissue extracts: growth factor production. J Neurosci Res. 2002;69(5):687–91. doi:10.1002/jnr.10334.
271. Kim HJ, Lee JH, Kim SH. Therapeutic effects of human mesenchymal stem cells on traumatic brain injury in rats: secretion of neurotrophic factors and inhibition of apoptosis. J Neurotrauma. 2010;27(1):131–8. doi:10.1089/neu.2008-0818.
272. Wang Z, Deng Q, Zhang X, Zhang J. Treatment of injured neurons with bone marrow stem cells cotransfected by hTERT and Ad-BDNF in vitro. J Mol Neurosci (MN). 2009;38(3):265–72. doi:10.1007/s12031-009-9208-5.
273. Wang Z, Yao W, Deng Q, Zhang X, Zhang J. Protective effects of BDNF overexpression bone marrow stromal cell transplantation in rat models of traumatic brain injury. J Mol Neurosci (MN). 2013d;49(2):409–16. doi:10.1007/s12031-012-9908-0.

274. Bedi SS, Walker PA, Shah SK, Jimenez F, Thomas CP, Smith P, Hetz RA, Xue H, Pati S, Dash PK, Cox CS Jr. Autologous bone marrow mononuclear cells therapy attenuates activated microglial/macrophage response and improves spatial learning after traumatic brain injury. J Trauma Acute Care Surg. 2013b. doi:10.1097/TA.0b013e31829617c6.
275. Walker PA, Bedi SS, Shah SK, Jimenez F, Xue H, Hamilton JA, Smith P, Thomas CP, Mays RW, Pati S, Cox CS Jr. Intravenous multipotent adult progenitor cell therapy after traumatic brain injury: modulation of the resident microglia population. J Neuroinflammation. 2012;9:228. doi:10.1186/1742-2094-9-228.
276. Scuteri A, Miloso M, Foudah D, Orciani M, Cavaletti G, Tredici G. Mesenchymal stem cells neuronal differentiation ability: a real perspective for nervous system repair? Curr Stem Cell Res Ther. 2011;6(2):82–92.
277. Parr AM, Tator CH, Keating A. Bone marrow-derived mesenchymal stromal cells for the repair of central nervous system injury. Bone Marrow Transplant. 2007;40(7):609–19. doi:10.1038/sj.bmt.1705757.
278. Tate MC, Shear DA, Hoffman SW, Stein DG, Archer DR, LaPlaca MC. Fibronectin promotes survival and migration of primary neural stem cells transplanted into the traumatically injured mouse brain. Cell Transplant. 2002;11(3):283–95.
279. Tate CC, Shear DA, Tate MC, Archer DR, Stein DG, LaPlaca MC. Laminin and fibronectin scaffolds enhance neural stem cell transplantation into the injured brain. J Tissue Eng Regen Med. 2009;3(3):208–17. doi:10.1002/term.154.
280. Tajiri N, Kaneko Y, Shinozuka K, Ishikawa H, Yankee E, McGrogan M, Case C, Borlongan CV. Stem cell recruitment of newly formed host cells via a successful seduction? Filling the gap between neurogenic niche and injured brain site. PloS one. 2013;8(9):e74857. doi:10.1371/journal.pone.0074857.
281. Yu B, Ma H, Kong L, Shi Y, Liu Y. Enhanced connexin 43 expression following neural stem cell transplantation in a rat model of traumatic brain injury. Arch Med Sci (AMS). 2013;9(1):132–8. doi:10.5114/aoms.2012.31438.
282. Weidenfeller C, Svendsen CN, Shusta EV. Differentiating embryonic neural progenitor cells induce blood-brain barrier properties. J Neurochem. 2007;101(2):555–65. doi:10.1111/j.1471-4159.2006.04394.x.
283. Cullen DK, Stabenfeldt SE, Simon CM, Tate CC, LaPlaca MC. In vitro neural injury model for optimization of tissue-engineered constructs. J Neurosci Res. 2007;85(16):3642–51. doi:10.1002/jnr.21434.
284. Delcroix GJ, Schiller PC, Benoit JP, Montero-Menei CN. Adult cell therapy for brain neuronal damages and the role of tissue engineering. Biomaterials. 2010;31(8):2105–20. doi:10.1016/j.biomaterials.2009.11.084.
285. Stabenfeldt SE, Irons HR, Laplaca MC. Stem cells and bioactive scaffolds as a treatment for traumatic brain injury. Curr Stem Cell Res Ther. 2011;6(3):208–20.
286. Tam RY, Fuehrmann T, Mitrousis N, Shoichet MS. Regenerative therapies for central nervous system diseases: a biomaterials approach. Neuropsychopharmacology (official publication of the American College of Neuropsychopharmacology). 2013. doi:10.1038/npp.2013.237.
287. Wang JY, Liou AK, Ren ZH, Zhang L, Brown BN, Cui XT, Badylak SF, Cai YN, Guan YQ, Leak RK, Chen J, Ji X, Chen L. Neurorestorative effect of urinary bladder matrix-mediated neural stem cell transplantation following traumatic brain injury in rats. CNS Neurol Disord Drug Targets. 2013c;12(3):413–25.
288. Guan J, Zhu Z, Zhao RC, Xiao Z, Wu C, Han Q, Chen L, Tong W, Zhang J, Gao J, Feng M, Bao X, Dai J, Wang R. Transplantation of human mesenchymal stem cells loaded on collagen scaffolds for the treatment of traumatic brain injury in rats. Biomaterials. 2013;34(24):5937–46. doi:10.1016/j.biomaterials.2013.04.047.
289. Bakshi A, Shimizu S, Keck CA, Cho S, LeBold DG, Morales D, Arenas E, Snyder EY, Watson DJ, McIntosh TK. Neural progenitor cells engineered to secrete GDNF show enhanced survival, neuronal differentiation and improve cognitive function following traumatic brain injury. Eur J Neurosci. 2006;23(8):2119–34. doi:10.1111/j.1460-9568.2006.04743.x.

290. Gao J, Prough DS, McAdoo DJ, Grady JJ, Parsley MO, Ma L, Tarensenko YI, Wu P. Transplantation of primed human fetal neural stem cells improves cognitive function in rats after traumatic brain injury. Exp Neurol. 2006;201(2):281–92. doi:10.1016/j.expneurol.2006.04.039.
291. Chang CP, Chio CC, Cheong CU, Chao CM, Cheng BC, Lin MT. Hypoxic preconditioning enhances the therapeutic potential of the secretome from cultured human mesenchymal stem cells in experimental traumatic brain injury. Clin Sci (Lond). 2013;124(3):165–76. doi:10.1042/CS20120226.
292. Chuang TJ, Lin KC, Chio CC, Wang CC, Chang CP, Kuo JR. Effects of secretome obtained from normoxia-preconditioned human mesenchymal stem cells in traumatic brain injury rats. J Trauma Acute Care Surg. 2012;73(5):1161–7. doi:10.1097/TA.0b013e318265d128.
293. Bakhtiary M, Marzban M, Mehdizadeh M, Joghataei MT, Khoei S, Pirhajati Mahabadi V, Laribi B, Tondar M, Moshkforoush A. Comparison of transplantation of bone marrow stromal cells (BMSC) and stem cell mobilization by granulocyte colony stimulating factor after traumatic brain injury in rat. Iran Biomed J. 2010;14(4):142–9.
294. Burgess A, Ayala-Grosso CA, Ganguly M, Jordao JF, Aubert I, Hynynen K. Targeted delivery of neural stem cells to the brain using MRI-guided focused ultrasound to disrupt the blood-brain barrier. PloS one. 2011;6(11):e27877. doi:10.1371/journal.pone.0027877.
295. Heile A, Brinker T. Clinical translation of stem cell therapy in traumatic brain injury: the potential of encapsulated mesenchymal cell biodelivery of glucagon-like peptide-1. Dialogues Clin Neurosci. 2011;13(3):279–86.
296. Krampera M, Galipeau J, Shi Y, Tarte K, Sensebe L. Immunological characterization of multipotent mesenchymal stromal cells–The International Society for Cellular Therapy (ISCT) working proposal. Cytotherapy. 2013;15(9):1054–61. doi:10.1016/j.jcyt.2013.02.010.
297. Cheng JL, Yang YJ, Li HL, Wang J, Wang MH, Zhang Y. In vivo tracing of superparamagnetic iron oxide-labeled bone marrow mesenchymal stem cells transplanted for traumatic brain injury by susceptibility weighted imaging in a rat model. Chin J Traumatol (Zhonghua chuang shang za zhi/Chinese Medical Association). 2010;13(3):173–7.
298. Darkazalli A, Levenson CW. Tracking stem cell migration and survival in brain injury: current approaches and future prospects. Histol Histopathol. 2012;27(10):1255–61.
299. Tang HL, Sun HP, Wu X, Sha HY, Feng XY, Zhu JH. Detection of neural stem cells function in rats with traumatic brain injury by manganese-enhanced magnetic resonance imaging. Chin Med J. 2011;124(12):1848–53.
300. Zhang H, Zheng X, Yang X, Fang S, Shen G, Zhao C, Tian M. 11C-NMSP/18F-FDG microPET to monitor neural stem cell transplantation in a rat model of traumatic brain injury. Eur J Nucl Med Mol Imaging. 2008;35(9):1699–708. doi:10.1007/s00259-008-0835-9.
301. McColgan P, Sharma P, Bentley P. Stem cell tracking in human trials: a meta-regression. Stem Cell Rev. 2011;7(4):1031–40. doi:10.1007/s12015-011-9260-8.
302. de Munter JP Wolters E. 70th Birthday symposium of Prof. Dr. Riederer: autologous adult stem cells in ischemic and traumatic CNS disorders. J Neural Transm. 2013;120(1):91–102. doi:10.1007/s00702-012-0868-x.
303. Han EY, Chun MH, Kim ST, Lim DP. Injection time-dependent effect of adult human bone marrow stromal cell transplantation in a rat model of severe traumatic brain injury. Curr Stem Cell Res Ther. 2013;8(2):172–81.

Chapter 16
Stem Cell Therapy for Neonatal Hypoxic–Ischemic Brain Injury

James Carroll

Various varieties of stem cells and methods of their administration are proposed as therapeutic modalities for neonatal hypoxic–ischemic (HI) brain injury. The widespread use of stem cells for this purpose and others, despite the lack of strong clinical evidence for their efficacy in most clinical situations, makes it incumbent to review the pathophysiology of the clinical condition of neonatal HI brain injury, the preclinical data dealing with animal HI injury, and the available clinical studies.

The central problem for the evaluation of treatment for neonatal HI injury is that two vastly different circumstances exist with respect to the condition: (1) the acute and (2) the chronic or long-standing result of the acute injury. The pathophysiology of the two variations, particularly with respect to the possible benefit of stem cell treatment, is quite different.

The situation is further complicated by the widespread interest in treating cerebral palsy (CP) with stem cells, an effort of great interest to parents and society at large. This effort has been given impetus by anecdotal stories carried on the Internet. However, only a small component of children with CP has neonatal HI brain injury as the cause for their disability. While the treatment of CP is a popular target for stem cell therapy, CP is a group of disorders of complex and varied cause. This chapter deals only with CP caused by HI injury in the neonatal period of the term infant.

Pathophysiology of Neonatal Hypoxic–Ischemic Injury

The pathophysiology of acute HI injury in term infants is well known. These processes are important for our discussion in that they relate to way stem cells interact with the tissue. The mechanism of injury has been summarized in detail by

J. Carroll (✉)
Augusta, GA, USA
e-mail: jcarroll@GRU.EDU

D. C. Hess (ed.), *Cell Therapy for Brain Injury*, DOI 10.1007/978-3-319-15063-5_16

Volpe [1]. Briefly, the key innate factors are the vulnerable vascular border zones, regional anaerobic mechanisms, and excitatory glutamate synapses. Hypercapnia, hypoxemia, and acidosis lead to loss of regional regulation of blood flow. The latter phenomenon, in combination with systemic hypotension, leads to a compounding of the preceding events with exacerbation of the abnormal blood flow, more anaerobic metabolism with increasing lactate, and a rise in the damaging glutamate. As a consequence, cellular adenosine triphosphate (ATP) is depleted, calcium intrudes into the cell, and excitatory amino acids bring about more cellular damage. Later reperfusion injury eventuates leading to more neuronal damage with cell necrosis followed later by apoptosis. Models of acute HI injury, discussed below, attempt to mimic these events.

Most of the studies of HI relate most closely to events in the term infant. The brain injuries in premature infants, however, comprise a significant portion of the pathology in patients with CP. While term infants tend to demonstrate more neuronal and cortical injuries, premature infants are more vulnerable to injury of the white matter. The influence of stem cell therapy in white matter injury is even less well understood and is not the subject here.

The additional molecular events occurring in the time period proximate to the injury are critical to the receptiveness of the tissue to stem cell repair. One of the factors important in this process is the chemokine SDF-1/CXCL12 system which contributes to the homing of stem cells to the site of injury [2]. These cells expressing the CXCR4 receptor migrate to the area of injury demonstrating upregulation of SDF-1. A number of other chemokine systems are activated and undoubtedly play a role in acute stem cell therapy [3].

On the other hand, the above molecular events have long been completed in chronic HI brain injury that has resulted in CP. The molecular elements in the tissue which “attract” stem cells are no longer upregulated, and there is little evidence to indicate that intrinsic cells are altered by the addition of stem cells. The chief idea for benefit in chronic HI injury, therefore, is replacement of neurons and supportive cell elements by the stem cells.

Types of Cells Which Could Be Used

There are a number of candidate stem cell types for transplantation. Only three have been employed clinically thus far in children: (1) umbilical cord blood, usually autologous, (2) mesenchymal cells derived from bone marrow, again usually autologous, and (3) in countries where the procedure is allowed, neural progenitor cells obtained from aborted fetal brain.

The most widely used cell source is human umbilical cord blood. The fraction of cells bearing the putative positive action is the CD34 negative or mesenchymal stem cell [4]. Mesenchymal stem cells are characterized by their plastic adherent properties under normal culture conditions. The cells possess a fibroblast-like morphology. These cells may exert a neurotrophic effect on endogenous brain cells. CD133 positive cells, another component of umbilical cord mesenchymal cells, may have the potential for axonal regrowth, reduction in apoptosis, and neuronal protection [5].

There are other mesenchymal cell candidates which could be used for treatment. Multipotent adult progenitor cells (MAPCs), discovered by Verfaille and coworkers [6, 7], and developed for possible commercial use by the biotech company, Athersys, Inc., could be used for treatment after appropriate safety testing. A clinical trial (NCT01436487) with MAPCs is in progress in adults who suffer stroke acutely. The advantage of this cell type is that it could be supplied as a shelf-ready product that does not require matching for the individual patient.

The other types of cells that could be employed in patients, after preclinical testing, include induced pluripotent stem (IPS) cells and embryonic stem cells. IPS cells may provide an excellent alternative for embryonic cells as they could be derived from the patient's own skin fibroblasts [8]. These cells are said to possess many of the biological characteristics of embryonic stem cells. IPS cells would require safety testing to ensure their lack of tumorigenic potential [9]. Embryonic stem cells are the most controversial of the group. They are derived from human embryonic tissue. Embryonic cells have been employed in several acute preclinical injury studies with positive results. However, it does not appear these cells will be candidates for use in children any time soon.

Summary of Acute Preclinical Studies

There are numerous preclinical studies which demonstrate the efficacy of various types of stem cells in acute neonatal HI injury. The following table shows some of the studies in neonatal rodents following the production of acute HI (Table 16.1).

The remarkable observation among the above studies (described in more detail below) is that the route of treatment does not seem to affect the generally favorable outcome. Most of the reports do not show significant incorporation of the transplanted cells into the brain substance. Therefore, the beneficial action of the cells seems to occur by neurotrophic or paracrine influences.

In the report of intra-arterial transplantation [10], the transplant did not restore stereological volume in the damaged tissue even though long-term adverse cognitive effects were reduced. This would suggest that there were positive influences of the transplant on remaining intrinsic cells, but not cell replacement.

Huang et al. [11] reported that even the route of intra-peritoneal administration was effective. Once again, the message from this study is the importance of trophic influences.

Wang et al. [12] showed that the cell loss in the CA1 sector of the hippocampus was reduced. The mechanism of cell preservation would also be invoked by these results.

Table 16.1 Stem cell treatment for acute neonatal rodent brain injury

Time after injury	Cell type	Route of administration	Reference
24 h	Umbilical cord blood	Intra-arterial	[10]
24 h	Umbilical cord blood	Intraperitoneal	[11]
24 h	Umbilical cord blood		[12]
24 h	Dental pulp stem cells	Intracerebral	[13]
24 h	Umbilical cord blood	Jugular vein	[14]
	Mesenchymal stem cells	Striatum	[15]
3 h	Umbilical cord blood	Intraperitoneal	[16]
	Umbilical cord blood	Intracerebral	[17]
7 days	Multipotent adult progenitor cells	Hippocampus	[18]
7 days	Multipotent adult progenitor cells	Intravenous	[19]
3 days	Fetal neural stem cells	Cerebral ventricle	[20]
24 h	Mesenchymal stem cells	Intraperitoneal	[21]

Dental pulp cells only rarely differentiated into neural cells, but neurological function was improved with the inhibition of proinflammatory cytokines, increased expression of anti-inflammatory substances, and reduced apoptosis [13].

De Paula et al. [14] reported the dose dependency of intravenous transplantation. This result suggests that dosing experiments would be an important component of clinical studies.

The cells reported by Part et al. [15] were derived from human placenta. With the striatal transplantation, they theorized that the improved motor behavior was related to dopaminergic differentiation. Similarly, Pimental-Coelho et al. [16] observed a neuroprotective effect was observed in the striatum.

In the work of Xia et al. [17] above, the damage had been inflicted prenatally demonstrating a similar effect to that in the more standard postnatal model.

The two reports by Yasuhara et al. [18, 19] demonstrated positive results using the potentially commercial preparation of MAPCs. The critical outcome from these two papers is that the result was favorable with both direct brain injection and intravenous administration.

Qu et al. [20] used human fetal neural stem cells by direct cerebral ventricle brain injection. They found extensive cell survival and migration, which generally was not noted by other investigators.

Meier et al. [21] reported the intraperitoneal injection of human umbilical cord blood-derived mononuclear cells postnatal day 8 following carotid ligation and hypoxia on day 7. A few of the cells were incorporated into injured brain.

We should point out, however, that at least one report has been negative. DePaula et al. [22] have reported that severe neonatal rodent HI was not altered by intravenous human umbilical cord blood.

Mechanism of Action of Stem Cells in Acute Studies

The mechanism of benefit with stem cell treatment in acute neonatal HI is uncertain, but there are several lines of thinking which are prominent, and all of the proposed mechanisms may play a role in an additive fashion.

The first possible beneficial action is the actual replacement of intrinsic cells by the transplanted stem cells. While a small percentage of stem cells engraft, and some may even take the form of neurons, their numbers do not appear to be adequate to support much enhancement of outcome [19, 23, 24].

The other prominent ideas include blood vessel regeneration, an effect on the spleen to reduce its release of inflammatory cells which might have a deleterious effect in brain, and finally, and most likely, improved survival of intrinsic brain cells promoted by various noncellular factors.

Blood vessel regeneration potentially occurs via adhesion of CXCR4-positive cells onto vascular endothelium [25], recruitment of endothelial progenitor cells [26], and formation of periendothelial vascular cells [27]. Also, crude bone marrow cells have been shown to form endothelial cells [28].

Vendrame et al. [29, 30] proposed that umbilical cord blood may reduce the brain's inflammatory response by acting on the spleen to reduce its output of inflammatory cells. These splenic inflammatory cells apparently increase the blood–brain barrier permeability, which produces an adverse effect. Walker et al. [31] showed that intravenous injection of MAPCs impeded the splenic response to injury by lessening the spleen's release of inflammatory cells in association with improved outcome. The work of Leonardo et al. [32] confirmed these findings.

There are numerous mechanisms by which increased intrinsic cell survival might occur. Wang et al. [12] reported that neural stem cell proliferation in the affected tissue may be promoted by the Sonic hedgehog signaling pathway. Rosenkranz et al. [33] noted that umbilical cord cell transplantation decreased the expected rise in pro-inflammatory cytokines (interleukin-1 alpha, interleukin-1beta, and tumor necrosis factor alpha) occurring after HI injury. A marker for activated microglia and macrophages was reduced. Inflammation would thereby be reduced in brain [33]. Bae et al. [34] agreed that paracrine influences were the primary influence on intrinsic cell survival. Human umbilical cord blood cells reduced upregulation of CX43 and thereby brought about attenuation of reactive gliosis [35]. Overall, inflammation was reduced. Rosenkranz et al. [36] found various increased proteins which promoted angiogenesis. Apoptosis was lessened. Vascular endothelial growth factor (VEGF) increased perhaps increasing the proliferation of blood vessels. The same group also reported that human umbilical cord blood cells increased SDF-1 expression acting in concert with the SDF-1/CXCR4 axis [2], which appears to be a major factor for cell homing. In a primate model, Li et al. [37] reported that cell transplantation caused upregulation of IL-10 expression. Neuronal apoptosis was decreased and neuronal proliferation increased. Dayer et al. [38] found that neural progenitor cells increase expression of fibroblast growth factor-2, thereby

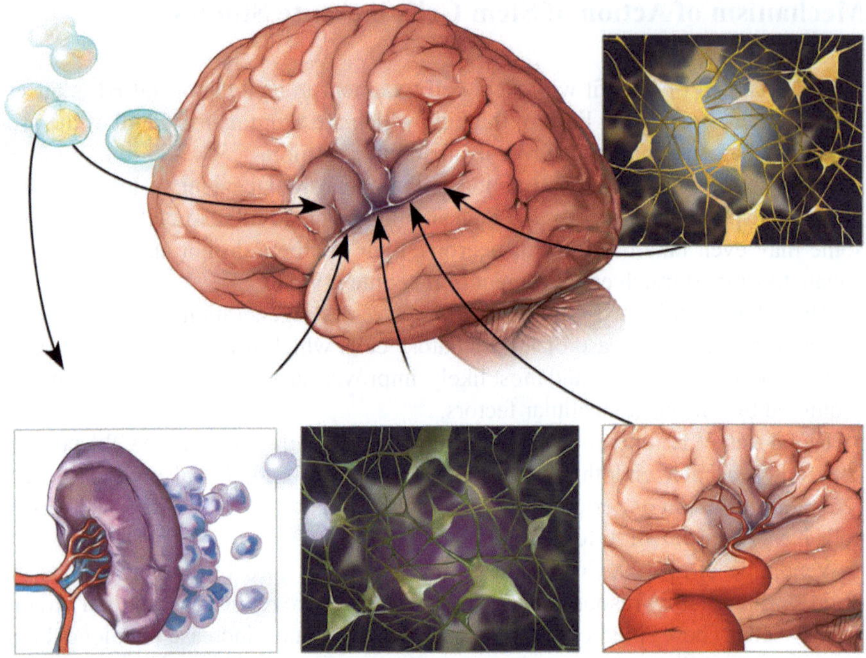

Fig. 16.1 The possible mechanisms of stem cell benefit are shown *above*. The drawing in the *upper right hand corner* depicts replacement of damaged cells. In the *left lower* drawing, the reduction in splenic release of inflammatory cells is shown. In the *right lower corner*, blood vessel regeneration is shown. In the *lower center*, the salvage of intrinsic neurons is depicted. (Courtesy of Pediatric Research)

improving HI outcome. These cells are thought to generate a pool of immature neurons available for repair (Fig. 16.1).

Summary of Acute Clinical Studies

There are no reports available as yet of controlled, double-blinded trials with stem cells in acute neonatal HI injury.

Luan et al. [39] reported a 75-day-old male who suffered severe HI injury at birth. The infant received transplantation of fetal neural stem cells into the cerebral ventricle. The infant showed improvement. There were no controls noted. In view of the relatively late transplant, it is unclear whether this child should be considered an example of acute or chronic HI injury.

Three trials in progress are listed in ClinicalTrials.gov. These are (1) Cord Blood for Neonatal Hypoxic-Ischemic Encephalopathy at Duke University, (2) Autologous Umbilical Cord Blood Transfusion for Preterm Neonates at Ain Shams University in Cairo, and (3) Autologous Stem Cells in Newborns with Oxygen Deprivation at

Hospital Universitario in Monterrey, Mexico. These trials appear to be confined to single institutions.

In any single medical center, there are only relatively small numbers of term infants with acute HI encephalopathy (HIE). Consequently, considerable time will be required to acquire the necessary numbers of subjects from one hospital system. In view of the fact there is ample preclinical laboratory evidence for benefit and a robust, multileveled rationale for the use of adult stem cells in the treatment of acute HIE, it is striking that multicentered, controlled trials have not been initiated. The cost of such a trial will be significant.

Cotton et al. [40] reported the treatment of 23 infants suffering from acute HI injury with intravenous autologous umbilical cord blood. This study demonstrated that, with considerable institutional preparation and coordination, it is feasible to conduct a trial with fresh, autologous umbilical cord blood in acute HI injury in newborns. The effort used to bring about this preliminary trial was considerable.

One of the problems inhibiting such a trial is the lack of a uniform, widely available cell preparation. Although autologous umbilical cord blood could be used for this purpose, there is some variability in the quality of preparation among the samples obtained, which would undoubtedly influence the outcome of the study. A widely available commercial preparation would be advantageous.

Characteristics of Chronic Clinical Cerebral Palsy

As we have noted, CP is a heterogeneous group of motor disorders of childhood. The condition is nonprogressive and may occur as a result of many specific disorders of brain. HI injury in the neonate comprises only a minority of the disorders among children with CP. While CP may manifest as hypotonia, ataxia, hemiparesis, dystonia or spasticity, the form resulting from neonatal HI brain injury is either spastic CP or spastic–dystonic CP. Therefore, in any clinical trial of chronic HI injury CP, the clinical inclusion criteria would need to be confined to spastic or spastic–dystonic CP.

In order to choose the subjects for such a trial the inclusion and exclusion criteria will need to be carefully constructed. The study should be confined to term or near term infants. The criteria composed by the American Congress of Obstetricians and Gynecologists [41] may prove useful. The four critical criteria are: (1) evidence of metabolic acidosis in arterial umbilical cord blood, (2) neonatal encephalopathy in infants born at 34 or more weeks of gestation, (3) CP of the spastic or dyskinetic type, and (4) exclusion of other causes. Other criteria that could be used as confirmatory would include a specific HI event, significant fetal bradycardia with abnormal decelerations, persistently low Apgar, evidence of adverse effect in other organs, and a confirmatory imaging study. Magnetic resonance imaging (MRI) is the best tool for this determination [42]. In the first 24 h, diffusion-weighted images may demonstrate increased signal intensity in central brain regions such as basal ganglia. This finding is seen in so-called severe asphyxia. T1 and T2 weighted images are more conveniently done at around 7 days of age when they diagnostically useful. Less severe or "partial" asphyxia results in a watershed pattern of injury with relative sparing of the basal ganglia.

Animal Models of Chronic HIE

Among the various animal models for CP perhaps the greatest difficulty is producing an injury which is sustained over a significant period of time. Rodents tend to die or recover quickly from brain injuries and consequently their motor manifestations of the injury fade rapidly. A sustained motor disability is essential to the use of the model in the experimental treatment of an injury designed to mimic long-standing brain injury.

Most of the models are variations of the Rice–Vanucci method [43] which impede delivery of oxygen to the animal brain either before or around the time of delivery. In the Rice–Vannucci method, the investigator ligates a unilateral carotid artery and then delivers a period of hypoxia. Variations on this theme have been conducted in rodents, piglets, and fetal sheep [44]. None of these variations are totally satisfactory.

Prenatal HI injury has been produced in the rabbit [45], but the injury is not sufficiently long-lasting to mimic CP in children. The induction of HI injury in prenatal rabbits by uterine ischemia results in motor disability in the animals, but once again the disability is not long lasting, perhaps less than 2 weeks. The motor disability may be assessed with a swim test [46]. The model is also amenable to study with MRI [47].

Maternal administration of lipopolysaccharide to pregnant rats late in gestation results in significant motor impairment of the offspring, but these findings tend to fade by 5 weeks [48]. A similar model has been produced in rabbits [49, 50]. This type of injury, however, does not necessarily mimic term infant HI injury and therefore would not be appropriate for a preclinical study of stem cell therapy for chronic HI injury. The model is more reminiscent of the periventricular leukomalacia of premature infants rather than term infant HI injury.

All these models suffer from the lack of similarity in structure to the human brain. The rodent models are particularly dissimilar in that the white matter is considerably less than in the mammal and differing in structure. The models overall are comparatively short-lived in terms of the motor abnormalities.

Clinical Trials in Chronic HIE

Stem cell therapy is widely proposed as a treatment for CP. Most of these therapies are offered for a fee in countries outside the USA and on a compassionate basis in at least one site in the USA. The treatment, in general, appears to be safe but the efficacy of the treatment is not clear. The problem is compounded considerably by the fact that CP is comprised by many different disorders only one of which (neonatal HI injury) is the subject of this chapter.

We are aware of several reports showing benefit. None of the reports address specifically the issue of the treatment of chronic neonatal HI injury.

Min et al. [51] reported benefit in a controlled study with allogeneic umbilical cord blood. To date, this is best study of the possible benefit of stem cell therapy in CP. In this study, erythropoietin was added to the cells. The three groups of patients were (1) the cells plus erythropoietin, (2) erythropoietin alone, and (3) the placebo group. They measured the gross motor performance measure, Bayley scales of infant development, F-fluorodeoxyglucose positron emission tomography, and diffusion tensor imaging. The first group had had significantly improved scores on the motor function measure and the Bayley scales. The diffusion tensor imaging and positron emission studies seemed to relate to the benefit seen with the cells plus erythropoietin. The subjects had a variety of diseases as a cause of their CP: periventricular leukomalacia, "diffuse encephalopathy", focal ischemia or hemorrhage, multicystic encephalomalacia, or cerebral malformations. It was not clear how many of the children sustained HI injury as a cause of their chronic disability. Because the children were not segregated as to the cause of their CP, we cannot draw conclusions about the possible benefit of the cells in chronic HI injury.

Chen et al. [52] reported an observer-blinded controlled study of 60 children with CP using autologous, bone-marrow derived neural stem cell like cells. The cells were implanted into the subarachnoid cavity. Thirty children received the treatment and 30 were controls. Motor function scores and language measurements were made on the children, and the motor function scores improved significantly in the treated group. No adverse effects were reported. Once again, the causes of CP among the subjects were not noted, so it is not possible to draw conclusions about the benefit of cell therapy in chronic HI injury.

Wang et al. [53] reported an uncontrolled study of 52 patients with CP who received transplantation of autologous bone marrow derived mesenchymal stem cells. The patients were reported to show improvement. The cells were delivered into the spinal fluid by lumbar puncture or into the brain by stereotactic surgery. Since the patients were not segregated as to cause, no conclusions can be drawn about benefit in chronic HI injury.

Li et al. [54] reported an 11-year-old male patient with CP who had sustained a brain injury from birth asphyxia. The child had severe visual impairment and was treated with four infusions of autologous bone marrow mesenchymal stem cells. His vision and motor abilities improved with the treatment. This type of study underlines the problem in the study of possible benefit in CP patients in that clinical findings may change as the child develops. In addition, new skills may be learned, albeit at a slower than normal rate.

Luan et al. [55] reported neural progenitor cell transplantation in children with CP. Forty-five children with CP received fetal tissue derived neural progenitor cells by injection into the lateral ventricle. Developmental improvement was noted to be significantly greater in the treated group as compared to the control subjects. The cause of the CP in the patients was not well defined.

These trials suffer from the deficiencies of either absent or inadequate controls or incomplete definition of the subjects' disease processes. Consequently, no conclusion can be drawn about the efficacy of stem cell therapy for the chronic form of neonatal HI injury. An undetermined number of the subjects likely had neonatal HI brain injury as the cause of their disability.

Our Trial

We are currently conducting a trial of autologous umbilical cord blood in children with CP (Project Title: [611187-2] "A Placebo-Controlled, Observer-Blinded, Crossover Study to Evaluate the Safety and Effectiveness of a Single Autologous Cord Blood Stem Cell Infusion for the Treatment of Cerebral Palsy in Children"). As yet, the trial is incomplete. The study protocol consists of administration of the cells on the first or second visit with the visits. The first and second visits are separated by 3 months. Half the children receive the cells on the first visit and half on the second visit in a double-blinded fashion. Saline placebo is given on the first or second visit. Follow-up has been designed to occur over the course of the year, but follow-up among the children recruited and treated has been difficult. Because of the anecdotal reports in the media of benefit of the treatment, parents are, of course, mainly interested in their child receiving the cells. Consequently, the parents are often not desirous of returning for follow-up after they are certain the cells have been administered. The second difficulty is that the construction of rigid but reasonable inclusion and exclusion criteria results in relatively small numbers of eligible patients. Finally, the study is not restricted to patients whose CP has resulted from neonatal HI injury. Consequently, most are not clearly in the neonatal HI category. Thus, the trial would not qualify as one for the assessment of chronic HI injury.

Methods of Assessment for Clinical HIE Injury

The prototype method of assessment is the general physical and neurological examination. Standardized examination protocols include the gross motor performance measure, gross motor functional measure, and the Bayley scales of infant development. MRI assessment may include routine scanning, diffusion tensor imaging, and fractional anisotropy. Fluorodeoxyglucose positron emission tomography may also be employed. MRI spectroscopy for lactate may provide a useful adjunct as lactate is increased with asphyxia.

What Should Be Done in Trials for Acute HIE

Acute neonatal HI injury is an ideal candidate for a clinical trial. The preclinical data supporting the treatment are robust. As noted above there are several plausible scientific explanations for the benefit of the cells noted in animal models.

The first matter to be decided is the choice of cell type for the trial. There are several candidates. One obvious choice is autologous umbilical cord cells. Another possibility would be MAPCs which could be supplied as a shelf-ready product that does not require cell matching. Other cell preparations, such IPS cells, could also be used. The choice among these cell types should be decided by head-to-head preclinical experiments comparing the cell types directly in animal experiments.

Second, the patient population should be characterized in a very specific fashion. Mild HIE and very severe HIE should be excluded. The trial would need to be an "add-on" to the now standardized cooling procedure.

Third, in order to insure adequate numbers of subjects, the trial would require a large number of centers with central coordination of the trial. No single center has enough patients with HIE to satisfy the numbers needed.

In the trial, much of the outcome evaluations would need to be done in the acute or subacute period, but in addition, long-term follow-up would be necessary until cognitive assessments could properly be done. These would include the testing mentioned above and perhaps more detailed intelligence quotient testing.

What Should Be Done in Trials for Chronic HIE

Possible trials for chronic HIE will be decidedly more difficult. First, the preclinical data supporting the scientific rationale for such a trial is quite minimal. Second, while there are anecdotal reports of clinical efficacy and several positive reports in the peer-reviewed literature, the controlled trials reporting efficacy deal with subjects with potentially different types of CP. As noted, CP has many causes and only a minority, perhaps 10–20 %, have resulted from HI injury.

First, a decision needs to be made about the best choice for a cell type. Given the problems noted above with the models for chronic HI injury, the choice may have to be made based on limited information.

Second, the clinical inclusion and exclusion criteria should be carefully constructed. These criteria should be based on the occurrence at birth of a well-documented HI injury, using criteria that would lead clinically in ideal circumstances to the use of the cooling protocol. The type of CP resulting should be identical among the subjects, that is, spastic quadriplegia with or without dystonic features. The clinical severity of the CP should be in the moderate to the severe range.

Finally, the trial should be multicentered in order to have sufficient numbers of subjects.

References

1. Volpe, J. Neurology of the newborn. 5th Ed. Philadelphia: Saunders/Elsevier Health Sciences; 2008. pp. 245–480.
2. Rosenkranz K, Kumbruch S, Lebermann K, et al. The chemokine SDF-1/CXCL12 contributes to the 'homing' of umbilical cord blood cells to a hypoxic-ischemic lesion in the rat brain. J Neurosci Res. 2010;88:1223–33.
3. Jiang L, Newman M, Saporta S, et al. MIP-1alpha and MCP-1 induce migration of human umbilical cord blood cells in models of stroke. Curr Neurovasc Res. 2008;5:118–24.
4. Lu X, Alshemali S, de Wynter EA, et al. Mesenchymal stem cells from CD34(-) human umbilical cord blood. Transfus Med. 2010;20:178–84
5. Tanaka N, Kamei N, Nakamae T, et al. CD133+ cells form human umbilical cord blood reduce cortical damage and promote axonal growth in neonatal rat organ co-cultures exposed to hypoxia. Int J Dev Neurosci. 2010;28:581–7.
6. Jiang T, Vaessen B, Lenvik T, et al. Multipotent progenitor cells can be isolated from postnatal murine bone marrow, muscle, and brain. Exp Hematol. 2002a;30:896–904.
7. Jiang Y, Jahagirdar B, Reinhardt RL, et al. Pluripotency of mesenchymal stem-cell-derived from adult marrow. Nature. 2002b;418:41–9.
8. Kazutoshi T, Yamanaka S. Induction of pluripotent stem cells from mouse embryonic and adult fibroblast cultures by defined factors. Cell. 2006;126:663–76.
9. Robbins RD, Prasain N, Maier BF, et al. Inducible pluripotent stem cells: not quite ready for prime time? Curr Opin Organ Transplant. 2010;15:61–7.
10. Greggio S, de Paula S, Azevedo PN, et al. Intra-arterial transplantation of human umbilical cord blood mononuclear cells in neonatal hypoxic-ischemic rats. Life Sci. 2014;96:33–9.
11. Huang HZ, Wen XH, Liu H, et al. Human umbilical cord blood mononuclear cell transplantation promotes long-term neurobehavioral functional development of newborn SD rats with hypoxic-ischemic brain injury. Zhonghua Er Ke Za Zhi. 2013;51:460–6.
12. Wang XL, Zhao YS, Hu MY, et al. Umbilical cord blood cells regulate endogenous neural stem cell proliferation via hedgehog signaling in hypoxic ischemic neonatal rats. Brain Res. 2013;1518:26–35.
13. Yamagata M, Yamamoto A, Kato E, et al. Human dental pulp-derived stem cells protect against hypoxic-ischemic brain injury in neonatal mice. Stroke. 2013;44:551–4.
14. de Paula S, Greggio S, Marinowic DR, et al. The dose-response effect of acute intravenous transplantation of human umbilical cord blood cells on brain damage and spatial memory deficits in neonatal hypoxia-ischemia. Neuroscience. 2012;210:431–41.
15. Park S, Koh SE, Maeng S, et al. Neural progenitors generated from the mesenchymal stem cells of first-trimester human placenta matured in the hypoxic-ischemic rat brain and mediated restoration of locomotor activity. Placenta. 2011;32:269–76.
16. Pimental-Coelho PM, Magalhaes ES, Lopes LM, et al. Human cord blood transplantation in a neonatal rat model of hypoxic-ischemic brain damage: functional outcome related to neuroprotection in the striatum. Stem Cells Dev. 2010;19:351–8.
17. Xia G, Hong X, Chen X, et al. Intracerebral transplantation of mesenchymal stem-cell-derived from human umbilical cord blood alleviated hypoxic ischemic brain injury in rat neonates. J Perinat Med. 2010;38:215–21.
18. Yasuhara T, Matsukawa N, Yu G, et al. Behavioral and histological characterization of intrahippocampal grafts of human bone mararrow-derived multipotent progenitor cells in neonatal rats with hypoxic-ischemic injury. Cell Transplant. 2006;15:231–8.
19. Yasuhara, T, Hara K, Maki M, Mays RW, et al. Intravenous grafts recapitulate the neurorestoration afforded by intracerebrally delivered multipotent adult progenitor cells in neonatal hypoxic-ischemic rats. J Cereb Blood Flow Metab. 2008;28:1804–10.
20. Qu SQ, Luan Z, Yin GC, et al. Transplantation of human fetal neural stem cells into cerebral ventricle of the neonatal rat following hypoxic-ischemic injury: survival, migration and differentiation. Zhongua Er Ke Za Zhi. 2005;43:576–9.

21. Meier CA, Middelanis J, Wasielewski B, et al. Spastic paresis after perinatal brain damage in rats is reduced by human cord blood mononuclear cells. Pediatr Res. 2006;59:244–9.
22. de Paula S, Vitola AS, Greggio S, et al. Hemispheric brain injury and behavioral deficits induced by severe neonatal hypoxia-ischemia in rats are not attenuated by intravenous administration of human umbilical cord blood cells. Pediatr Res. 2009;65:631–5.
23. Riess P, Zhang C, Saatman KE, et al. Transplanted neural stem cells survive, differentiate, and improve neurological motor function after experimental traumatic brain injury. Neurosurgery. 2002;51:1043–52.
24. Zhao LR, Duan WM, Reyes M, et al. Human bone marrow stem cells exhibit neural phenotypes and ameliorate neurological deficits after grafting into the ischemic brain of rats. Exp Neurol. 2002;174:11–20.
25. Peled A, Kollet O, Ponomaryov T, et al. The chemokine SDF-1 activates the integrins LFA-1, VLA-4, and VLA-5 on immature human CD34(+) cells: role in transendothelial/stromal migration and engraftment of NOD/SCID mice. Blood. 2000;95:3289–96.
26. Yamaguchi J, Kusano KF, Masuo O, et al. Stromal cell-derived factor-1effects on ex vivo expanded endothelial progenitor cell recruitment for ischemic neovascularization. Circulation. 2003;107:1322–8.
27. Rajantie I, Ilmonen M, Alminaite A, et al. Adult bone marrow-derived cells recruited during angiogenesis comprise precursors for periendothelial vascular mural cells. Blood. 2004;104:2084–6.
28. Borlongan CV, Lind JG, Dillon-Carter O, et al. Bone marrow grafts restore cerebral blood flow and blood brain barrier in stroke rats. Brain Res. 2004;1010:108–16.
29. Vendrame M, Gemma C, de Mesquita D, et al. Anti-inflammatory effects of human cord blood cells in a rat model of stroke. Stem Cells Dev. 2005;14:595–604.
30. Vendrame M, Gemma C, Pennypacker KR, et al. Cord blood rescues stroke-induced changes in splenocyte phenotype and function. Exp Neurol. 2006;199:191–200.
31. Walker PA, Shah SK, Jimenez F, et al. Intravenous multipotent adult progenitor cell therapy for traumatic brain injury: preserving the blood brain barrier via an interaction with splenocytes. Exp Neurol. 2010;225:341–52.
32. Leonardo CC, Hall AA, Collier LA, et al. Human umbilical cord blood cell therapy blocks the morphological change and recruitment of CD11b-expressing, isolectin-binding proinflammatory cells after middle cerebral artery occlusion. J Neurosci Res. 2010;88: 1213–22.
33. Rosenkranz K, Tenbruch M, May C, et al. Changes in interleukin-1 alpha serum levels after transplantation of umbilical cord blood cells in a model of perinatal hypoxic-ischemic brain damage. Ann Anat. 2013;195:122–7.
34. Bae SH, Kong TH, Lee HS, et al. Long-lasting paracrine effects of human cord blood cells on damaged neocortex in an animal model of cerebral palsy. Cell Transplant. 2012;21: 2497–515.
35. Waisielewski B, Jensen A, Roth-Harer A, et al. Neuroglial activation and Cx43 expression are reduced upon transplantation of human umbilical cord blood cells after perinatal hypoxic-ischemic injury. Brain Res. 2012;1487:39–53.
36. Rosenkranz K, Krumbach S, Tenbusch M, Marcus K, Marschner K, Dermietzel R, Meier C. Transplantation of human umbilical cord blood cells mediated beneficial effects on apoptosis, angiogenesis and neuronal survival after hypoxic-ischemic brain injury in rats. Cell Tissue Res. 2012;348:429–38.
37. Li J, Zhu H, Liu Y, et al. Human mesenchymal stem cell transplantation protects against cerebral ischemic injury and upregulates interleukin-10 expression in Macaca fascicularis. Brain Res. 2010;1334:65–72.
38. Dayer AG, Jenny B, Sauvain MO, et al. Expression of FGF-2 in neural progenitor cells enhances their potential for cellular brain repair in the rodent cortex. Brain. 2007;130: 2962–76.
39. Luan Z, Yin GC, Hu XH, et al. Treatment of an infant with severe neonatal hypoxic-ischemic encephalopathy sequelae with transplantation of human neural stem cells into cerebral ventricle. Zhonghua Er Ke Za Zhi. 2005;43:580–3.

40. Cotten MG, Murtha AP, Goldberg RN, et al. Feasibility of autologous cord blood cells for infants with hypoxic-ischemic encephalopathy. J Pediatr. 2014;164:973–979.e1. doi:10.1016/j.peds.2013.11.036.
41. Committee on Obstetrics Practice: Inappropriate use of the terms fetal distress and birth Asphyxia. Obstet Gynecol. 2005;106:1469–70.
42. Huang BY, Castillo M. Hypoxic-ischemic brain injury: imaging findings from birth to adulthood. Radiographics. 2008;28:417–39.
43. Rice JE, Vanucci RC, Brierly JB. The influence of immaturity on hypoxic-ischemic brain damage in the rat. Ann Neurol. 1981;9:131–41.
44. Johnston MV, Ferriero DM, Vannucci SJ, et al. Models of cerebral palsy; which ones are better. J Child Neurol. 2005;20:984–7.
45. Derrick M, Drobyshevsky A, Ji X, et al. A model of cerebral palsy from fetal hypoxia-ischemia. Stroke. 2007;38(2 Suppl):731–5.
46. Derrick M, Drobyshevsky A, Ji X, et al. Hypoxia-ischemic causes persistent movement deficits in a perinatal rabbit model of cerebral palsy: assessed by a new swim test. Int J Dev Neurosci. 2009;27:549–57.
47. Droboshevsky A, Derrick M, Luo K, et al. Near-term fetal hypoxia-ischemia in rabbits: MRI can predict muscle tone abnormalities and deep brain injury. Stroke. 2012;43:2757–63.
48. Rousset CL, Kassem J, Aubert A, et al. Maternal exposure to lipopolysaccharide leads to transient motor dysfunction in neonatal rats. Dev Neurosci. 2013;35:172–81.
49. Kannan S, Saadani-Makki F, Balakrishnan B, et al. Magnitude of [(11)C]PK11195 binding is related to severity of motor deficits in a rabbit model of cerebral palsy induced by intrauterine endotoxin exposure. Dev Neurosci. 2011;33:231–40.
50. Stigger F, Felizzola AL, Kronbauer GA, et al. Effects of fetal exposure to lipolysaccharide, perinatal anoxia and sensorimotor restriction on motor skills and musculoskeletal tissue: implication for an animal model of cerebral palsy. Exp Neurol. 2011;228:183–91.
51. Min K, Song J, Kang JY, et al. Umbilical cord blood therapy potentiated with erythropoietin for children with cerebral palsy: a double-blind, randomized, placebo-controlled trial. Stem Cells. 2013;31:581–91.
52. Chen G, Wang Y, Xu Z, et al. Neural stem-cell-lie cells derived from autologous bone mesenchymal stem cells for the treatment of patients with cerebral palsy. J Transl Med. 2013;11:21.
53. Wang X, Cheng H, Hua R, et al. Effects of bone marrow mesenchymal stromal cells on gross motor function measure scores of children with cerebral palsy: a preliminary clinical study. Cytotherapy. 2013;15:1549–62.
54. Li M, Yu A, Zhang F, Dai G, et al. Treatment of one case of cerebral palsy combined with posterior visual pathway injury using autologous bone marrow mesenchymal stem cells. J Transl Med. 2012;10:100.
55. Luan Z, Liu W, Qu S, Du K, et al. Effects of neural progenitor cell transplantation in children with severe cerebral palsy. Cell Transplant. 2012;21(Suppl 1):591–8.

Chapter 17
Cell-Based Therapies in Neonatal Stroke

Masahiro Tsuji and Michael V. Johnston

Neonatal Stroke

Definition

Neonatal stroke encompasses a range of focal and multifocal ischemic and hemorrhagic cerebral tissue injuries, and the terms neonatal stroke and perinatal stroke are often used interchangeably [1]. There is no consensus on terminology, definition, or classification of neonatal/perinatal stroke [2]. A workshop sponsored by the National Institute of Child Health and Human Development and the National Institute of Neurological Disorders and Stroke (Maryland, USA) defined ischemic perinatal stroke as "a group of heterogeneous conditions in which there is focal disruption of cerebral blood flow secondary to arterial or cerebral venous thrombosis or embolization, between 20 weeks of fetal life through the 28th postnatal day, confirmed by neuroimaging or neuropathologic studies" [2]. This definition distinguishes perinatal stroke from neonatal hypoxic–ischemic encephalopathy (HIE). However, those two disease entities share common risk factors and mechanisms, and can coexist in the same newborn [3, 4]. Neonatal encephalopathy is a neonatal neurological syndrome with clinical features with acute brain dysfunction, which covers all of HIE cases and some portion of neonatal stroke cases [3, 5, 6] (Fig. 17.1). The two disorders present somewhat differently in the newborn period and have different appearances on magnetic resonance imaging (MRI) [7, 8].

M. Tsuji (✉)
Department of Regenerative Medicine and Tissue Engineering, National Cerebral and Cardiovascular Center, 5-7-1, Fujishiro-dai, 565–8565 Suita, Osaka, Japan
e-mail: tsuji.masahiro.ri@ncvc.go.jp

M. V. Johnston
Departments of Neurology and Pediatrics, Kennedy Krieger Institute and Johns Hopkins University, School of Medicine, 707 North Broadway, 21205 Baltimore, MD, USA
e-mail: Johnston@kennedykrieger.org

D. C. Hess (ed.), *Cell Therapy for Brain Injury,* DOI 10.1007/978-3-319-15063-5_17

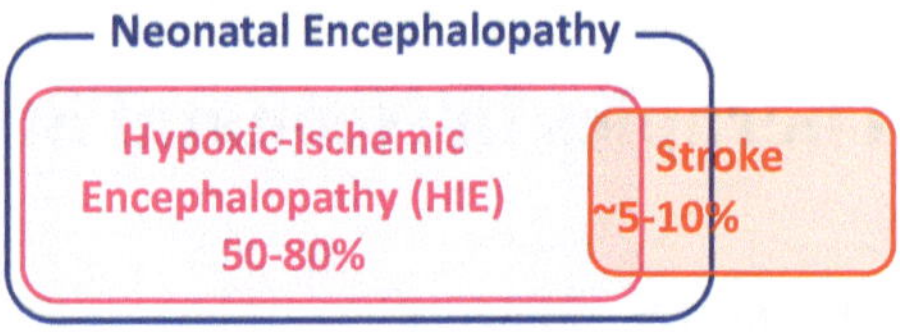

Fig. 17.1 Neonatal stroke account for up to 5–10 % of neonatal encephalopathy cases. Some neonatal stroke cases present with seizures without encephalopathy during the neonatal period. Approximately, 40 % of neonatal stroke cases do not present symptoms in the neonatal period

Incidence

Neonatal stroke is not a rare condition. The estimated incidence of ischemic perinatal arterial ischemic stroke ranges between 1 in 2300 and 1 in 5000 births, places the perinatal period second only to elderly age groups with respect to the incidence of stroke [2]. The estimated incidence of neonatal cerebral sinus venous thrombosis ranges between 1 and 2.7 in 100,000 births [1]. The incidence of neonatal encephalopathy in developed countries is 1–6 per 1000 live births [6]. From 50 to 80 % of neonatal encephalopathy cases are considered to have HIE, and up to 5–10 % of such cases are considered to have stroke [6] (Fig. 17.1).

Pediatric strokes occurring beyond the neonatal period are a serious health problem as well. The incidence of pediatric stroke is approximately 10–13 per 100,000 children per year, and cerebrovascular disorders are among the top ten causes of death in children in the USA [9]. This chapter, however, mostly focuses on neonatal stroke because of almost lack of publications in cell therapies in pediatric stroke.

Clinical Presentation

The cardinal clinical feature of neonatal encephalopathy is depressed level of consciousness, often associated with seizures [6]. Approximately, 60 % of the infants with perinatal/neonatal stroke present early symptoms, of which clonic and/or tonic seizures are most frequent (approximately 90 %) [4]. Other early symptoms include recurrent apnea/desaturation, persistently altered tone, and decreased level of consciousness. The remaining 40 % of the children with perinatal stroke do not present specific symptoms in the neonatal period, and are only recognized later with the emergence of the symptoms such as hemiplegia and seizures [4].

Treatment

The current treatment for infants with stroke is predominantly supportive, as there is no evidence-based specific treatment available [4, 10]. The onset of neonatal stroke is antenatal in some cases and is unknown in others. Hence, treatments that have a narrow therapeutic window, such as tissue plasminogen activator (tPA), are not feasible for perinatal/neonatal stroke. Cell-based-therapy has attracted attention as a novel treatment for a number of neurological diseases, including neonatal encephalopathy [11]. This is not only because of its possible regenerative properties but also because of the long therapeutic time window for the effect of stem cells. More than 1000 of therapeutic treatments for ischemic brain injury have reported neuroprotective in studies in neonatal and adult animals [12]. Although more than 100 treatments among more than 1000 candidates have been tested in clinical trials, as few as two treatments have proven clinical efficacy: tPA for adult stroke and therapeutic hypothermia for neonatal encephalopathy. Even in well-controlled animal studies, therapeutic time windows of those treatments are disappointingly short from a clinical standpoint. Most treatments are neuroprotective only when started before the insult. Although some therapies are neuroprotective with posttreatment, the therapeutic time windows are confined to first several hours after the insult. In contrast, cell therapies have been shown to have a neuroprotective effect in animal studies even when administered days after the insult [13, 14].

Preclinical Studies on Cell Therapies

Overview

More than 50 research articles on cell-based therapies for perinatal/neonatal brain injury have been published in English literature to date since the first report in this field by Elsayed et al. in 1996 [15] (Table 17.1, 17.2, 17.3). The vast majority of those studies have used rodent models of neonatal HIE. Only four studies in rodent models of neonatal stroke have been published [16–19] (Table 17.2). No study on the effects of cell-based therapy for perinatal/neonatal brain injury in large animal models has been reported, although some groups are doing research on it (abstracts of conferences and personal communications). No study in animal models of childhood stroke has been reported.

During the first decade since 1996 in this research arena, intracerebral transplantation of either the neocortical block of fetal brain or neural stem cells (NSCs) was investigated [20, 21] (Table 17.1). Systemic administration by intraperitoneal injection of cells was first reported by Guan et al. in non-English literature in 2004 [22], and by Meier et al. in English literature in 2006 [23]. Also, Guan et al. was the first to report the effects of mesenchymal stem cells (MSCs), and Meier et al. was the first to report the effects of mononuclear cells (MNCs) in neonatal models. A clinically feasible route of systemic administration by intravenous injection, was first reported by Yasuhara et al. [24]. During the second decade of this research arena, a similar number of studies using intracranial transplantation or systemic administration of stem cells have been reported. With regard to the donor cells, the MNC fraction of human umbilical cord blood (hUCB) and MSCs derived from rodent bone marrow (BM) have been most extensively explored. In many of those studies, MNCs are transfused systematically and MSCs are transplanted intracranially.

Table 17.1 Review of reported studies in models with neonatal hypoxic–ischemic brain damage

	Research group	Animal	Cell			Timing[a]	Delivery route	Follow-up	Improvement		Author
			Source	Type	Dose				Histology	Behavior	
1		P7 rat	Fetal brain	Neocortical tissue	block	7d	i.c.	6wk	no	NA	Elsayed et al. [15]
2		Neonatal rat	Fetal brain	Neocortical tissue	block	3d	i.c.	12wk	NA	yes	Jansen et al. [20]
3	A	P7 mouse	Mouse	NSC	1×10^7	7d	i.c.	12wk	yes	NA	Park et al. [21]
4	A	P7 mouse	Mouse	NSC	$0.4-1.6 \times 10^5$	3d	i.c.	4wk	NA	NA	Park et al. [25]
5	A	P7 mouse	Mouse	NT3-NSC	$\sim 3 \times 10^5$	3d	i.c.		NA	NA	Park et al. [26]
6	A	P7 mouse	Human, mouse	NSC	$\sim 3 \times 10^5$	3d	i.c.	10wk	NA	NA	Imitola et al. [27]
7		P7 rat	Neo-natal mouse	MASC	5×10^4	24 h or 5d	i.c.	3wk	NA	NA	Zheng et al. [28]
8	B	P7 rat	Rat BM	MAPC	2×10^5	7d	i.c.	3wk	NA	yes	Yasuhara et al. [29]
9	B	P7 rat	Mouse BM	MAPC	2×10^5	7d	i.c. or i.v.	3wk	yes	yes	Yasuhara et al. [24]
10	B	P7 rat	hUCB	MNC +mannitol	1.5×10^4	7d	i.v.	3wk	NA	yes	Yasuhara et al. [13]
11	C	P7 rat	hUCB	MNC	1×10^7	24 h	i.p.	2wk	NA	yes	Meier et al. [23]
12	C	P7 rat	hUCB	MNC	1×10^7	24 h	i.p.	2wk	NA	NA	Rosenkranz et al. [30]
13	C	P7 rat	hUCB	MNC	1×10^7	24 h	i.p.	6wk	yes	NA	Geißler et al. [31]
14	C	P7 rat	hUCB	MNC	1×10^7	24 h	i.p. or i.t.	7wk	yes	yes	Wasielewski et al. [32]

Table 17.1 (continued)

	Research group	Animal	Cell			Timing[a]	Delivery route	Follow-up	Improvement		Author
			Source	Type	Dose				Histology	Behavior	
15	C	P7 rat	hUCB	MNC	1×10^7	24 h	i.p.	2wk	yes	NA	Rosenkranz et al. [33]
16	C	P7 rat	hUCB	MNC	1×10^7	24 h	i.p.	2wk	NA	NA	Rosenkranz et al. [34]
17		P7 mouse	Mouse ES cell	ES cell-derived cell	1×10^4	2–3d	i.c.	8m	yes	yes	Ma et al. [35]
18		P7 rat	Fetal rat brain	NSC +ChABC	2×10^5	24 h	i.c.	8d	yes	NA	Sato et al. [36]
19	D	P7 rat	hUCB	MNC	1×10^7	24 h	i.v.	3wk	no	no	de Paula et al. [37]
20	D	P7 rat	hUCB	MNC	1×10^6, 10^7, 10^8	24 h	i.v.	8wk	yes	yes	de Paula et al. [38]
21	D	P7 rat	hUCB	MNC	1×10^6, 10^7	24 h	Intraarterial	9wk	no	yes	Greggio et al. [39]
22		P7 rat	Human BM	MSC	1×10^6	72 h	Intracardiac	6wk	no	yes	Lee et al. [40]
23		P7 rat	Human ES	NSC	1.5×10^5	24 h	i.c.	4wk	no	yes	Daadi et al. [41]
24		P7 rat	hUCB	MSC	1×10^5	3d	i.c.	4wk	yes	yes	Xia et al. [42]
25		P7 rat	hUCB	MNC	2×10^6	3h	i.p.	7d	yes	yes	Pimentel-Coelho et al. [43]
26	E	P9 mouse	Mouse BM	MSC	1×10^5	3 or 10d	i.c.	4wk	yes	yes	van Velthoven et al. [44]
27	E	P9 mouse	Mouse BM	MSC	5×10^5	10d	Intranasal	4wk	yes	yes	van Velthoven et al. [45]
28	E	P9 mouse	Mouse BM	MSC	1×10^5	3 or 3 + 10d	i.c.	4wk	yes	yes	van Velthoven et al. [46]

Table 17.1 (continued)

	Research group	Animal	Cell			Timing[a]	Delivery route	Follow-up	Improvement		Author
			Source	Type	Dose				Histology	Behavior	
29	E	P9 mouse	Mouse BM	MSC	1×10^5	3, 10, or 3+10d	i.c.	4wk	yes	NA	van Velthoven et al. [47]
30	E	P9 mouse	Mouse BM	MSC	1×10^5	3d+10d	i.c.	4wk	yes	yes	van Velthoven et al. [48]
31	E	P9 mouse	Mouse BM	MSC	0.25, 0.5, 1×10^6	3, 10, 17, or 3+10d	Intranasal	9wk	yes	yes	Donega et al. [14]
32	E	P9 mouse	Mouse BM	GM-MSC	5×10^5	10d	Intranasal	4wk	yes	yes	van Velthoven et al. [49]
33	E	P9 mouse	Mouse BM	MSC	1×10^6	10d	Intranasal	2wk	yes	NA	Donega et al. [50]
34	F	P10 rat	Mouse	NSC	5×10^5	3d	i.c.	58wk	NA	NA	Obenaus et al. [51]
35	F	P10 rat	Human	NSC	$\sim 2.5 \times 10^5$	3d	i.c.	13wk	yes	yes	Ashwal et al. [52]
36		P7 rat	Rat embryo	NSC, VEGF-NSC	1×10^5	3d	i.c.	5wk	yes	yes	Zheng et al. [53]
37		P7 rat	hUCB	MNC	1×10^7	24 h	i.v.	10wk	yes	yes	Bae et al. [54]
38		P2 mouse	Mouse ES	NPC	2×10^5	48 h	i.c.	3wk	NA	yes	Shinoyama et al. [55]
39		P5 mouse	hDP	SHED	2×10^5	24 h	i.c.	8wk	yes	yes	Yamagata et al. [56]
40		P7 rat	hDP	hDP stem cell	1×10^5	3d	i.c.	5wk	yes	yes	Fang et al. [57]
41	G	P7 rat	hUCB	MNC	3×10^6	24h	i.c.	2wk	yes	NA	Wang et al. [58]
42	G	P7 rat	hUCB	MNC	3×10^6	24h	i.c.	4wk	yes	NA	Wang et al. [59]

Table 17.1 (continued)

	Research group	Animal	Cell			Timing[a]	Delivery route	Follow-up	Improvement		Author
			Source	Type	Dose				Histology	Behavior	
43		P3 rat	hUCB	MSC	1×10^6	0, 1, and 2d	i.p.	4wk	yes	yes	Zhu et al. [60]
44		P7 mouse	New-born mouse	splenocyte	5×10^6	3wk	i.v.	3wk	NA	no	Wang et al. [61]
45		P8 mouse	Rat BM	MSC, MNC	1×10^5	48 h	i.p. or i.v.	24h	NA	NA	Ohshima et al. [62]

Studies in which cells are administered before the brain injury are not included here. Non-English literatures are not included.

A–G each alphabet indicates the same research group, *P* postnatal day, *BM* bone marrow, *hUCB* human umbilical cord blood, *ES* embryonic stem, *hDP* human dental pulp, *NSC* neural stem cell, *NT3-NSC* neurotrophin-3-expressing NSC, *MASC* multipotent astrocytic stem cell, *MAPC* multipotent adult progenitor cell, *MNC* mononuclear cell, *ChABC* chondroitinase ABC, *MSC* mesenchymal stem cell, *GM-MSC* genetically modified MSC, *VEGF-NSC* vascular endothelial growth factor transfected NSC, *SHED* stem cell from human exfoliated deciduous teeth, *i.c.* intracranial, *i.t.* intrathecal, *i.v.* intravenous, *i.p.* intraperitoneal, *NA* not assessed, *yes*, at least one test among several tests examined is improved by the cell therapy, or at least one treatment protocol among several protocols tested is beneficial

[a] timing of administration after injury

Table 17.2 Review of reported studies in models with neonatal stroke

	Animal	Model	Cell			Timing[a]	Delivery route	Follow-up	Improvement		Author
			Source	Type	Dose				Histology	Behavior	
1	P12 mouse	permanent CCAO	mouse ES cell	NSC	1×10^5	2d or 7d	into striatum	4wk	yes	no	Comi et al. [16]
2	P10 rat	permanent MCAO	hUCB	MSC	1×10^5	6 h	intraven-tricular	4wk	yes	yes	Kim et al. [17]
3	P10 rat	transient MCAO	rat BM	MSC[b]	1×10^6	3d	intranasal	4wk	yes	yes	van Velthoven et al. [18]
4	P12 mouse	permanent MCAO	hUCB	$CD34^+$ cell	1×10^5	48 h	i.v.	7wk	yes	no	Tsuji et al. [19]

P postnatal day, *CCAO* common carotid artery occlusion, *MCAO* middle cerebral artery occlusion, *ES* embryonic stem, *hUCB* human umbilical cord blood, *BM* bone marrow, *NSC* neural stem cell, *MSC* mesenchymal stem cell, *i.v.* intravenous, *yes*, at least one test among several tests examined is improved by the cell therapy, or at least one treatment protocol among several protocols tested is beneficial

[a] Timing of administration after injury

[b] MSC, or MSC overexpressing brain-derived neurotrophic factor

Table 17.3 Review of reported studies in models with excitotoxic brain damage

	Research group	Animal	Cell			Timing[a]	Delivery route	Follow-up	Improvement		Author
			Source	Type	Dose				Histology	Behavior	
1		P5–7 mouse	Human EG cell	NSC	1.5×10^5	3d	brain paren-chymal, intraventricular	10d	yes	NA	Mueller et al. [63]
2		P7 rat	Mouse ES cell	ES cell	1×10^6	8d	into injured striatum	2wk	NA	NA	Vadivelu et al. [64]
3		P5 rat	Neonatal rat	MSC		1d	near the lesion		yes	yes	Chen et al. [65]
4	A	P5 rat	Mouse embryo's brain	NDP	3×10^5	4 h or 72 h	into contralat-eral hemisphere	5m	yes	yes	Titomanlio et al. [66]
5	A	P5 rat	hUCB	MNC	$1, 3 \times 10^6$, 1×10^7	0 h or 24 h	i.p. or i.v.	5d	no	NA	Dalous et al. [67]

A same research group, *P* postnatal day, *EG* embryonic germ, *ES* embryonic stem, *hUCB* human umbilical cord blood, *NSC* neural stem cell, *MSC* mesenchymal stem cell, *NDP* neurosphere-derived precursor, *MNC* mononuclear cell, *i.p.* intraperitoneal, *i.v.* intravenous, *NA* not assessed, *yes*, at least one test among several tests examined is improved by the cell therapy, or at least one treatment protocol among several protocols tested is beneficial

[a] Timing of administration after injury

hUCB-MNCs The MNC fraction of hUCB contains many stem cell types, including hematopoietic stem cells, endothelial progenitor cells, and MSCs [68–70]. A subpopulation of cells within human UCB-MNC fraction has the potential to become neural cells [71]. hUCB-MNCs secrete a higher level of trophic factors, such as brain-derived neurotrophic factor (BDNF) and neurotrophin-4/5, than adult peripheral blood MNCs [72].

MSCs MSCs are present in BM, adipose tissue, amniotic tissue, and UCB. MSCs are easy to harvest, and are capable of differentiating into mesodermal cell lineages such as adipocytes, skeletal myoblasts, chondroblasts and osteoblasts, and neuroglial cells [69, 73]. MSCs have several favorable characteristics such as low immunogenicity in allogeneic (nonself) transplantation and no tumorigenicity [74].

Animal Models There is no widely used model of neonatal stroke. Some rodent models of stroke subject animals to transient or permanent occlusion of unilateral middle cerebral artery (MCAO) [75–78]. Other stroke models use occlusion of the common carotid artery (CCA) only [79, 80] or a combination of CCA and MCA ligation [81]. Rodent models of neonatal brain injury use postnatal day 7–12 (P7–12) rat or mouse pups, as those pups are considered comparable to human term P0 newborns with regard to brain maturation [82].

This chapter focuses on neonatal stroke, but introduces a brief summary of study data obtained in rodent models of neonatal HIE. That is because rodent models of neonatal stroke and neonatal HIE form a continuum of hypoxia–ischemia ranging from a more hypoxic to an ischemic insult with respect to pathophysiology, which also occurs in clinical settings. An extensively used rodent model of neonatal HIE, the Rice–Vannucci model, has mixed histopathology and exhibits a focal stroke in approximately half of the pups with the HI insult [83]. The rodent model of HIE subjects animals to permanent unilateral CCA occlusion (CCAO) followed by transient exposure to systemic hypoxia (30 min to 4 h, 8–10 % O_2) [83, 84].

Studies in Neonatal Stoke Models: Four Reports

The four studies that examined cell therapies in rodent models of neonatal stroke are summarized in Table 17.2.

Comi et al. reported the effects of stem-cell-based therapy in a neonatal stroke model for the first time [16]. They used P12 CD1 mice with permanent unilateral CCAO. They prepared murine embryonic stem (ES) cell-derived NSCs, and injected a suspension of 1×10^5 cells into ipsilateral striatum 2 or 7 days after the occlusion. Pups with the NSC treatment administered at 2 days, but not at 7 days, had less severe hemispheric brain atrophy compared with either nontreated or vehicle-treated pups 28 days after the occlusion. Three out of ten pups injected with the NSCs developed local tumors.

Kim et al. reported effects of MSC transplantation in P10 Sprague-Dawley rats with permanent MCAO [17]. hUCB-derived MSCs (1×10^5 cells) were administered into the ipsilateral lateral ventricle at 6 h after the occlusion. MSC transplantation improved the survival rate as well as the body weight gain after MCAO. MSC transplantation attenuated infarct volume measured by MRI at three time points examined; 3, 7, and 28 days after MCAO. MRI demonstrated the presence of transplanted super-paramagnetic iron oxide-tagged MSCs until 28 days after the transplantation. Functional deficits measured by the rotarod and cylinder tests after MCAO were partially ameliorated by the transplantation. An increased number of cells with markers for apoptotic cell death, reactive microglia, and astrogliosis in the penumbra of MCAO were also reduced by the MSC transplantation. Only a few transplanted MSCs were labeled with markers for neurons or astrocytes. The authors consider that anti-inflammatory effects mediated by increased expression of trophic factors may be the primary mechanism underlying the effects of transplanted MSCs on injured brain.

Van Velthoven and colleagues published several studies on the effects of mouse BM-derived MSCs in a mouse model of HIE, and recently examined the effects of MSCs in a rat model of neonatal stroke [18]. They used P10 Sprague-Dawley rats with transient MCAO. At 3 days post-MCAO, 1×10^6 rat BM-derived MSCs were delivered intranasally. They demonstrated that intranasally applied MSCs can migrate into the brain in the HIE model [45]. Both treatments of MSCs alone or MSCs overexpressing BDNF equally reduced infarct size and motor deficits (cylinder test and adhesive removal test) at 28 days after MCAO. Although the ischemic brain injury did not induce cell proliferation in the ipsilateral striatum and subventricular zone (SVZ), MSC treatment induced long-lasting cell proliferation up to 3 weeks after MCAO. Together with their observations in the HIE model, the authors consider that transplanted MSCs induce repair by secreting, as well as stimulating secretion of, several growth and differentiation factors, thereby stimulating endogenous neurogenesis and angiogenesis [47].

One of the authors of this chapter studied the effects of hUCB $CD34^+$ cells (hematopoietic stem cells/endothelial progenitor cells) in P12 immunocompromised mice (severe combined immunodeficiency (SCID) mice) with permanent MCAO [19]. We injected 1×10^5 hUCB $CD34^+$ cells intravenously 48 h after MCAO. The cell treatment ameliorated cerebral blood flow (CBF) in the peri-infarct area at 24 h, but not at 7 days or 7 weeks, after MCAO. Hemispheric volume loss was reduced, and the average diameter of blood vessels in the peri-infarct area was increased by the cell treatment assessed at 7 weeks after MCAO. The protective effect of cell therapy on cerebral tissue loss is significant but relatively small, and none of the cell-treated mice exhibited outstanding improvement suggesting that there may be a limit to the improvement. This implies that the cell treatment can rescue and/or restore ischemic penumbra but not the ischemic core. Behavioral deficits examined by rotarod and open-field tests were not significantly improved by hUCB $CD34^+$ cells. Few hUCB cells were identified in the brain 24 h after the intravenous injection. We believe that transient augmentation of CBF in the peri-infarct area is one of the mechanisms of action of hUCB $CD34^+$ cells.

It seems difficult to draw conclusions from the four studies discussed above, and an optimal protocol for clinical use of cell therapies against neonatal stroke has not been developed. Nevertheless, those four studies suggest that cell therapies have potential to exert neuroprotective effects even though cells are administered hours or days after brain injury.

Studies in Neonatal Models with Excitotoxicity: Five Reports

Five studies have reported the effects of cell-based therapies in rodent models with excitotoxicity, in which brain lesions are induced by intracerebral injection of excitotoxic compounds such as a glutamate analogue [63–67]. Those studies are summarized in Table 17.3. Excitotoxic neuronal injury is involved in the cascade of injury activated in the pathophysiology of both neonatal stroke and HIE [85].

Mueller et al. transplanted human embryonic germ cell-derived NSCs intracerebrally 3 days after injection of *N*-methyl-D-aspartate (NMDA) receptor agonist quinolinic acid injection [63]. Subsets of transplanted cells expressed neuronal and glial cell markers, and partially restored the complement of striatal neurons. Interestingly, the survival of hNSCs was five times greater in the lesioned brains compared with sham-control brains.

Vadivelu et al. transplanted murine ES-derived progenitor cells into injured striatum 8 days after the NMDA injection [64]. Transplanted cells differentiated into neural cells, and generated endothelial cells, which integrated with host cells to form chimeric vasculature. Unlike in adult animals, endothelial cell proliferation and vascular density in the peri-ischemic regions were limited in immature rats with transient MCAO [86]. Therefore, enhancement of angiogenesis after neonatal stroke may be a promising strategy for recovery.

Chen et al. transplanted MSCs intracerebrally 1 day after the glutamate analogue ibotenate injection [65]. MSC treatment increased anti-myelin immunoreactivity in the corpus callosum and improved reaching and retrieval skills. The study showed little evidence of differentiation into neural phenotype.

Titomanlio et al. transplanted neurosphere-derived precursors intracerebrally either 4 h or 72 h after an ibotenate injection [66]. Both early (4 h) and late (72 h) cell transplantation reduced the extent of brain lesion and memory impairments. The transplanted cells differentiated into oligodendrocyte and neurons although the transplanted cells finally died.

Dalous et al., in the same group as Titomanlio, examined the effects of hUCB-MNCs [67]. When MNCs were administered intraperitoneally either immediately or 24 h after the insult, a low-cell dose (1×10^6 cells) ameliorated the lesion size in the cortex, but a high-cell dose (1×10^7 cells) exacerbated the lesion size in the white matter at 5 days after the insult. When MNCs were administered intravenously, neither cell dose caused significant effects on the lesion size.

Mechanisms of Action

Cell-based therapies for brain injury can be divided into two categories from the standpoint of the postulated mechanisms of action: those aiming at cell replacement and those aiming at other beneficial effects. Cell therapies in the first category are designed so that administered cells survive in the brain, differentiate into neural cell lineage, and reconstruct adequate neural network with synapse formation. Those studies include all studies with brain tissue block [15, 20] and NSCs derived from ES cells or induced pluripotent stem (iPS) cells [21, 55], and some with non-NSCs, such as MSCs. Tissues and cells are directly injected into the brain in most of these studies, into brain parenchyma, cerebral ventricle, or cerebrospinal fluid space. Cell therapies in the second category are designed so that administered cells can stimulate endogenous neurogenesis and/or angiogenesis, secrete several trophic factors, modulate inflammation/immune response, and/or increase CBF. The studies in the second category include many experiments with MNCs, MSCs, multipotent adult progenitor cells (MAPCs) [24], and $CD34^{+}$ cells. Cells are injected either intracerebrally or systemically. Accumulating evidence from studies in animal models of neonatal HIE and of adult stroke shows that cell therapies have all of the abovementioned capacities as mechanisms of action. A cell therapy of any kind seems to have multiple effects for injured brain, and this multipotency may be the basis of its long therapeutic time window, thereby enabling it to translate into clinical use.

In order to develop cell-based therapies, whatever the postulated mechanism of action is, understanding the mechanisms of neurogenesis and angiogenesis after ischemic injury in the immature brain is important. Several studies in immature rodents reveal that HI injury promotes cell proliferation in the sub-ventricular zone (SVZ) [87, 88]. Studies show contradicting results with regard to the differentiation rate of proliferating cells into neurons after neonatal HI [89]. Only one study explored neurogenesis after neonatal stroke [90]. Kadam et al. labeled newly generated cells with bromodexyuridine (BrdU) injected 6–8 days after permanent unilateral CCAO in P12 CD1 mice. Neonatal ischemia significantly reduced the cell counts of new cells in the dentate gyrus of the hippocampus which are derived from sub-granular zone (SGZ), and increased the counts in the striatum and neocortex which are presumably derived from SVZ compared with sham controls. Cell-lineage commitment patterns, which are predominantly neuronal in the hippocampus and nonneuronal in the striatum and neocortex in sham controls, are maintained after neonatal ischemia. Those observations are mostly in accordance with the data presented by Zhu and colleagues in P9 C57Bl/6 mice with HI [91, 92].

Treatment Protocol

The optimal protocol for a cell therapy is not known in neonatal HIE models or adult stroke models, in which far more studies have been done compared with neonatal stroke models. Cell type, dose, timing after insult, and administration route are the crucial issues when designing the most effective and feasible protocol for clinical application.

Cell Type No study has compared effects of different cell types in models of neonatal brain injury. A study in an adult mouse model of permanent MCAO compared the effects of BM-MSCs and BM-MNCs [93]. Cells were transplanted into ipsilateral striatum. MSCs, but not MNCs, extensively migrated into the peri-infarct area. Approximately, 20 % of transplanted MSCs expressed a neuronal marker, NeuN, while 1.4 % of transplanted MNCs expressed NeuN. Those results may indicate that MSCs are more advantageous than MNCs. However, MNCs have other potential benefits for clinical use because MNCs can easily be isolated by a density gradient separation after correcting from UCB or BM, while MSCs need to be cultured in vitro in most of the cases with bovine serum.

Cell Dose A few studies in models of neonatal brain injury have compared different cell doses. Some studies demonstrate that a higher dose is more beneficial [14, 38, 39], while a study does not confirm it [67].

Timing After Insult Limited number of studies in models of neonatal brain injury have examined therapeutic time window. Van Velthoven and colleagues demonstrated that MSC treatment at either 3 days or 10 days, but not 17 days, after HI insult is similarly beneficial [14, 44, 46, 47]. Comi et al. reported NSC transplantation at 2 days, but not 7 days, after CCAO is beneficial [16]. Two studies in an excitotoxicity model showed no time-dependent effects comparing 4 h versus 72 h after insult [66] and immediate versus 24 h after insult [67]. Another study showed no difference in characteristics of grafted mutlipotent astrocytic stem cells between 24 h and 5 days after HI injury [28].

Administration Route Few studies in models of neonatal brain injury have examined the influence of different administration routes. Effects of intravenously administered UCB-MNCs and intraperitoneally administered UBC-MNCs are different [67]. In contrast, two studies showed that cells administered by different routes are equally beneficial: intravenous versus intracerebral [24], and intraperitoneal versus intrathecal [60]. Quite a few studies demonstrated beneficial effects of cell-based therapies despite the limited number and duration of donor cell survival in the brain [19, 24, 43, 47]. Two studies, one in neonatal excitotoxic and the other in neonatal HI brain injury model, compared systemic cell distributions after intravenous and intraperitoneal injections [62, 67]. Intravenously administered cells are detected in the brain more than intraperitoneally administered cells. However, the number of cell detected in the brain is relatively low even if infused intravenously. Instead, a large number of cells accumulate in the lungs, liver, and spleen. The same pattern of systemic cell distribution is noted in a juvenile primate with MCAO [62].

In addition to the limited amount of evidence, study results are often contradictory to each other as mentioned above. That may be because the studies used different models, strains, degrees and extents of injury, donor cells, evaluation methods, and time points. More studies are warranted to find optimal protocols for clinical application.

Age-Dependent Differences in Cell Therapies

It is well known that neonatal brains and adult brains are different with respect to neurodegeneration and neural repair, and this notion is also true in cell therapies. There are dynamic age-dependent changes in characteristics of both donor cells (administered cells) and recipient brains. For instance, transplanted hUCB-NSCs survived longer time in uninjured neonatal rat brain than in uninjured adult rat brain [94]. Therefore, caution should be exercised when translating evidence in adult stroke studies into neonatal stroke.

Clinical Studies on Cell Therapies

Overview

Only one clinical trial is listed in the website of the US National Institute of Health (ClinicalTrials.gov) as a cell-based therapy for neonatal or childhood stroke; ClinicalTrials.gov Identifier: NCT01700166 (Table 17.5). It is a phase 1 trial in children aged 6 weeks to 6 years who have prenatal or perinatal stroke. Children with evidence of HIE were excluded from the trial. Their own (autologous) UCB was intravenously injected to the children. This trial, however, has been withdrawn prior to enrollment due to relocation of the principal investigator.

Caution should be exercised when translating results obtained from preclinical and clinical studies in adult stroke into neonatal and childhood stroke practice. Neonatal stroke and adult stroke are different in many aspects, such as background risk factors, pathophysiology, and repair processes [10]. Neonates are generally considered to be more resistant to cerebral ischemic insults and have higher potential to repair the insult. Such assumptions, however, are not always true. For instance, children with early stroke perform more poorly compared with children with late stroke [95].

Clinical Trials During Acute Phase of Brain Injury

Clinical trials of an acute-phase treatment for childhood stroke are extremely difficult to conduct. A phase 1 clinical study with tPA for pediatric stroke (NCT01591096) was terminated due to lack of patient accrual. The study design was to administer tPA intravenously for children aged 2–17 years within 4.5 h from the onset of acute arterial ischemic stroke. Several inherent factors in childhood stroke, such as its low incidence and ambiguous time of onset, make those clinical trials difficult to conduct.

Clinical trials of an acute-phase treatment for neonatal stoke may be less difficult to conduct than those for childhood stroke. That is because its incidence is higher and concentrated in the perinatal period, and patients are already in medical institutes. However, the trials for neonatal stroke are more difficult to conduct than those for neonatal HIE. The onset of neonatal stroke is often ambiguous compared with that of neonatal HIE. Symptoms of neonatal stroke are generally milder than those of neonatal HIE. As many as 40 % of newborns with stroke are unrecognized during the neonatal period [4].

No clinical trial of cell therapy during the acute phase of neonatal stroke is listed in ClinicalTrials.gov. As mentioned above (see the section "Definition"), it is often difficult to clearly distinguish between neonatal stroke and neonatal HIE [96]. Some percentage of participants of a clinical trial for "neonatal HIE" may have neonatal stroke. Interestingly, when the landmark clinical trial of whole-body hypothermia was first published, the participating infants were referred as neonates with HIE [97]. Several years later, the follow-up study of the same cohort was published under the title of "Childhood outcomes after hypothermia for neonatal encephalopathy" [98]. A portion of infants presenting as neonatal encephalopathy has a focal stroke [3]. Hence, clinical studies for neonatal HIE are briefly reviewed here.

Three clinical trials are listed in ClinicalTrials.gov as trials for newborns with neonatal HIE; NCT00593242, NCT01506258, NCT01649648 (Table 17.4). All of three trials use intravenous injection of autologous UCB within a few days after birth. The trial NCT00593242 (principal investigator, Dr. Cotten at Duke University, USA) uses noncryopreserved volume- and red blood cell-reduced UCBCs, while NCT01506258 (Dr. Mancias-Guerra at Hospital Universitario Dr. Jose E. Gonzalez, Mexico) uses noncryopreserved $CD34^+$ UCBCs. Details of treatment protocol of NCT01649648 (Dr. Lee at National University Hospital, Singapore) are not documented in ClinicalTrials.gov. Those three trials are small nonrandomized trials, therefore efficacy of the cell therapies has not yet been known. Only NCT00593242 has been completed, and the results have been published [99]. Cotten and colleagues enrolled 23 infants who were cooled for HIE and intravenously infused UCBCs (up to 4 doses, $1–5 \times 10^7$ cells/dose, ~ 72 postnatal hours). No significant infusion reactions were noted. Of the 23 infants, 1-year neurodevelopmental outcomes were assessed utilizing the Bayley Scales of Infant and Toddler Development, third edition (Bayley III) in 18 infants, and 13 infants (72 %) had Bayley scores ≥ 85. During the study period, 82 infants did not have available UCB and were cooled for HIE. Of the 82 infants, 46 were assessed and 19 (41 %) had Bayley scores ≥ 85. Of note, 26 % of UCBC-treated infants were outborn (transported from an outside hospital after delivery), while 88 % of cooled infants without UCBC treatment were outborn. Outborn infants tend to have poor outcome than inborn infants, hence the benefit of UCBCs may be exaggerated. Nevertheless, the trial suggests that autologous UCBC infusion therapy for neonatal HIE is safe and feasible, and may improve the outcome. A randomized double-blind study is needed. A phase 1 trial almost same as the one at Duke University is about to start in Japan (principal investigator: Dr. Shintaku at Osaka City University).

A clinical trial (NCT01121328) is focusing neonates born premature, less than 34 weeks (Table 17.4). Autologous UCB mononuclear cells are infused in the first 14 days after birth.

The properties of hUCBCs may be altered by several factors, such as the gestational age and perinatal asphyxia [100, 101]; For example, Aly et al. reported that although UCB-MNC count does not differ between healthy-term newborns and term newborns with perinatal asphyxia, neuronal differentiation of hUCB-MSCs is

Table 17.4 Review of clinical trials of acute phase treatment in newborns or children with acute brain injury

	ClinicalTrials. gov	Phase	Status	Participants			Arms	Random-ization	Cell			Timing	Route	Loca-tion
	Identifier			Diagnosis	Number	Age		–	Source	Type	Dose			
1	NCT00593242	I	Com-pleted, published	HIE[a]	23	NB	1	–	auto	UCB	$1\text{-}5 \times 10^7 \times 4$	~72h	i.v.	USA[e]
2	NCT01506258	NS	NA	Asphyxia[b]	20	NB	2	–	auto	UCB $CD34^+$	NA	~48h	i.v.	Mexico
3	NCT01649648	I	Recruiting	HIE[c]	10	NB	1	–	auto	UCB	NA	~3d	NA	Singa-pore
4	NCT01121328	I	Recruiting	Preterm[d]	60	NB	1	–	auto	UCB–MNC	NA	~14d	i.v.	Egypt
5	NCT00254722	I	Com-pleted, published	TBI	10	5–14y	1	–	auto	BM-MNC	6×10^6/kg	~36h	i.v.	USA[f]
6	NCT01851083	II	Recruiting	TBI	50	5–17y	2	+	auto	BM-MNC	6×10^6/kg	~36h	i.v.	USA[f]

NS not specified, *NA* not available, *HIE* hypoxic–ischemic encephalopathy, *TBI* traumatic brain injury, *NB* newborn, *auto* autologous, *UCB* umbilical cord blood, *BM* bone marrow, *MNC* mononuclear cell, *i.v.*: intravenous

Inclusion criteria:

[a] ≥ 35wk gestation, and met cooling criteria

[b] 37–42wk, Apger <5 at 5min, UCB pH<7.0, signs of HIE

[c] >36wk, met cooling criteria

[d] <34wk, birth weight <1500g

[e] Duke Univ

[f] Univ. of Texas Health Science Center, Houston

more pronounced in the cells derived from newborns with asphyxia [102]. Apoptosis of neutrophils is impaired in cord blood compared with adult peripheral blood, and the apoptosis is reduced by hypoxia [103]. Lymphocyte counts are elevated in term infants with HIE, although the counts rapidly normalized [104]. Those alterations may become beneficial or detrimental to infants receiving UCB therapy.

Two clinical trials with the same protocol and the same principal investigator, one phase 1 and the other phase 2, are listed in ClinicalTrials.gov as an acute-phase cell-based treatment for traumatic brain injury (TBI) in childhood; NCT00254722, NCT01851083 (Table 17.4). The phase 1 trial has completed, and the results have been published [105]. Ten children with severe TBI aged 5–14 years were treated with autologous BM-MNCs, 6×10^6 cells/kg body weight delivered intravenously within 48 h after TBI. There were no episodes of BM harvest- and infusion-related adverse reactions. Efficacy of the treatment was not assessed as it was a phase 1 trial with no control group.

Luan et al. reported allogeneic transplantations of human neural precursor cells in six newborns with severe brain injury [106]. One of them had severe carbon monoxide poisoning at fifth day of birth; another had severe hypoglycemia; the others had severe neonatal HIE. Neural precursor cells were obtained from the forebrain of a spontaneously aborted fetus. The transplantations were performed at day 4 to day 20 after birth. They were followed up for 12 months. Four patients were normal in psychomotor development, and two patients developed cerebral palsy (CP).

Clinical Trials During Chronic Phase of Brain Injury

The most frequent sequela of neonatal stroke is CP, and approximately half of survivors develop CP [107, 108]. CP is a group of permanent disorders affecting motor development and posture resulting from various brain injuries that occurred during prenatal or neonatal period. Ischemic perinatal stroke accounts for 30% of hemiplegic CP [109], which accounts for 18–41% of all CP cases [110, 111]. Therefore, clinical studies for CP inevitably include children with neonatal stroke, unless type and/or etiology of CP are specified.

Thirteen trials from nine research groups are listed in ClinicalTrials.gov as cell-based therapies for children with CP (Table 17.5). Two clinical trials by the same investigators (NCT01404663, NCT01763255) set its inclusion criteria for the participant as spastic quadriplegic CP. This criterion excludes most of CP children resulting from neonatal stroke as they are generally hemiplegic. Hence, 11 trials from eight research groups are related with children with neonatal stroke.

Two of the 11 trials have been published (NCT01193660, NCT01019733). NCT01193660 is a randomized trial with 105 participants conducted in CHA University in South Korea [112]. HLA-matched allogeneic UCB consisting of $>3 \times 10^7$/kg total nucleated cells (TNCs) was intravenously injected into CP children along with erythropoietin and immunosuppressive treatment. Compared with the control group and the group treated with erythropoietin only, the UCB-treated

Table 17.5 Review of clinical trials of chronic phase treatment in children with neonatal stroke and cerebral palsy

		ClinicalTrials. gov Identifier	Phase	Status	Participants			Arms	Random-ization	Cell			Route	Loca-tion
					Diagnosis	Number	Age			Source	type	dose		
[1]		NCT01700166	I	Withdrawn	pre-, perinatal stroke	10	6wk–6y	1	–	Auto	UCB	NA	i.v.	USA[a]
1		NCT01019733	NS	Completed, published	CP with perinatal HI	18	2–5y	1	–	auto	BM-TNC	$5\text{-}54\times10^8$	i.t. and i.v.	Mex-ico
2		NCT01072370	I, II	Recruiting	CP	40	1–12y	2	+	Auto	UCB-MNC	$>1\times10^7$/kg	i.v.	USA[b]
3		NCT01147653	II	Recruiting	spastic CP	120	1–6y	2	+	Auto	UCB	$>1\times10^7$/kg	i.v.	USA[c]
4	A	NCT01193660	NS	Completed, published	CP	105	10m–10y	3	+	Allo	UCB	$>3\times10^7$/kg	i.v.	Korea
5	A	NCT01528436	II	Completed	CP	37	6m–20y	2	+	Allo	UCB	NA	i.v. or i.a.	Korea
6	A	NCT01639404	NS	Completed	CP	17	6m–20y	1	–	Allo	UCB	NA	i.v. or i.a.	Korea
7	A	NCT01991145	NS	Recruiting	CP	120	10m–6y	4	+	Allo	UCB	NA	i.v.	Korea
8		NCT01832454	II, III	Recruiting	CP	100	3–5y	1	–	Auto	BM-MNC	1×10^8	i.t.	India
9		NCT01929434	III	Before recruiting	CP	300	1–14y	3	+	NA	UCB-MSC	NA	i.t.	China
10		NCT01978821	I	Completed	CP	40	17m–22y	1	–	Auto	BM-MNC	NA	NA	India
11		NCT01988584	II	Recruiting	CP	30	2–10y	3	+	Auto	UCB or BM-MNC	NA	i.v.	USA[a]

Table 17.5 (Continued)

		ClinicalTrials. gov Identifier	Phase	Status	Participants			Arms	Randomization	Cell			Route	Location
					Diagnosis	Number	Age			Source	type	dose		
12	B	NCT01404663	I	Completed	Spastic quadriplegic CP	12	4–12y	1	–	Auto	BM $CD133^+$	NA	i.t.	Iran
13	B	NCT01763255	I, II	Completed	Spastic quadriplegic CP	8	4–12y	2	+	NA	BM $CD133^+$	NA	i.t.	Iran

A, B The same alphabet indicates the same institute
NS not specified, *CP* cerebral palsy, *HI* hypoxia-ischemia, *auto* autologous, *allo* allogeneic, *NA* not available, *UCB* umbilical cord blood, *BM* bone marrow, *TNC* total nucleated cell, *MNC* mononuclear cell, *MSC* mesenchymal stem cell, *i.v.* intravenous, *i.t.* intrathecal, *i.a.* intraarterial
[a] Univ. of Texas Health Science Center, Houston
[b] Georgia Regents Univ.
[c] Duke Univ.

group had significantly higher scores on the Gross Motor Performance Measure and Bayley II Mental and Motor scales at 6 months. The incidence of serious adverse events did not differ between groups. The same group compared allogeneic (three patients) and autologous (four patients) UCB transplantation in children with CP [113]. The allogeneic transplantation showed better outcome than autologous transplant. NCT01019733 is a phase 1 trial in 18 pediatric patients with CP associated with perinatal HI brain injury [114]. The principal investigator is the same as NCT01506258 (an acute-phase treatment for neonatal HIE; see the section "Clinical Trials During Acute Phase of Brain Injury"). After stimulation with granulocyte colony-stimulating factor, autologous BM–TNCs (a median of 13×10^8 cells) including $CD34^+$ cells (10×10^6 cells) were injected intrathecally, and the remaining cells from the BM aspiration were administered intravenously. Early adverse effects including headache, vomiting, and fever occurred in three patients. No serious adverse effects occurred. No MRI changes were found at 6 months of follow-up.

Among remaining nine trials, one trial (NCT01988584) used either autologous UCB or BM-MNCs, two trials (NCT01832454, NCT01978821) use autologous BM-MNCs, and six trials (NCT01072370, NCT01147653, NCT01193660, NCT01528436, NCT01639404, NCT01991145, NCT01929434) used either autologous or allogeneic UCB. Cells are administered intravenously in many of the trials, but intra-arterially or intrathecally in some trials. NCT01929434 is the only one phase 3 trial among them, and the only trial using MSCs, while others using MNCs or TNCs.

Safety and feasibility of intravenous infusion of cryopreserved autologous UCB have been reported in a study in children with neurological disorders, most of whom had CP [115]. Among 184 study participants (median age 27 months, range 6 days–9 years), three patients experienced infusion reaction, which resolved after discontinuation of the infusion and medical therapy. No other adverse events have been reported during the 12-month follow-up.

Apart from the clinical studies listed in ClinicalTrials.gov, there are a few case reports of cell therapies in infants with brain injury. Jensen and Hamelmann reported a boy with HI brain injury due to cardiac arrest at 2 years of age [116]. He received autologous intravenous UCB transfusion 9 weeks after the cardiac arrest. At 2-months follow-up, he demonstrated remarkable neurofunctional recovery from a vegetative state. Jansen et al. attribute the recovery to the cell therapy. Chen et al. conducted a randomized controlled trial of allogeneic transplantations of olfactory ensheathing cells (OECs) in children with CP [117]. OECs were isolated from aborted human fetal olfactory bulb, and 2×10^6 cells were injected into the corona radiata of the frontal lobe of the HLA-matched patients with no immunosuppressant therapy. Motor function scores in cell-treated group ($n=6$) were significantly better than those in the control group ($n=8$) 6 months after the intervention. They reported that no patients experienced side effects. Lee et al. conducted a pilot study of the intravenous infusion of autologous UCB in 20 children with CP [118]. There was no control group. Five children showed more improvements in neurodevelopmental evaluations than would normally be expected during a 6-month period after the infusion. The improvements occurred significantly in children with hemiplegia or diplegia rather than quadriplegia. Although those results seem promising, they are limited by the fact that they are reported in a case report and a clinical trial with very small sample size or with no control group. Therefore, it is difficult to interpret the results.

Conclusion

There is a paucity of preclinical studies and a lack of clinical data on cell-based therapies for neonatal stroke. The preclinical studies suggest that cell therapies have a potential for ameliorating infant brain injury even when the treatment is started days after the insult. Rigorous preclinical studies are needed before these therapies are applied clinically, especially when the therapy uses cells that may have a tumorigenic risk and invasive administration route. However, many patients and their parents are desperately seeking opportunities to receive cell therapies, as the current therapies for neonatal stroke as well as other neonatal brain injuries do not appear to be effective. It will be important to stratify cell therapies according to the risks involved in clinical application, and to proceed to the clinical trial without delay under the approval and monitoring of regulatory authorities when the therapy has low risk.

References

1. Rutherford MA, Ramenghi LA, Cowan FM. Neonatal stroke. Arch Dis Child Fetal Neonatal Ed. 2012;97:F377–84.
2. Raju TN, Nelson KB, Ferriero D, Lynch JK, Participants N-NPSW. Ischemic perinatal stroke: summary of a workshop sponsored by the National Institute of Child Health and Human Development and the National Institute of Neurological Disorders and Stroke. Pediatrics. 2007;120:609–16.
3. Ramaswamy V, Miller SP, Barkovich AJ, Partridge JC, Ferriero DM. Perinatal stroke in term infants with neonatal encephalopathy. Neurology. 2004;62:2088–91.
4. Chabrier S, Husson B, Dinomais M, Landrieu P, Nguyen The Tich S. New insights (and new interrogations) in perinatal arterial ischemic stroke. Thromb Res. 2011;127:13–22.
5. Ferriero DM. Neonatal brain injury. N Engl J Med. 2004;351:1985–95.
6. Volpe JJ. Neonatal encephalopathy: an inadequate term for hypoxic-ischemic encephalopathy. Ann Neurol. 2012;72:156–66.
7. Cowan F, Rutherford M, Groenendaal F, Eken P, Mercuri E, Bydder GM, Meiners LC, Dubowitz LM, de Vries LS. Origin and timing of brain lesions in term infants with neonatal encephalopathy. The Lancet. 2003;361:736–42.
8. Johnston MV. MRI for neonatal encephalopathy in full-term infants. The Lancet. 2003;361:713–4.
9. de Veber GA. Cerebrovascular disease. In: Swaiman KF, Ashwal S, Ferriero DM, editors. Pediatric Neurology, principles & practice. 4th ed. Philadelphia: Mosby; 2006. pp. 1759–801.
10. Roach ES, Golomb MR, Adams R, Biller J, Daniels S, Deveber G, Ferriero D, Jones BV, Kirkham FJ, Scott RM, Smith ER, American Heart Association Stroke C, Council on Cardiovascular Disease in the Y. Management of stroke in infants and children: a scientific

statement from a special writing group of the American Heart Association Stroke Council and the Council on Cardiovascular Disease in the Young. Stroke. 2008;39:2644–91.

11. Bennet L, Tan S, Van den Heuij L, Derrick M, Groenendaal F, van Bel F, Juul S, Back SA, Northington F, Robertson NJ, Mallard C, Gunn AJ. Cell therapy for neonatal hypoxia-ischemia and cerebral palsy. Ann Neurol. 2012;71:589–600.
12. O'Collins VE, Macleod MR, Donnan GA, Horky LL, van der Worp BH, Howells DW. 1,026 experimental treatments in acute stroke. Ann Neurol. 2006;59:467–77.
13. Yasuhara T, Hara K, Maki M, Xu L, Yu G, Ali MM, Masuda T, Yu SJ, Bae EK, Hayashi T, Matsukawa N, Kaneko Y, Kuzmin-Nichols N, Ellovitch S, Cruz EL, Klasko SK, Sanberg CD, Sanberg PR, Borlongan CV. Mannitol facilitates neurotrophic factor up-regulation and behavioural recovery in neonatal hypoxic-ischaemic rats with human umbilical cord blood grafts. J Cell Mol Med. 2010;14:914–21.
14. Donega V, van Velthoven CT, Nijboer CH, van Bel F, Kas MJ, Kavelaars A, Heijnen CJ. Intranasal mesenchymal stem cell treatment for neonatal brain damage: long-term cognitive and sensorimotor improvement. PloS one. 2013a;8:e51253.
15. Elsayed MH, Hogan TP, Shaw PL, Castro AJ. Use of fetal cortical grafts in hypoxic-ischemic brain injury in neonatal rats. Exp Neurol. 1996;137:127–41.
16. Comi AM, Cho E, Mulholland JD, Hooper A, Li Q, Qu Y, Gary DS, McDonald JW, Johnston MV. Neural stem cells reduce brain injury after unilateral carotid ligation. Pediatr Neurol. 2008;38:86–92.
17. Kim ES, Ahn SY, Im GH, Sung DK, Park YR, Choi SH, Choi SJ, Chang YS, Oh W, Lee JH, Park WS. Human umbilical cord blood-derived mesenchymal stem cell transplantation attenuates severe brain injury by permanent middle cerebral artery occlusion in newborn rats. Pediatr Res. 2012;72:277–84.
18. van Velthoven CT, Sheldon RA, Kavelaars A, Derugin N, Vexler ZS, Willemen HL, Maas M, Heijnen CJ, Ferriero DM. Mesenchymal stem cell transplantation attenuates brain injury after neonatal stroke. Stroke. 2013;44:1426–32.
19. Tsuji M, Taguchi A, Ohshima M, Kasahara Y, Sato Y, Tsuda H, Otani K, Yamahara K, Ihara M, Harada-Shiba M, Ikeda T, Matsuyama T. Effects of intravenous administration of umbilical cord blood $CD34^+$ cells in a mouse model of neonatal stroke. Neuroscience. 2014;263:148–58.
20. Jansen EM, Solberg L, Underhill S, Wilson S, Cozzari C, Hartman BK, Faris PL, Low WC. Transplantation of fetal neocortex ameliorates sensorimotor and locomotor deficits following neonatal ischemic-hypoxic brain injury in rats. Exp Neurol. 1997;147:487–97.
21. Park KI, Teng YD, Snyder EY. The injured brain interacts reciprocally with neural stem cells supported by scaffolds to reconstitute lost tissue. Nat Biotechnol. 2002;20:1111–7.
22. Guan XQ, Yu JL, Li LQ, Liu GX. Study on mesenchymal stem cells entering the brain through the blood-brain barrier. Zhonghua er ke za zhi. 2004;42:920–3.
23. Meier C, Middelanis J, Wasielewski B, Neuhoff S, Roth-Haerer A, Gantert M, Dinse HR, Dermietzel R, Jensen A. Spastic paresis after perinatal brain damage in rats is reduced by human cord blood mononuclear cells. Pediatr Res. 2006;59:244–9.
24. Yasuhara T, Hara K, Maki M, Mays RW, Deans RJ, Hess DC, Carroll JE, Borlongan CV. Intravenous grafts recapitulate the neurorestoration afforded by intracerebrally delivered multipotent adult progenitor cells in neonatal hypoxic-ischemic rats. J Cereb Blood Flow Metab. 2008;28:1804–10.
25. Park KI, Hack MA, Ourednik J, Yandava B, Flax JD, Stieg PE, Gullans S, Jensen FE, Sidman RL, Ourednik V, Snyder EY. Acute injury directs the migration, proliferation, and differentiation of solid organ stem cells: evidence from the effect of hypoxia-ischemia in the CNS on clonal "reporter" neural stem cells. Exp Neurol. 2006a;199:156–78.
26. Park KI, Himes BT, Stieg PE, Tessler A, Fischer I, Snyder EY. Neural stem cells may be uniquely suited for combined gene therapy and cell replacement: Evidence from engraftment of Neurotrophin-3-expressing stem cells in hypoxic-ischemic brain injury. Exp Neurol. 2006b;199:179–90.

27. Imitola J, Raddassi K, Park KI, Mueller FJ, Nieto M, Teng YD, Frenkel D, Li J, Sidman RL, Walsh CA, Snyder EY, Khoury SJ. Directed migration of neural stem cells to sites of CNS injury by the stromal cell-derived factor 1alpha/CXC chemokine receptor 4 pathway. Proc Natl Acad Sci U S A. 2004;101:18117–22.
28. Zheng T, Rossignol C, Leibovici A, Anderson KJ, Steindler DA, Weiss MD. Transplantation of multipotent astrocytic stem cells into a rat model of neonatal hypoxic-ischemic encephalopathy. Brain Res. 2006;1112:99–105.
29. Yasuhara T, Matsukawa N, Yu G, Xu L, Mays RW, Kovach J, Deans RJ, Hess DC, Carroll JE, Borlongan CV. Behavioral and histological characterization of intrahippocampal grafts of human bone marrow-derived multipotent progenitor cells in neonatal rats with hypoxic-ischemic injury. Cell Transplant. 2006;15:231–8.
30. Rosenkranz K, Kumbruch S, Lebermann K, Marschner K, Jensen A, Dermietzel R, Meier C. The chemokine SDF-1/CXCL12 contributes to the 'homing' of umbilical cord blood cells to a hypoxic-ischemic lesion in the rat brain. J Neurosci Res. 2010;88:1223–33.
31. Geißler M, Dinse HR, Neuhoff S, Kreikemeier K, Meier C. Human umbilical cord blood cells restore brain damage induced changes in rat somatosensory cortex. PloS one. 2011;6:e20194.
32. Wasielewski B, Jensen A, Roth-Harer A, Dermietzel R, Meier C. Neuroglial activation and Cx43 expression are reduced upon transplantation of human umbilical cord blood cells after perinatal hypoxic-ischemic injury. Brain Res. 2012;1487:39–53.
33. Rosenkranz K, Kumbruch S, Tenbusch M, Marcus K, Marschner K, Dermietzel R, Meier C. Transplantation of human umbilical cord blood cells mediated beneficial effects on apoptosis, angiogenesis and neuronal survival after hypoxic-ischemic brain injury in rats. Cell Tissue Res. 2012;348:429–38.
34. Rosenkranz K, Tenbusch M, May C, Marcus K, Meier C. Changes in Interleukin-1 alpha serum levels after transplantation of umbilical cord blood cells in a model of perinatal hypoxic-ischemic brain damage. Ann Anat. 2013;195:122–7.
35. Ma J, Wang Y, Yang J, Yang M, Chang KA, Zhang L, Jiang F, Li Y, Zhang Z, Heo C, Suh YH. Treatment of hypoxic-ischemic encephalopathy in mouse by transplantation of embryonic stem-cell-derived cells. Neurochem Int. 2007;51:57–65.
36. Sato Y, Nakanishi K, Hayakawa M, Kakizawa H, Saito A, Kuroda Y, Ida M, Tokita Y, Aono S, Matsui F, Kojima S, Oohira A. Reduction of brain injury in neonatal hypoxic-ischemic rats by intracerebroventricular injection of neural stem/progenitor cells together with chondroitinase ABC. Reprod Sci. 2008;15:613–20.
37. de Paula S, Vitola AS, Greggio S, de Paula D, Mello PB, Lubianca JM, Xavier LL, Fiori HH, Dacosta JC. Hemispheric brain injury and behavioral deficits induced by severe neonatal hypoxia-ischemia in rats are not attenuated by intravenous administration of human umbilical cord blood cells. Pediatr Res. 2009;65:631–5.
38. de Paula S, Greggio S, Marinowic DR, Machado DC, DaCosta JC. The dose-response effect of acute intravenous transplantation of human umbilical cord blood cells on brain damage and spatial memory deficits in neonatal hypoxia-ischemia. Neuroscience. 2012;210:431–41.
39. Greggio S, de Paula S, Azevedo PN, Venturin GT, Dacosta JC. Intra-arterial transplantation of human umbilical cord blood mononuclear cells in neonatal hypoxic-ischemic rats. Life Sci. 2014;96:33–9.
40. Lee JA, Kim BI, Jo CH, Choi CW, Kim EK, Kim HS, Yoon KS, Choi JH. Mesenchymal stem cell transplantation for hypoxic-ischemic brain injury in neonatal rat model. Pediatr Res. 2010;67:42–6.
41. Daadi MM, Davis AS, Arac A, Li Z, Maag AL, Bhatnagar R, Jiang K, Sun G, Wu JC, Steinberg GK. Human neural stem cell grafts modify microglial response and enhance axonal sprouting in neonatal hypoxic-ischemic brain injury. Stroke. 2010;41:516–23.
42. Xia G, Hong X, Chen X, Lan F, Zhang G, Liao L. Intracerebral transplantation of mesenchymal stem cells derived from human umbilical cord blood alleviates hypoxic ischemic brain injury in rat neonates. J Perinat Med. 2010;38:215–21.
43. Pimentel-Coelho PM, Magalhaes ES, Lopes LM, deAzevedo LC, Santiago MF, Mendez-Otero R. Human cord blood transplantation in a neonatal rat model of hypoxic-ischemic

brain damage: functional outcome related to neuroprotection in the striatum. Stem Cells Dev. 2010;19:351–8.

44. van Velthoven CT, Kavelaars A, van Bel F, Heijnen CJ. Mesenchymal stem cell treatment after neonatal hypoxic-ischemic brain injury improves behavioral outcome and induces neuronal and oligodendrocyte regeneration. Brain Behav Immun. 2010a;24:387–93.
45. van Velthoven CT, Kavelaars A, van Bel F, Heijnen CJ. Nasal administration of stem cells: a promising novel route to treat neonatal ischemic brain damage. Pediatr Res. 2010b;68:419–22.
46. van Velthoven CT, Kavelaars A, van Bel F, Heijnen CJ. Repeated mesenchymal stem cell treatment after neonatal hypoxia-ischemia has distinct effects on formation and maturation of new neurons and oligodendrocytes leading to restoration of damage, corticospinal motor tract activity, and sensorimotor function. J Neurosci. 2010c;30:9603–11.
47. van Velthoven CT, Kavelaars A, van Bel F, Heijnen CJ. Mesenchymal stem cell transplantation changes the gene expression profile of the neonatal ischemic brain. Brain Behav Immun. 2011;25:1342–8.
48. van Velthoven CT, Kavelaars A, van Bel F, Heijnen CJ. Nasal administration of stem cells: a promising novel route to treat neonatal ischemic brain damage. Pediatr Res. 2012;68:419–22.
49. van Velthoven CT, Braccioli L, Willemen HL, Kavelaars A, Heijnen CJ. Therapeutic potential of genetically modified mesenchymal stem cells after neonatal hypoxic-ischemic brain damage. Mol Ther. 2014;22:645–54.
50. Donega V, Nijboer CH, van Tilborg G, Dijkhuizen RM, Kavelaars A, Heijnen CJ. Intranasally administered mesenchymal stem cells promote a regenerative niche for repair of neonatal ischemic brain injury. Exp Neurol. 2014;261:53–64.
51. Obenaus A, Dilmac N, Tone B, Tian HR, Hartman R, Digicaylioglu M, Snyder EY, Ashwal S. Long-term magnetic resonance imaging of stem cells in neonatal ischemic injury. Ann Neurol. 2011;69:282–91.
52. Ashwal S, Ghosh N, Turenius CI, Dulcich M, Denham CM, Tone B, Hartman R, Snyder EY, Obenaus A. Reparative effects of neural stem cells in neonatal rats with hypoxic-ischemic injury are not influenced by host sex. Pediatric Res. 2014;75:603–11.
53. Zheng XR, Zhang SS, Yin F, Tang JL, Yang YJ, Wang X, Zhong L. Neuroprotection of VEGF-expression neural stem cells in neonatal cerebral palsy rats. Behav Brain Res. 2012;230:108–15.
54. Bae SH, Kong TH, Lee HS, Kim KS, Hong KS, Chopp M, Kang MS, Moon J. Long-lasting paracrine effects of human cord blood cells on damaged neocortex in an animal model of cerebral palsy. Cell Transplant. 2012a;21:2497–515.
55. Shinoyama M, Ideguchi M, Kida H, Kajiwara K, Kagawa Y, Maeda Y, Nomura S, Suzuki M. Cortical region-specific engraftment of embryonic stem-cell-derived neural progenitor cells restores axonal sprouting to a subcortical target and achieves motor functional recovery in a mouse model of neonatal hypoxic-ischemic brain injury. Front Cell Neurosci. 2013;7:128.
56. Yamagata M, Yamamoto A, Kako E, Kaneko N, Matsubara K, Sakai K, Sawamoto K, Ueda M. Human dental pulp-derived stem cells protect against hypoxic-ischemic brain injury in neonatal mice. Stroke. 2013;44:551–4.
57. Fang CZ, Yang YJ, Wang QH, Yao Y, Zhang XY, He XH. Intraventricular injection of human dental pulp stem cells improves hypoxic-ischemic brain damage in neonatal rats. PloS one. 2013;8:e66748.
58. Wang XL, Zhao YS, Hu MY, Sun YQ, Chen YX, Bi XH. Umbilical cord blood cells regulate endogenous neural stem cell proliferation via hedgehog signaling in hypoxic ischemic neonatal rats. Brain Res. 2013;1518:26–35.
59. Wang X, Zhao Y, Wang X. Umbilical cord blood cells regulate the differentiation of endogenous neural stem cells in hypoxic ischemic neonatal rats via the hedgehog signaling pathway. Brain Res. 2014;1560:18–26.
60. Zhu LH, Bai X, Zhang N, Wang SY, Li W, Jiang L. Improvement of human umbilical cord mesenchymal stem cell transplantation on glial cell and behavioral function in a neonatal model of periventricular white matter damage. Brain Res. 2014;1563:13–21.

61. Wang F, Shen Y, Tsuru E, Yamashita T, Baba N, Tsuda M, Maeda N, Sagara Y. Syngeneic transplantation of newborn splenocytes in a murine model of neonatal ischemia-reperfusion brain injury. J Matern Fetal Neonatal Med. 2014;18:1–19.
62. Ohshima M, Taguchi A, Tsuda H, Sato Y, Yamahara K, Harada-Shiba M, Miyazato M, Ikeda T, Iida H, Tsuji M. Intraperitoneal and intravenous deliveries are not comparable in terms of drug efficacy and cell distribution in neonatal mice with hypoxia-ischemia. Brain Dev. 2014. http://dx.doi.org/10.1016/j.braindev.2014.06.010.
63. Mueller D, Shamblott MJ, Fox HE, Gearhart JD, Martin LJ. Transplanted human embryonic germ cell-derived neural stem cells replace neurons and oligodendrocytes in the forebrain of neonatal mice with excitotoxic brain damage. J Neurosci Res. 2005;82:592–608.
64. Vadivelu S, Platik MM, Choi L, Lacy ML, Shah AR, Qu Y, Holekamp TF, Becker D, Gottlieb DI, Gidday JM, McDonald JW. Multi-germ layer lineage central nervous system repair: nerve and vascular cell generation by embryonic stem cells transplanted in the injured brain. J Neurosurg. 2005;103:124–35.
65. Chen A, Siow B, Blamire AM, Lako M, Clowry GJ. Transplantation of magnetically labeled mesenchymal stem cells in a model of perinatal brain injury. Stem Cell Res. 2010a;5:255–66.
66. Titomanlio L, Bouslama M, Le Verche V, Dalous J, Kaindl AM, Tsenkina Y, Lacaud A, Peineau S, El Ghouzzi V, Lelievre V, Gressens P. Implanted neurosphere-derived precursors promote recovery after neonatal excitotoxic brain injury. Stem Cells Dev. 2011;20:865–79.
67. Dalous J, Pansiot J, Pham H, Chatel P, Nadaradja C, D'Agostino I, Vottier G, Schwendimann L, Vanneaux V, Charriaut-Marlangue C, Titomanlio L, Gressens P, Larghero J, Baud O. Use of human umbilical cord blood mononuclear cells to prevent perinatal brain injury: a preclinical study. Stem Cells Dev. 2012;22:169–79.
68. Ingram DA, Mead LE, Tanaka H, Meade V, Fenoglio A, Mortell K, Pollok K, Ferkowicz MJ, Gilley D, Yoder MC. Identification of a novel hierarchy of endothelial progenitor cells using human peripheral and umbilical cord blood. Blood. 2004;104:2752–60.
69. Lee OK, Kuo TK, Chen WM, Lee KD, Hsieh SL, Chen TH. Isolation of multipotent mesenchymal stem cells from umbilical cord blood. Blood. 2004;103:1669–75.
70. Verina T, Fatemi A, Johnston MV, Comi AM. Pluripotent possibilities: human umbilical cord blood cell treatment after neonatal brain injury. Pediatr Neurol. 2013;48:346–54.
71. Chen N, Hudson JE, Walczak P, Misiuta I, Garbuzova-Davis S, Jiang L, Sanchez-Ramos J, Sanberg PR, Zigova T, Willing AE. Human umbilical cord blood progenitors: the potential of these hematopoietic cells to become neural. Stem Cells. 2005;23:1560–70.
72. Fan CG, Zhang QJ, Tang FW, Han ZB, Wang GS, Han ZC. Human umbilical cord blood cells express neurotrophic factors. Neurosci Lett. 2005;380:322–5.
73. Kopen GC, Prockop DJ, Phinney DG. Marrow stromal cells migrate throughout forebrain and cerebellum, and they differentiate into astrocytes after injection into neonatal mouse brains. Proc Natl Acad Sci U S A. 1999;96:10711–6.
74. Phillips AW, Johnston MV, Fatemi A. The potential for cell-based therapy in perinatal brain injuries. Transl Stroke Res. 2013;4:137–48.
75. Ashwal S, Cole DJ, Osborne S, Osborne TN, Pearce WJ. A new model of neonatal stroke: reversible middle cerebral artery occlusion in the rat pup. Pediatr Neurol. 1995;12:191–6.
76. Derugin N, Ferriero DM, Vexler ZS. Neonatal reversible focal cerebral ischemia: a new model. Neurosci Res. 1998;32:349–53.
77. Wen TC, Rogido M, Gressens P, Sola A. A reproducible experimental model of focal cerebral ischemia in the neonatal rat. Brain Res Brain Res Protoc. 2004;13:76–83.
78. Tsuji M, Ohshima M, Taguchi A, Kasahara Y, Ikeda T, Matsuyama T. A novel reproducible model of neonatal stroke in mice: Comparison with a hypoxia-ischemia model. Exp Neurol. 2013;247C:218–25.
79. Mitsufuji N, Yoshioka H, Okano S, Nishiki T, Sawada T. A new model of transient cerebral ischemia in neonatal rats. J Cereb Blood Flow Metab. 1996;16:237–43.
80. Comi AM, Weisz CJ, Highet BH, Johnston MV, Wilson MA. A new model of stroke and ischemic seizures in the immature mouse. Pediatr Neurol. 2004;31:254–7.

81. Renolleau S, Aggoun-Zouaoui D, Ben-Ari Y, Charriaut-Marlangue C. A model of transient unilateral focal ischemia with reperfusion in the P7 neonatal rat: morphological changes indicative of apoptosis. Stroke. 1998;29:1454–60.
82. Hagberg H, Peebles D, Mallard C. Models of white matter injury: comparison of infectious, hypoxic-ischemic, and excitotoxic insults. Ment Retard Dev Disabil Res Rev. 2002;8:30–8.
83. Rice JE 3rd, Vannucci RC, Brierley JB. The influence of immaturity on hypoxic-ischemic brain damage in the rat. Ann Neurol. 1981;9:131–41.
84. Johnston MV, Ferriero DM, Vannucci SJ, Hagberg H. Models of cerebral palsy: which ones are best? J Child Neurol. 2005;20:984–7.
85. Johnston MV, Nakajima W, Hagberg H. Mechanisms of hypoxic neurodegeneration in the developing brain. Neuroscientist. 2002;8:212–20.
86. Fernandez-Lopez D, Faustino J, Derugin N, Vexler ZS. Acute and chronic vascular responses to experimental focal arterial stroke in the neonate rat. Transl Stroke Res. 2013;4:179–88.
87. Hayashi T, Iwai M, Ikeda T, Jin G, Deguchi K, Nagotani S, Zhang H, Sehara Y, Nagano I, Shoji M, Ikenoue T, Abe K. Neural precursor cells division and migration in neonatal rat brain after ischemic/hypoxic injury. Brain Res. 2005;1038:41–9.
88. Ong J, Plane JM, Parent JM, Silverstein FS. Hypoxic-ischemic injury stimulates subventricular zone proliferation and neurogenesis in the neonatal rat. Pediatr Res. 2005;58:600–6.
89. Donega V, van Velthoven CT, Nijboer CH, Kavelaars A, Heijnen CJ. The endogenous regenerative capacity of the damaged newborn brain: boosting neurogenesis with mesenchymal stem cell treatment. J Cereb Blood Flow Metab. 2013b;33:625–34.
90. Kadam SD, Mulholland JD, McDonald JW, Comi AM. Neurogenesis and neuronal commitment following ischemia in a new mouse model for neonatal stroke. Brain Res. 2008;1208: 35–45.
91. Qiu L, Zhu C, Wang X, Xu F, Eriksson PS, Nilsson M, Cooper-Kuhn CM, Kuhn HG, Blomgren K. Less neurogenesis and inflammation in the immature than in the juvenile brain after cerebral hypoxia-ischemia. J Cereb Blood Flow Metab. 2007;27:785–94.
92. Zhu C, Qiu L, Wang X, Xu F, Nilsson M, Cooper-Kuhn C, Kuhn HG, Blomgren K. Age-dependent regenerative responses in the striatum and cortex after hypoxia-ischemia. J Cereb Blood Flow Metab. 2009;29:342–54.
93. Shichinohe H, Kuroda S, Maruichi K, Osanai T, Sugiyama T, Chiba Y, Yamaguchi A, Iwasaki Y. Bone marrow stromal cells and bone marrow-derived mononuclear cells: which are suitable as cell source of transplantation for mice infarct brain? Neuropathology. 2010;30: 113–22.
94. Jablonska A, Kozlowska H, Markiewicz I, Domanska-Janik K, Lukomska B. Transplantation of neural stem cells derived from human cord blood to the brain of adult and neonatal rats. Acta Neurobiol Exp (Wars). 2010;70:337–50.
95. Max JE, Bruce M, Keatley E, Delis D. Pediatric stroke: plasticity, vulnerability, and age of lesion onset. J Neuropsychiatry Clin Neurosci. 2010;22:30–9.
96. Dammann O, Ferriero D, Gressens P. Neonatal encephalopathy or hypoxic-ischemic encephalopathy? Appropriate terminology matters. Pediatr Res. 2011;70:1–2.
97. Shankaran S, Laptook AR, Ehrenkranz RA, Tyson JE, McDonald SA, Donovan EF, Fanaroff AA, Poole WK, Wright LL, Higgins RD, Finer NN, Carlo WA, Duara S, Oh W, Cotten CM, Stevenson DK, Stoll BJ, Lemons JA, Guillet R, Jobe AH, National Institute of Child H, Human Development Neonatal Research N. Whole-body hypothermia for neonates with hypoxic-ischemic encephalopathy. N Engl J Med. 2005;353:1574–84.
98. Shankaran S, Pappas A, McDonald SA, Vohr BR, Hintz SR, Yolton K, Gustafson KE, Leach TM, Green C, Bara R, Petrie Huitema CM, Ehrenkranz RA, Tyson JE, Das A, Hammond J, Peralta-Carcelen M, Evans PW, Heyne RJ, Wilson-Costello DE, Vaucher YE, Bauer CR, Dusick AM, Adams-Chapman I, Goldstein RF, Guillet R, Papile LA, Higgins RD. Childhood outcomes after hypothermia for neonatal encephalopathy. N Engl J Med. 2012;366:2085–92.

99. Cotten CM, Murtha AP, Goldberg RN, Grotegut CA, Smith PB, Goldstein RF, Fisher KA, Gustafson KE, Waters-Pick B, Swamy GK, Rattray B, Tan S, Kurtzberg J. Feasibility of autologous cord blood cells for infants with hypoxic-ischemic encephalopathy. J Pediatr. 2014;164(973–979):e971.
100. Javed MJ, Mead LE, Prater D, Bessler WK, Foster D, Case J, Goebel WS, Yoder MC, Haneline LS, Ingram DA. Endothelial colony forming cells and mesenchymal stem cells are enriched at different gestational ages in human umbilical cord blood. Pediatr Res. 2008;64:68–73.
101. Ligi I, Simoncini S, Tellier E, Vassallo PF, Sabatier F, Guillet B, Lamy E, Sarlon G, Quemener C, Bikfalvi A, Marcelli M, Pascal A, Dizier B, Simeoni U, Dignat-George F, Anfosso F. A switch toward angiostatic gene expression impairs the angiogenic properties of endothelial progenitor cells in low birth weight preterm infants. Blood. 2011;118:1699–709.
102. Aly H, Mohsen L, Badrawi N, Gabr H, Ali Z, Akmal D. Viability and neural differentiation of mesenchymal stem cells derived from the umbilical cord following perinatal asphyxia. J Perinat. 2012;32:671–6.
103. Hanna N, Graboski S, Laskin DL, Weinberger B. Effects of ibuprofen and hypoxia on neutrophil apoptosis in neonates. Biol Neonate. 2004;86:235–9.
104. Phelan JP, Korst LM, Ahn MO, Martin GI. Neonatal nucleated red blood cell and lymphocyte counts in fetal brain injury. Obstet Gynecol. 1998;91:485–9.
105. Cox CS Jr, Baumgartner JE, Harting MT, Worth LL, Walker PA, Shah SK, Ewing-Cobbs L, Hasan KM, Day MC, Lee D, Jimenez F, Gee A. Autologous bone marrow mononuclear cell therapy for severe traumatic brain injury in children. Neurosurgery. 2011;68:588–600.
106. Luan Z, Liu WP, Qu SQ, Qu SQ, Hu XH, Wang ZY, He S, Liu CQ, Xiao M. Treatment of newborns with severe injured brain with transplantation of human neural precursor cells. Zhonghua er ke za zhi. 2011;49:445–9.
107. Sreenan C, Bhargava R, Robertson CM. Cerebral infarction in the term newborn: clinical presentation and long-term outcome. J Pediatr. 2000;137:351–5.
108. Lee J, Croen LA, Lindan C, Nash KB, Yoshida CK, Ferriero DM, Barkovich AJ, Wu YW. Predictors of outcome in perinatal arterial stroke: a population-based study. Ann Neurol. 2005;58:303–8.
109. Raju TN. Ischemic perinatal stroke: challenge and opportunities. Int J Stroke. 2008;3:169–72.
110. Menkes JH, Sarnat HB. Perinatal asphyxias and Trauma. In: Menkers JH, Sarnat HB, Maria BL, editors. Child neurology. 7th ed. Philadelphia: Lippincott Williams & Wilkins; 2000. pp. 367–431.
111. Suzuki J, Miyajima T, Fujii T. Epidemiological study of cerebral palsy in Shiga Prefecture, Japan, during 1977–2000. Part 3: clinical features of cerebral palsy at six years of age. No to Hattatsu 2009;41:289–93.
112. Min K, Song J, Kang JY, Pt JK, Ryu JS, Kang MS, Jang S, Kim SH, Oh D, Kim MK, Soo KS, Kim M. Umbilical cord blood therapy potentiated with erythropoietin for children with cerebral palsy: a double-blind, randomized, placebo-controlled trial. Stem Cells. 2013;31:581–91.
113. Bae SH, Lee HS, Kang MS, Strupp BJ, Chopp M, Moon J. The levels of pro-inflammatory factors are significantly decreased in cerebral palsy patients following an allogeneic umbilical cord blood cell transplant. Int J Stem Cells. 2012b;5:31–8.
114. Mancias-Guerra C, Marroquin-Escamilla AR, Gonzalez-Llano O, Villarreal-Martinez L, Jaime-Perez JC, Garcia-Rodriguez F, Valdes-Burnes SL, Rodriguez-Romo LN, Barrera-Morales DC, Sanchez-Hernandez JJ, Cantu-Rodriguez OG, Gutierrez-Aguirre CH, Gomez-De Leon A, Elizondo-Riojas G, Salazar-Riojas R, Gomez-Almaguer D. Safety and tolerability of intrathecal delivery of autologous bone marrow nucleated cells in children with cerebral palsy: an open-label phase I trial. Cytotherapy. 2014;16:810–20.
115. Sun J, Allison J, McLaughlin C, Sledge L, Waters-Pick B, Wease S, Kurtzberg J. Differences in quality between privately and publicly banked umbilical cord blood units: a pilot study of autologous cord blood infusion in children with acquired neurologic disorders. Transfusion. 2010;50:1980–7.

116. Jensen A, Hamelmann E. First autologous cell therapy of cerebral palsy caused by hypoxic-ischemic brain damage in a child after cardiac arrest-individual treatment with cord blood. Case Rep Transplant. 2013;2013:951827.
117. Chen L, Huang H, Xi H, Xie Z, Liu R, Jiang Z, Zhang F, Liu Y, Chen D, Wang Q, Wang H, Ren Y, Zhou C. Intracranial transplant of olfactory ensheathing cells in children and adolescents with cerebral palsy: a randomized controlled clinical trial. Cell Transplant. 2010b;19:185–91.
118. Lee YH, Choi KV, Moon JH, Jun HJ, Kang HR, Oh SI, Kim HS, Um JS, Kim MJ, Choi YY, Lee YJ, Kim HJ, Lee JH, Son SM, Choi SJ, Oh W, Yang YS. Safety and feasibility of countering neurological impairment by intravenous administration of autologous cord blood in cerebral palsy. J Transl Med. 2012;10:58.

Chapter 18
Issues in Clinical Trial Design in Stem Cell Trials After Stroke

Steven C. Cramer

Introduction

Stroke remains a major cause of human disability. Currently approved therapies for a new stroke reach a limited fraction of patients, largely because of the narrow time window. For example, in the USA, IV tissue plasminogen activator (tPA) is given to approximately 5 % of patients [1], and note that half or more of those so treated nonetheless have significant long-term disability [2, 3]. Some studies suggest that stem-cell-based therapies have the potential to improve outcomes after stroke via acute neuroprotective mechanisms. A larger body of research has focused on stem cells as restorative therapies, which have the potential to improve outcomes in a large fraction of patients in part because the therapeutic time window is measured in days– weeks; the latter stem cell application is the focus of the current chapter.

Restorative stroke trials have received increased attention in recent years. Increasing experience has identified a number of issues that are important to consider when testing the efficacy of restorative therapies such as stem cells. Some issues are shared with acute stroke trials while some are not. Some issues are common to many different restorative therapies, while others are most relevant to cell-based therapies.

Several key principles emerge across restorative stroke trials. First, unlike the other organs in the body, the brain is really a collection of dozens or hundreds of different functional units. Appreciation of this point is important to topics such as endpoints to measure treatment effect. Second, many restorative therapies achieve maximum effect when coupled with behavioral training. A brain galvanized for repair requires behavioral shaping, i.e., restorative therapies improve outcomes on the basis of experience-dependent plasticity. This is a major distinction as compared

S. C. Cramer (✉)
Department of Neurology, Anatomy and Neurobiology, and Physical Medicine and Rehabilitation, Institute for Clinical and Translational Science, Sue and Bill Gross Stem Cell Research Center, UC Irvine Medical Center, University of California, Irvine, 200 S. Manchester Ave. Suite 206, Orange, CA 92868, USA
e-mail: scramer@uci.edu

D. C. Hess (ed.), *Cell Therapy for Brain Injury*, DOI 10.1007/978-3-319-15063-5_18

to acute reperfusion or neuroprotective therapies, where treatment effects do not make any behavioral demands upon subjects. Third, many sources of heterogeneity reduce study power but can be minimized. The price of variance is no cheaper in the context of restorative stroke therapies. This issue pertains to many aspects of clinical trial design, particularly training of study staff and stratification of enrollees.

Issues in Restorative Stroke Trials That Are Similar to Issues in Acute Stroke Trials

Some issues in clinical trial design are shared between restorative and acute stroke trials. A critical example in this regard relates to effective translation of preclinical findings to a clinical investigation [4–6]. Review of the methodological quality of preclinical studies has received increased attention [6–8]. Lower-quality preclinical studies, such as those that lack blinding or that do not randomize treatment assignments, are less likely to directly translate to successful human clinical trials. Although major differences exist between the species used in preclinical studies (such as rats) and humans [9], the nature of the endpoints found significant in preclinical studies may nonetheless be of guiding value for selection of endpoints in human investigations. While allometric scaling from rodent to human is difficult [10–12], preclinical studies can nonetheless provide some information regarding choice of dose(s) to study in humans. Stroke recovery is a four-dimensional target, with efficacy changing according to time post stroke when the therapy is introduced [13–17]. As such, translation of preclinical stem cell findings to human studies must consider when treatment was initiated in animals in relation to time post stroke.

Variance in endpoint assessment represents a large source of variance that can be strategically reduced. Acute stroke trialists are familiar with certification methods for endpoints such as the NIH stroke scale and the modified Rankin scale. Recent restorative studies have pursued similar approaches [18], and a number of publications are providing key tools for recovery-related endpoints [19, 20]. One study [20] provided detailed methods and training materials for the Fugl-Meyer scale and found that training with this method improved accuracy and reduced variance of Fugl-Meyer scoring. Training with this method reduced variance in the Fugl-Meyer scale scoring by 20 %, which would decrease sample size requirements from 137 to 88 in a theoretical trial aiming to detect a 7-point difference in this scale.

The entry criteria used in preclinical stem cell stroke studies are valuable for defining the target population in a human translational trial. For example, anatomical and physiological features of patient enrollees should be aligned with those present in preclinical models. This issue was raised regarding translation of epidural motor cortex stimulation studies, where rodent and primate studies showing efficacy of this intervention required preserved motor evoked responses [21–24] but a phase III human trial [25] did not. In a post hoc analysis, Nouri and Cramer found that human patients in the phase III trial who were randomized to epidural motor cortex stimulation and who (like the animals selected for preclinical studies) had a preserved motor evoked response were 2.5 times more likely to achieve the primary efficacy endpoint as compared to human patients randomized to epidural motor cortex stimulation who did not have a preserved motor-evoked response (67% vs. 27%, $p < 0.05$) [26].

Issues Related to Outcome Measures

Choice of outcome measures is a central issue in the design of stem cell trials after stroke. Unlike many acute stroke studies in which patient enrollment by design must be very rapid, many studies of repair-based therapies have the potential to perform more detailed measurements of patients at baseline, prior to treatment; this enables within-subject analyses, which have potential statistical advantages over cross-sectional outcome assessments [27].

Earlier-phase studies might emphasize safety-based measures or biomarkers, while later-phase studies often rely on specific scales to demonstrate efficacy [28]. Many different scales are available [29–33]. An important consideration is which dimension an endpoint falls within on the World Health Organization (WHO) international classification of functioning, disability, and health (ICF) [34, 35]. Loss of body functions and structures, formerly referred to as impairments, includes deficits that are a consequence of stroke. Activity limitations, formerly referred to as disabilities, reflect difficulties patients with stroke experience in the performance of functional tasks. Participation restrictions, formerly referred to as handicaps, refer to difficulties patients with stroke encounter in societal roles.

Each stroke scale has its limitations as an outcome measure [36–38]. Some are insensitive to small deficits [39] or require special skills to administer. The utility of some scales depends on the population under study. Some scales have been criticized for having a ceiling effect in patients with mild stroke; for example, a normal score on the functional independence measure or the Barthel index can be reached despite persistence of deficits [40–42]. Patient-reported outcomes may be useful in stem cell trials after stroke. These measures are sensitive to change, reveal disability with high accuracy, are considered the gold standard for many social and emotional consequences of brain injury, and have been the subject of several recent initiatives [43–46].

Many, if not most, scales are neither continuous nor linear. For some clinical scales, values are treated as a continuous variable but are actually the sum of several ordinal variables, the latter being a comparatively weaker form of measurement [47]. Scales that are not continuous have reduced granularity. While many global scales reduce patient outcomes to a single number, or sometimes a single digit, humans are very complex creatures, and improved resolution of measurement can increase the likelihood of identifying meaningful treatment gains [39, 48, 49]. This issue might be overcome by devices such as digital sensors. The higher resolution of such sensors might increase sensitivity and reliability; however, such measures may have limited relevance to a patient's function and independence [35, 50]. Few stroke outcome measures are truly linear. For example, a gain of 5 points in NIH stroke scale score from 8 to 3 does not have the same meaning as a gain from 19 to 14. More assumptions must be made when interpreting data from scales that are not linear.

Dichotomous outcome measures are commonly used in stroke trials, but concerns such as those listed above make this approach particularly problematic in the setting of restorative stroke trials. A dichotomous outcome measure classifies treatment response as successful or not; this reduces the human condition from a single digit to a single binary digit. One potential solution in the design of such clinical trials is to define a successful outcome in a manner that varies according to a patient's baseline status. This approach is known as a sliding dichotomous outcome, or responder analysis. A recent analysis of acute stroke trial outcome measures emphasized the utility of this approach [51], and noted its ability to increase study power. With this approach, patient subgroups are specified before the trial on the basis of established prognostic measures such as age, baseline behavioral status, or extent of injury. Successful response to therapy is defined differently for each subgroup. A sliding outcomes approach has been used in several acute stroke trials [51], e.g., the AbESTT-II trial of Abciximab for acute stroke defined good outcome as modified Rankin scale (mRS) score of 0 for patients with baseline NIHSS score of 4–7, mRS score of 0–1 for baseline NIHSS of 8–14, and mRS score of 0–2 for baseline NIHSS 15–22 [52]. A sliding outcomes approach has also been used in trials enrolling patients with chronic stroke. For example, the locomotor experience applied poststroke (LEAPS) trial of locomotor training [53] defined success in the primary outcome measure (proportion of participants with improved functional walking level) as gait velocity ≥ 0.4 m/s for enrollees with baseline gait velocity <0.4 m/s and as gait velocity ≥ 0.8 m/s for enrollees with baseline gait velocity 0.4–0.8 m/s.

Sliding dichotomous outcomes are attractive candidates for assessing therapeutic efficacy in restorative stroke trials enrolling patients in either the acute or the chronic phase. For example, in a study examining motor recovery, in a patient with severe baseline motor deficits, return of dexterous hand movements might be very unlikely, while an increase in grip force from 10 to 40 N might be readily achievable; such a boost in force of squeezing would likely be highly relevant to function in such a patient. On the other hand, in a patient with mild baseline motor deficits, return of dexterous hand movements might be attainable and relevant to functional gains, while the same 30-N boost in grip force would be of trivial significance.

An alternative approach for dealing with the heterogeneity of the stroke population when selecting outcome measures is to use a composite endpoint. This approach was employed in a phase III trial of epidural motor cortex stimulation [25, 26], where the primary outcome measure required specific gains on both the Fugl-Meyer arm motor scale (which measures loss of body functions and structures) and the Arm Motor Ability Test (which measures activities limitations). One recent study examining the effect of robot-based therapy on arm motor status after stroke generated a composite endpoint by using a principal components analysis of two endpoints, the Fugl-Meyer scale and the Action Research Arm Test (which measures activities limitations) [54].

One way to classify endpoints in clinical stroke trials is global versus modality-specific. A global endpoint takes an ombnibus approach, summing many neurological considerations into a single measurement. The NIH stroke scale and the modified Rankin stroke scale are generally classified as global outcome measures. On the other hand, a modality-specific endpoint examines individual neural systems separately. Examples include the Fugl-Meyer motor scale and the Western Aphasia Battery.

The choice between a global outcome measure and a modality-specific outcome measure may be guided by key features of a clinical trial. The fact that the brain is functionally and anatomically an agglomeration of many different neural systems suggests that trials of restorative agents might benefit from use of modality-specific outcome measures. This is because restorative therapies achieve their effect by improving the function of specific neural systems. Improvement in a neural system can occur when sufficient substrate survives and is available to be repaired. A behavior for which the underlying brain regions are utterly destroyed by stroke is less likely to improve with therapy than a behavior for which the underlying brain regions remain accessible to a restorative therapy. For example, a patient whose stroke partially spares the language system but destroys the left hemisphere motor system may show treatment-related gains in language function but negligible change in motor function in response to a restorative therapy. These treatment effects would be captured by many different aphasia scales but may or may not be in evidence when assessed with a global outcome measure. On the other hand, a global outcome measure is a powerful tool for establishing substantial treatment-related gains. The beneficial effect of therapies such as IV tPA that salvage massive volumes of brain tissue across many different neural systems might be apparent with a global outcome measure. The choice between a modality-specific outcome measure and a global outcome measure depends on factors such as the mechanisms underlying treatment-related gains.

The importance of modality-specific measures to restorative therapies takes on added dimension when one considers that many of the therapies provided to patients as standard of care after stroke have features that are modality-specific. Examples include occupational therapy, physical therapy, and speech therapy. Evidence suggests that many features of stroke recovery are experience-dependent (see below), and so the nature of such concomitant therapy is often modality-specific, a feature that is better captured by endpoints than ignored. Furthermore, the rate and degree of behavioral gains often vary widely across different behavioral modalities [55–57]. Global endpoints in essence compress behavioral outcomes by averaging across these various recovery curves, while modality-specific outcomes retain the granularity that is characteristic of a neurologist's assessment of stroke outcomes.

Regulatory perspectives might also influence choice of outcome measures. Approval of IV tPA was based on clinical trials that relied on global endpoints, and uncertainty exists as to regulatory perspectives on modality-specific outcome measures in clinical stroke trials [58]. Recently, the FDA-approved 4-aminopyridine for patients with multiple sclerosis to improve walking ability [59], a modality-specific endpoint.

Improvement in global clinical status is of course a goal of paramount importance, but a treatment that provides gains by promoting neuroplasticity might demonstrate maximum effect in only those brain networks that have sustained subtotal injury, underscoring the complementary value of endpoints that measure these modality-specific treatment effects.

Concomitant Experience as an Adjuvant Therapy

As noted above, neural repair occurs on the basis of experience-dependent plasticity, and this issue is generally not a factor in acute reperfusion or neuroprotection stroke therapies. Evidence suggests that benefits from a restorative therapy require specific training—an adjuvant therapy of sorts. By contrast, a patient need not engage in any specific behavior in order to derive gains from IV tPA. A landmark study on this topic was published by Feeney et al. [60], who found that in rodents with an experimental stroke, amphetamine therapy improved motor outcome, but only if drug dosing was paired with training. Subsequent studies have confirmed this principle across many other classes of post-stroke restorative therapy [21, 61–64].

Clinical trials examining cellular therapies that aim to promote neural repair after stroke will likely benefit from attention to patient experiences that occur concomitant with the therapy of interest, as these experiences can influence therapeutic efficacy. Issues of interest might include the timing, content, dosage, and intensity of such experiences [37, 65–71]. A patient's psychosocial experiences and environment may also be important along this axis [72–77], issues that stress the limits by rodent studies model human stroke recovery.

There are several potential options by which a clinical trial might be designed to address these issues. Some studies may be able to control such issues, e.g., via entry criteria, or by strictly dictating details of concomitant therapy. This approach may be particularly relevant to studies planned in countries with health systems that have as consistent approach to post-stroke rehabilitation therapy. Studies that are not able to control such measures, e.g., due to cost, might measure the amount of relevant experience. Such an approach provided useful insights in one recent repair-based stroke clinical trial, where the amount of outside physiotherapy (i.e., physiotherapy occurring in parallel with trial participation, but prescribed by private physicians, outside of trial jurisdiction) was found to differ significantly between active and placebo treatment arms [78]. Such measures can then be treated as planned covariates of interest in statistical analyses.

Identifying and Enrolling the Ideal Target Population

The heterogeneity of stroke makes prediction of treatment responders from nonresponders a great challenge. The ability to predict response to therapy and prospectively separate subgroups could be useful for stratifying patients [79] in order to maximize behavioral gains, efficiently utilize rehabilitation and financial resources, and reduce variance to increase power in clinical trials. Acute reperfusion therapies generally target blood clots or arteries, whereas restorative therapies usually target the brain. Studies need to identify and enroll those patients whose brain state is ideal for responding to the restorative therapy of interest.

This need is complicated by the fact that many different variables can influence the likelihood that a patient can respond to a given restorative therapy after stroke. Predictors of response to a restorative therapy after stroke can be grouped into three main categories: (1) measures of neural injury such as extent of injury to white matter or gray matter [26, 80–83]; (2) measures of neural function such as functional activation [84–87], functional connectivity [88, 89], and neurophysiological status [26, 85]; and (3) clinical measures such as demographics [90], baseline behavioral status [80, 86, 91, 92], and affective disorders [74, 93]. Some data in human subjects also suggest that genetic variation might contribute to variance in response to a restorative therapy [94–97].

A major challenge for restorative trials is to understand which measure, or measures, is most important to patient selection, as such insights can be used to guide entry criteria or stratification variables. A more selective approach to patient recruitment might reduce variance and so increase study power, but this can carry important costs. Selectively enrolling a narrow fraction of the stroke population slows enrollment, which likely increases total study costs; encourages larger numbers of enrollment sites, which likely increases variance in outcome measures; and limits the extent to which study findings generalize across the broader stroke population. There is some convergence of findings that, at least among patients with chronic stroke, optimal prediction of response to restorative therapies comes from combining measures of neural function and neural injury [54, 80].

Integration of Study Therapy and Procedures with Standard of Care

As above, restorative trials will often need to consider therapy provided to patients outside of study procedures. This need must be implemented in the context of a patient who is receiving standard of care treatment after stroke. In the LEAPS trial [53], where functional walking ability was the primary endpoint, 81.9 % of subjects were also receiving physical therapy outside of study procedures—an average of 25 such sessions. Further complicating all of this is the fact that post-stroke rehabilitation therapy is highly variable across sites in some countries [98].

Biomarkers

Biomarkers have the potential to strengthen clinical trials of brain repair after stroke [99]. A biomarker can be defined as a measure that provides insight into a tissue state or disease state, and in a clinical trial context would provide information beyond that available from bedside exam. Biomarkers have the potential to identify patients most likely to respond to a treatment, and so might reduce variance and increase study power [100–102]. A biomarker might also provide insight into a treatment's mechanism of action [103–105], which can provide useful insights at the stage of protocol development, or to refine the target patient population. There are important caveats in the selection of any biomarker [106], for example, the utility of a biomarker is highest when its relationships with the disease process and with the therapy are well understood [107, 108].

A number of specific measures are potentially available to serve as biomarkers in the context of a restorative stroke trial. Simple measures derived from blood testing have been proposed [109, 110]. Imaging-based methods can provide anatomical measures of injury [82, 111], tissue status such as cortical thickness [112], white matter tract integrity [80, 82, 113–115], regional brain function [86, 116, 117], network interactions [118], or chemical state [119, 120]. Physiological assessments might also be useful [26, 80, 121]. Measures of injury to a predefined functional brain region, such as the extent of insult to the hand region of primary motor cortex [122], white matter cholinergic projections [123], corticospinal tract [54], or left temporal language areas [124], might provide useful insights into the likelihood that a particular therapy will be able to promote repair in a specific target region.

Complexities of Restorative Clinical Trials Related to the Study of Stem Cells

Many different types of cellular product are under study in the treatment of stroke. These therapeutic strategies sometimes raise a number of issues that are encountered less often with other approaches to stroke recovery such as pharmacological. Some stem cell therapies represent a combined approach, with cells combined with gene therapy, exposure to neurotrophic factors or hypoxia, or inclusion of a bioscaffold. The long-term fate of the therapeutic product can be an issue of particular concern with certain forms of cellular therapy, and many years of patient follow-up may be indicated. Stem cell therapies are sometimes given in parallel with an immunosuppressant regimen, a nontrivial consideration among patients with stroke. Some patients, scientists, and health-care providers have ethical concerns with use of certain cells [125]. Some protocols for delivery of cell-based products involve neurointerventional or neurosurgical procedures, increasing the complexity of the intervention and assessment of its safety. Stem cells are living organisms, and in some cases their biological activity or biological identity could change with time or with storage, and so protocols for product testing and release are often required prior to a patient being treated [126]. While none of these issues is prohibitive, each represents a level of complexity that may be uncommon in studies of classes of restorative agent.

References

1. Adeoye O, Hornung R, Khatri P, Kleindorfer D. Recombinant tissue-type plasminogen activator use for ischemic stroke in the United States: a doubling of treatment rates over the course of 5 years. Stroke. 2011;42:1952–5.
2. Hacke W, Kaste M, Bluhmki E, Brozman M, Davalos A, Guidetti D, et al. Thrombolysis with alteplase 3–4.5 hours after acute ischemic stroke. N Engl J Med. 2008;359:1317–29.
3. Tissue plasminogen activator for acute ischemic stroke. The national institute of neurological disorders and stroke rt-pa stroke study group. N Engl J Med. 1995;333:1581–7.
4. Grotta J, Bratina P. Subjective experiences of 24 patients dramatically recovering from stroke. Stroke. 1995;26:1285–8.
5. Fisher M, Ratan R. New perspectives on developing acute stroke therapy. Ann Neurol. 2003;53:10–20.
6. Philip M, Benatar M, Fisher M, Savitz SI. Methodological quality of animal studies of neuroprotective agents currently in phase ii/iii acute ischemic stroke trials. Stroke. 2009;40:577–81.
7. Savitz SI, Fisher M. Future of neuroprotection for acute stroke: in the aftermath of the saint trials. Ann Neurol. 2007;61:396–402.
8. Vu Q, Xie K, Eckert M, Zhao W, Cramer SC. Meta-analysis of preclinical studies of mesenchymal stromal cells for ischemic stroke. Neurology. 2014;82:1277–86.
9. Cramer S. Clinical issues in animal models of stroke and rehabilitation. ILAR J/Nat Res Counc, Inst Lab Anim Resour. 2003;44:83–4.
10. Watanabe K, Bois FY, Zeise L. Interspecies extrapolation: a reexamination of acute toxicity data. Risk Anal. 1992;12:301–10.
11. Davidson IW, Parker JC, Beliles RP. Biological basis for extrapolation across mammalian species. Regul Toxicol Pharmacol. 1986;6:211–37.
12. Guidance for industry estimating the maximum safe starting dose in initial clinical trials for therapeutics in adult healthy volunteers USDoHaHSFaDA. 2005.
13. Li S, Carmichael ST. Growth-associated gene and protein expression in the region of axonal sprouting in the aged brain after stroke. Neurobiol Dis. 2006;23:362–73.
14. Stroemer R, Kent T, Hulsebosch C. Enhanced neocortical neural sprouting, synaptogenesis, and behavioral recovery with d-amphetamine therapy after neocortical infarction in rats. Stroke. 1998;29:2381–95.
15. Jones T, Schallert T. Overgrowth and pruning of dendrites in adult rats recovering from neocortical damage. Brain Res. 1992;581:156–60.
16. Ren J, Kaplan P, Charette M, Speller H, Finklestein S. Time window of intracisternal osteogenic protein-1 in enhancing functional recovery after stroke. Neuropharmacology. 2000;39:860–5.
17. Biernaskie J, Chernenko G, Corbett D. Efficacy of rehabilitative experience declines with time after focal ischemic brain injury. J Neurosci. 2004;24:1245–54.
18. Winstein CJ, Wolf SL, Dromerick AW, Lane CJ, Nelsen MA, Lewthwaite R, et al. Interdisciplinary comprehensive arm rehabilitation evaluation (icare): a randomized controlled trial protocol. BMC Neurol. 2013;13:5.
19. Yozbatiran N, Der-Yeghiaian L, Cramer SC. A standardized approach to performing the action research arm test. Neurorehabil Neural Repair. 2008;22:78–90.

20. See J, Dodakian L, Chou C, Chan V, McKenzie A, Reinkensmeyer DJ, et al. A standardized approach to the fugl-meyer assessment and its implications for clinical trials. Neurorehabil Neural Repair. 2013;27:732–41.
21. Adkins-Muir D, Jones T. Cortical electrical stimulation combined with rehabilitative training: enhanced functional recovery and dendritic plasticity following focal cortical ischemia in rats. Neurol Res. 2003;25:780–8.
22. Kleim J, Bruneau R, VandenBerg P, MacDonald E, Mulrooney R, Pocock D. Motor cortex stimulation enhances motor recovery and reduces peri-infarct dysfunction following ischemic insult. Neurol Res. 2003;25:789–93.
23. Plautz E, Barbay S, Frost S, Friel K, Dancause N, Zoubina E, et al. Post-infarct cortical plasticity and behavioral recovery using concurrent cortical stimulation and rehabilitative training: a feasibility study in primates. Neurol Res. 2003;25:801–10.
24. Teskey G, Flynn C, Goertzen C, Monfils M, Young N. Cortical stimulation improves skilled forelimb use following a focal ischemic infarct in the rat. Neurol Res. 2003;25:794–800.
25. Levy R, Benson R, Winstein C, for the Everest Study Investigators. Cortical stimulation for upper-extremity hemiparesis from ischemic stroke: Everest study primary endpoint results. International Stroke Conference. 2008.
26. Nouri S, Cramer SC. Anatomy and physiology predict response to motor cortex stimulation after stroke. Neurology. 2011 Sep 13;77(11):1076–83.
27. Calautti C, Baron J. Functional neuroimaging studies of motor recovery after stroke in adults: a review. Stroke. 2003;34:1553–66.
28. Cramer SC. Repairing the human brain after stroke. Ii. Restorative therapies. Ann Neurol. 2008;63:549–60.
29. Lyden P, Lau G. A critical appraisal of stroke evaluation and rating scales. Stroke. 1991;22:1345–52.
30. Gresham G, Duncan P, Stason W, Adams H, Adelman A, Alexander D, et al. Post-stroke rehabilitation. Rockville: U.S. Department of Health and Human Services. Public Health Service, Agency for Health Care Policy and Research; 1995.
31. Barak S, Duncan PW. Issues in selecting outcome measures to assess functional recovery after stroke. Neuro Rx. 2006;3:505–24.
32. Baker K, Cano SJ, Playford ED. Outcome measurement in stroke: a scale selection strategy. Stroke. 2011;42:1787–94.
33. Lees KR, Bath PM, Schellinger PD, Kerr DM, Fulton R, Hacke W, et al. Contemporary outcome measures in acute stroke research: choice of primary outcome measure. Stroke. 2012;43:1163–70.
34. World Health Organization. International classification of functioning, disability and health (icf). Geneva: World Health Organization; 2008.
35. Miller EL, Murray L, Richards L, Zorowitz RD, Bakas T, Clark P, et al. Comprehensive overview of nursing and interdisciplinary rehabilitation care of the stroke patient: a scientific statement from the American heart association. Stroke. 2010;41:2402–48.
36. Wood-Dauphinee S, Williams J, Shapiro S. Examing outcome measures in a clinical study of stroke. Stroke. 1990;21:731–9.
37. Dobkin B. The clinical science of neurologic rehabilitation. New York: Oxford University Press; 2003.
38. Cramer S, Nelles G, Schaechter J, Kaplan J, Finklestein S. Computerized measurement of motor performance after stroke. Stroke. 1997;28:2162–8.
39. Stewart JC, Cramer SC. Patient-reported measures provide unique insights into motor function after stroke. Stroke. 2013;44:1111–6.
40. Duncan PW, Samsa GP, Weinberger M, Goldstein LB, Bonito A, Witter DM, et al. Health status of individuals with mild stroke. Stroke. 1997;28:740–5.
41. Edwards DF, Hahn M, Baum C, Dromerick AW. The impact of mild stroke on meaningful activity and life satisfaction. J Stroke Cerebrovasc Dis. 2006;15:151–7.
42. Carlsson GE, Moller A, Blomstrand C. Consequences of mild stroke in persons <75 years–a 1-year follow-up. Cerebrovas Dis. (Basel, Switzerland). 2003;16:383–8.

43. Barrett AM. Rose-colored answers: neuropsychological deficits and patient-reported outcomes after stroke. Behav Neurol. 2010;22:17–23.
44. Snyder CF, Aaronson NK, Choucair AK, Elliott TE, Greenhalgh J, Halyard MY, et al. Implementing patient-reported outcomes assessment in clinical practice: a review of the options and considerations. Qual Life Res. 2012 Oct;21(8):1305–14.
45. Department of Health. Equity and excellence: liberating the NHS. London: Department of Health; 2010.
46. Promis. Dynamic tools to measure health outcomes from the patient perspective. http://nihpromis.Org/. Zugegriffen: 01. Oct. 2014.
47. Luce R, Narens L. Measurement scales on the continuum. Science. 1987;236:1527–32.
48. Cramer SC, Koroshetz WJ, Finklestein SP. The case for modality-specific outcome measures in clinical trials of stroke recovery-promoting agents. Stroke. 2007;38:1393–5.
49. Cramer SC, Fitzpatrick C, Warren M, Hill MD, Brown D, Whitaker L, et al. The beta-hcg+erythropoietin in acute stroke (betas) study: a 3-center, single-dose, open-label, noncontrolled, phase iia safety trial. Stroke. 2010;41:927–31.
50. Roth EJ, Heinemann AW, Lovell LL, Harvey RL, McGuire JR, Diaz S. Impairment and disability: their relation during stroke rehabilitation. Arch Phys Med Rehabil. 1998;79:329–35.
51. Bath PM, Lees KR, Schellinger PD, Altman H, Bland M, Hogg C, et al. Statistical analysis of the primary outcome in acute stroke trials. Stroke. 2012;43:1171–8.
52. Adams HP, Jr., Leclerc JR, Bluhmki E, Clarke W, Hansen MD, Hacke W. Measuring outcomes as a function of baseline severity of ischemic stroke. Cerebrovasc Dis (Basel, Switzerland). 2004;18:124–9.
53. Duncan PW, Sullivan KJ, Behrman AL, Azen SP, Wu SS, Nadeau SE, et al. Body-weight-supported treadmill rehabilitation after stroke. N Engl J Med. 2011;364:2026–36.
54. Burke Quinlan E, Dodakian L, See J, McKenzie A, Le V, Wojnowicz M, et al. Neural function, injury, and stroke subtype predict treatment gains after stroke. Ann Neurol. 2015 Jan;77(1):132–45.
55. Hier D, Mondlock J, Caplan L. Recovery of behavioral abnormalities after right hemisphere stroke. Neurology. 1983;33:345–50.
56. Marshall R, Perera G, Lazar R, Krakauer J, Constantine R, DeLaPaz R. Evolution of cortical activation during recovery from corticospinal tract infarction. Stroke. 2000;31:656–61.
57. Markgraf C, Green E, Hurwitz B, Morikawa E, Dietrich W, McCabe P, et al. Sensorimotor and cognitive consequences of middle cerebral artery occlusion in rats. Brain Res. 1992;575:238–46.
58. Fisher M, Hanley DF, Howard G, Jauch EC, Warach S. Recommendations from the stair v meeting on acute stroke trials, technology and outcomes. Stroke. 2007;38:245–8.
59. Traynor K. Dalfampridine approved for ms. Am J Health Syst Pharm. 2010;67:335
60. Feeney D, Gonzalez A, Law W. Amphetamine, halperidol, and experience interact to affect the rate of recovery after motor cortex injury. Science. 1982;217:855–7.
61. Fang PC, Barbay S, Plautz EJ, Hoover E, Strittmatter SM, Nudo RJ. Combination of nep 1–40 treatment and motor training enhances behavioral recovery after a focal cortical infarct in rats. Stroke. 2010;41:544–9.
62. Starkey ML, Schwab ME. Anti-nogo-a and training: Can one plus one equal three? Exp Neurol. 2012 May;235(1):53–61.
63. Hovda D, Feeney D. Amphetamine with experience promotes recovery of locomotor function after unilateral frontal cortex injury in the cat. Brain Res. 1984;298:358–61.
64. Adkins DL, Hsu JE, Jones TA. Motor cortical stimulation promotes synaptic plasticity and behavioral improvements following sensorimotor cortex lesions. Exp Neurol. 2008;212:14–28.
65. Kwakkel G. Impact of intensity of practice after stroke: issues for consideration. Disabil Rehabil. 2006;28:823–30.
66. Kwakkel G, Wagenaar R, Twisk J, Lankhorst G, Koetsier J. Intensity of leg and arm training after primary middle-cerebral-artery stroke: a randomised trial. Lancet. 1999;354:191–6.
67. Van Peppen RP Kwakkel G Wood-Dauphinee S Hendriks HJ Van der Wees PJ Dekker J. The impact of physical therapy on functional outcomes after stroke: What's the evidence? Clin Rehabil. 2004;18:833–62.

68. Cicerone KD, Dahlberg C, Malec JF, Langenbahn DM, Felicetti T, Kneipp S, et al. Evidence-based cognitive rehabilitation: updated review of the literature from 1998 through 2002. Arch Phys Med Rehabil. 2005;86:1681–92.
69. Bhogal S, Teasell R, Speechley M. Intensity of aphasia therapy, impact on recovery. Stroke. 2003;34:987–93.
70. Jones T, Chu C, Grande L, Gregory A. Motor skills training enhances lesion-induced structural plasticity in the motor cortex of adult rats. J Neurosci. 1999;19:10153–63.
71. Johansson B. Brain plasticity and stroke rehabilitation. The willis lecture. Stroke. 2000;31:223–30.
72. Smith J, Forster A, Young J. Cochrane review: information provision for stroke patients and their caregivers. Clin Rehabil. 2009;23:195–206.
73. Glass TA, Matchar DB, Belyea M, Feussner JR. Impact of social support on outcome in first stroke. Stroke. 1993;24:64–70.
74. Lai SM, Duncan PW, Keighley J, Johnson D. Depressive symptoms and independence in badl and iadl. J Rehabil Res Dev. 2002;39:589–96.
75. Jonsson AC, Lindgren I, Hallstrom B, Norrving B, Lindgren A. Determinants of quality of life in stroke survivors and their informal caregivers. Stroke. 2005;36:803–8.
76. Mukherjee D, Levin RL, Heller W. The cognitive, emotional, and social sequelae of stroke: Psychological and ethical concerns in post-stroke adaptation. Top Stroke Rehabil. 2006;13:26–35.
77. McFadden E, Luben R, Wareham N, Bingham S, Khaw KT. Social class, risk factors, and stroke incidence in men and women: a prospective study in the European prospective investigation into cancer in norfolk cohort. Stroke. 2009;40:1070–7.
78. Cramer S, Dobkin B, Noser E, Rodriguez R, Enney L. A randomized, placebo-controlled, double-blind study of ropinirole in chronic stroke. Stroke. 2009 Sep;40(9):3034–8.
79. Cramer SC. Stratifying patients with stroke in trials that target brain repair. Stroke. 2010;41:S114–6.
80. Stinear CM, Barber PA, Smale PR, Coxon JP, Fleming MK, Byblow WD. Functional potential in chronic stroke patients depends on corticospinal tract integrity. Brain. 2007;130:170–80.
81. Lindenberg R, Zhu LL, Ruber T, Schlaug G. Predicting functional motor potential in chronic stroke patients using diffusion tensor imaging. Hum Brain Mapp. 2012;33:1040–51.
82. Riley JD, Le V, Der-Yeghiaian L, See J, Newton JM, Ward NS, et al. Anatomy of stroke injury predicts gains from therapy. Stroke. 2011;42:421–6.
83. Stinear CM, Barber PA, Petoe M, Anwar S, Byblow WD. The prep algorithm predicts potential for upper limb recovery after stroke. Brain: J Neurol. 2012;135:2527–35.
84. Dong Y, Dobkin BH, Cen SY, Wu AD, Winstein CJ. Motor cortex activation during treatment may predict therapeutic gains in paretic hand function after stroke. Stroke. 2006;37:1552–5.
85. Milot MH, Spencer SJ, Chan V, Allington JP, Klein J, Chou C, et al. Corticospinal excitability as a predictor of functional gains at the affected upper limb following robotic training in chronic stroke survivors. Neurorehabil Neural Repair. 2014 Nov-Dec;28(9):819–27.
86. Cramer SC, Parrish TB, Levy RM, Stebbins GT, Ruland SD, Lowry DW, et al. Predicting functional gains in a stroke trial. Stroke. 2007;38:2108–14.
87. Laible M, Grieshammer S, Seidel G, Rijntjes M, Weiller C, Hamzei F. Association of activity changes in the primary sensory cortex with successful motor rehabilitation of the hand following stroke. Neurorehabil Neural Repair. 2012;26:881–8.
88. Sergi F, Krebs HI, Groissier B, Rykman A, Guglielmelli E, Volpe BT, et al. Predicting efficacy of robot-aided rehabilitation in chronic stroke patients using an mri-compatible robotic device. Conference proceedings:… annual international conference of the IEEE engineering in medicine and biology society. IEEE engineering in medicine and biology society. Conference. 2011;2011:7470–3.
89. Varkuti B, Guan C, Pan Y, Phua KS, Ang KK, Kuah CW, et al. Resting state changes in functional connectivity correlate with movement recovery for bci and robot-assisted upper-extremity training after stroke. Neurorehabil Neural Repair. 2013;27:53–62.

90. Graham JE, Ripsin CM, Deutsch A, Kuo YF, Markello S, Granger CV, et al. Relationship between diabetes codes that affect medicare reimbursement (tier comorbidities) and outcomes in stroke rehabilitation. Arch Phys Med Rehabil. 2009;90:1110–6.
91. Dam M, Tonin P, Casson S, Ermani M, Pizzolato G, Iaia V, et al. The effects of long-term rehabilitation therapy on poststroke hemiplegic patients. Stroke. 1993;24:1186–91.
92. Kononen M, Tarkka IM, Niskanen E, Pihlajamaki M, Mervaala E, Pitkanen K, et al. Functional mri and motor behavioral changes obtained with constraint-induced movement therapy in chronic stroke. Eur J Neurol. 2012;19:578–86.
93. Gillen R, Tennen H, McKee TE, Gernert-Dott P, Affleck G. Depressive symptoms and history of depression predict rehabilitation efficiency in stroke patients. Arch Phys Med Rehabil. 2001;82:1645–9.
94. McCarron MO, Muir KW, Nicoll JA, Stewart J, Currie Y, Brown K, et al. Prospective study of apolipoprotein e genotype and functional outcome following ischemic stroke. Arch Neurol. 2000;57:1480–4.
95. Cramer SC, Procaccio V. Correlation between genetic polymorphisms and stroke recovery: analysis of the gain Americas and gain international studies. Eur J Neurol. 2012;19:718–24.
96. Pearson-Fuhrhop KM, Burke E, Cramer SC. The influence of genetic factors on brain plasticity and recovery after neural injury. Curr Opin Neurol. 2012 Dec;25(6):682–8.
97. Siironen J, Juvela S, Kanarek K, Vilkki J, Hernesniemi J, Lappalainen J. The met allele of the bdnf val66met polymorphism predicts poor outcome among survivors of aneurysmal subarachnoid hemorrhage. Stroke. 2007;38:2858–60.
98. Freburger JK, Holmes GM, Ku LJ, Cutchin MP, Heatwole-Shank K, Edwards LJ. Disparities in postacute rehabilitation care for stroke: an analysis of the state inpatient databases. Arch Phys Med Rehabil. 2011;92:1220–9.
99. Burke E, Cramer SC. Biomarkers and predictors of restorative therapy effects after stroke. Curr Neurol Neurosci Rep. 2013;13:329.
100. Toth G, Albers GW. Use of mri to estimate the therapeutic window in acute stroke: is perfusion-weighted imaging/diffusion-weighted imaging mismatch an epithet for salvageable ischemic brain tissue? Stroke. 2009;40:333–5.
101. Donnan GA, Baron JC, Ma H, Davis SM. Penumbral selection of patients for trials of acute stroke therapy. Lancet Neurol. 2009;8:261–9.
102. Feuerstein GZ, Zaleska MM, Krams M, Wang X, Day M, Rutkowski JL, et al. Missing steps in the stair case: a translational medicine perspective on the development of nxy-059 for treatment of acute ischemic stroke. J Cereb Blood Flow Metab. 2008;28:217–9.
103. Carey J, Kimberley T, Lewis S, Auerbach E, Dorsey L, Rundquist P, et al. Analysis of fmri and finger tracking training in subjects with chronic stroke. Brain. 2002;125:773–88.
104. Johansen-Berg H, Dawes H, Guy C, Smith S, Wade D, Matthews P. Correlation between motor improvements and altered fmri activity after rehabilitative therapy. Brain. 2002;125:2731–42.
105. Koski L, Mernar T, Dobkin B. Immediate and long-term changes in corticomotor output in response to rehabilitation: correlation with functional improvements in chronic stroke. Neurorehabil Neural Repair. 2004;18:230–49.
106. Milot MH, Cramer SC. Biomarkers of recovery after stroke. Curr Opin Neurol. 2008;21:654–9.
107. Fleming T, DeMets D. Surrogate end points in clinical trials: are we being misled? Ann Intern Med. 1996;125:605–13.
108. Bucher H, Guyatt G, Cook D, Holbrook A, McAlister F. Users' guides to the medical literature: Xix. Applying clinical trial results. A. How to use an article measuring the effect of an intervention on surrogate end points? Evidence-based medicine working group. JAMA. 1999;282:771–8.
109. Geiger S, Holdenrieder S, Stieber P, Hamann GF, Bruening R, Ma J, et al. Nucleosomes as a new prognostic marker in early cerebral stroke. J Neurol. 2007;254:617–23.

110. Yip HK, Chang LT, Chang WN, Lu CH, Liou CW, Lan MY, et al. Level and value of circulating endothelial progenitor cells in patients after acute ischemic stroke. Stroke. 2008;39:69–74.
111. Brott T, Marler J, Olinger C, Adams H, Tomsick T, Barsan W, et al. Measurements of acute cerebral infarction: lesion size by computed tomography. Stroke. 1989;20:871–5.
112. Schaechter JD, Moore CI, Connell BD, Rosen BR, Dijkhuizen RM. Structural and functional plasticity in the somatosensory cortex of chronic stroke patients. Brain. 2006;129:2722–33.
113. Ding G, Jiang Q, Li L, Zhang L, Zhang ZG, Ledbetter KA, et al. Magnetic resonance imaging investigation of axonal remodeling and angiogenesis after embolic stroke in sildenafil-treated rats. J Cereb Blood Flow Metab. 2008;28:1440–8.
114. Lindenberg R, Zhu LL, Ruber T, Schlaug G. Predicting functional motor potential in chronic stroke patients using diffusion tensor imaging. Hum Brain Mapp. 2012 May;33(5):1040–51.
115. Marchina S, Zhu LL, Norton A, Zipse L, Wan CY, Schlaug G. Impairment of speech production predicted by lesion load of the left arcuate fasciculus. Stroke. 2011;42:2251–6.
116. Hodics T, Cohen LG, Cramer SC. Functional imaging of intervention effects in stroke motor rehabilitation. Arch Phys Med Rehabil. 2006;87:36–42.
117. Richards LG, Stewart KC, Woodbury ML, Senesac C, Cauraugh JH. Movement-dependent stroke recovery: a systematic review and meta-analysis of tms and fmri evidence. Neuropsychologia. 2008;46:3–11.
118. Carter AR, Astafiev SV, Lang CE, Connor LT, Rengachary J, Strube MJ, et al. Resting interhemispheric functional magnetic resonance imaging connectivity predicts performance after stroke. Ann Neurol. 2010;67:365–75.
119. Parsons M, Li T, Barber P, Yang Q, Darby D, Desmond P, et al. Combined (1)h mr spectroscopy and diffusion-weighted mri improves the prediction of stroke outcome. Neurology. 2000;55:498–505.
120. Pendlebury S, Blamire A, Lee M, Styles P, Matthews P. Axonal injury in the internal capsule correlates with motor impairment after stroke. Stroke. 1999;30:956–62.
121. Talelli P, Greenwood RJ, Rothwell JC. Arm function after stroke: neurophysiological correlates and recovery mechanisms assessed by transcranial magnetic stimulation. Clin Neurophysiol. 2006;117:1641–59.
122. Crafton K, Mark A, Cramer S. Improved understanding of cortical injury by incorporating measures of functional anatomy. Brain. 2003;126:1650–9.
123. Bocti C, Swartz RH, Gao FQ, Sahlas DJ, Behl P, Black SE. A new visual rating scale to assess strategic white matter hyperintensities within cholinergic pathways in dementia. Stroke. 2005;36:2126–31.
124. Hillis AE, Gold L, Kannan V, Cloutman L, Kleinman JT, Newhart M, et al. Site of the ischemic penumbra as a predictor of potential for recovery of functions. Neurology. 2008;71:184–9.
125. Hyun I. The bioethics of stem cell research and therapy. J Clin Invest. 2010;120:71–5.
126. Food and Drug Administration Center for Biologics Evaluation and Research. Guidance for industry potency tests for cellular and gene therapy products. http://www.Fda.Gov/downloads/biologicsbloodvaccines/guidancecomplianceregulatoryinformation/guidances/cellularandgenetherapy/ucm243392.Pdf. 2011. Accessed 1 March 2015.

Index

D. C. Hess (ed.), *Cell Therapy for Brain Injury,* DOI 10.1007/978-3-319-15063-5

MIX
Papier aus verantwortungsvollen Quellen
Paper from responsible sources
FSC® C105338

If you have any concerns about our products,
you can contact us on
ProductSafety@springernature.com

In case Publisher is established outside the EU,
the EU authorized representative is:
Springer Nature Customer Service Center GmbH
Europaplatz 3, 69115 Heidelberg, Germany

Printed by Libri Plureos GmbH
in Hamburg, Germany